Compendium of Hydrogen Energy

Related titles

Solid-state hydrogen storage
(ISBN 978-1-84569-270-4)

Solid oxide fuel cell technology
(ISBN 978-1-84569-628-3)

Handbook of membrane reactors Volume 1: Fundamental materials science, design and optimisation
(ISBN 978-0-85709-414-8)

Woodhead Publishing Series in Energy: Number 84

Compendium of Hydrogen Energy

Volume 2: Hydrogen Storage, Distribution and Infrastructure

Edited by

Ram B. Gupta, Angelo Basile and T. Nejat Veziroğlu

ELSEVIER

AMSTERDAM • BOSTON • CAMBRIDGE • HEIDELBERG
LONDON • NEW YORK • OXFORD • PARIS • SAN DIEGO
SAN FRANCISCO • SINGAPORE • SYDNEY • TOKYO
Woodhead Publishing is an imprint of Elsevier

WP WOODHEAD PUBLISHING

Woodhead Publishing is an imprint of Elsevier
80 High Street, Sawston, Cambridge, CB22 3HJ, UK
225 Wyman Street, Waltham, MA 02451, USA
Langford Lane, Kidlington, OX5 1GB, UK

Notices
Knowledge and best practice in this field are constantly changing. As new research and experience broaden our understanding, changes in research methods, professional practices, or medical treatment may become necessary.

Practitioners and researchers must always rely on their own experience and knowledge in evaluating and using any information, methods, compounds, or experiments described herein. In using such information or methods they should be mindful of their own safety and the safety of others, including parties for whom they have a professional responsibility.

To the fullest extent of the law, neither the Publisher nor the authors, contributors, or editors, assume any liability for any injury and/or damage to persons or property as a matter of products liability, negligence or otherwise, or from any use or operation of any methods, products, instructions, or ideas contained in the material herein.

British Library Cataloguing in Publication Data
A catalogue record for this book is available from the British Library.

Library of Congress Control Number: 2015937059

ISBN 978-1-78242-362-1 (print)
ISBN 978-1-78242-384-3 (online)

Contents

List of contributors

P. Agnolucci UCL Energy Institute, University College London, London, United Kingdom

R.K. Ahluwalia Argonne National Laboratory, Argonne, IL, USA

C.C. Ahn California Institute of Technology, Pasadena, CA, USA

O. Akgul Centre for Process Systems Engineering, University College London, London, United Kingdom

A.-M. Beaulieu Institut de recherche sur l'hydrogène, Université du Québec à Trois-Rivières, Trois-Rivières, QC, Canada

N. Bento DINÂMIA'CET, ISCTE – University Institute of Lisbon, Lisboa, Portugal

P. Bénard Institut de recherche sur l'hydrogène, Université du Québec à Trois-Rivières, Trois-Rivières, QC, Canada

R. Chahine Institut de recherche sur l'hydrogène, Université du Québec à Trois-Rivières, Trois-Rivières, QC, Canada

A. Chakraborty Indian Institute of Technology, Kharagpur, India

P.K. Chattaraj Indian Institute of Technology, Kharagpur, India

R. Das Indian Institute of Technology, Kharagpur, India

R.K. Dixon Global Environment Facility, Washington, DC, USA

D. Durette Institut de recherche sur l'hydrogène, Université du Québec à Trois-Rivières, Trois-Rivières, QC, Canada

R. Gerboni Politecnico di Torino, Torino, Italy

I.A. Gondal National University of Sciences & Technology, Rawalpindi, Pakistan

T.Q. Hua Argonne National Laboratory, Argonne, IL, USA

J.-E. Kim Virginia Tech, Blacksburg, VA, USA

H.W. Langmi HySA Infrastructure Centre of Competence, Council for Scientific and Industrial Research, Pretoria, South Africa

F.G.N. Li UCL Energy Institute, University College London, London, United Kingdom

J. Li Tsinghua University, Beijing, China

W. McDowall UCL Energy Institute, University College London, London, United Kingdom

S. McWhorter Savannah River National Laboratory, Aiken, SC, USA

N.M. Musyoka HySA Infrastructure Centre of Competence, Council for Scientific and Industrial Research, Pretoria, South Africa

K. Ohira Institute of Fluid Science, Tohoku University, Sendai, Japan

S. Pan Indian Institute of Technology, Kharagpur, India

M. Panfilov LEMTA - Université de Lorraine/CNRS, Nancy, France

L.G. Papageorgiou Centre for Process Systems Engineering, University College London, London, United Kingdom

J.-K. Peng Argonne National Laboratory, Argonne, IL, USA

Y.-H. Percival Zhang Virginia Tech; Cell-Free Bioinnovations Inc., and Institute for Critical Technology and Applied Science (ICTAS), Virginia Tech, Blacksburg, VA, USA

J. Ren HySA Infrastructure Centre of Competence, Council for Scientific and Industrial Research, Pretoria, South Africa

N.T. Stetson U.S. Department of Energy, Washington, DC, USA

G. Valenti Politecnico di Milano, Milano, Italy

M.Q. Wang Argonne National Laboratory, Chicago, IL, USA

N.K. Zhevago National Research Centre, Kurchatov Institute, Moscow, Russia

Woodhead Publishing Series in Energy

1 **Generating power at high efficiency: Combined cycle technology for sustainable energy production**
Eric Jeffs

2 **Advanced separation techniques for nuclear fuel reprocessing and radioactive waste treatment**
Edited by Kenneth L. Nash and Gregg J. Lumetta

3 **Bioalcohol production: Biochemical conversion of lignocellulosic biomass**
Edited by Keith W. Waldron

4 **Understanding and mitigating ageing in nuclear power plants: Materials and operational aspects of plant life management (PLiM)**
Edited by Philip G. Tipping

5 **Advanced power plant materials, design and technology**
Edited by Dermot Roddy

6 **Stand-alone and hybrid wind energy systems: Technology, energy storage and applications**
Edited by John K. Kaldellis

7 **Biodiesel science and technology: From soil to oil**
Jan C. J. Bart, Natale Palmeri and Stefano Cavallaro

8 **Developments and innovation in carbon dioxide (CO_2) capture and storage technology Volume 1: Carbon dioxide (CO_2) capture, transport and industrial applications**
Edited by M. Mercedes Maroto-Valer

9 **Geological repository systems for safe disposal of spent nuclear fuels and radioactive waste**
Edited by Joonhong Ahn and Michael J. Apted

10 **Wind energy systems: Optimising design and construction for safe and reliable operation**
Edited by John D. Sørensen and Jens N. Sørensen

11 **Solid oxide fuel cell technology: Principles, performance and operations**
Kevin Huang and John Bannister Goodenough

12 **Handbook of advanced radioactive waste conditioning technologies**
Edited by Michael I. Ojovan

13 **Membranes for clean and renewable power applications**
Edited by Annarosa Gugliuzza and Angelo Basile

14 **Materials for energy efficiency and thermal comfort in buildings**
Edited by Matthew R. Hall

15 **Handbook of biofuels production: Processes and technologies**
Edited by Rafael Luque, Juan Campelo and James Clark

16 **Developments and innovation in carbon dioxide (CO_2) capture and storage technology Volume 2: Carbon dioxide (CO_2) storage and utilisation**
Edited by M. Mercedes Maroto-Valer

17 **Oxy-fuel combustion for power generation and carbon dioxide (CO_2) capture**
Edited by Ligang Zheng
18 **Small and micro combined heat and power (CHP) systems: Advanced design, performance, materials and applications**
Edited by Robert Beith
19 **Advances in clean hydrocarbon fuel processing: Science and technology**
Edited by M. Rashid Khan
20 **Modern gas turbine systems: High efficiency, low emission, fuel flexible power generation**
Edited by Peter Jansohn
21 **Concentrating solar power technology: Principles, developments and applications**
Edited by Keith Lovegrove and Wes Stein
22 **Nuclear corrosion science and engineering**
Edited by Damien Féron
23 **Power plant life management and performance improvement**
Edited by John E. Oakey
24 **Electrical drives for direct drive renewable energy systems**
Edited by Markus Mueller and Henk Polinder
25 **Advanced membrane science and technology for sustainable energy and environmental applications**
Edited by Angelo Basile and Suzana Pereira Nunes
26 **Irradiation embrittlement of reactor pressure vessels (RPVs) in nuclear power plants**
Edited by Naoki Soneda
27 **High temperature superconductors (HTS) for energy applications**
Edited by Ziad Melhem
28 **Infrastructure and methodologies for the justification of nuclear power programmes**
Edited by Agustín Alonso
29 **Waste to energy conversion technology**
Edited by Naomi B. Klinghoffer and Marco J. Castaldi
30 **Polymer electrolyte membrane and direct methanol fuel cell technology Volume 1: Fundamentals and performance of low temperature fuel cells**
Edited by Christoph Hartnig and Christina Roth
31 **Polymer electrolyte membrane and direct methanol fuel cell technology Volume 2: *In situ* characterization techniques for low temperature fuel cells**
Edited by Christoph Hartnig and Christina Roth
32 **Combined cycle systems for near-zero emission power generation**
Edited by Ashok D. Rao
33 **Modern earth buildings: Materials, engineering, construction and applications**
Edited by Matthew R. Hall, Rick Lindsay and Meror Krayenhoff
34 **Metropolitan sustainability: Understanding and improving the urban environment**
Edited by Frank Zeman
35 **Functional materials for sustainable energy applications**
Edited by John A. Kilner, Stephen J. Skinner, Stuart J. C. Irvine and Peter P. Edwards
36 **Nuclear decommissioning: Planning, execution and international experience**
Edited by Michele Laraia
37 **Nuclear fuel cycle science and engineering**
Edited by Ian Crossland
38 **Electricity transmission, distribution and storage systems**
Edited by Ziad Melhem
39 **Advances in biodiesel production: Processes and technologies**
Edited by Rafael Luque and Juan A. Melero

40 **Biomass combustion science, technology and engineering**
Edited by Lasse Rosendahl
41 **Ultra-supercritical coal power plants: Materials, technologies and optimisation**
Edited by Dongke Zhang
42 **Radionuclide behaviour in the natural environment: Science, implications and lessons for the nuclear industry**
Edited by Christophe Poinssot and Horst Geckeis
43 **Calcium and chemical looping technology for power generation and carbon dioxide (CO_2) capture: Solid oxygen- and CO_2-carriers**
Paul Fennell and E. J. Anthony
44 **Materials' ageing and degradation in light water reactors: Mechanisms, and management**
Edited by K. L. Murty
45 **Structural alloys for power plants: Operational challenges and high-temperature materials**
Edited by Amir Shirzadi and Susan Jackson
46 **Biolubricants: Science and technology**
Jan C. J. Bart, Emanuele Gucciardi and Stefano Cavallaro
47 **Advances in wind turbine blade design and materials**
Edited by Povl Brøndsted and Rogier P. L. Nijssen
48 **Radioactive waste management and contaminated site clean-up: Processes, technologies and international experience**
Edited by William E. Lee, Michael I. Ojovan, Carol M. Jantzen
49 **Probabilistic safety assessment for optimum nuclear power plant life management (PLiM): Theory and application of reliability analysis methods for major power plant components**
Gennadij V. Arkadov, Alexander F. Getman and Andrei N. Rodionov
50 **The coal handbook: Towards cleaner production Volume 1: Coal production**
Edited by Dave Osborne
51 **The coal handbook: Towards cleaner production Volume 2: Coal utilisation**
Edited by Dave Osborne
52 **The biogas handbook: Science, production and applications**
Edited by Arthur Wellinger, Jerry Murphy and David Baxter
53 **Advances in biorefineries: Biomass and waste supply chain exploitation**
Edited by Keith Waldron
54 **Geological storage of carbon dioxide (CO_2): Geoscience, technologies, environmental aspects and legal frameworks**
Edited by Jon Gluyas and Simon Mathias
55 **Handbook of membrane reactors Volume 1: Fundamental materials science, design and optimisation**
Edited by Angelo Basile
56 **Handbook of membrane reactors Volume 2: Reactor types and industrial applications**
Edited by Angelo Basile
57 **Alternative fuels and advanced vehicle technologies for improved environmental performance: Towards zero carbon transportation**
Edited by Richard Folkson
58 **Handbook of microalgal bioprocess engineering**
Christopher Lan and Bei Wang
59 **Fluidized bed technologies for near-zero emission combustion and gasification**
Edited by Fabrizio Scala

60 **Managing nuclear projects: A comprehensive management resource**
Edited by Jas Devgun

61 **Handbook of Process Integration (PI): Minimisation of energy and water use, waste and emissions**
Edited by Jiří J. Klemeš

62 **Coal power plant materials and life assessment**
Edited by Ahmed Shibli

63 **Advances in hydrogen production, storage and distribution**
Edited by Ahmed Basile and Adolfo Iulianelli

64 **Handbook of small modular nuclear reactors**
Edited by Mario D. Carelli and Dan T. Ingersoll

65 **Superconductors in the power grid: Materials and applications**
Edited by Christopher Rey

66 **Advances in thermal energy storage systems: Methods and applications**
Edited by Luisa F. Cabeza

67 **Advances in batteries for medium and large-scale energy storage**
Edited by Chris Menictas, Maria Skyllas-Kazacos and Tuti Mariana Lim

68 **Palladium membrane technology for hydrogen production, carbon capture and other applications**
Edited by Aggelos Doukelis, Kyriakos Panopoulos, Antonios Koumanakos and Emmanouil Kakaras

69 **Gasification for synthetic fuel production: Fundamentals, processes and applications**
Edited by Rafael Luque and James G. Speight

70 **Renewable heating and cooling: Technologies and applications**
Edited by Gerhard Stryi-Hipp

71 **Environmental remediation and restoration of contaminated nuclear and NORM sites**
Edited by Leo van Velzen

72 **Eco-friendly innovation in electricity networks**
Edited by Jean-Luc Bessede

73 **The 2011 Fukushima nuclear power plant accident: How and why it happened**
Yotaro Hatamura, Seiji Abe, Masao Fuchigami and Naoto Kasahara. Translated by Kenji Iino

74 **Lignocellulose biorefinery engineering: Principles and applications**
Hongzhang Chen

75 **Advances in membrane technologies for water treatment: Materials, processes and applications**
Edited by Angelo Basile, Alfredo Cassano and Navin Rastogi

76 **Membrane reactors for energy applications and basic chemical production**
Edited by Angelo Basile, Luisa Di Paola, Faisal Hai and Vincenzo Piemonte

77 **Pervaporation, vapour permeation and membrane distillation: Principles and applications**
Edited by Angelo Basile, Alberto Figoli and Mohamed Khayet

78 **Safe and secure transport and storage of radioactive materials**
Edited by Ken Sorenson

79 **Reprocessing and recycling of spent nuclear fuel**
Edited by Robin Taylor

80 ***Advances in battery technologies for electric vehicles***
Edited by Bruno Scrosati, Juergen Garche and Werner Tillmetz

81 **Rechargeable lithium batteries: From fundamentals to applications**
Edited by Alejandro A. Franco

82 **Calcium and chemical looping technology for power generation and carbon dioxide (CO_2) capture**
Edited by Paul Fennell and Ben Anthony

83 **Compendium of Hydrogen Energy Volume 1: Hydrogen Production and Purificiation**
Edited by Velu Subramani, Angelo Basile and T. Nejat Veziroğlu

84 **Compendium of Hydrogen Energy Volume 2: Hydrogen Storage, Distribution and Infrastructure**
Edited by Ram B. Gupta, Angelo Basile and T. Nejat Veziroğlu

85 **Compendium of Hydrogen Energy Volume 3: Hydrogen Energy Conversion**
Edited by Frano Barbir, Angelo Basile and T. Nejat Veziroğlu

86 **Compendium of Hydrogen Energy Volume 4: Hydrogen Use, Safety and the Hydrogen Economy**
Edited by Michael Ball, Angelo Basile and T. Nejat Veziroğlu

Part One

Hydrogen Storage in Pure Form

Introduction to hydrogen storage

1

N.T. Stetson[1], S. McWhorter[2], C.C. Ahn[3]
[1]U.S. Department of Energy, Washington, DC, USA; [2]Savannah River National Laboratory, Aiken, SC, USA; [3]California Institute of Technology, Pasadena, CA, USA

Abbreviations

ΔH	enthalpy of reaction
ΔS	entropy of reaction
AGA	American Gas Association
ANSI	American National Standards Institute
CF	carbon fiber
DOE	Department of Energy
HHV	higher heating value
ISO	International Organization for Standardization
KHK	Kouatsu-Gas Hoan Kyoukai (High Pressure Gas Safety Institute of Japan)
LHV	lower heating value
LLNL	Lawrence Livermore National Laboratory
MH_x	metal hydrides
MLI	multilayer insulation
mpg	miles per gallon
NASA	National Aeronautics and Space Administration
NGV2	American National Standard for Natural Gas Vehicle Containers
PEM	polymer electrolyte membrane
R	gas constant
SAE	Society of Automotive Engineers
STP	standard temperature and pressure (273.15 K, 0.100 MPa)
TUV	Technischer Überwachungsverrein

1.1 Introduction

There is significant interest in the use of hydrogen as an energy carrier. Since the 1990s this interest has been driven by geo-political and climate change concerns as well as advancements in technology. For instance, the United States imported over four billion barrels of crude oil and petroleum products annually from 2000 to 2011, exceeding five billion barrels in each of 2005 and 2006 (U.S. Imports of Crude Oil and Petroleum Products, 2014). With the price of crude oil ranging up to over $100 per barrel, these import rates resulted in expenditures of up to US$1 billion per day being sent to foreign entities to import nondomestic oil. During this timeframe there was also increasing evidence that the use of fossil fuels and the incumbent

Compendium of Hydrogen Energy. http://dx.doi.org/10.1016/B978-1-78242-362-1.00001-8
2016 Published by Elsevier Ltd.

release and accumulation of carbon dioxide and other greenhouse gases in the atmosphere was causing significant climate change. Therefore, development of clean, sustainable sources of energy that are widely available throughout the world was considered a high priority. An example was in 2003 when US President George W. Bush announced the "Hydrogen Fuel Initiative" to invest US$1.2 billion in research and development to make hydrogen competitive for powering vehicles and generating electricity (Selected Speeches of President George W. Bush, 2001–2008).

Advances in polymer electrolyte membrane (PEM) fuel cell technology have also contributed to the interest in use of hydrogen as an energy carrier. The generation of electricity from hydrogen and oxygen through fuel cell technology was first demonstrated in 1838 by William Robert Grove (Fuel Cell History Project, 2015). In the 1960s General Electric invented ion-exchange membrane fuel cells that were successfully used on seven of NASA's Gemini space missions in 1965 and 1966 (Warshay and Prokopius, 1989). Later space missions, such as the Apollo and Space Shuttle programs, used alkaline fuel cells instead of ion-exchange membrane fuel cell technology due to performance issues and the high amounts of platinum catalyst that were used in ion-exchange fuel cells. Throughout the 1990s developments to improve power density while lowering the required amount of platinum catalysts, particularly at Los Alamos National Laboratory and Ballard Power Systems, spurred renewed interest in PEM fuel cells for automotive propulsion and other power applications (Koppel, 1999). Fuel cells can provide several benefits compared to conventional power sources. When hydrogen is used directly as the fuel, the reaction product is water, thus no greenhouse gases are produced at the point of use. The chemical to electrical energy conversion efficiency of complete automotive PEM fuel cell systems have been shown to approach 60%, significantly higher than for devices such as internal combustion engines (Wipke et al., 2012). Higher total efficiencies can be realized in combined heat and power applications when both the electrical and heat energy are utilized.

PEM and alkaline fuel cells operate at relatively low temperatures, making them more suitable for applications that undergo frequent on/off cycles and require quick start-up, such as mobile, portable, and back-up power. Power requirements for these applications can range from as low as a watt or so for portable applications (e.g., portable electronics) up to about 100 kW or so for mobile applications (e.g., passenger cars and buses). Many of these applications are currently in the domain of battery systems. Hydrogen fuel cell systems have both advantages and disadvantages when compared with battery systems. Typical batteries are redox systems with an oxidant and a reductant (fuel) stored within the battery electrodes. In typical hydrogen fuel cell systems oxygen from the air is used as the oxidant and stored hydrogen is the reductant (fuel). Since the fuel is separately stored in hydrogen fuel cell systems, refueling/recharging is dependent on the rate hydrogen can be charged into the storage system. Many hydrogen storage technologies are capable of rapid recharging without any detrimental impact on system performance. Since the fuel storage is separate from the conversion device, in hydrogen fuel cell systems the fuel cell is sized for the required power while the hydrogen storage is sized for the amount of energy required. In battery systems, the energy and power are coupled within the electrode design and are not independent. The separation of fuel storage from the energy conversion device also

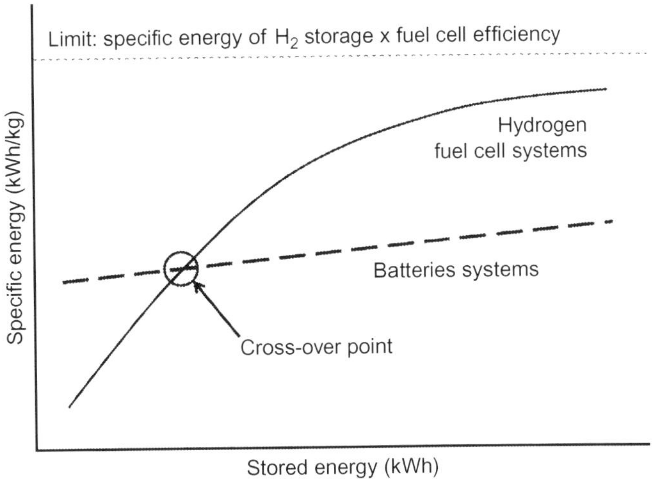

Figure 1.1 Relationship of specific energy versus stored energy for batteries and hydrogen fuel cell systems.

means that hydrogen fuel cell systems become more advantageous as the amount of energy required increases. This is obvious in plots of system specific energy (kWh/kg) versus stored energy (kWh) (see Figure 1.1) or energy density (kWh/L) versus stored energy (kWh). For applications that do not require much energy (i.e., short runtimes), batteries tend to have lower mass and volume than hydrogen fuel cell systems, but when greater amounts of energy are required (i.e., long runtimes), the fuel cell mass and volume remain essentially constant, and the specific energy and energy density continuously increases and approaches the specific energy and energy density of the fuel storage times the conversion efficiency of the fuel cell. For a given battery chemistry, the specific energy and energy density of the battery system is relatively constant with only incremental improvements with larger battery systems.

Hydrogen is appropriately referred to as an energy carrier, like electricity, instead of as a fuel. This is due to the fact that on earth hydrogen is not normally found in the elemental state, but is instead bound to other elements in compounds. Energy has to be expended to liberate or produce elemental hydrogen from being bound within compounds. At standard temperature and pressure (STP) conditions, elemental hydrogen exists as a diatomic molecule, with a density of 0.0001 kg/L. The normal boiling point for hydrogen is at 20.3 K, where it has a normal density of 0.071 kg/L. Each hydrogen nuclei contains a single proton, therefore, two spin isomers are possible for the diatomic molecule: para with the spins antiparallel or ortho with the spins parallel. While the para isomer is the lowest energy configuration, at room temperature about 75% of hydrogen exists as the ortho isomer. Near the normal boiling point, the conversion energy from the ortho to para isomers is similar to the latent heat of vaporization and thus near total conversion to the para isomer is of critical importance during hydrogen liquefaction to extend the dormancy of liquid hydrogen storage (Stetson et al., 2015).

Hydrogen reacts with oxygen to produce water according to:

$$H_{2(g)} + \tfrac{1}{2}O_{2(g)} \rightarrow H_2O_{(l\ or\ g)} \quad (1.1)$$

When the product water is taken to its standard liquid state, the specific energy of combustion is the higher heating value of 141.8 MJ/kg. However, in typical use, as with fuel cells or internal combustion engines, the product water is in the gaseous state and the lower heating value of 120.0 MJ/kg is appropriate. The specific energy of hydrogen is greater than all other conventional fuels, approximately three times greater than that of typical liquid fuels such as gasoline and diesel. However, due to hydrogen's very low density, the energy density of hydrogen gas at STP is over 2500 times lower than the liquid fuels. The energy density of liquid hydrogen at 20 K is still only one quarter of gasoline and diesel. To overcome the low energy density of hydrogen gas, it is normally stored as a compressed gas, typically at pressures ranging from about 15 to 70 MPa. Table 1.1 compares the specific energy and energy density of several conventional fuels and hydrogen as a compressed gas and liquid. These values are for the substance only and do not account for the mass and volume of the containment vessel or other accessories required for the storage system. The comparison shows that even when compressed to pressures as high as 70 MPa, hydrogen still suffers from low relative energy density. Additionally, neither high-pressure compressed nor liquid hydrogen are suitable for the range of applications that hydrogen fuel cells are being considered. Therefore, significant efforts have been devoted to developing advanced alternative hydrogen storage technologies.

Table 1.1 Specific energy and energy density for several common fuels

Fuel	Gravimetric (MJ/kg)	Volumetric (MJ/L)
Diesel[a]	43.4	36.8
JP-8/jet kerosene[b]	43.4	34.7
Gasoline (E10) (288 K)[c]	41.6	30.8
E85 (288 K)[c]	30	23.4
Liquid propane (10 bar) (298.15 K)[d]	46	22.8
Ethanol (288 K)[d]	26.8	21.2
Methanol[d]	19.9	15.7
CH_4 (liq) (111 K)[d]	50	21.2
CH_4 (250 bar) (298.15 K)[d]	50	9.4
H_2 (liq) (20 K)[d]	120	8.6
H_2 (700 bar) (298.15 K)[d]	120	4.7
H_2 (350 bar) (298.15 K)[d]	120	2.8

[a]Based on Chevron-Phillips surrogate composition from Mueller et al. (2012).
[b]Based on surrogate composition from Agency for Toxic Substances and Disease Registry of CDC (1998), Table 3-3 with the original source at http://www.dtic.mil/cgibin/GetTRDoc?Location=U2&doc=GetTRDoc.pdf&AD=ADA247835.
[c]https://greet.es.anl.gov/carbon_footprint_calculator and Schremp and Janusch (2009), Fig. 13, p. 39.
[d]Derived from thermochemical data at http://webbook.nist.gov/.

There are a number of potential storage technologies for hydrogen that may offer advantages for different applications. The traditional methods of storing hydrogen are either as a compressed gas at ambient temperatures or as a cryogenic liquid at low pressure. Cryogenic vessels for liquid hydrogen typically employ multilayered vacuum super insulation sandwiched between a double-wall. Compressed hydrogen is typically stored in high-pressure cylinders that include: Type I—all metal; Type II—metal liner with composite overwrap of the cylindrical section; Type III—metal liner with full composite overwrap; or Type IV—nonmetallic liner with a full composite overwrap. Type I all metal cylinders tend to be the lowest cost but also have the greatest mass and are therefore used primarily at lower pressures and for stationary applications. For high-pressure and mobile applications, such for passenger vehicles, Type IV composite overwrapped pressure vessels are predominant since they offer the lowest mass, but tend to be the highest cost. To improve the energy density, the storage of compressed hydrogen at subambient temperatures has been investigated. These investigations have considered the applicability of the various pressure vessel types at the target storage temperature, the performance of the insulation required, and the trade-offs between storage pressure, temperature, and cost. The physical storage of hydrogen as a liquid, or as an ambient or subambient compressed gas, will be discussed in more depth later in this chapter as well as by authors Valenti, Ohira, Ahluwalia, and Zhevago in later chapters.

In addition to the physical methods of storing elemental hydrogen as either a compressed gas or cryogenic liquid, hydrogen can also be stored bound to other elements within a material. The three primary classes of materials for materials-based hydrogen storage are adsorbents, reversible metal hydrides, and chemical hydrogen storage materials. Hydrogen adsorbents are high-surface area, porous materials in which the dihydrogen molecule weakly bonds (adsorbs) to the surface. Reversible metal hydrides are materials, where monatomic hydrogen is bound to other elements, typically metals or metalloids, within a solid. The two types of metal hydrides most commonly investigated for hydrogen storage are intermetallic (also called interstitial) metal hydrides where the hydrogen atoms occupy interstitial sites within an alloy, and complex hydrides where the hydrogen is covalently bound to another atom to form a complex anion, where charge is balanced by the presence of cations. For both adsorbents and reversible metal hydrides, a temperature/pressure equilibrium condition is used to charge and release hydrogen to and from the material. In chemical hydrogen storage materials, the hydrogen is typically bound to another element through either covalent or ionic bonds. The hydrogen is released from the material through nonequilibrium processes and is considered "nonreversible" in that the dehydrogenated material cannot be hydrogenated simply by applying pressurized hydrogen. Methods to release hydrogen from chemical hydrogen storage materials include thermolysis and chemical reactions. A typical chemical reaction is hydrolysis as shown in Equation (1.2):

$$MH_x + xH_2O \rightarrow x/2H_2 + M(OH)_x \tag{1.2}$$

When the hydrogen interacts with other elements, the hydrogen atoms can be brought closer together and therefore higher energy densities are possible. For instance, atomic hydrogen can be accommodated in sub Å interstices in intermetallic hydrides (Westlake, 1983), whereas the nearly spheroidal dimensions of molecular hydrogen are on the order of 3.6 Å (Koresh and Soffer, 1980). In adsorbents, the density of hydrogen physisorbed on the surface can approach the density of liquid hydrogen at temperatures significantly above the critical temperature of 33 K and at relatively low pressures. For metal hydrides and chemical hydrogen storage materials, hydrogen densities twice that of liquid hydrogen can be obtained at ambient temperatures and low pressures. For instance, the hydrogen density within Mg_2FeH_6 (Didisheim et al., 1984) and AlH_3 (Graetz et al., 2011) is approximately 150 g/L, but the actual hydrogen density that can be realized in a storage system will be less since the materials are typically used as powders with lower bulk densities. Materials-based hydrogen storage will be discussed in more depth later in this chapter, as well as in later chapters by Benard, Langmi, Zhang, and Chattaraj. In addition, several reviews have been published on engineering aspects of materials-based hydrogen storage systems; e.g., see Bowman et al. (2013) and Brooks et al. (2011).

When consideration is given to the hydrogen storage requirements for the various applications hydrogen fuel cells devices are being developed for, it becomes obvious that certain storage technologies may not be suitable for some applications but preferred for other applications. Therefore, it is important to understand the application performance requirements and the attributes of the storage technology to determine which storage technology is suited for which application (McWhorter et al., 2011; Stetson et al., 2013). For instance, liquid hydrogen requires double-walled vessels with multilayered vacuum superinsulation to minimize heat leakage that will boil-off the stored hydrogen. It is therefore most suitable for large bulk stationary storage and not suitable for small portable applications. Portable power applications typically favor compact storage that operates at ambient temperature and preferably low pressure and therefore reversible metal hydrides and chemical hydrogen storage systems may be most suitable. Automotive applications require high energy density and high specific energy systems that are capable of frequent start/stop cycles, good dormancy (i.e., little to no loss of hydrogen when not in use) and low cost. The initial launch of commercial hydrogen fuel cell vehicles use Type IV 70 MPa compressed hydrogen systems since that technology is the most mature and currently able to best meet performance requirements. However, significantly higher energy density and lower costs are needed for to obtain wide-scale commercial success.

1.2 Physical storage

Physical storage of hydrogen fundamentally refers to the utilization of a structural vessel to contain the hydrogen while altering the density through variations in the pressure and/or temperature. Figure 1.2 shows the calculated density of H_2 (kg/m^3) as a function of temperature at a few common storage pressures. What is apparent from Figure 1.2 is the ability to dynamically change the density of H_2 over a range of temperature and pressure combinations, especially with decreasing temperature and

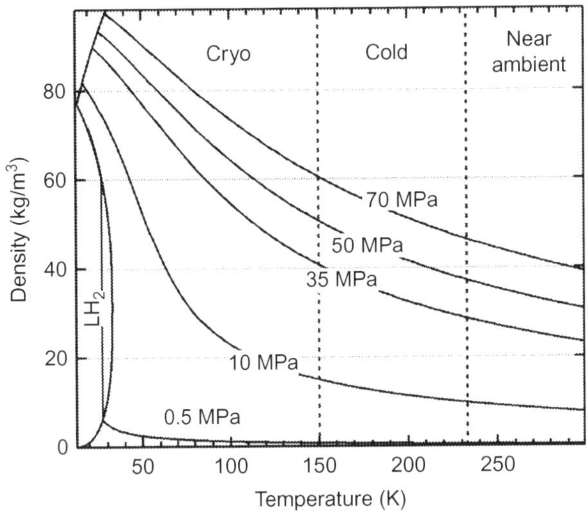

Figure 1.2 Hydrogen density versus temperature for several storage pressures.

increasing pressure or combinations thereof. Thus, the four most common approaches to storing hydrogen, such as for automotive applications, are easily defined by the pressure–temperature operating regimes, including:

1. Compressed hydrogen vessels operating at high pressures, as high as 70 MPa, and near ambient temperature;
2. Cold-compressed hydrogen vessels operating at high pressures (ca., 50 MPa) but lower temperatures (i.e., <273 K and >150 K);
3. Cryo-compressed hydrogen vessels typically operating at ≤ 35 MPa and temperatures less than 150 K; and
4. Liquid hydrogen vessels operating at low pressures (i.e., <0.6 MPa) and low temperatures near the normal boiling point of H_2, 20 K.

However, when considering these approaches there are a host of key performance characteristics such as density, weight, volume, cost, dormancy, refueling, safety, and environmental impact that must be evaluated in the context of the application.

1.2.1 Compressed hydrogen

The current near-term technology for onboard automotive hydrogen storage is 35 and 70 MPa nominal working-pressure compressed gas vessels. The main advantage of employing a compressed hydrogen storage system is the ability to rapidly refuel the vehicles in approximately 3–5 min. However, given that the density of hydrogen at elevated pressures deviates from ideal behavior and increases much more slowly than the pressure, robust pressure vessels are required to withstand the higher pressures, which can incur significant weight and cost penalties. In addition, safety of the compressed hydrogen storage is a concern, whether real or perceived. Accordingly, the design and implementation of the cylinders is regulated by government agencies and requires specific design standards and certifications. There are four standard types of cylinders that are used for hydrogen storage: Type I—all-metal cylinders, Type II—all-metal hoop-wrapped composite cylinders, Type III—fully wrapped composite cylinders with metallic liners (e.g., Al-6061), and Type IV—fully wrapped composite cylinders with nonload bearing nonmetallic liners (i.e., usually a polymer such as high density polyethylene). For Type I and II cylinders, the mass of the metal required to contain the pressure generally only allows storage of 1% or 2% hydrogen compared to the cylinder mass, which drops to less than 1% hydrogen by mass at pressures above 35 MPa. Consequently, Type I and II cylinders are impractical for automotive applications where weight and volume play a critical role in defining the overall efficiency of the vehicle. As such, most recent efforts at developing high-pressure, compressed cylinders for light-duty vehicles has been devoted to Type III and IV cylinders, which can deliver hydrogen capacities that are at least four times that of Type I cylinders, as indicated by analyses through the US Department of Energy's (DOE) Office of Energy Efficiency and Renewable Energy's Fuel Cell Technologies Office (Department of Energy—Office of Energy Efficiency and Renewable Energy, 2015). For example, analysis of 35 MPa and 70 MPa Type IV composite cylinders that use high-strength carbon fiber (CF) reinforcement to reduce cylinder weight for automotive applications indicates hydrogen system capacities of 5.4% and 4.4% hydrogen by mass and 17.7 and 25.0 g hydrogen per liter volume, at a cost of ~$13 and $17/kWh, respectively (McWhorter and Ordaz, 2013).

Compressed hydrogen storage systems have been demonstrated in hundreds of prototype fuel cell vehicles and are commercially available at low production volumes. The cylinders within these systems have been certified worldwide according to ISO-11439 (Europe), ANSI/AGA NGV2 (US), and Reijikijun Betten (Iceland) standards, and approved by TUV (Germany) and KHK (Japan). In fact, the United Nations recently adopted a Global Technical Regulation that unifies the regulation requirements of these automotive systems based on the guidance from SAE J2579 hydrogen system standard (United Nations Global Technical Regulations, 2015).

1.2.2 Cold-compressed hydrogen

By logically tracking the increase in density from right to left in Figure 1.2, we would deduce that utilizing colder temperatures combined with higher pressures would be a reasonable next approach to increasing storage density. In fact, the cold-compressed hydrogen pathway is a relatively new concept being considered by the U.S. DOE as a means to reduce the costs of onboard hydrogen storage while increasing gas density by reduction in mass and volume of the storage vessel, thus, reducing the amount of expensive CF required for vessel reinforcement. In this scenario, the hydrogen gas would be cooled to liquid nitrogen temperatures (i.e., 90 K) at a terminal or station and then compressed to a higher pressure (e.g., 50 MPa) for storage onboard a vehicle. In this case, the final gas temperature in the compressed gas vessel would warm to approximately 190 K as a result of the heat of compression and isoenthalpic expansion (i.e., the reverse Joule–Thompson effect) between the delivery source and storage vessel. The storage vessel would be designed to withstand higher pressures and also require vacuum insulation to maintain the cold temperature when the vehicle is not in use. The Pacific Northwest National Laboratory and Ford Motor Company have been evaluating this concept of enhanced operating conditions and have identified conditions (i.e., 50 MPa and 200 K) where the hydrogen gas density approaches 42 g hydrogen per liter and an estimated Type IV tank mass of 48.2 kg to store 5.6 kg of usable hydrogen, which is roughly a 45% reduction over the current Type IV 70 MPa vessel mass for storing an equivalent amount of hydrogen, as reported by the US DOE (Simmons, 2015). In 2011, Paster et al. (2011) presented a case study on several pathways for delivery and onboard storage of hydrogen that highlighted the attributes of the cold-compressed hydrogen pathway. Their results indicated that the cold-compressed pathway could reach the lowest cost for hydrogen on a mass basis, while the overall storage system cost was higher than all other technologies excluding the 70 MPa case. However, considering the improvements identified by Simmons (2015) this route could potentially show real promise for improving the storage system costs as well as volumetric and gravimetric storage efficiencies for automotive applications.

1.2.3 Liquid hydrogen

Historically, the liquefaction of hydrogen and cryogenic liquid storage have been the preferred method to increase hydrogen density for bulk transport and storage. With the advent of regeneratively cooled refrigeration devices by Hampson and Linde in 1895, James Dewar first liquefied molecular hydrogen in 1898 using a regeneratively cooled liquid nitrogen system, storing it in silver glazed, doubled-walled glass flasks (i.e., Dewar flasks). Oddly enough, Dewar never patented the design of these flasks but the flasks continue to serve as an important tool in the utilization of cryogenic hydrogen. Commercial scale production of liquid hydrogen would not be forthcoming until the 1950s, as large quantities of liquid hydrogen would be required for launching spacecraft, such as in NASA's Centaur rocket stage.

Hydrogen is a liquid at its boiling point of 20 K with a density of approximately 71 g per liter at atmospheric pressure. Further increases in density are expected with the application of cryogenic temperatures and pressure as observed in Figure 1.2. However, each technology requires well-insulated and expensive cryogenic storage vessels to prevent boil-off and maximize dormancy (i.e., length of time the hydrogen can be stored without loss through venting). Moreover, the liquefaction process is energy intensive and consumes approximately ~25–30% of the energy content of the stored hydrogen (Ahluwalia et al., 2007). While the initial energy expenditures for producing liquid hydrogen are considerable, the boil-off loss due to heat leakage through the storage vessel walls poses perhaps the greatest challenge to the liquid hydrogen storage for long-term storage and storage of nonbulk quantities of hydrogen, such as for automotive applications.

In the early 1970s, cryogenic hydrogen storage was regarded as a viable option for automotive applications and was demonstrated in several prototype vehicles operating internal combustion engines. Perhaps the most well-known and sustained efforts were by the BMW AG (Wolf, 2013). These efforts were focused on lengthening the storage dormancy by reducing liquid boil-off as well as on facilitating safe filling of the tank

Figure 1.3 Image of a liquid hydrogen storage vessel designed by Linde for BMW. The vessel has a storage volume of 143-L and can be operated up to 5 MPa pressure. The total mass of the vessel including valves and piping is about 100 kg.

with minimal losses and advancing the design for a more compact efficient storage system. Figure 1.3 shows an image of an early version designed by Linde for BMW to fit between the boot and rear seats of the car. The capacities of these tanks were mostly between 7 and 12 kg of hydrogen, with maximum fill pressures usually below 1 MPa. Advancement in the design and materials of construction of the liquid hydrogen vessels, however, showed marked improvement in gravimetric system capacity (i.e., including the valve and piping box) from approximately 9% to 15% H_2 by mass, following a 50% reduction in weight of the vessel mainly due to switching to almost all aluminum rather than steel for most components as used in the preceding vessel designs (Wolf, 2013).

Even though cryogenic hydrogen approaches can offer the potential for large capacity and compact-size storage vessels, evaporative losses due to system heat transport could be the Achilles heel. Heat flow into the interior of the storage tank will unavoidably cause the liquid hydrogen fuel to gradually evaporate or boil-off, requiring relief valves to vent the hydrogen vapor as the internal tank pressure rises above 0.4–0.6 MPa (or corresponding temperatures of 26–28 K). As these systems are required to operate at 20–25 K, there is a large temperature differential to the environment, thus almost unavoidable heat input from thermal conduction, convection, and radiation. Among these, thermal conduction through pipes, cables, mounts, etc. (i.e., parasitic losses) and thermal radiation from the environment are the main contributors.

Maintaining a low surface-to-volume (S/V) ratio of the tank is key to minimizing heat flow and is normally accomplished by working with cylindrical tank structures as shown in Figure 1.3. These tank structures still require evacuated metallic double-walled containers and multiple layers of vacuum super insulation, known as multi-layer insulation (MLI), which is normally either alternating metallic and/or thermally insulated polymeric or glass films chosen to minimize thermal conduction and reduce evaporative losses to between 0.3% and 3% per day. In fact, to minimize heat transfer through the insulation, 15 or more alternating layers of highly reflective foil and low-conductivity spacers are placed in the inner shell of a typical MLI installation as

Figure 1.4 Multilayer insulation from the Space Station Mir's docking module.

shown in Figure 1.4. In some extremely high-performance NASA applications, the effective mean apparent conductivity of MLI can reach as low as 0.01 mW/mK (Barron, 1985). While high-performance MLI can go a long way toward reducing heat transfer, even a few watts can result in evaporative losses when a vehicle is not operated frequently. Therefore, it is imperative that the technology be matched to the appropriate automotive application (i.e., consider driving habits) to optimize storage system efficiency.

Additionally, a related issue is concerned with thermal losses during refilling the vessels. Since intimate contact of the transfer line (i.e., refueling line) and the tank must occur, the transfer line must be cooled to approximately 20 K using a helium purge, i.e., establishing a "cold finger" connection point. Considering both internal and external effects of the potential for thermal losses, these systems usually lead to unacceptable hydrogen losses, especially for terrestrial applications. The success of liquid hydrogen storage ultimately lies in improving insulation, vessel design, and liquefaction infrastructure design and efficiency.

1.2.4 Cryo-compressed hydrogen

One approach that originated at Lawrence Livermore National Laboratory (LLNL) and was pursued by BMW combines high-pressure and cryogenic storage vessels to increase gravimetric and volumetric capacities while overcoming some of the shortcomings of both compressed hydrogen and pure liquid hydrogen storage. At cryogenic temperatures, gaseous hydrogen is much more dense—at atmospheric pressure, hydrogen at 80 K is 3.73 times denser than hydrogen at 298 K. At higher pressures and cryogenic temperatures, similar increases in density are realized; thus, delivering an increase in vehicle range of the same factor. Assuming a 60-mpg equivalent hydrogen fuel cell vehicle with an insulated storage tank designed to hold 5 kg of usable hydrogen at 298 K and 70 MPa (sufficient for a 300-mile range), the same tank would be able to hold 16.5 kg of hydrogen at 80 K (sufficient for a 1295-mile range, far exceeding current standards). Therefore, with the use of an insulated pressure vessel built to minimize heat leakage and withstand moderate pressures, the compact benefits of liquid hydrogen storage and the simplicity of ambient-temperature compressed hydrogen storage can be exploited. However, the relative benefits of cryo-compressed hydrogen dramatically decrease at pressures greater than 20 MPa. In fact, at 40 MPa, the advantage factor decreases by 25%. Nonetheless, a notable advantage of cryo-compressed hydrogen over liquid hydrogen is that it requires less energy to produce where the theoretical compression to 20 MPa and cooling work to 80 K is approximately 10 MJ/kg, which is similar to the work required to compress hydrogen to roughly 50 MPa while remaining at 20–30% of the liquefaction work. However, most current refueling stations do not take advantage of this aspect as they deliver from either liquid hydrogen or compressed sources.

One of the most significant advantages of the cryo-compressed approach is the concept of a flexibly fueled system. Since the vessel is designed with the ability to operate at cryogenic temperatures and moderate to high pressures (20 MPa and higher), the insulated pressure vessel can be flexibly fueled with liquid hydrogen, cold-compressed hydrogen, ambient-temperature compressed hydrogen, or cryo-compressed hydrogen. Additionally, the ability to operate at moderate pressures (i.e., 35 MPa) significantly extends the dormancy period, allowing for containment of gas phase hydrogen as the temperature of the gas increases due to heat transfer. LLNL has performed extensive testing on insulated pressure vessels to address thermodynamic considerations such as dormancy and concerns of mechanical and thermal stresses due to the broad operating ranges of this approach. A schematic of an advanced cryo-compressed vessel designed by LLNL and Spencer Composites is shown in Figure 1.5. The vessel contains a Type III composite inner vessel designed to maintain hydrogen gas at 35 MPa surrounded by vacuum space filled with numerous layers of MLI and then a metallic jacket. An

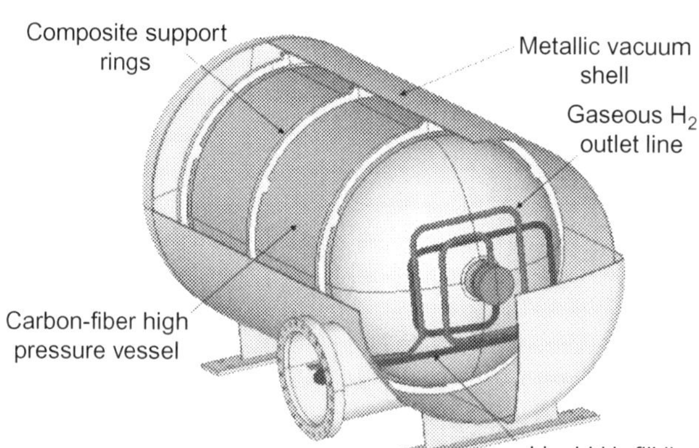

Figure 1.5 Schematic of a Gen 2 cryo-compressed hydrogen storage vessel designed by LLNL. The vessel contains a Type III composite inner vessel designed to maintain hydrogen gas at 35 MPa surrounded by vacuum space filled with numerous layers of MLI and then a (including all internal piping and valves) are 187 kg and 323 L, respectively.

advanced prototype of this vessel has been demonstrated onboard an experimental hybrid hydrogen Toyota Prius (Aceves et al., 2010). While packaging of the vessel onboard the Prius was not optimal, it delivered over 1000 km driving range from 10.7 kg of hydrogen stored and an estimated 2 weeks of evaporation free hydrogen hold time resulting in an estimated heat transfer rate of 3–4 W. Ongoing studies from LLNL in conjunction with BMW are addressing cyclic fatigue, burst testing, advanced superinsulation designs as well as improving the volumetric size of the vessel for optimal automotive packaging.

Cryo-compressed hydrogen storage is discussed in more detail in the chapter by Ahluwalia.

1.3 Material-based hydrogen storage

While compressed gas technology satisfies the immediate needs for onboard hydrogen storage in presently deployed PEM fuel cell applications, improvements gained by adopting lower containment pressures and higher volumetric density can offer the possibility of better well-to-power plant efficiency gained through the use of materials-based systems. In the following sections, we delineate some of the materials-based storage categories that have been studied recently. A number of excellent recent review articles exist that address the many specific systems that have been investigated (Klebanoff et al., 2014; Sakintuna et al., 2007; Orimo et al., 2007; Klebanoff, 2013), and we therefore only note some of the principles behind these material categories and cite only particular compounds that exemplify the categories discussed. Note that as with compressed gas systems where the exterior tank, balance-of-plant components, and diagnostic components that make up the total system, add substantial mass, volume, and cost, these quantities should also be considered in the metrics of materials-based storage systems. Because the system make-up will have different engineering and balance of plant requirements depending on the nature of the material being used (Didisheim et al., 1984; Graetz et al., 2011; Pasini et al., 2013; Felderhoff et al., 2007), in the rest of this chapter the volumetric or gravimetric density descriptions apply to values calculated on the basis of material densities only. Note also that a number of different conventions have arisen in defining hydrogen capacity. In this chapter the mass percent is defined as the mass of hydrogen divided by the mass of the dehydrogenated ab(d)sorbent material only (H/M), whereas weight percent (wt%) refers to the hydrogen mass the mass of ad(b)sorbent material plus stored hydrogen ($H/(M+H)$).

While theoretical density values can be of some use for comparative purposes, under conditions where the ultimate goal is to supply hydrogen to a PEM fuel cell, the quantity of hydrogen that a material-based storage system can supply will necessarily be smaller than that the total quantity of hydrogen contained within the material. This is due in part to the requirements of meeting minimum pressure and delivery rate requirements. The total amount of hydrogen that can be provided under minimum delivery requirements is the usable hydrogen capacity.

1.3.1 Transition metal hydrides

Metal hydrides (MH_x) continue to be the most technologically relevant class of hydrogen storage materials given their utility in a range of applications including neutron moderation (Vetrano, 1970), electrochemical cycling (Young and Nei, 2013), thermal storage (Felderhoff and Bogdanovic, 2009), heat pumps (Lototskyy et al., 2010), and purification/separation (Wang et al., 2005). While many alkali or *sp* metals also form saline or covalent hydrides the recognition that transition metal hydrides, in particular, are in fact distinct compounds as distinguished from solid solutions of hydrogen is attributable to the band-structure calculations of Switendick (1978). However, the attribute of relevance where application solutions are sought is the solid solution region of the phase diagram, between metal and hydride phases. In *d*-electron transition metals and intermetallic alloy hydrides, stability against cycling over this solubility range can yield a long flat plateau pressure region of the isotherm. Given the small size of hydrogen and the number of available interstices, some intermetallic systems can have solubility limits with a volumetric hydrogen density with values that exceed that of liquid hydrogen (Sandrock, 1999). This high level of solubility arises from the high jump frequency that allows atomic hydrogen to diffuse readily through the metal lattice through phonon dominated tunneling at technologically relevant temperatures. While it is tempting to assume that the ultimate volumetric density of hydrogen is accurately represented by the stoichiometry of the metal hydride phase itself, for practical applications where high cycle life is required, the quantity of hydrogen available to an engineering system is actually limited to the quantity across the solid-solution range between the alloy and hydride phases (Wolf, 2013).

While there is continued and steady research in metal hydrides, most recently on possible applications for solar thermal storage (Harries et al., 2012), excellent earlier reviews (Sandrock et al., 1992) and texts (Mueller et al., 1968) provide pertinent information and insight on this class of materials that are still relevant to modern studies. A compendium on somewhat simpler binary metal hydride systems that include thermodynamic, crystallographic, and phase equilibria diagrams illustrate the important principles behind hydrogen metal interactions (Manchester, 2000), even though it is the intermetallic forming hydrides such as A_2B, AB, AB_2, and AB_5 alloys that are used for the most demanding present-day applications. A comprehensive database of published hydrogen alloys with relevant properties has also been compiled by Sandrock and Thomas (Hydride Information Center (Hypark), 2015) as part of what has evolved into a more general hydrogen storage materials database.

The thermodynamics of hydrogen adsorption/desorption govern their temperature and pressure range of applicability. The change in enthalpy and entropy for the adsorption/desorption reaction over the constant pressure region of the phase diagram can be described by application of the isothermal Gibbs free energy to the van't Hoff equation (Equation (1.3)) to yield the linear form, which is

$$\ln P = \frac{\Delta H}{RT} - \frac{\Delta S}{R} \tag{1.3}$$

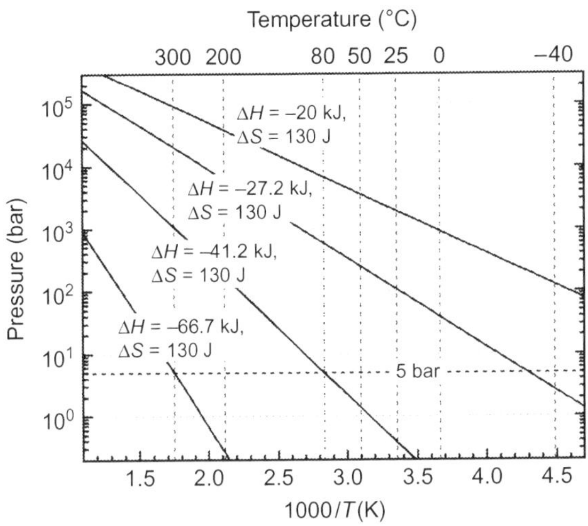

Figure 1.6 Calculated van't Hoff data for several enthalpy values assuming a constant entropy value.

It is therefore possible to relate the thermodynamics over the solid solution region shown schematically in Figure 1.6, to ln(P). The enthalpy of formation (ΔH) is negative for this category, which reflects the endothermic release of hydrogen for the reaction. ΔS is the entropy of formation, R is the gas constant, and T is the absolute temperature. An analysis of the thermodynamic behavior is especially critical for application specific equilibrium pressure requirements. While many articles cite a 0.1 MPa equilibrium value in determining an appropriate value or range of ΔH, PEM fuel cells may require pressure higher than 0.5 MPa. Thus, the van't Hoff equation can be used for the initial assessment of the viability of a hydride for a particular application given an operating temperature and pressure range. It is also necessary to know the entropy value as the slope and enthalpy at a particular pressure and temperature can vary depending on the value of ΔS. While it is generally a given that ΔS is dominated by the change in gas phase entropy when hydrogen transforms from diatomic gas to atomic hydrogen into the metal lattice, ΔS_{H_2} Rudman and Sandrock (1982) noted that a large change in the gas phase entropy value should accompany a change in temperature. However, cognizant of the nearly constant value of ΔS that is determined from empirical work Oates and Flanagan (1981) noted that for a given plateau pressure, a nearly constant ΔS would be attributed to a similarly sloped increasing hydride entropy as determined from an Einstein oscillator model. Interestingly, assuming that the entropy change ΔS is nearly equivalent to the change in gas entropy at ~130 J/mol K, requires an enthalpy for a 0.5 MPa equilibrium pressure over a temperature range of −40 to 80 °C between 27 and 41 kJ/mol of hydrogen.

AB_5 alloys remain the primary material of choice when high hydrogen cyclability is paramount under technologically modest conditions. When costs are a concern, minimally processed mischmetal is often adequate, but the most demanding of applications will rely on $LaNi_{5-x}M_x$ (M = e.g., Al, Mn, Sn). The use of $LaNi_{4.8}Sn_{0.2}$, for instance, provided the first stage of closed-loop cooling for the bolometers used in the Planck mission, and far exceeded the 18-month life time to operate for over 4 years,

Figure 1.7 Difference in jump frequency for H, C, and Fe in α-Fe. Adapted from Hirth (1980).

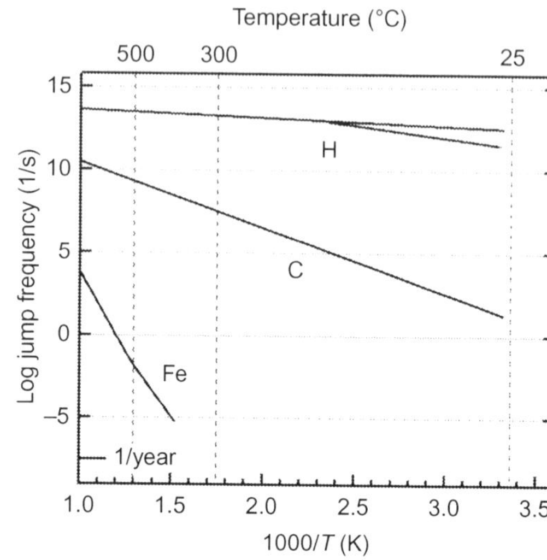

collecting high angular resolution, high dynamic range images of the cosmic microwave background (New York Times, 2015).

While the thermodynamics of hydrides play a crucial role in assessing the suitability of a particular system, the kinetics can also be critical. Metal hydrides typically can be expected to have reasonably fast kinetics, brought on by a jump frequency, that for instance in α-Fe, can reach 10^{12}/s at 298 K as shown in Figure 1.7 (Hirth, 1980). This can be compared to the jump frequency of C, which is on the order of 1/10 s. The split in H jump frequency is attributed to defects that act as H traps.

1.3.2 Complex hydrides

Unlike the solid solutions of reversible interstitial metal hydrides described in the previous section, the materials referred to as "complex hydrides" consist typically of alkali or alkaline earth elements that are ionically bonded to a complex anion. The anions themselves can consist of central atoms that are typically transition or main-group metals or metalloids (e.g., Fe, Ni, B, Al) or N, to which hydrogen is covalently bonded. As with metal hydrides, the desorption reaction in this series of compounds can be endothermic upon hydrogen release. With the pursuit of higher gravimetric and volumetric hydrogen density materials, the use of complex hydrides that consist of low Z cations offers possibilities for use given a complete dehydrogenation of the stoichiometric hydride. The invigorated level of effort in investigating this class of materials can be traced back to the original work of Bogdanovic and Schwickardi (1997) when it was discovered that Ti-mediated dehydrogenation could be effected in $NaAlH_4$, prior to melting (Reilly et al., n.d.). However, as with a number of complex hydride dehydrogenation reactions, the actual reaction pathway and diffusion mechanism seldom consists of a simple route where all of the bound hydrogen is released over a solid solution range that consists solely of hydride and nonhydride phases of these materials (Callini et al., 2014). In systems where the ionic character of bonding of the complex hydride plays a role, electrostatic forces can dominate and impose limits on the diffusion mechanisms that may be required to effect the atom mobility needed for hydrogen hydrogenation/dehydrogenation within a bulk crystal. Because the magnitude of electrostatic effects is invariably large, mobile defects like interstitials and vacancies with energies that are on the order of an eV should be present, the paucity of which can result in the relatively poor kinetics that are endemic to this class of hydrogen storage materials. But for this category of materials no single diffusion mechanism can be used to describe hydrogenation/dehydrogenation behavior given the number of possible pathways that may be dependent on defect densities, grain boundaries, surface tension, concentration gradients, as well as intermediate, metastable, or stable phases that can lie along the reaction pathway. The following is a discussion of several specific examples that illustrate the issues accompanying the consideration of complex hydride materials as a hydrogen storage medium.

Given the aggressive gravimetric and volumetric targets established for light-duty vehicles, and while somewhat high from a thermodynamic standpoint, a compound like $LiBH_4$ appears to be a possibility. While the formation enthalpy of 190.5 kJ/mol is high, the decomposition enthalpy of ~66.7 kJ/mol hydrogen, given the reaction $LiBH_4 \leftrightarrow LiH + B + 3/2\ H_2$, would require ~300°C to effect a 0.5 MPa equilibrium pressure. However, a dissociation reaction such as this one that produces elemental boron would appear to preclude the possibility of a simple rehydrogenation reaction.

In addition to the class of single compound materials that have just been discussed, destabilization reactions that rely on altering the normally high enthalpy of dehydrogenation of a single hydride by reacting a combination of compounds has been pursued both empirically (Vajo et al., 2005) and computationally (Alapati et al., 2007). While the reaction enthalpies suggest dehydrogenation/rehydrogenation under modest conditions, systems based on destabilization reactions still rely on solid-state diffusion so relatively high temperatures are still required to overcome the activation barriers to atom mobility. For instance, the most compelling of the destabilized systems that again uses $LiBH_4$ in the reaction, $MgH_2 + 2LiBH_4 \rightarrow 2MgB_2 + 2LiH + 4H_2$ has a reaction enthalpy of 46 kJ/mol hydrogen and would thus normally be expected to have an equilibrium pressure of 0.5 MPa at 122°C but actual temperatures >300°C (Felderhoff and Bogdanovic, 2009) are required before appreciable dehydrogenation occurs. Still, for higher temperature applications, material-based hydrogen using solid-state reactions can offer a storage solution.

1.3.3 Chemical hydrogen storage

This category generally refers to covalently bound hydrogen in either solid or liquid form and consists of compounds that generally have the highest density of hydrogen. Hydrogen release from chemical hydrogen systems is usually exothermic or has a small endothermic enthalpy, thus rehydrogenation typically requires the dehydrogenated product to be rehydrogenated offboard by processes other than through the application of hydrogen pressure. For single-use, high value-added applications, this approach can be useful but consideration of chemical hydrogen for large-scale transportation use will require offboard rehydrogenation of the spent product, which at present greatly increases the well-to-powerplant fuel cost. Moreover, as with complex hydrogen compounds, release from chemical hydrogen compounds can have reaction pathways that may not be easily predicted.

The dehydrogenation of chemical hydrogen itself can be accomplished either hydrolytically by reaction with water (Kojima et al., 2004), or thermolytically, where the compound is heated, as with ammonia borane (NH_3BH_3) (Gutowska et al., 2005) and alane (AlH_3) (Graetz and Reilly, 2007). Given their high hydrogen capacity, ammonia borane and amide and amine compounds and variants remain systems of interest. As in the case of complex hydrides, the reaction pathway for dehydrogenation of chemical hydrogen is not necessarily straightforward and may be governed to the extent to which the dehydrogenation product remains as a liquid phase or slurry that can be handled easily.

In addition to covalently bound hydrogen as solids, compounds that are capable of binding hydrogen as liquids have been studied. Examples of systems based on liquid carriers include n-ethylcarbazole (Cooper et al., 2005) and methyl-cyclopentane (Luo et al., 2013) as shown in Figure 1.8. In addition to the need for offboard rehydrogenation of the spent product, some of the difficulty in working with these liquids is that the dehydrogenated product can solidify, complicating the handling of the spent materials and thus affecting their overall utility.

1.3.4 Hydrogen sorbents

Unlike other forms of solid-state storage, one of the advantages in the use of adsorbents as a storage medium is that dihydrogen retains its molecular form throughout the adsorption/desorption cycle with minimal activation energy. The primary disadvantage in the use of sorbents is the relatively weak adsorption enthalpies that are typical of gas–solid interactions when compared to that of bond formation with chemical hydrogen, or interstitial atomic hydrogen in metal hydrides. Additionally, the van der Waals dimension of molecular hydrogen is large in comparison to atomic hydrogen, putting limits on the overall volumetric density that systems based on dihydrogen can achieve.

Under many conditions, the nature of the adsorption process can still offer volumetric density advantages over the use of compressed gas storage (Chahine and Bénard, 1998), although not as high as intermetallic hydride densities. For dihydrogen that adsorbs onto a substrate, the London dispersion forces that can be regarded as transient nonpolar dipoles of polarizable substances (Parasegian, 2006) provides a

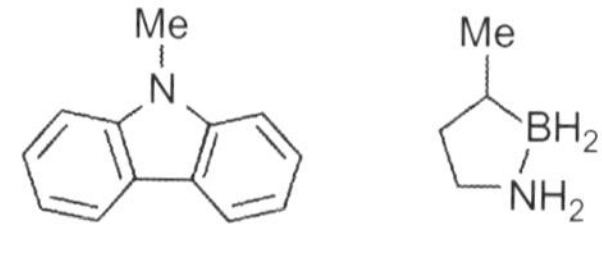

Figure 1.8 Structures of n-ethyl carbazole on the left and methyl cyclopentane on the right.

convenient framework in appreciating the nature of adsorption above the triple-point temperature and pressure.

Unlike chemical hydrogen or metal hydrides, the data that is measured either gravimetrically or volumetrically is the Gibbs surface excess (Gibbs, 1993). This quantity is a true measurement of increased concentration of gas at the surface of an adsorbent over the concentration that might be expected from the gas law and is the only data that is determined empirically.

The enthalpy range for hydrogen adsorbents can run typically from 5 to 10 kJ/mol. At this point, we should note that the term "enthalpy" for sorption *is not* a single valued quantity, although it is often referenced this way. When reported, a single enthalpy value generally refers to the "differential enthalpy of adsorption at zero coverage," more commonly referred to as the Henry's Law value, where the gas concentration varies linearly with pressure (Rouquerol et al., 1999). The Henry's Law value is essentially the enthalpy or potential assigned to the initial molecule adsorbed onto a surface. This value will generally be the highest quantity as the initial molecule adsorbs onto the highest potential site and adsorbs free of the influence of neighboring adsorbed gas molecules. Of more general technological interest is the isosteric enthalpy of adsorption, which traces the change in enthalpy over as large a range of coverage as pressure measurements allow. In all known instances for hydrogen sorption, the isosteric enthalpy of adsorption is expected to decay monotonically.

The sorbents that have been investigated recently range from coordination polymers (Robson, 2008) to activated carbons. These classes of materials have the initial requirements for high gravimetric density as adsorption relies on having a high number of sites, upon which gas molecules can adsorb. For activated carbons, we can generally expect the absolute quantity of adsorbed gas molecules to vary as a function of surface area and micropore volume (Chahine and Bose, 1996). As the gravimetric adsorption is highly correlated with surface area, maximizing surface area consistent with retaining a high micropore volume is the geometric combination that will best satisfy the design criteria for successful adsorbents.

References

Aceves, S.M., Espinosa-Loza, F., Ledesma-Orozco, E., Ross, T.O., Weisberg, A.H., Brunner, T.B., Kircher, O., 2010. High-density automotive hydrogen storage with cryogenic capable pressure vessels. Int. J. Hydrogen Energy 35, 1219–1226.

Agency for Toxic Substances and Disease Registry of CDC, 1998. Toxicological Profile for Jet Fuels JP-5 and JP-8. August, CAS#: 8008-20-6, http://www.atsdr.cdc.gov/ToxProfiles/tp.asp?id=773&tid=150.

Ahluwalia, R.K., Hua, T.Q., Peng, J.-K., 2007. Fuel cycle efficiencies of different automotive on-board hydrogen storage options. Int. J. Hydrogen Energy 32, 3592–3602.

Alapati, S.V., Johnson, J.K., Sholl, D.S., 2007. Using first principles calculations to identify new destabilized metal hydride reactions for reversible hydrogen storage. Phys. Chem. Chem. Phys. 9, 1438–1452.

Barron, R.F., 1985. Multilayer insulations, Cryogenic Systems, second ed. Oxford University Press, New York, p. 397.

Bogdanovic, B., Schwickardi, M.J., 1997. Ti-doped alkali metal aluminium hydrides as potential novel reversible hydrogen storage materials. J. Alloys Compd. 253–254, 1–9.
Bowman, B., Anton, D., Stetson, N., 2013. Engineering assessments of condensed-phase hydrogen storage systems. In: Klebanoff, L. (Ed.), Hydrogen Storage Technology: Materials and Applications. CRC Press, Boca Raton, pp. 385–403.
Brooks, K., Devarakonda, M., Rassat, S., Holladay, J., 2011. Systems modeling of chemical hydride hydrogen storage materials for fuel cell applications. J. Fuel Cell Sci. Technol. 8 (6), 061021. http://dx.doi.org/10.1115/1.4004477 (6 pages).
Callini, E., Borgschulte, A., Hugelshofer, C.L., Ramirez-Cuesta, A.J., Züttel, A., 2014. The role of Ti in alanates and borohydrides: catalysis and metathesis. J. Phys. Chem. C 118, 77–84. http://dx.doi.org/10.1021/jp407999r.
Chahine, R., Bénard, P., 1998. Adsorption storage of gaseous hydrogen at cryogenic temperatures. In: Kittel, P. (Ed.), Advances in Cryogenic Engineering, vol. 43. Plenum Press, New York, pp. 1257–1264.
Chahine, R., Bose, T.K., 1996. Characterization and optimization of adsorbents for hydrogen storage. In: Hydrogen Energy Progress XI: Proceedings of the 11th World Hydrogen Energy Conference. Intl. Assoc. of Hydrogen Energy, Stuttgart, pp. 1259–1263.
Cooper, A.C., Fowler, D.E., Scott, A.R., et al., 2005. Hydrogen storage and delivery by reversible hydrogenation of liquid-phase hydrogen carriers. Abstr. Pap. Am. Chem. Soc. 229, U868, Part: 1 Meeting Abstract: 113-FUEL.
Department of Energy—Office of Energy Efficiency and Renewable Energy. Fuel cell technologies program multi-year research, development and demonstration plan: planned activities for 2011–2020. Avaliable from: http://energy.gov/sites/prod/files/2014/11/f19/fcto_myrdd_storage.pdf (17.02.15.).
Didisheim, J.J., Zolliker, P., Yvon, K., Fischer, P., Schefer, J., Gubelmann, M., Williams, A.F., 1984. Dimagnesium iron(II) hydride, Mg_2FeH_6, containing octahedral FeH_6^- anions. Inorg. Chem. 23, 1953.
Felderhoff, M., Bogdanovic, B., 2009. High temperature metal hydrides as heat storage materials for solar and related applications. Int. J. Mol. Sci. 10 (1), 325–344. http://dx.doi.org/10.3390/ijms10010325.
Felderhoff, M., Weidenthaler, C., von Helmolt, R., Eberle, U., 2007. Hydrogen storage: the remaining scientific and technological challenges. Phys. Chem. Chem. Phys. 9, 2643–2653.
Fuel Cell History Project. Smithsonian National Museum of American History. Available from: http://americanhistory.si.edu/fuelcells/origins/origins.htm (11.01.15.).
Gibbs, J.W., 1993. The Scientific Papers of J. Willard Gibbs. Ox Bow Press, Woodbridge, p. 219 (reprint).
Graetz, J., Reilly, J.J., 2007. Kinetically stabilized hydrogen storage materials. Scr. Mater. 56 (10), 835–839.
Graetz, J., Reilly, J.J., Yartys, V.A., Maehlen, J.P., Bulychev, B.M., Antonov, V.E., Tartsov, B.P., Gabis, I.E., 2011. Aluminum hydride as a hydrogen and energy storage material: past, present and future. J. Alloys Compd. 509S, S517–S528.
Gutowska, A., Li, L., Shin, Y., Wang, C.M., Li, X.S., Linehan, J.C., Smith, R.S., Kay, B.D., Schmid, B., Shaw, W., Gutowski, M., Autrey, T., 2005. Nanoscaffold mediates hydrogen release and the reactivity of ammonia borane. Angew. Chem. Int. Ed. 44, 3578–3582.
Harries, D.N., Paskevicius, M., Sheppard, D.A., Price, T.E.C., Buckley, C.E., 2012. Concentrating solar thermal heat storage using metal hydrides. Proc. IEEE 20 (2), 539–549. http://dx.doi.org/10.1109/JPROC.2011.2158509.

Hirth, J.P., 1980. Effects of hydrogen on the properties of iron and steel. Metall. Trans. A 11A, 861–890.

Hua, T.Q., Ahluwalia, R.K., Peng, J.-K., Kromer, M., Lasher, S., McKenney, K., Law, K., Sinha, J., 2011. Technical assessment of compressed hydrogen storage tank systems for automotive applications. Int. J. Hydrogen Energy 36, 3037–3049.

Hydride Information Center (Hypark) database incorporated into the US DOE Hydrogen Storage Materials Database. Available from: http://www.hydrogenmaterialssearch.govtools.us/ (16.02.15.).

Klebanoff, L. (Ed.), 2013. Hydrogen Storage Technology: Materials and Applications. CRC Press, Boca Raton.

Klebanoff, L.E., Ott, K.C., Simpson, L.J., O'Malley, K., Stetson, N.T., 2014. Accelerating the understanding and development of hydrogen storage materials: a review of the five-year efforts of the three DOE Hydrogen Storage Materials Centers of Excellence. Metall. Mater. Trans. E 1 (2), 81–117.

Kojima, Y., Kawai, Y., Kimbara, M., Nakanishi, H., Matsumoto, S., 2004. Hydrogen generation by hydrolysis reaction of lithium borohydride. Int. J. Hydrogen Energy 29, 1213–1217.

Koppel, T., 1999. Powering the Future: The Ballard Fuel Cell and the Race to Change the World. John Wiley & Sons Canada Ltd, USA, pp. 89–103 (Chapter 6).

Koresh, J., Soffer, A., 1980. Study of molecular-sieve carbons, 2. Estimation of cross-sectional diameters of non-spherical molecules. J. Chem. Soc. Farad. Trans. I 76, 2472–2485.

Lototskyy, M.V., Yartys, V.A., Pollet, B.G., Bowman Jr., R.C., 2010. Metal hydride hydrogen compressors: a review. Int. J. Hydrogen Energy 39, 5815–5851.

Luo, W., Campbell, P.G., Zakharov, L.N., Liu, S.Y., 2013. A single-component liquid-phase hydrogen storage material. J. Am. Chem. Soc. 135 (23), 8760.

Manchester, F.D. (Ed.), 2000. Phase Diagrams of Binary Hydrogen Alloys. In: Monograph Series on Alloy Phase Diagrams, 13. ASM International, OH, USA.

McWhorter, S., Ordaz, G., 2013. Onboard Type IV Compressed Hydrogen Storage Systems - Current Performance and Cost, US DOE Fuel Cell Technologies Office Record. Available from http://hydrogen.energy.gov/pdfs/13010_onboard_storage_performance_cost.pdf (29.07.15.).

McWhorter, S., Read, C., Ordaz, G., Stetson, N., 2011. Materials-based hydrogen storage: attributes for near-term, early market PEM fuel cells. Curr. Opin. Solid State Mater. Sci. 15, 29–38.

Mueller, W.M., Blackledge, J.P., Libowitz, G.G. (Eds.), 1968. Metal Hydrides. Academic Press, New York.

Mueller, C.J., Cannella, W.J., Bruno, T.J., Bunting, B., Dettman, H.D., Franz, J.A., Huber, M.L., Natarajan, M., Pitz, W.J., Ratcliff, M.A., Wright, K., 2012. Methodology for formulating diesel surrogate fuels with accurate compositional, ignition-quality, and volatility characteristics. Energy Fuels 26, 3284–3303. http://dx.doi.org/10.1021/ef300303e.

Oates, W.A., Flanagan, T.B., 1981. The solubility of hydrogen in transition metals and their alloys. Prog. Solid State Chem. 13, 193–283.

Orimo, S.-I., Nakamori, Y., Eliseo, J.R., Züttel, A., Jensen, C.M., 2007. Complex hydrides for hydrogen storage. Chem. Rev. 107 (10), 4111–4132.

Parasegian, V.A., 2006. van der Waals Forces. Cambridge University Press, New York.

Pasini, J.-M., Corgnale, C., van Hassel, B.A., Motyka, T., Kumar, S., Simmons, K.L., 2013. Metal hydride material requirements for automotive hydrogen storage systems. Int. J. Hydrogen Energy 38 (23), 9755–9765.

Paster, M.D., Ahluwalia, R.K., Berry, G., Elgowainey, A., Lasher, S., McKenney, K., Gardiner, M., 2011. Technical assessment of compressed hydrogen storage tank systems for automotive applications. Int. J. Hydrogen Energy 36, 14534–14551.

Reilly, J.J., Wiswall, R.H., Waide, C.H. Motor vehicle storage of hydrogen using metal hydrides, EPA report TEC-75/001.

Robson, R., 2008. Design and its limitations in the construction of bi- and poly-nuclear coordination complexes and coordination polymers (aka MOFs): a personal view. Dalton Trans. 38, 5113–5131.

Rouquerol, F., Rouquerol, J., Sing, K., 1999. Adsorption by Powders and Porous Solids. Academic Press, London.

Rudman, P.S., Sandrock, G.D., 1982. Metallurgy of rechargeable hydrides. Annu. Rev. Mater. Sci. 12, 271–294.

Sakintuna, B., Lamari-Darkrim, F., Hirscher, M., 2007. Metal hydride materials for solid hydrogen storage: a review. Int. J. Hydrogen Energy 32, 1121–1140.

Sandrock, G., 1999. A panoramic overview of hydrogen storage alloys from a gas reaction point of view. J. Alloys Compd. 293–295, 877–888.

Sandrock, G., Suda, S., Schlapbach, L., 1992. Ch. 5. Applications. Hydrogen in Intermetallic Compounds II, Topics in Applied Physics, vol. 67. Springer-Verlag, Berlin, pp. 197–258, Dantzer, P., 1997. Metal hydride technology: a critical review. Hydrogen in Metals III, Topics in Applied Physics, vol. 73. Springer-Verlag, Berlin, pp. 279–340.

Schremp, G., Janusch, N., 2009. Fuel Delivery Temperature Study. California Energy Commission, USA, CEC-600–2009–002-CMF.

Selected Speeches of President George W. Bush, 2001–2008, pp. 152–153. Available from: http://georgewbush-whitehouse.archives.gov/infocus/bushrecord/documents/Selected_Speeches_George_W_Bush.pdf or Whitehouse Press Release, January 28, 2003. Available from Whitehouse historical site: http://georgewbush-whitehouse.archives.gov/news/releases/2003/01/20030128-25.html (23.11.14.).

Simmons, K. Synergistically enhanced materials and design parameters for reducing the costs of hydrogen storage tanks, DOE Fuel Cell Technologies Office 2014 annual progress report. Available from: http://www.hydrogen.energy.gov/pdfs/progress14/iv_f_3_simmons_2014.pdf (17.02.15.).

Stetson, N.T., Ordaz, G., Adams, J., Randolph, K., McWhorter, S., 2013. The use of application specific performance targets and engineering considerations to guide hydrogen storage materials development. J. Alloys Compd. 580, S333–S336.

Stetson, N.T., Bowman, R.C., Olson, G.L., 2015. Overview of hydrogen storage, transportation, handling and distribution. In: Sheriff, S.A., Goswami, D.Y., Stefanakos, E.K., Steinfeld, A. (Eds.), Handbook of Hydrogen Energy. CRC Press, Boca Raton, pp. 567–592.

Switendick, A.C., 1978. The change in electronic properties on hydrogen alloying and hydride formation. In: Alefeld, G., Völkl, J. (Eds.), Hydrogen in Metals I. Topics in Applied Physics, vol. 28. Springer, Berlin, pp. 101–129.

The New York Times. Universe as an infant: fatter than expected and kind of lumpy. Available from: http://www.nytimes.com/2013/03/22/science/space/planck-satellite-shows-image-of-infant-universe.html?pagewanted=all&module=Search&mabReward=relbias%3Ar%2C%7B%221%22%3A%22RI%3A8%22%7D (16.02.15.).

U.S. Imports of Crude Oil and Petroleum Products, U.S. Energy Information Administration. Available from: http://www.eia.gov/dnav/pet/hist/LeafHandler.ashx?n=pet&s=mttimus1&f=a (23.11.14.).

United Nations Global Technical Regulations. Global technical regulation on hydrogen and fuel cell vehicles. Available from: http://www.unece.org/fileadmin/DAM/trans/main/wp29/wp29wgs/wp29gen/wp29registry/ECE-TRANS-180a13e.pdf (17.02.15.).

Vajo, J.J., Skeith, S.L., Mertens, F., 2005. Reversible storage of hydrogen in destabilized $LiBH_4$. J. Phys. Chem. B 109 (9), 3719–3722.

Vetrano, J.B., 1970. Hydrides as neutron moderator and reflector materials. Nucl. Eng. Des. 14 (3), 390.

Wang, Z.M., Li, V., Chan, S.L.I., 2005. Review of alloy membranes/film for hydrogen separation or purification. J. Rare Earths 23, 611–616.

Warshay, M., Prokopius, P.R., 1989. Fuel cells in space: yesterday, today and tomorrow. NASA Technical Memorandum 102366.

Westlake, D.G., 1983. Hydrides of intermetallic compounds: a review of stabilities, stoichiometries and preferred hydrogen sites. J. Less Common Met. 91, 1–20.

Wipke, K., et al., 2012. National fuel cell electric vehicle learning demonstration final report. Technical report NREL/TP-5600–54860, July, p. 13. Available from: http://www.nrel.gov/hydrogen/pdfs/54860.pdf (23.11.14.).

Wolf, J., 2013. Liquid-hydrogen technology for vehicles. MRS Bull. 27 (9), 684–687.

Young, K.-H., Nei, J., 2013. The current status of hydrogen storage alloy development for electrochemical applications. Materials 6, 4574–4608. http://dx.doi.org/10.3390/ma6104574.

Hydrogen liquefaction and liquid hydrogen storage

2

G. Valenti
Politecnico di Milano, Milano, Italy

Abbreviation

ASU	air separation unit
EOS	equation of state
EQHHPP	Euro-Quebec Hydro-Hydrogen Pilot Project
LPG	liquefied petroleum gas
NASA	US National Aeronautics and Space Administration
NIST	US National Institute of Standards and Technology
WE-NET	World Energy Network

Symbols and constants

α	para-to-ortho concentration ratio
β	coefficient of (isobaric) thermal expansion (see Equation (2.3)) (K^{-1})
μ	Joule–Thomson coefficient (see Equation (2.1)) ($K\ Pa^{-1}$)
E_J	J-th quantum state of energy (J)
J	rotational quantum number, integer
I	molecular hydrogen moment of inertia, $4.67e^{-48}$ ($kg\ m^{-2}$)
M	hydrogen molar weight, 2.01588 ($kg\ kmol^{-1}$)
P	pressure (Pa)
R	universal gas constant, 8314.472 ($J\ kmol^{-1}\ K^{-1}$)
T	temperature (K)
T_i	inversion temperature (K)
c_p	specific heat at constant pressure (see Equation (2.4)) ($J\ kg^{-1}\ K^{-1}$)
c_p^0	specific heat at constant pressure in the hypothetical ideal gas state ($J\ kg^{-1}\ K^{-1}$)
g_J	statistical weight of J-th quantum level
g_e	statistical weight of an electronic level
h	specific enthalpy ($J\ kg^{-1}$)
$\tilde{h}$	Planck constant, $6.62606896e^{-34}$ (J s)
k	Boltzmann constant, $1.3806504e^{-23}$ ($J\ K^{-1}$)
i	nuclear spin
s	specific entropy ($J\ kg^{-1}\ K^{-1}$)
v	specific volume ($m^3\ kg^{-1}$)
x	molar fraction ($kmol\ kmol^{-1}$)

Compendium of Hydrogen Energy. http://dx.doi.org/10.1016/B978-1-78242-362-1.00002-X

Subscripts

eH_2 equilibrium-hydrogen
nH_2 normal-hydrogen
oH_2 orthohydrogen
pH_2 parahydrogen

2.1 Introduction: Why liquefying hydrogen?

In view of a vast hydrogen infrastructure, very large quantities of hydrogen may be distributed and stored as a cryogenic liquid. Hydrogen is being considered as a mobility fuel due to the ease of its use in rather conventional internal combustion engines or, more efficiently, in fuel cells. A comparison with common fuels can highlight the advantages and disadvantages of its adoption from the perspective of energy content. Figure 2.1 shows the lower heating value on mass and volume bases for hydrogen, methane, liquefied petroleum gas (LPG), gasoline, and diesel.[1] Hydrogen, methane, and LPG are represented for different pressure conditions as well as gaseous and liquid phases.

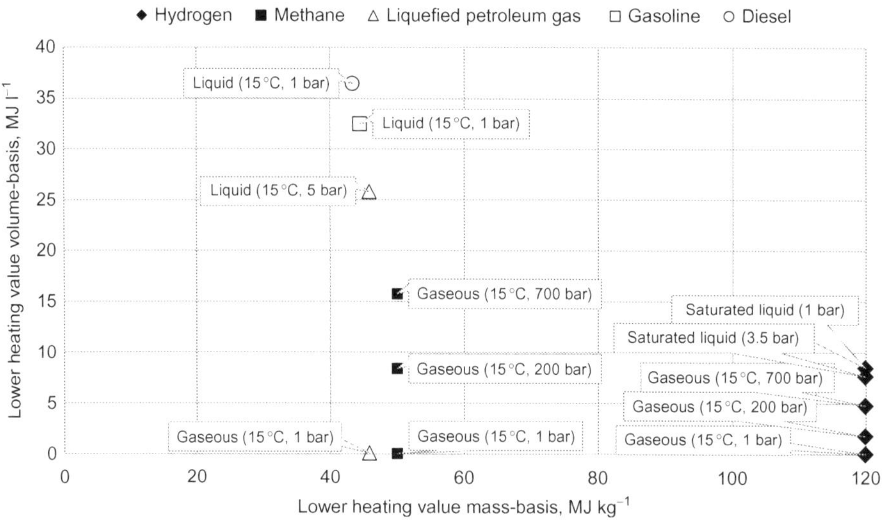

Figure 2.1 Lower heating values on a mass- and on a volume-basis for a number of mobility fuels (hydrogen, methane, liquefied petroleum gas—LPG, gasoline, and diesel). Solid markers refer to pure substances, while hollow to mixtures. Properties of hydrogen, methane, and LPG are computed with the code Refprop ver.9.1; LPG composition is taken to be simply 30% propane and 70% *iso*-butane. Properties of gasoline and diesel are representative, as they may vary appreciably in reality.

[1]The heating value is sometimes referred to as the calorific value, and the lower and higher heating values as net and gross heating values.

Hydrogen and methane are pure substances (indicated by solid markers), whereas LPG, gasoline, and diesel are mixtures (hollow markers). The adopted values for the mixtures are only representative as they may vary in reality. On a mass-basis, hydrogen shows the largest heating value, equal to 120 MJ kg^{-1}, which is more than twice as much as those of all other fuels. Methane shows 50.0 MJ kg^{-1}, LPG around 46 MJ kg^{-1}, gasoline slightly more than 44 MJ kg^{-1}, and diesel slightly less than 44 MJ kg^{-1}. Hence, hydrogen has a very high energy content compared to other common fuels of the same weight. In contrast, hydrogen is characterized by small heating values on a volume basis. At the ambient conditions of 15 °C and 1 bar, at which hydrogen exists as a gas phase, the lower heating values are as small as 0.01 MJ l^{-1} due to a very small density of just 0.084 kg m^{-3}, since hydrogen is a molecule with a very low molar mass. After pressurizing gaseous hydrogen to 200 or to 700 bar, the density increases to 15.0 and 40.2 kg m^{-3} and, hence, the heating value increases to 1.80 and 4.82 MJ l^{-1}, respectively. A pressure of 200 bar is common for gaseous hydrogen bottles, whereas the pressure of 700 bar is a target for hydrogen tanks that are under development. For comparison, compressed methane at the same pressures of 200 and 700 bar and at 15 °C has lower heating values of 8.39 and 15.8 MJ l^{-1}, which are greater than the analogs of hydrogen. In saturated liquid conditions at 1 and 3.5 bar, the hydrogen heating value doubles up to 8.50 and 7.68 MJ l^{-1}, respectively, due to the densities of 70.9 and 64.0 kg m^{-3}. LPG, which is actually a gas at 1 bar but a liquid at 5 bar, is characterized by a heating value of about 26 MJ l^{-1} at 5 bar. Gasoline and diesel are even better, with values of around 32 and 36 MJ l^{-1} Hence, hydrogen has a low energy content compared to other common fuels for the same volume (i.e., for the same capacity tank). Similar considerations hold true for higher heating values, noting that on a mass-basis the hydrogen value is 142 MJ kg^{-1} (almost 20% greater than its lower heating value), methane 55.6 MJ kg^{-1} (11% greater than the lower), LPG about 50 MJ kg^{-1} (+8%), and gasoline and diesel around 47 MJ kg^{-1} (+6%), and 45 MJ kg^{-1} (+3%), respectively.

A comparison with common mobility fuels shows one of the major, if not *the* major, issue with the use of hydrogen as an energy vector: its low energy content on a volume basis, i.e., on the basis of a tank capacity. As observed previously, this issue may be mitigated by liquefying hydrogen. However, hydrogen exists as a saturated liquid at 1 bar at a cryogenic temperature as low as around 20 K, i.e., around −253 °C. Hydrogen liquefaction is an energy-intensive process and liquid hydrogen storage is an advanced technology. Nevertheless, hydrogen liquefiers and liquid hydrogen vessels do exist around the world. Relatively large quantities of hydrogen are liquefied in central plants, distributed by cryogenic truck trailers or even train carriages and pipelines, and stored in vacuum-insulated vessels at the end-using sites. This common infrastructure of liquid hydrogen is depicted in Figure 2.2.

In view of a mobility based on hydrogen, the distribution and storage of hydrogen as a liquid is one of the most feasible options from energy, technical, and economic perspectives. In such a scenario, the liquefaction capacity required for powering the transportation sector, in particular the road transportation, is orders of magnitudes larger than the current infrastructure. A numerical example can help illustrate the size of the infrastructure that would be required. According to Eurostat, the statistical office of the European Union, there were about 260–280 million vehicles on the roads

Figure 2.2 An illustrated overview of a common liquid hydrogen infrastructure: liquefaction in a central plant and storage in a large vacuum-insulated vessel acting as a buffer (top, courtesy of Air Products) and distribution from the plant to the end users by a cryogenic truck trailer (bottom, courtesy of Air Products).

of the EU-28 countries in the years 2008–2012 (Eurostat, 2014). In the same time period, the final energy consumption of the road transport was 300–310 million tons of oil equivalent (Eurostat, 2013). On average, the yearly power consumption of the road transport was around 400–410 GW,[2] i.e., 1 million vehicles would require an average power consumption of 1.5 GW. Because the hydrogen lower heating value is 120 MJ kg^{-1}, to meet the demand of 1 million vehicles a hydrogen liquefier would need to process 12.5 kg s^{-1} (or equivalently 1080 t day^{-1}). Moreover, 260–280 such liquefiers would have to be located throughout Europe. Both the single liquefier capacity and the total number of liquefiers exceed by orders of magnitudes the current hydrogen liquefaction scenario, which will be outlined in Section 2.4.

[2]The ton of oil equivalent is a conventional standardized unit defined on the basis of a ton of oil with a lower heating value of 41,868 kJ kg^{-1} (Eurostat, 2013).

2.2 Basics of cryogenic liquefaction

Cryogenics literally means the production of icy cold from ancient Greek. This term is used today as a synonym for temperatures lower than, indicatively, −150 °C (123 K). *Cryogenic liquefaction* is the process of turning a gaseous species at atmospheric conditions into a liquid at atmospheric pressure but at cryogenic temperature. Cryogenic liquefaction applies to the so-called permanent gases: helium, hydrogen, neon, nitrogen, and oxygen (as well as air in a more general sense).

The era of cryogenic liquefaction started in the 1840s when Gorrie manufactured an air-expansion engine for the production of ice (Barron, 1985; Flynn, 2005). Of course, a series of events have occurred since then. Importantly, Cailletet and Pictet were the first to independently produce a fog of liquid oxygen in 1877 by precooling it and expanding it through a valve, while Wroblewski and Olszewski completely liquefied oxygen and nitrogen in 1883 in a similar process. In 1892, Dewar developed a vacuum-insulated vessel for the effective storage of cryogenic liquids. Soon after, in 1895, Linde was granted a patent on air liquefaction based on the expansion through a valve and established the company known today by his name. Although Linde was not the first to liquefy air, he was the first to understand its industrial importance. Claude developed a practical system for air liquefaction employing an expander in 1902 and established the company known today as Air Liquide. Onnes succeeded in liquefying helium in 1908, while Collins developed an efficient helium liquefier around 1947. From the 1910s on, larger and larger cryogenic liquefiers were constructed in North America and in Europe. In the 1930s and 1940s, the development of rocket propulsion based on liquid oxygen and liquid hydrogen provided an exceptional thrust to the sector. Current applications of cryogenic technologies include not only rocket propulsion, but also studies in high-energy physics, electronics, metallurgy, and food processing to name a few important ones.

The following sections describe the fundamental cooling effects on which cryogenic liquefiers are based and the fundamental liquefaction cycles adopted for large-scale applications.

2.2.1 Fundamental cooling effects

Cryogenic liquefiers are based on the thermodynamic concept that the cooling effect is obtained by expanding adiabatically a fluid from a proper initial condition. The adiabatic expansion can be executed either with or without mechanically extracting energy from the fluid being expanded. Expansion without energy extraction, which is an isenthalpic process, is realized by a throttling valve, while expansion with energy extraction, which is ideally an isentropic process, is realized by an expanding machine. Because liquefiers are based on the expansion of a fluid, they require a compression process to complete the cycle from the pressure perspective. This compression is executed at ambient temperature and, commonly, in an intercooled manner. Additionally, liquefiers require heat exchangers to complete the cycle from the temperature perspective.

In general, throttling of any fluid leads to an appreciable temperature change, either positive or negative. This behavior is described by the thermodynamic property called the Joule–Thomson coefficient, μ (K Pa^{-1}), after the homonymous scientists and experiment. The coefficient is by definition

$$\mu := \left(\frac{\partial T}{\partial p}\right)_h \tag{2.1}$$

where T is the temperature (K), p is pressure (Pa), and h is the specific enthalpy (J kg^{-1}). The Joule–Thomson coefficient shows whether an infinitesimal throttling causes an infinitesimal increase, decrease, or even invariability of the fluid temperature from a given temperature and pressure condition. The locus of temperature and pressure conditions from which the infinitesimal throttling results in the invariability of the temperature is called *inversion curve*. Similarly, the temperature at which for a given pressure the infinitesimal throttling results in the invariability is called the *inversion temperature* (likewise, an *inversion pressure* may be defined). The following identity can be proved to hold true

$$\left(\frac{\partial T}{\partial p}\right)_h \equiv \frac{v(\beta T - 1)}{c_p} \tag{2.2}$$

where v is the specific volume (m^3 kg^{-1}), β is the coefficient of (isobaric) thermal expansion of the fluid (K^{-1}), and c_p is its specific heat at constant pressure (J kg^{-1} K^{-1}). The coefficient of thermal expansion and the specific heat at constant pressure are defined as follows

$$\beta := \frac{1}{v}\left(\frac{\partial v}{\partial T}\right)_p \tag{2.3}$$

and

$$c_p := \left(\frac{\partial h}{\partial T}\right)_p \tag{2.4}$$

Therefore, the inversion temperature, T_i (K), turns to be

$$T_i \equiv \frac{1}{\beta} \tag{2.5}$$

which may have two, one, or no solution, meaning that for a given pressure there may be two, one, or no inversion temperature. The inversion curve divides the pressure and temperature region into two areas: one area is characterized by a positive Joule–Thomson coefficient, where throttling yields a cooling effect, and the other by a negative coefficient, where throttling yields a warming effect. (Section 2.3 will describe the hydrogen thermodynamic properties, depicting in particular its inversion curve on a pressure–temperature diagram.) The maximum cooling effect of throttling is obtained from an initial condition laying on the inversion curve. Furthermore, throttling of a liquid may generate a two-phase state, with partly evaporated liquid, at a lower temperature. The latter process is sometimes called *flash evaporation*. Gas liquefiers that adopt throttling employ always the flash evaporation option to generate from a relatively warm liquid a colder saturated liquid as final product. The low-pressure vapor that is generated through the flash evaporation is recompressed either at low temperature or at ambient temperature, as will be shown later.

Unlike throttling, expanding a fluid with energy extraction leads ideally to a lower temperature. This behavior is described by the following identity, which can be proved to hold true

$$\left(\frac{\partial T}{\partial p}\right)_s \equiv \frac{v\beta T}{c_\mathrm{p}} \tag{2.6}$$

which is never negative. It is null exclusively for the hypothetical case of an incompressible fluid, for which β is null because v is constant. Furthermore, the larger β the larger the cooling effect, but the lower the temperature T the lower the cooling effect. Consequently, gas liquefiers that adopt expanding with energy extraction, i.e., an expanding machine, expand a gas, not a liquid, and they expand that gas from an intermediate temperature between the temperature of the ambient and the temperature of the liquefied fluid.

2.2.2 Fundamental liquefaction cycles

The most fundamental liquefaction scheme is the (simple or single-pressure) *Linde–Hampson cycle*, illustrated in Figure 2.3a, which employs exclusively a throttling valve to achieve the cooling effect (Barron, 1985; Flynn, 2005). The gaseous inlet is mixed with a vapor return from the flash evaporation and compressed in an intercooled compressor. The compressed stream is cooled down recuperatively by the vapor return in a heat exchanger and, ultimately, throttled. The colder saturated liquid is extracted as the product, while the saturated vapor returns to the compressor. The performance of the plant, which is usually never high, can be improved by the splitting of the throttling over two pressure levels. This new scheme is called a *dual-pressure Linde–Hampson cycle* as depicted in Figure 2.3b. When the fluid to be liquefied has an inversion curve below the ambient temperature for any pressure, such as hydrogen, the process requires a precooling system, which cools the stream prior to the recuperative heat exchanger and to the throttling valve (Figure 2.3c).

The performance of the liquefier can be increased remarkably by the adoption of an expanding machine. The fundamental scheme comprising the expander is the *Claude cycle* (Figure 2.4a), which also employs three heat exchangers, instead of one as in the Linde–Hampson cycle, the throttling valve, and the intercooled compressor. The heat exchangers operate over three different temperature intervals while the expander

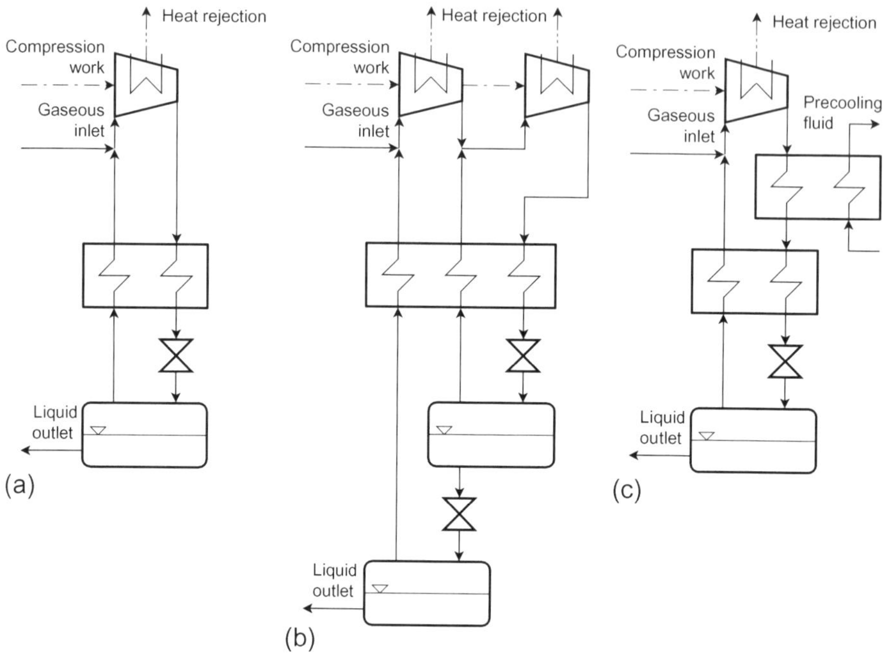

Figure 2.3 Linde–Hampson cycle in the simple or single-pressure configuration (left), dual-pressure (center), and single-pressure with precooling (right). In these cycles, the processed fluid is also the working fluid. (a) Simple or single-pressure, (b) dual-pressure, and (c) single-pressure with precooling.

operates in the mid interval. The compressor operates at the ambient temperature and the throttling valve at the cold temperature. Like the Linde–Hampson, the Claude cycle can operate over an additional pressure level. In this case, the throttling valve works between maximum and minimum pressures while the expanding machine between mid and minimum (Figure 2.4b). Moreover, the Claude cycle can adopt a precooling system, even though it is not strictly necessary for fluids that like hydrogen have a low inversion curve. The Claude cycle has diverse modifications: the *Heylandt cycle* that does not employ the high-temperature exchanger or the *Kapitza cycle* that does not employ the low-temperature exchanger. A third modification is the *Collins cycle*, which extends the use of one expander to two or more expanders, interposing a heat exchanger between any two machines. As in the previous schemes, the Heylandt, Kapitza, and Collins cycles have dual-pressure architecture and a precooling configuration. These three cycles are not shown here because of modest interest in the large-scale liquefaction of hydrogen.

All the cycles described so far also use the processed fluid, which is the fluid to be liquefied, as a working fluid, which is the refrigerant. Technically, it is possible to separate the process fluid from the working fluid. A sequence of vapor compression

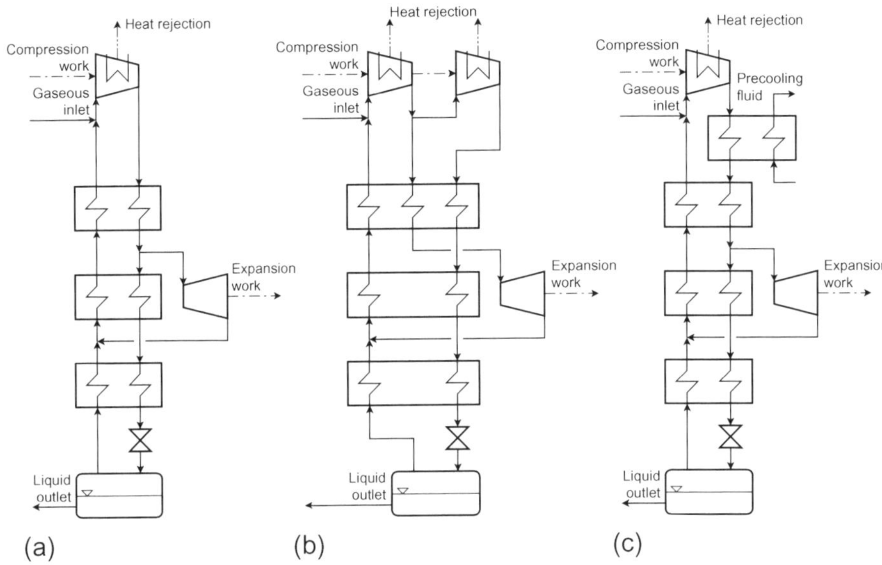

Figure 2.4 Claude cycle in the simple or single-pressure configuration (left), dual-pressure (center), and single-pressure with precooling (right). In these cycles, the processed fluid is also the working fluid. Modifications of the Claude cycle are the Heylandt and the Kapitza cycle, whereas an evolution is the Collins cycles, which are not shown here. (a) Simple or single-pressure, (b) dual-pressure, and (c) single-pressure with precooling.

systems operating on different pure fluids, and thus at different temperature ranges, can be organized such that the evaporator of a cycle works as the condenser of the next bottoming cycle. This deriving scheme is called the *cascade cycle* and is shown in Figure 2.5a, where each pure fluid operates in a separate closed loop and at a given temperature range. Alternatively, the refrigerant fluid can be a mixture of a specified composition operating in one single loop over the entire temperature interval. The scheme is called the *mixed-refrigerant cycle* (not shown here). In the cascade cycle and in the mixed-refrigerant cycle, the refrigerant operates in both the vapor and liquid phases. In constrast, an additional cycle, called the reversed *Joule–Brayton cycle*, uses a pure or a mixture refrigerant, but operates exclusively in the gas phase and employs solely expanders to achieve the cooling effect (Figure 2.5b).

In all liquefaction cycles, the stream of the fluid to be liquefied is always at a pressure higher than its critical pressure. Indeed, at supercritical pressure, there is no phase change within any of the heat exchangers, which would lead to small temperature variations on large energy transfers. Furthermore, a subcritical (pure) fluid would require extracting the condensation energy at a constant temperature, which is an adverse situation for achieving high effectiveness of the heat exchangers (or low entropy generation from a second-law perspective).

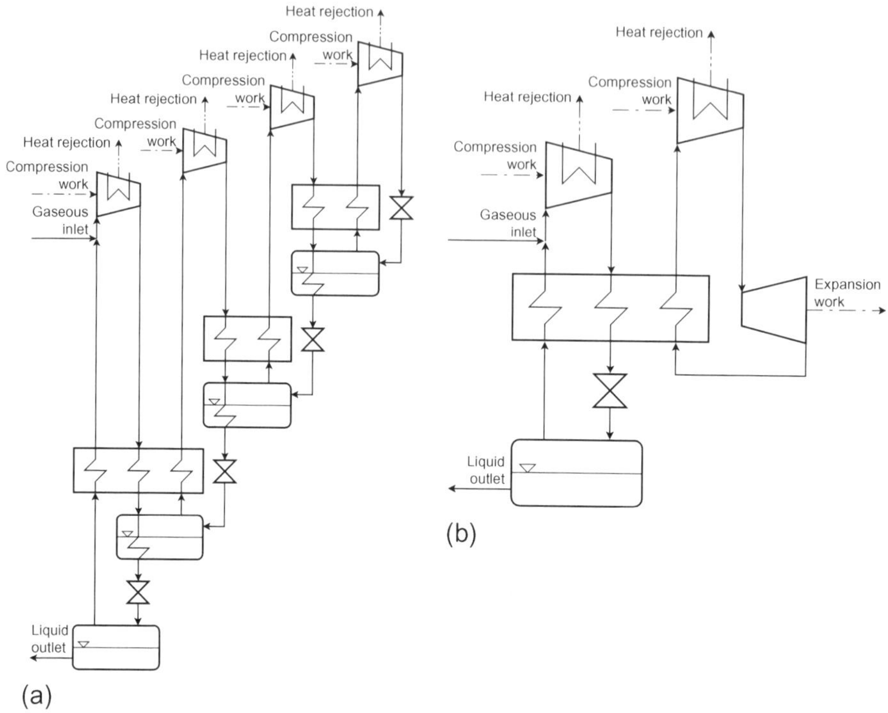

Figure 2.5 Cascade cycle in a three-loop configuration (left) and reversed Joule–Brayton cycle in a single-loop configuration (right). In these cycles, the processed fluid is different from the working fluid. The cascade cycle can employ pure fluids or mixture of desired composition. In the latter case, the cycle is called mixed-refrigerant cycle, which is not shown here. (a) Cascade and (b) reversed Joule–Brayton.

2.3 Hydrogen thermodynamic properties at ambient and low temperatures

Since the dawn of the universe, hydrogen has been the most abundant element. According to current knowledge, hydrogen and helium were the first elements to form about 300,000 years after the Big Bang because of the inherent stability of their nuclei (Ridgen, 2003). Today, hydrogen constitutes 90% of the total atoms or 75% of the total matter preceding by far helium, the second most abundant element. In space, hydrogen exists predominantly in the atomic and ionic states, while partly in molecular form. On Earth, hydrogen is bonded mostly to other atoms in chemical compounds, such as water and hydrocarbons.

The following section describes hydrogen as an element and as a molecule, highlighting in particular the peculiar thermodynamic behavior of the molecular hydrogen. Such behavior has a direct effect on the design of the hydrogen liquefaction cycle.

2.3.1 Elemental hydrogen

Hydrogen is the smallest element in nature with an atomic number equal to 1 and an atomic weight of 1.00794. It exists naturally in two stable and one unstable radioactive isotopes of mass number ranging from 1 to 3.[3] In mid-1931, Birge and Menzel suggested the existence of the two stable isotopes, while Urey, Brickwedde, and Menzel proved it in late 1931. Urey et al. proposed the names *protium* and *deuterium*, which are derived from the ancient Greek words meaning *first* and *second* (Farkas, 1935). These names are still in common use along with the other two terms of *light hydrogen* and *heavy hydrogen*. Rutherford, Oliphant, and Harteck observed an unstable isotope in 1934 and gave it the name *tritium*, meaning *third*. The IUPAC-compliant nomenclature of the three isotopes is in the order ^1_1H, ^2_1H, and ^3_1H, while the symbols D and T are still often employed to indicate deuterium and tritium. The three isotopes have a relative composition of approximately 99.98%, 0.02%, and traces on an atomic basis. In recent years, isotopes with atomic masses up to 7 have been synthesized in the laboratories.

2.3.2 Molecular hydrogen

Molecular hydrogen was observed accidentally for the first time by Theophrastus von Hohenheim, also known as Paracelsus, who noted that the attack of strong acids against metals generated a flammable gas. Other chemists and physicists repeated his experience, including Robert Boyle who described the properties of this flammable gas in 1671. The credit of the discovery that the generated gas was made of a new element is attributed commonly to Henry Cavendish because he was able to isolate it and measure its relevant properties in 1776. Shortly later, in 1783, Antoine Lavoisier gave the new element the name *hydrogen* after he discovered in collaboration with Pierre-Simon Laplace that burning it in air produced water. Etymologically, hydrogen means forming, *genes*, water, *hydro*, both words derived from ancient Greek.

Hydrogen isotopes can combine and form diatomic molecules. Since tritium is unstable and radioactive, only molecules comprising protium and deuterium, H_2, HD, D_2, are of practical interest. Molecular hydrogen is thus a mixture of those three molecules. As will be explained in the following, H_2 and D_2 exist in two modifications each, so that molecular hydrogen is naturally a mixture of five compounds (HD and the two modifications for H_2 and the two for D_2). However, being deuterium scarcely abundant with respect to protium, molecular hydrogen can be approximated to be only H_2 for industrial applications.

2.3.3 Modifications of molecular hydrogen

Hydrogen, as any other substance, can be described by *statistical thermodynamics*, the branch of physics that studies matter combining the probability theory and quantum mechanics with a microscopic mechanical model of the constituents of the matter

[3]The atomic number is the number of protons in the nucleus, while the mass number is the number of protons and neutrons in the nucleus.

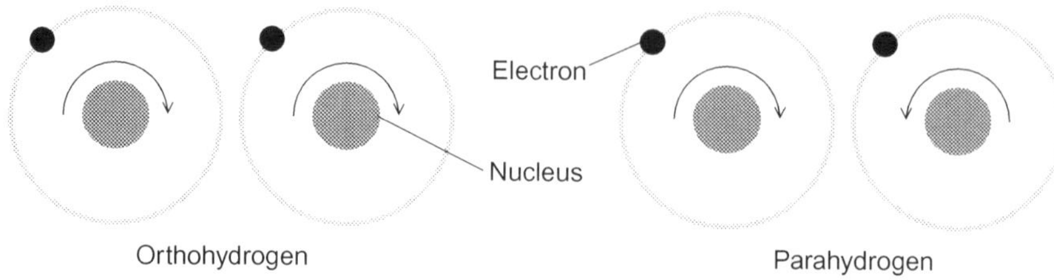

Figure 2.6 Pictorial schematic of the parallel and antiparallel nuclear spins of the two molecular hydrogen modifications: orthohydrogen and parahydrogen, respectively.

itself. Its aim is to predict the properties that are measurable at the macroscopic level based on the microscopic behavior of the measured substance (Tien and Lienhard, 1988).

According to the statistical thermodynamic theory, for homonuclear diatomic molecules, which are molecules composed of two like atoms, for example, H_2 and D_2 as well as O_2 and N_2, where there is a division of the *quantized rotational energy levels* of their nuclear spins into two groups referred to as the *ortho* and the *para* series. One series is composed of the even-numbered and the other of the odd-numbered levels. For the case of H_2, the para series is comprised of even numbered, while ortho of odd. *Parahydrogen* and *orthohydrogen* represents two different types, namely *modifications*, of molecular hydrogen that have their own thermodynamic behaviors (Woolley et al., 1948). In simpler words, the distinction between the two modifications of hydrogen is the relative spin of the two nuclei. If the nuclear spin is in the same direction (parallel) the modification is called orthohydrogen; in contrast, if it is in the opposite direction (antiparallel) it is called parahydrogen, as shown in Figure 2.6. The ratio of the average number of all hydrogen molecules in the para series and that in the ortho series varies with temperature. The ambient temperature equilibrium composition is called *normal-hydrogen*, whereas the equilibrium composition at any temperature is called simply *equilibrium-hydrogen* (at ambient temperature, normal and equilibrium hydrogen are the same mixture).

2.3.4 Thermodynamics of molecular hydrogen modifications

A classical thermodynamic model of a generic fluid comprises an equation of state (EOS) to calculate the properties of that fluid away from the ideal gas behavior. In the past, formulations of EOS used to be defined explicitly in the pressure, whereas in recent decades they have been mostly defined explicitly in the Helmholtz free energy because of the consequent advantage that all thermodynamic properties can be calculated as derivatives without requiring integrals as in the former approach (Span, 2000). Both pressure- and Helmholtz free energy-explicit formulations necessitate a separate equation for the specific heat in the hypothetical ideal gas state. In regards to hydrogen, McCarty et al. (1981) prepared an extensive and comprehensive report on a large set of hydrogen properties, while Jacobsen et al. (2007) and Leachman et al. (2007) provided two updates on the thermodynamic and transport properties of hydrogen. Hydrogen shows a peculiar calorimetric behavior, represented by the specific heat, and a conventional volumetric behavior, represented by the EOS.

The specific heats at constant pressure in the hypothetical ideal gas state, c_p^0 (J kg^{-1} K^{-1}), for orthohydrogen, parahydrogen, normal-hydrogen, and equilibrium-hydrogen up to 600 K can be computed considering only the nuclei rotational energy levels and neglecting their vibrational ones (Valenti et al., 2012). The nuclear spin, denoted by i, is equal to 1/2. The statistical weight of an electronic level, g_e, is equal to 1. Let k indicate the Boltzmann constant (1.3806504e^{-23} J K^{-1}), $\tilde{h}$ the Planck constant (6.62606896e^{-34} J s), and I the molecular hydrogen moment of inertia (4.67e^{-48}). The statistical weight of the J-th quantum level, g_J, is

$$g_J = \begin{cases} g_e(2i+1)i(2J+1), & J\,\text{even}\,(\text{para}) \\ g_e(2i+1)(i+1)(2J+1), & J\,\text{odd}\,(\text{ortho}) \end{cases} \tag{2.7}$$

while the J-th quantum energy state, E_J, is

$$E_J = J(J+1)\frac{\tilde{h}^2}{8\pi^2 I} \tag{2.8}$$

The equilibrium ratio of the average number of all molecules in the para series and that in the ortho series, i.e., the para-to-ortho fraction α, is

$$\alpha = \frac{\sum_{J=\text{even}} g_J e^{-\frac{E_J}{kT}}}{\sum_{J=\text{odd}} g_J e^{-\frac{E_J}{kT}}} \tag{2.9}$$

and thus the concentration of parahydrogen x_{pH_2} in the equilibrium mixture turns out to be

$$x_{pH_2} = \frac{\alpha}{1+\alpha} \tag{2.10}$$

The orthohydrogen heat capacity at constant pressure in the hypothetical ideal gas state due to translational and rotational modes, with respect to universal gas constant R (8314.472 J kmol^{-1} K^{-1}), is

$$\frac{c^0_{p,oH_2}}{R} = \frac{\sum_{J=\text{odd}} g_J \left(\frac{E_J}{kT}\right)^2 e^{-\frac{E_J}{kT}}}{\sum_{J=\text{odd}} g_J e^{-\frac{E_J}{kT}}} - \left(\frac{\sum_{J=\text{odd}} g_J \frac{E_J}{kT} e^{-\frac{E_J}{kT}}}{\sum_{J=\text{odd}} g_J e^{-\frac{E_J}{kT}}}\right)^2 + \frac{5}{2} \tag{2.11}$$

while that for parahydrogen is

$$\frac{c^0_{p,pH_2}}{R} = \frac{\sum_{J=\text{even}} g_J \left(\frac{E_J}{kT}\right)^2 e^{-\frac{E_J}{kT}}}{\sum_{J=\text{even}} g_J e^{-\frac{E_J}{kT}}} - \left(\frac{\sum_{J=\text{even}} g_J \frac{E_J}{kT} e^{-\frac{E_J}{kT}}}{\sum_{J=\text{even}} g_J e^{-\frac{E_J}{kT}}}\right)^2 + \frac{5}{2} \tag{2.12}$$

and for equilibrium-hydrogen is

$$\frac{c^0_{\mathrm{p,eH_2}}}{R} = \frac{\sum_{J=\mathrm{all}} g_J \left(\frac{E_J}{kT}\right)^2 \mathrm{e}^{-\frac{E_J}{kT}}}{\sum_{J=\mathrm{all}} g_J \mathrm{e}^{-\frac{E_J}{kT}}} - \left(\frac{\sum_{J=\mathrm{all}} g_J \frac{E_J}{kT} \mathrm{e}^{-\frac{E_J}{kT}}}{\sum_{J=\mathrm{all}} g_J \mathrm{e}^{-\frac{E_J}{kT}}}\right)^2 + \frac{5}{2} \tag{2.13}$$

Finally, the ideal gas heat capacity of normal hydrogen is evaluated as the proper weighted average of ortho- and parahydrogen

$$c^0_{\mathrm{p,nH_2}} = \frac{1}{4} c^0_{\mathrm{p,pH_2}} + \frac{3}{4} c^0_{\mathrm{p,oH_2}} \tag{2.14}$$

The parahydrogen concentration in the equilibrium composition and the heat capacities in the hypothetical ideal gas state of all modifications and mixtures resulting from this procedure, truncating all summations to J equal to 10, are illustrated in Figures 2.7 and 2.8, respectively. From the ambient temperature on, the equilibrium composition, which is normal-hydrogen by definition, shows a 3:1 ortho-to-parahydrogen ratio, or in other words parahydrogen concentration in normal-hydrogen is 25%. As the temperature decreases, the parahydrogen content increases: at 20 K it is 99.84% and approaching 0 K it tends to 100%.

More recently, Le Roy et al. (1990) reported new and more complex calculations of the thermodynamic properties of the modifications of the diatomic molecules made by hydrogen, deuterium, and tritium. These calculations are still considered the most accurate source of data for hydrogen. The relative error on the specific heat computed from the described procedure with respect to the most accurate calculation by Le Roy et al. is at most 1%.

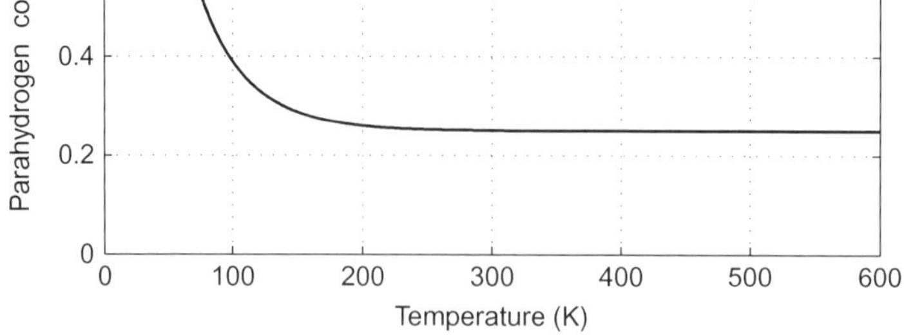

Figure 2.7 Parahydrogen concentration at equilibrium, $x_{\mathrm{pH_2}}$, as a function of temperature, T (K).

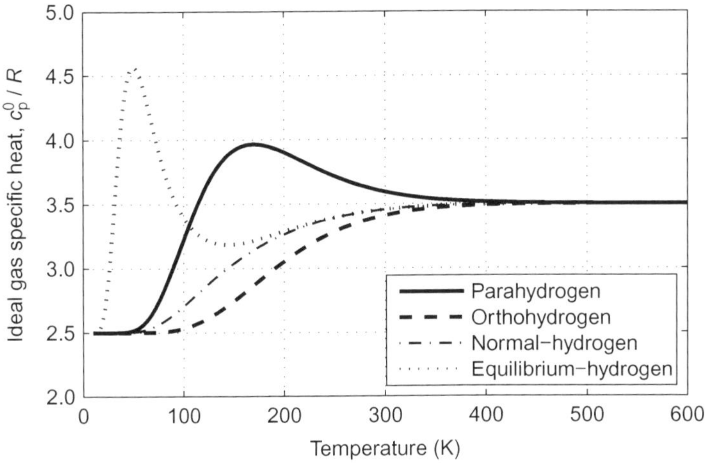

Figure 2.8 Para-, ortho-, normal-, and equilibrium-hydrogen heat capacity at constant pressure in the hypothetical state of ideal gas, c_p^0, as a function of temperature, T (K), defined with respect to the universal gas constant R.

If on one hand orthohydrogen and parahydrogen have a highly dissimilar calorimetric behavior, on the other they share similar volumetric behavior but in the cryogenic region where any differences are insignificant (McCarty et al., 1981). The state-of-the-art Helmholtz free energy-explicit EOS for orthohydrogen, parahydrogen, and normal-hydrogen is that proposed by Leachman et al. (2009). It appears though that no EOS is available already for equilibrium-hydrogen. However, because equilibrium-hydrogen is primarily made of parahydrogen in the low-temperature region, its volumetric behavior can be approximated by that of parahydrogen or, in other words, the EOS of parahydrogen can be taken for equilibrium-hydrogen (Valenti et al., 2012).

Since the volumetric behavior of hydrogen modifications is relatively similar and the calorimetric behavior are remarkably dissimilar, the enthalpy of conversion from normal-hydrogen to equilibrium-hydrogen in the real fluid state at any pressure can be computed with very good approximation from the difference in their heat capacities in the hypothetical ideal gas state. Figure 2.9 shows the enthalpy of conversion from normal-hydrogen to equilibrium-hydrogen in the ideal gas state as a function of temperature.[4] From ambient temperature on, the enthalpy of conversion is null because normal hydrogen and equilibrium hydrogen coincide. However, as the temperature decreases, the enthalpy of conversion increases rapidly to values well over 100 times

[4]Other authors, e.g., McCarty et al. (1981), report the enthalpy of conversion of normal-hydrogen to parahydrogen instead of to equilibrium-hydrogen as in this text. However, the enthalpy of conversion to equilibrium-hydrogen is more important from the perspective of the liquefaction process because hydrogen liquefiers adopt technologies to promote reaching the equilibrium composition in the hydrogen stream while cooling that stream.

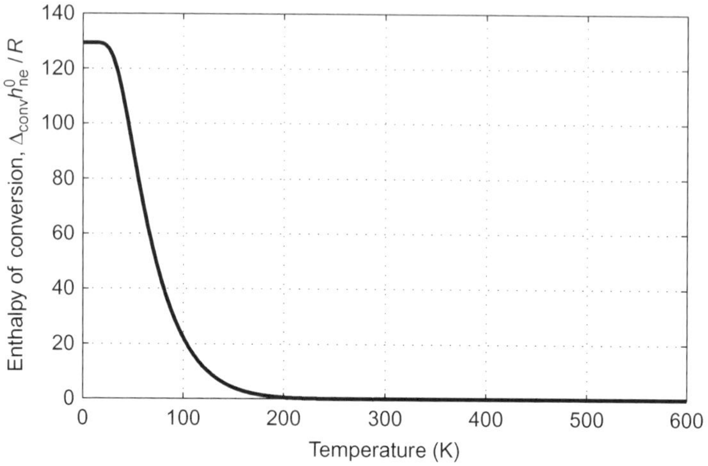

Figure 2.9 Normal-hydrogen to equilibrium-hydrogen enthalpy of conversion in the hypothetical state of ideal gas, $\Delta_{conv}h^0_{ne}$, as a function of temperature, T (K), defined with respect to the universal gas constant R. This enthalpy of conversion in the hypothetical state of ideal gas can be considered as a very good approximation for the conversion in the real state at any pressure.

the universal gas constant. Numerically, since the hydrogen molar weight M equals to 2.01588 kg kmol^{-1}, the enthalpy of conversion from normal- to equilibrium-hydrogen at 20 K is about 532 kJ kg^{-1}. For comparison, the enthalpy of vaporization of para-hydrogen at 20 K is about 447 kJ kg^{-1}. Therefore, the enthalpy of conversion is exothermic at decreasing values of temperature and it is higher than the enthalpy of vaporization in the cryogenic region. From the perspective of the liquefaction process, the enthalpy of conversion is an additional cooling load that greatly increases the liquefaction work.

In the absence of a catalyst, the orthohydrogen to parahydrogen conversion is a second-order reaction and, thus, the rate of change of the orthohydrogen composition is

$$\frac{dx_{oH_2}}{dt} = C_2 x_{oH_2}^2 \tag{2.15}$$

where C_2 is the reaction-rate constant equal to 0.0114 h^{-1} at the normal boiling point (Barron, 1985). In such a case, the time t (h) required for reaching a certain orthohydrogen composition, x_{oH_2}, starting from the normal composition of 75% is

$$t = \frac{1}{0.0114}\left(\frac{1}{x_{oH_2}} - \frac{1}{0.75}\right) \tag{2.16}$$

For instance, the time required for the noncatalyzed conversion from 75% to 5% of orthohydrogen requires over 2 months. Hence, the noncatalyzed conversion is an extremely low process. However, in the presence of a catalyst that is mixed well with the hydrogen stream, the conversion within the gas phase approaches a first-order reaction

$$\frac{dx_{oH_2}}{dt} = C_1 x_{oH_2} \tag{2.17}$$

where C_1 depends on the specific catalyst as well as the operating pressure and temperature. Similarly, the conversion within the liquid phase approaches a zero-order reaction

$$\frac{dx_{oH_2}}{dt} = C_0 \tag{2.18}$$

where C_0 depends also on the specific catalyst as well as the operating pressure and temperature. Because the normal- to equilibrium-hydrogen conversion is strongly exothermic in the cryogenic region, hydrogen liquefiers always catalytically promote the conversion of orthohydrogen to parahydrogen in the hydrogen stream while cooling that stream.

The inversion curve of parahydrogen, plotted from the data by McCarty et al. (1981), is depicted in Figure 2.10. This curve shows that the maximum inversion temperature, T_i (K), is around 200 K, which is well below the ambient temperature. Consequently, all liquefaction cycles based on the sole throttling cooling effect require precooling of the hydrogen stream by other means to operate.

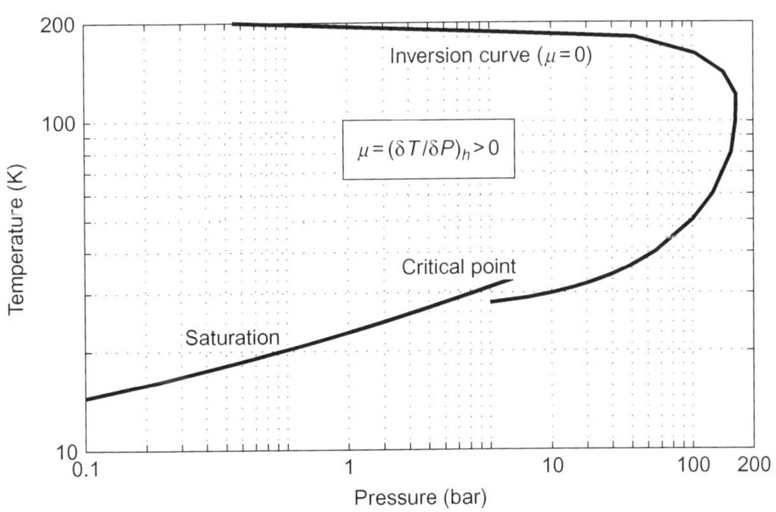

Figure 2.10 Inversion curve and saturation curve of parahydrogen (plot of the data from McCarty et al., 1981). It shows that the maximum inversion temperature, T_i (K), is around 200 K, which is well below the ambient temperature.

Today, the most accurate computation tool for the thermodynamic and transport properties of hydrogen is likely the software REFPROP by the U.S. National Institute of Standards and Technology (known as NIST). Version 9 of this code contains detailed models for orthohydrogen, parahydrogen, and normal-hydrogen. Valenti et al. (2012) prepared an additional model for the computation of the thermodynamic properties of equilibrium-hydrogen. This model is not as accurate as the others, but it is sufficiently accurate for practical purposes.

2.4 Large-scale hydrogen liquefaction and storage

In the early 1880s, Wroblewski and Olszewski established the first low-temperature laboratory, the Cracow University Laboratory in Poland, and succeeded in liquefying oxygen and nitrogen in sufficient quantities to investigate their properties. However, when they attempted to liquefy hydrogen in 1884, they obtained only a mist of liquid droplets. About a decade later, in 1898, it was Dewar, the inventor of the vacuum-insulated vessel, who obtained liquid hydrogen as a bulk, instead of a fog, at the Royal Institute of London. While up until the late 1940s, hydrogen liquefiers for continuous production were more and more common in laboratories, they never reached large scales and high efficiencies. The drive for the construction of industrial-size systems came in the early 1950s from the increasing demand of liquid hydrogen requested by the aerospace sector, for airplane and rocket propulsion, and by the nuclear sector, for deuterium production via cryogenic distillation. Until now, larger and larger liquefiers have been constructed to meet the increasing demand of bulk hydrogen deliveries from diverse sectors, such as semiconductor, electronic, metallurgical, and chemical industries to name a few. Today, both small- and large-scale liquefiers are a relatively common technology characterized by a high industrial maturity (Haselden, 1971; Barron, 1985; Flynn, 2005; McIntosh, 2004).

In general terms, a hydrogen liquefier may be classified as small if its capacity is up to 0.5 t day^{-1}, while large if greater than 5.0 t day^{-1}. The measurement unit of tons, also called metric tons, per day is customary in the field of hydrogen liquefaction, despite other units, such as liters per hour, in common use for smaller systems. Small- and large-scale hydrogen liquefiers do not differ just in capacity but also in the manner by which liquefaction is accomplished with current technology. Small-scale hydrogen liquefaction is realized by cryogenic refrigerators, which are systems operating with a refrigerant in a closed-loop thermodynamic cycle that cools and liquefies the hydrogen stream. In the case of hydrogen, the adopted refrigerant is helium. Hydrogen, which is the processed fluid, and helium, which is the working fluid, do not mix in any point of the process, and always remain separated. In contrast, large-scale liquefaction is accomplished by open-loop thermodynamic cycles in which hydrogen is both the processed fluid and the working fluid. This distinction is only technological because from a theoretical perspective closed-loop refrigerators can be resized to large scales, as well as downsized open-loop plants. In addition, for very small-scale hydrogen liquefaction, magnetic refrigerators can be employed. Such refrigerators work based on the *adiabatic demagnetization* of *paramagnetic material*, such as iron ammonium alum, to refrigerate substances even below 1 K (Barron, 1985; Flynn, 2005).

Large-scale hydrogen liquefiers must employ catalytic ortho-to-parahydrogen conversion reactors because, as seen in Section 2.3.4, the reaction is highly exothermic and otherwise very slow. The reaction is so exothermic that the associated enthalpy of conversion is greater than the enthalpy of vaporization at storage tank pressure. Briefly, if orthohydrogen is not converted during liquefaction, the conversion occurs in the storage tank releasing an amount of energy large enough to vaporize the entire content of that tank.

The hydrogen-liquefaction work strongly depends on inlet conditions (pressure and temperature), outlet conditions (saturation pressure and parahydrogen content), and ambient temperature. Let us assume a gaseous hydrogen feed at 1 bar and 288 K, a saturated liquid product at 1 bar, and the ambient at 288 K. The resulting ideal liquefaction work is 11.5 MJ kg^{-1} (3.19 kWh kg^{-1}), if the outlet composition is the normal 75% ortho- and 25% parahydrogen, and 13.6 MJ kg^{-1} (3.78 kWh kg^{-1}), if the composition is at equilibrium. Hence, the hydrogen liquefaction work is large even in ideal terms, around 10% of the lower heating value of 120 MJ kg^{-1}, and it is much larger in the case of ortho-to-parahydrogen conversion, around 18% greater than the case without the conversion. If the inlet pressure is 20 bar instead of 1 bar, as is typical for a number of hydrogen production processes, the ideal liquefaction work is 7.89 MJ kg^{-1} (2.19 kWh kg^{-1}) and 9.97 MJ kg^{-1} (2.77 kWh kg^{-1}) for the two cases with and without conversion. Hence, increasing the inlet pressure up to 20 bar reduces the ideal liquefaction work by 25–30%.

The following sections describe today's and likely tomorrow's technology for hydrogen liquefaction on a large scale and, as it will be shown, also on the *mega* scale.

2.4.1 Today's technology

The world's hydrogen liquefaction capacity was estimated to be about 290 t day^{-1} in the year 2003 (West, 2003) and 355 t day^{-1} in the year 2009 (Krasae-in et al., 2010a). The largest capacity, more than 84% of the total, is located in North America, whereas Asia accounts for over 8% and Europe for almost 7%. The capacity of the other continents is either negligible or null. The largest operating plants, rated at over 30 t day^{-1}, are located in North America. Considering the lower heating value of hydrogen (120 MJ kg^{-1}), the total world capacity is approximately 500 MW. This figure, 500 MW for the whole world, is much lower than the requirement set in Section 2.1, 1500 MW for a single liquefier, demonstrating that a mobility largely based on hydrogen would require an infrastructure greater than the current one by at least two orders of magnitude.

Today's large-scale hydrogen liquefiers are based on modifications of the original precooled Claude cycle (see Section 2.2.2). Typically, hydrogen expansion through a machine is split into two processes separated by a heat exchanger for additional cooling, as shown in Figure 2.11a. Precooling is obtained by a dedicated refrigeration cycle or, when the liquefier is located nearby an air separation unit (ASU), by liquid

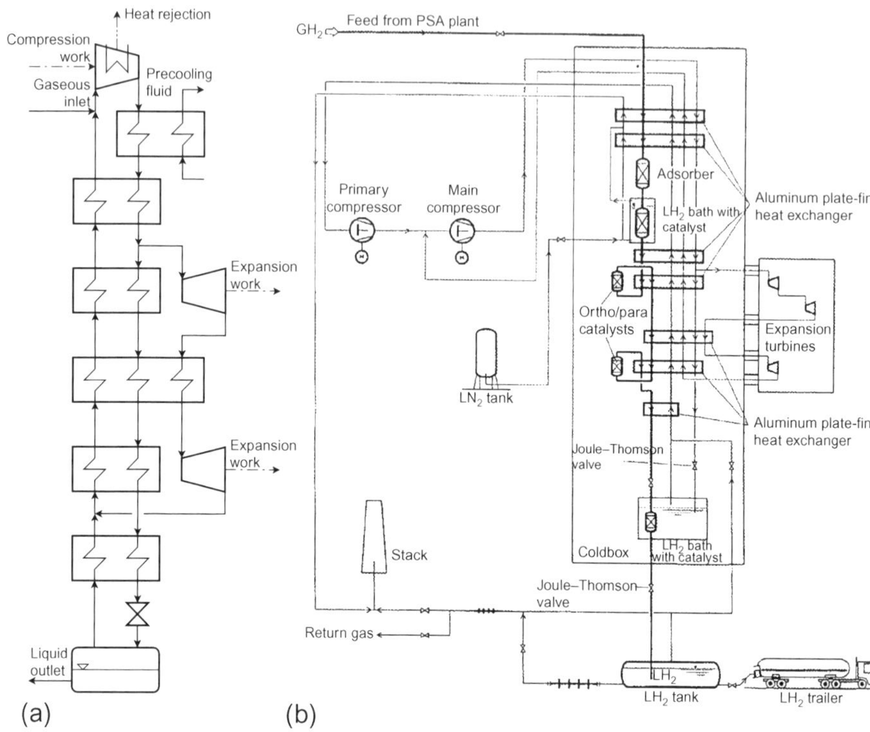

Figure 2.11 Modern large-scale hydrogen liquefiers are based on modifications of the fundamental Claude cycle (left) and employ catalytic orthohydrogen to parahydrogen conversion reactors, either in a batch mode (right, Bracha et al., 1994) or in continuous mode. (a) Modified Claude and (b) Recent large-scale hydrogen liquefaction scheme.

nitrogen from that separation unit. Catalytic conversion of orthohydrogen to parahydrogen can be executed in a *batch mode* or in a *continuous mode*. If in batch mode, the reactors can be either adiabatic or isothermal. Isothermal reactors are immerged in liquid nitrogen or in liquid hydrogen baths. If in continuous mode, the catalyst is placed within the heat exchangers. Considering the liquefaction work requirement, continuous conversion is definitely more efficient because the energy released by the conversion itself is extracted at the highest possible temperature requiring the lowest possible liquefaction work. In contrast, the energy released by a batch conversion is extracted at the temperature of the reactor that is lower than in the continuous case yielding a higher liquefaction work (Lipman et al., 1963).[5] Typically, hydrogen liquefiers are designed to produce a liquid with parahydrogen content greater than 95%. Figure 2.11b shows the process scheme of a modern hydrogen liquefier, i.e., the plant in Ingolstadt owed and operated by Linde AG until recently. This process scheme

[5]This reasoning is based on the thermodynamic concept that the work required to transfer a given amount of energy from a cold system to the ambient increases as the system temperature decreases.

shows all the precooling the hydrogen feed, modifying the fundamental Claude cycle, and using the batch conversion concept.

The hydrogen compressors in large-scale liquefiers are commonly positive-displacement reciprocating machines, either oil-flooded or, better, oil-free. If oil-flooded, an accurate oil separation is required to prevent contamination of the liquid hydrogen product. The hydrogen expanders are instead dynamic centripetal turbines with oil bearings or, better yet, with gas bearings (Ohlig and Bischoff, 2012). Turbines often have an oil-brake circuit to dissipate the expander output, which is a fraction of the compressor input, even though theoretically the extracted mechanical energy may be recuperated and used within the process. Aluminum fin-plate heat exchangers are usually employed in hydrogen liquefiers due to their compactness and to the capability to allow very low minimal internal temperature approaches, as low as few Kelvins. Expanders and heat exchangers are placed within vacuum-insulated vessels, called *cold boxes*, to reduce the energy leakage from the ambient to the process, which as at a much lower temperature.

The work requirement of liquefiers currently in operation is in the range 30–45 MJ kg^{-1} (about 8–12 kWh kg^{-1}), depending on the size and age of the specific plant, for an inlet at about 20 bar and a parahydrogen content of at least 95%. Compared to an ideal requirement of 9.97 MJ kg^{-1} (2.77 kWh kg^{-1}), the liquefiers in operation have a second-law efficiency in the range 22–33%. New designs of hydrogen liquefiers are claimed to have a work requirement as low as 25 MJ kg^{-1} (7.0 kWh kg^{-1}), resulting in a 40% second-law efficiency, which is actually a remarkable achievement.

Once liquefied, hydrogen is stored in cryogenic tanks up to 300 m^3 in size. Tanks are generally cylindrical, placed horizontally or even vertically to reduce the footprint, but they may be spherical for larger applications because the sphere has the lowest area-to-volume ratio. For instance, NASA uses spherical vessels to store liquid hydrogen and liquid hydrogen for fueling rockets. Cryogenic vessels adopt sophisticated materials to reduce conductive and radiant energy flow from the environment to the stored fluid as well as vacuum-insulated jackets to eliminate convection flow. Modern large-scale cryogenic tanks are characterized by a daily evaporation rate, namely *boil-off*, that may be even below 0.1% day^{-1} of the total stored amount.

2.4.2 Future technologies

A number of researchers have suggested major evolutions of conventional large-scale hydrogen liquefiers with the primary goal of minimizing the liquefaction work requirement while increasing the plant size by even an order of magnitude. In a recent work, Quack (2002) proposed a conceptual design of a large-scale liquefier comprised of three sections: (i) precooling by a propane refrigeration cycle organized over three pressure levels. Then, (ii) cooling by a helium–neon reversed Brayton cycle. Ultimately, (iii) throttling of the pressurized hydrogen stream to storage conditions and recompressing in the cold region the fraction evaporated during the isenthalpic expansion. The predicted second-law efficiency exceeds 50%. Valenti and Macchi (2008) report the numerical results of a detailed process based on four helium reversed Brayton cycles, all of which are organized in a cascade and share the same maximum pressure. The proposed process does not adopt precooling and it employs an expanding machine instead of a throttling valve to reach the liquid storage condition. The expander allows avoiding the formation of an evaporated fraction that must be either recompressed in the cold end or heated recuperatively to ambient temperature (see Section 2.2.2). The predicted second-law efficiency of this process approaches 50%. Berstad et al. (2010), Krasae-in et al. (2010b), and Krasae-in (2014) suggest instead liquefiers based on first (i) precooling by a mixed-refrigerant cycle, similarly to the technology of the liquefaction of natural gas, and then (ii) cooling further by either a helium–neon cycle, in the former case, or four hydrogen reversed Brayton cycle, in the latter. The predicted second-law efficiencies of these processes are in the range 45–48%.

A few international projects have focused on improving liquefaction and transport of liquid hydrogen across countries and even continents. The Japanese World Energy Network (WE-NET) project (1993–2002) focused on enabling the introduction of a worldwide network for the development of abundantly available renewable resources and their distribution by way of liquid hydrogen. A number of liquefaction schemes were analyzed in the project, including modified Claude, helium-reversed Brayton, mixed-refrigerant, neon with a cryogenic compressor, and nelium (a mixture of neon and helium) cycles (Fukano et al., 2000). Around the same time, the Euro-Québec Hydro-Hydrogen Pilot Project (EQHHPP) assessed the techno-economic feasibility of producing hydrogen in Canada from renewable electricity, then liquefying and shipping it overseas to Europe (Drolet et al., 1996). More recently, from 2008 until 2012, the U.S. Department of Energy, within the Hydrogen and Fuel Cells Program, supported research activities to develop cost- and energy-effective liquefaction processes based on active magnetic regenerative liquefiers and on once-through helium-reversed Brayton cycles.[6] From 2011 until 2013, the European Community, under the Fuel Cells and Hydrogen Joint Technology Initiative, has supported the Integrated Design for Efficient Advanced Liquefaction of Hydrogen (IDEALHY) project aimed at designing a large-scale high-efficiency liquefier, investigating safety issues, and assessing the overall techno-economic feasibility.[7] In 2012, an industrial consortium, led by Japanese companies, initiated the HyGrid project to investigate the viability of producing hydrogen from coal in Australia, liquefying it and shipping it to Japan.[8]

2.5 Advantages and disadvantages

Hydrogen is a valuable energy vector because of the ease and cleanness of its use. It is characterized by a very high energy content on a mass-basis yet quite low content on a volume-basis. Liquid hydrogen has the advantage of mitigating the problem of low volumetric energy content even at ambient pressure. Moreover, today hydrogen liquefaction is a mature and proven technology. Nevertheless, the disadvantages are

[6]http://www.hydrogen.energy.gov.

[7]http://www.idealhy.eu.

[8]http://www.hygrid.jp.

related to the fact that hydrogen liquefaction work is high even (10 MJ kg^{-1} or more in ideal term, 25 MJ kg^{-1} or more in practical terms) and that hydrogen exists as a liquid at cryogenic temperatures (the saturation temperature at 1 bar is as low as 20 K). Therefore, managing a cryogenic fluid may require care and technologies that are not in common practice today.

2.6 Current uses of liquid hydrogen

Currently, hydrogen is used mostly in the gaseous form by the petrochemical, pharmaceutical, metallurgical, electronic, and food industries. It is used directly in liquid form by aerospace agencies as a propellant. For instance, the new Space Launch System by the U.S. National Aeronautics and Space Administration (known as NASA) is based on liquid hydrogen and liquid oxygen propulsion. Moreover, liquid hydrogen is also employed directly in the electronics and metallurgical sector for specific production processes. Nevertheless, despite the end use, when large consumption is present, hydrogen is more economically transported and stored as a liquid over mid-range distances (Sherif et al., 1997; Ramachandran and Menon, 1998).

2.7 Sources of further information and advice

The interested reader may refer first to the two well-known textbooks by Barron (1985) and Flynn (2005). Moreover, the reader can also locate the other reports and articles cited in this work. In general, detailed information may be derived from two reference journals of the sector: (i) the *International Journal of Hydrogen Energy* and (ii) *Cryogenics*. Furthermore, the proceedings of (iii) the *International conference on cryogenics*, (iv) the *World Hydrogen Energy Conference*, and (v) the *World Hydrogen Technologies Convention* may be also valuable sources of data.

Acknowledgments

The author kindly thanks Stefano Foresti from Politecnico di Milano for proofreading the text and offering suggestions. Moreover, he gratefully acknowledges Linde Kryotechnik and, in particular, Lars Blum for sharing information. He is also very thankful to Air Products, especially Diana Raine and Peter Bout, as well as to Sapio, in particular, Michela Capoccia and Andrea Seghezzi, for providing both data and pictures.

References

Barron, R.F., 1985. Cryogenic Systems, second ed. Oxford University Press, New York.

Berstad, D., Stang, J.H., Neksa, P., 2010. Large-scale hydrogen liquefier utilising mixed-refrigerant pre-cooling. Int. J. Hydrog. Energy 35 (10), 4512–4523. http://dx.doi.org/10.1016/j.ijhydene.2010.02.001.

Bracha, M., Lorenz, G., Patzelt, A., Wanner, M., 1994. Large-scale hydrogen liquefaction in Germany. Int. J. Hydrog. Energy 19 (1), 53–59. http://dx.doi.org/10.1016/0360-3199(94)90177-5.

Drolet, B., Gretz, J., Kluyskens, D., Sandmann, F., Wurster, R., 1996. The Euro-Québec Hydro-Hydrogen Pilot Project [EQHHPP]: demonstration phase. Int. J. Hydrog. Energy 21 (4), 305–316. http://dx.doi.org/10.1016/0360-3199(95)00083-6.

Eurostat, 2013. Energy, transport and environment indicators. In: Eurostat Pocketbooks. http://dx.doi.org/10.2785/4663.

Eurostat, 2014. Statistics database. Available from: http://epp.eurostat.ec.europa.eu/portal/page/portal/statistics/search_database (accessed 14.11.14.).

Farkas, A., 1935. Orthohydrogen, Parahydrogen and Heavy Hydrogen. Cambridge University Press, London.

Flynn, T.M., 2005. Cryogenic Engineering, second ed. Marcel Dekker, New York.

Fukano, T., Yamashita, N., Ohira, K., 2000. A study of the large hydrogen liquefaction process. J. High Press. Inst. Jpn. 38 (5), 298–305. http://dx.doi.org/10.11181/hpi1972.38.298.

Haselden, G.G., 1971. Cryogenic Fundamentals. Academic Press, London.

Jacobsen, R.T., Leachman, J.W., Penoncello, S.G., Lemmon, E.W., 2007. Current status of thermodynamic properties of hydrogen. Int. J. Thermophys. 28 (3), 758–772. http://dx.doi.org/10.1007/s10765-007-0226-7.

Krasae-in, S., 2014. Optimal operation of a large-scale liquid hydrogen plant utilizing mixed fluid refrigeration system. Int. J. Hydrog. Energy 39 (13), 7015–7029. http://dx.doi.org/10.1016/j.ijhydene.2014.02.046.

Krasae-in, S., Stang, J.H., Neksa, P., 2010a. Development of large-scale hydrogen liquefaction processes from 1898 to 2009. Int. J. Hydrog. Energy 35 (10), 4524–4533. http://dx.doi.org/10.1016/j.ijhydene.2010.02.109.

Krasae-in, S., Stang, J.H., Neksa, P., 2010b. Simulation on a proposed large-scale liquid hydrogen plant using a multi-component refrigerant refrigeration system. Int. J. Hydrog. Energy 35 (22), 12531–12544. http://dx.doi.org/10.1016/j.ijhydene.2010.08.062.

Le Roy, R.J., Chapman, S.G., McCourt, F.R.W., 1990. Accurate thermodynamic properties of the six isotopomers of diatomic hydrogen. J. Phys. Chem. 94 (2), 923–929. http://dx.doi.org/10.1021/j100365a077.

Leachman, J.W., Jacobsen, R.T., Penoncello, S.G., Huber, M.L., 2007. Current status of transport properties of hydrogen. Int. J. Thermophys. 28 (3), 773–795. http://dx.doi.org/10.1007/s10765-007-0229-4.

Leachman, J.W., Jacobsen, R.T., Penoncello, S.G., Lemmon, E.W., 2009. Fundamental equations of state for parahydrogen, normal hydrogen, and orthohydrogen. J. Phys. Chem. Ref. Data 38 (3), 721–748. http://dx.doi.org/10.1063/1.3160306.

Lipman, M.S., Cheung, H., Roberts, O.P., 1963. Continuous conversion hydrogen liquefier. Chem. Eng. Prog. 59 (8), 49–54.

McCarty, R.D., Hord, J., Roder, H.M., 1981. Selected properties of hydrogen (Engineering design data), Monograph 168, National Bureau of Standards.

McIntosh, G.E., 2004. Hydrogen liquefiers since 1950. In: Proc. 2003 Cryogenic Engineering Conference, vol. 49A, pp. 9–15. http://dx.doi.org/10.1063/1.1774661.

Ohlig, K., Bischoff, S., 2012. Dynamic gas bearing turbine technology in hydrogen plants. In: AIP Conference Proceedings, vol. 1434, p. 814. http://dx.doi.org/10.1063/1.4706994.

Quack, H., 2002. Conceptual design of high efficiency large capacity hydrogen liquefier. In: Proceeding of Cryogenic Engineering Conference, vol. 47, pp. 255–263. http://dx.doi.org/10.1063/1.1472029.

Ramachandran, R., Menon, R.K., 1998. An overview of industrial uses of hydrogen. Int. J. Hydrog. Energy 23 (7), 593–598. http://dx.doi.org/10.1016/S0360-3199(97)00112-2.

Ridgen, J., 2003. Hydrogen: The Essential Element. Harvard University Press, Cambridge, MA.

Sherif, S.A., Zeytinoglu, N., Veziroglu, T.N., 1997. Liquid hydrogen: potential, problems, and a proposed research program. Int. J. Hydrog. Energy 22 (7), 683–688. http://dx.doi.org/10.1016/S0360-3199(96)00201-7.

Span, R., 2000. Multiparameter Equations of State—An Accurate Source of Thermodynamic Property Data. Springer, Berlin.

Tien, C.L., Lienhard, J.H., 1988. Statistical Thermodynamic. Hemisphere Publishing, Washington (revised printing).

Valenti, G., Macchi, E., 2008. Proposal of an innovative, high-efficiency, large-scale hydrogen liquefier. Int. J. Hydrog. Energy 33 (12), 3116–3121. http://dx.doi.org/10.1016/j.ijhydene.2008.03.044.

Valenti, G., Macchi, E., Brioschi, S., 2012. The influence of the thermodynamic model of equilibrium-hydrogen on the simulation of its liquefaction. Int. J. Hydrog. Energy 37 (14), 10779–10788. http://dx.doi.org/10.1016/j.ijhydene.2012.04.050.

West, J.W., 2003. The economics of small to medium liquid hydrogen facilities. In: CryoGas International, May, 28-33, Available from: www.rmwsolutions.net/pub3.pdf (November 2014).

Woolley, H.W., Scott, R.B., Brickedde, F.G., 1948. Compilation of thermal properties of hydrogen in its various isotopic and ortho-para modifications. J. Res. Natl. Bur. Stand. 41 (5), 379–475.

Slush hydrogen production, storage, and transportation

3

K. Ohira
Institute of Fluid Science, Tohoku University, Sendai, Japan

Abbreviations

A	corrugated pipe with an inner diameter of 12 mm
B	corrugated pipe with an inner diameter of 15 mm
CHF	critical heat flux
GGG	gadolinium gallium garnet
GHe	gaseous helium
GH_2	gaseous hydrogen
LCR	inductance (L), capacitance (C), and resistance (R)
LH_2	liquid hydrogen
LN_2	liquid nitrogen
NBP	normal boiling point
PIV	particle image velocimetry
% Carnot efficiency	percent of Carnot efficiency
SLH_2	slush hydrogen
SLN_2	slush nitrogen
SLUSH	slush fluid
SMES	superconducting magnetic energy storage
TP	triple point

List of symbols

A	coefficient in Equation (3.13)
B	magnetic flux density (T, Wb/m^2)
C	capacitance (F)
C_0	sensitivity to capacitance in Equation (3.1) (constant) (F)
C_d	invalid capacitance in Equation (3.1) (constant) (F)
C_1	coefficient in Equation (3.5) (constant) (s^{-1})
C_2	constant in Equation (3.5) (s^{-1})
C_{sf}	factor in Equation (3.15)
c	light speed in a vacuum in Equation (3.4) (m/s)
c_l	specific heat of liquid in Equation (3.15) (J/kg K)
c_p	specific heat at constant pressure (J/kg K)
d	distance between flat-plate and cylinder (m)
d	pipe inner diameter (m)
F	coefficient in Equation (3.3)

Compendium of Hydrogen Energy. http://dx.doi.org/10.1016/B978-1-78242-362-1.00003-1

f_c cut-off frequency in Equation (3.4) (s^{-1})
g gravitational acceleration (m/s^2)
h heat transfer coefficient (W/m^2 K)
h specific enthalpy (J/kg)
h_{lv} latent heat of vaporization (J/kg)
Δh specific enthalpy difference (J/kg)
K Kutateladze factor in Equation (3.14)
K pressure drop coefficient in Equation (3.10)
k thermal conductivity (W/m K)
L distance between antennas in Equation (3.2) (m)
L distance between density meters (m)
L pipe length (m)
n exponent in Equation (3.13)
P pressure (Pa)
P_c critical pressure (Pa)
ΔP pressure drop (Pa)
q heat flux (W/m^2)
q^* critical heat flux (W/m^2)
r pressure drop ratio of slush to liquid nitrogen
r radius (m)
r_h heat transfer coefficient ratio of slush to liquid nitrogen
s corrugation pitch in Equation (3.11) (m)
s exponent in Equation (3.15)
s specific entropy (J/kg K, J/mol K)
T temperature (K)
ΔT temperature difference (superheat) (K)
t corrugation height (m)
t delay time (s)
u flow velocity (m/s)
x volumetric solid fraction

Greeks

γ empirical constant (= 0.438) in Equation (3.11)
ε specific dielectric constant
$\Delta\varepsilon$ change in specific dielectric constant
λ microwave wavelength in Equation (3.2) (m)
λ pipe friction factor
μ viscosity (Pa s)
μ_{SL} apparent viscosity of slush nitrogen (Pa s)
ρ density (kg/m^3)
σ surface tension (N/m)
$\Delta\phi$ phase-shift difference between slush and liquid hydrogen
$\Delta\phi_1$ phase shift in liquid hydrogen
$\Delta\phi_2$ phase shift in slush hydrogen

Dimensionless numbers

Fr	Froude number
Nu	Nusselt number
Pr	Prandtl number
Ra	Rayleigh number
Re	Reynolds number
Re_{SL}	slush Reynolds number

Subscripts

h	high temperature
l	liquid
l	low temperature
s	solid
SL	slush
v	vapor

3.1 Introduction: What is slush hydrogen?

Slush hydrogen is a cryogenic solid–liquid two-phase fluid, wherein solid hydrogen particles having a particle diameter of several mm are contained in liquid hydrogen, featuring greater density, and refrigerant heat capacity than liquid hydrogen. Because liquid hydrogen's density and latent heat of vaporization are approximately 1/14 and 1/5 of those of water, respectively, fuel storage tank capacities become accordingly greater for rockets and fuel cells. Also, due to boil-off resulting from heat inleak during transport and storage, the reduction in transport and storage efficiency is a practical problem.

In comparison, slush hydrogen is characterized by greater density and refrigerant heat capacity (enthalpy). That is, the utilization of slush hydrogen not only enables more efficient hydrogen transport and storage, but in pipe flow affected by heat inleak or heat generation due to superconductor quenching, heat is absorbed by the solid particles' heat of fusion, thus reducing the fluid temperature rise and the phase change from liquid to vapor–liquid two-phase flow.

Taking advantage of the solid particles' higher density and heat of fusion, slush hydrogen offers superior characteristics as a functional thermal fluid, and various applications are anticipated for hydrogen transport and storage. Furthermore, practical development is proceeding for MgB_2 (magnesium diboride) metallic superconducting material having a superconductive transition temperature of 39 K, for which slush or liquid hydrogen can be used as the refrigerant required for high-temperature superconducting equipment.

3.2 Hydrogen energy system using slush hydrogen

In light of dramatic growth in the use of fuel cells, expanded electrical power requirements associated with information technology and the need to reduce greenhouse gas emissions, a hydrogen energy system using slush hydrogen as shown in Figure 3.1 has been proposed (Ohira, 2011).

When transporting hydrogen in the form of slush hydrogen through long-distance pipelines, it can also be used as the refrigerant for superconductive power transmission using the metallic high-temperature superconducting material MgB_2. If the slush or liquid hydrogen is also used as the refrigerant for superconducting magnetic energy storage (SMES) using MgB_2 at the destination point of the pipeline, hydrogen fuel, and electrical power can be simultaneously transported and stored, resulting in a synergistic effect.

The production, storage, and transport technologies enclosed by squares in Figure 3.1 form the basis of the slush hydrogen energy system; Table 3.1 presents the major technology items needed for application.

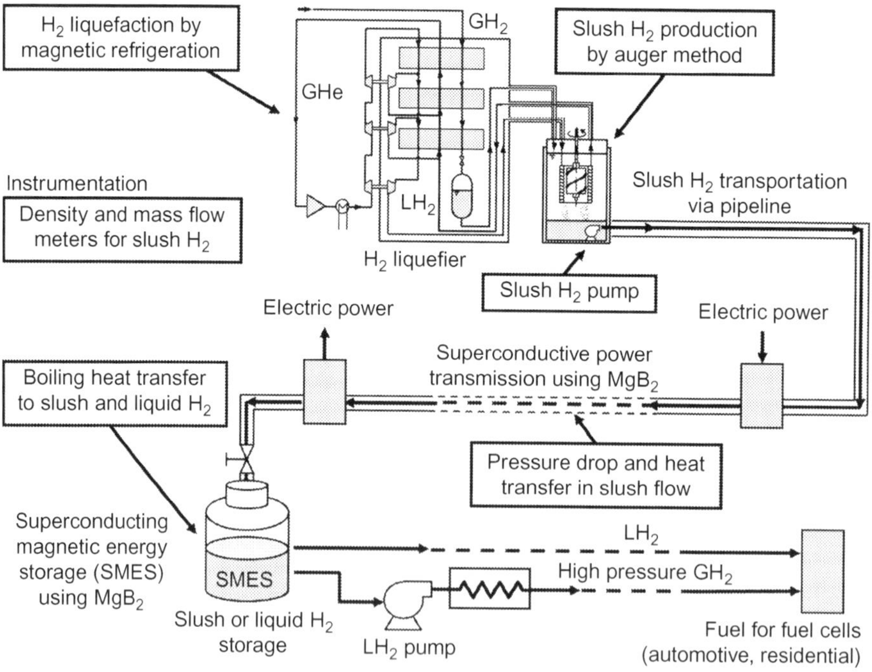

Figure 3.1 Hydrogen energy system using slush hydrogen.

Table 3.1 Major technology needed for applications of slush hydrogen

Technology	Content		References
Production	Spray method		Ohira (2004b)
	Freeze–thaw method		Sindt and Ludtke (1970) and Ohira (2011)
	Auger method (He refrigeration method)		Voth (1985), Daney et al. (1990), Ohira et al. (1994), and Ohira (2004b)
	Magnetic refrigeration for H_2 liquefaction		Waynert et al. (1989) and Ohira et al. (2000)
Storage	Aging effect of solid particles		Sindt (1970)
	Thermal stratification		Park (2010)
Transportation	Transfer equipment	Pump	Vaniman et al. (1969)
	Flow characteristics	Flow pattern	Sindt and Ludtke (1970) and Ohira (2011)
		Pressure drop	Sindt and Ludtke (1970), Ohira (2011), and Ohira et al. (2011, 2012b)
	Numerical analysis		Hardy (1990), Hardy and Whalen (1992), and Ohira et al. (2012a, 2013)
Heat transfer	Pool boiling		Sindt (1974) and Ohira (2003b)
	Forced convection		Ohira et al. (2011, 2012a)
Instrumentation	Density	Gamma ray, beta ray	Weitzel et al. (1971)
		Capacitance	Weitzel et al. (1971) Ohira and Nakamichi (2000)
		Microwave	Ellerbruch (1971) and Ohira et al. (2003b)
	Mass flow rate	Capacitance	Ohira et al. (2003a) and Ohira (2004c)
		Microwave	Ellerbruch (1971) and Ohira et al. (2005)

3.3 Thermophysical properties of slush hydrogen

As illustrated in Figure 3.2, there are two different molecular forms of hydrogen: ortho-hydrogen and para-hydrogen (Barron, 1985). In the case of the former, the two protons of the hydrogen molecule spin in the same direction, while in the latter case they spin in opposite directions.

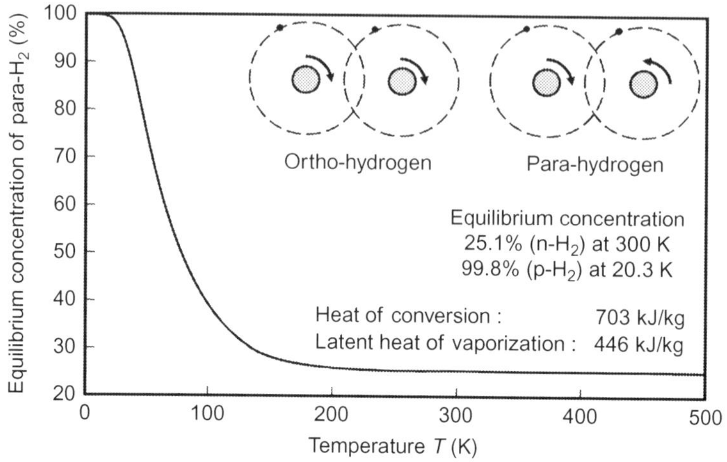

Figure 3.2 Para-hydrogen concentration in equilibrium-hydrogen.

The equilibrium concentrations of ortho- and para-hydrogen differ according to temperature. That of para-hydrogen is approximately 25% at room temperature, but 99.8% at liquid temperature. The changeover of ortho to para-hydrogen generates heat of conversion of 703 kJ/kg at the normal boiling point of hydrogen. Thus, when hydrogen gas with a para-hydrogen concentration of 25% is liquefied and stored, ortho–para conversion gradually occurs, generating a heat of conversion of 527 kJ/kg. Since liquid hydrogen's latent heat of vaporization is 446 kJ/kg, substantial boil-off occurs during long-term storage, and storage efficiency becomes poor.

Hydrogen liquefiers normally use a catalyst to speed up the conversion during the liquefaction process so as to produce liquid hydrogen for storage and transport with a para-hydrogen concentration of nearly 100%.

The pressure–temperature (*P–T*) diagram for para-hydrogen, Figure 3.3, shows the gas, liquid, and solid phases, and indicates the critical point (1.28 MPa, 33.0 K), the normal boiling point (0.10 MPa, 20.3 K), and the triple point (0.0070 MPa, 13.8 K).

The temperature–entropy (*T–s*) diagram for para-hydrogen is presented in Figure 3.4 (Johnson, 1961; McCarty, 1975; DeWitt et al., 1990). The diagram shows the quantity of state for saturated vapor and liquid hydrogen at atmospheric pressure, and for liquid hydrogen and slush hydrogen having a mass solid fraction (referred to hereafter as solid fraction) of 0.50 at the triple-point pressure, together with the respective enthalpy differences (Δh).

Table 3.2 indicates the thermophysical properties of para-hydrogen at the normal boiling point and the triple point, and the properties of slush hydrogen (13.8 K) having a solid fraction of 0.50. In the case of the latter, density is 15% higher than for normal boiling liquid hydrogen (20.3 K), and the refrigerant heat capacity (enthalpy) through change to gaseous state (20.3 K) increases by 18%.

Another representative slush fluid is slush nitrogen (63.2 K), which is being considered for use as a refrigerant for high-temperature superconducting equipment. In this case, density is 16% higher than for liquid nitrogen (77.3 K), while the refrigerant heat capacity (enthalpy) increases by 22%.

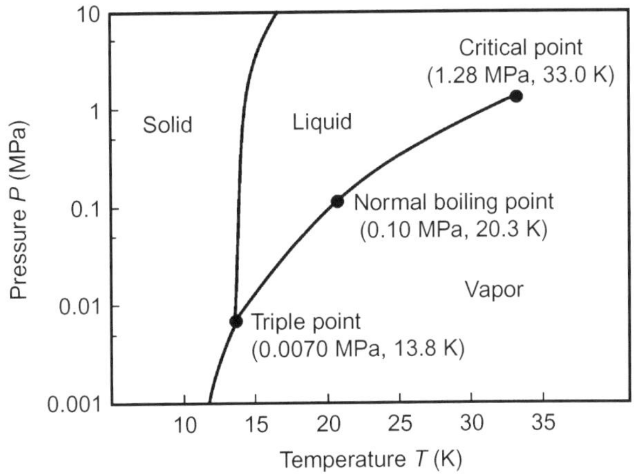

Figure 3.3 Pressure–temperature diagram for para-hydrogen.

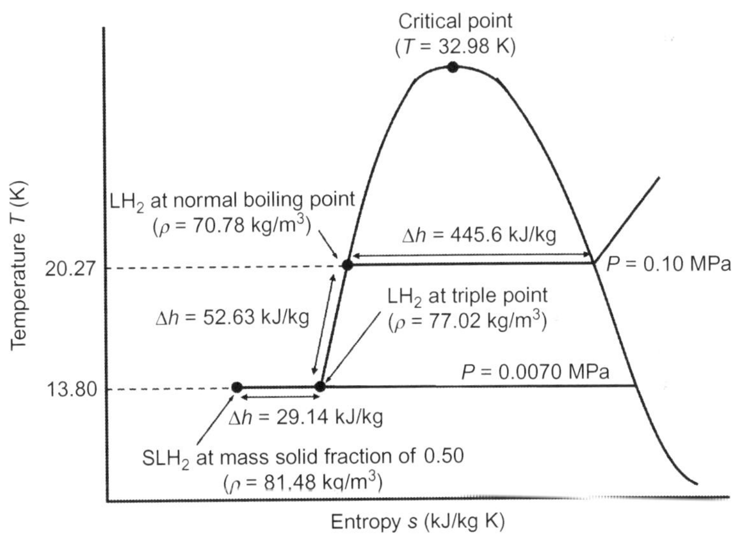

Figure 3.4 Temperature–entropy diagram for para-hydrogen.

3.4 Process of producing and storing slush hydrogen

3.4.1 Hydrogen liquefaction by magnetic refrigeration

In hydrogen liquefaction by magnetic refrigeration, the adiabatic demagnetization method to generate temperatures below 1 K is extended to the high-temperature region. As shown in Figure 3.5, magnetic liquefaction uses an external magnetic field to magnetize and demagnetize paramagnetic material in repeated cycles, thus producing low temperatures through the magnetocaloric effect. Since magnetic refrigeration can ideally realize a reversed Carnot cycle, it is possible to achieve theoretically high

Table 3.2 **Thermophysical properties of para-hydrogen at the normal boiling point and the triple point**

	Normal boiling point (0.10 MPa, 20.3 K)		Triple point (0.0070 MPa, 13.8 K)			
	Vapor	**Liquid**	**Vapor**	**Liquid**	**Solid**	**SLH_2 (solid fraction of 0.5)**
Density ρ (kg/m^3)	1.338	70.78	0.1256	77.02	86.50	81.49
Enthalpy h (kJ/kg)	189.34	−256.26	140.28	−308.89	−367.17	−338.03
Specific heat c_p (kJ/kg K)	12.153	9.688	10.516	6.513	–	–
Thermal conductivity k (mW/m K)	16.3	100	10.4	73	900	–
Viscosity μ (mPa s)	0.0011	0.0133	0.00073	0.026	–	–
Specific dielectric constant ε (–)	1.004	1.230	1.0004	1.252	1.286	1.269

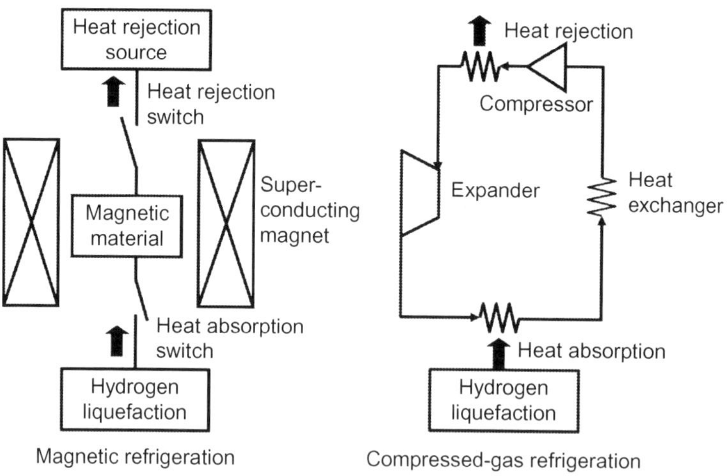

Figure 3.5 Comparison of hydrogen liquefaction between magnetic refrigeration and compressed-gas refrigeration.

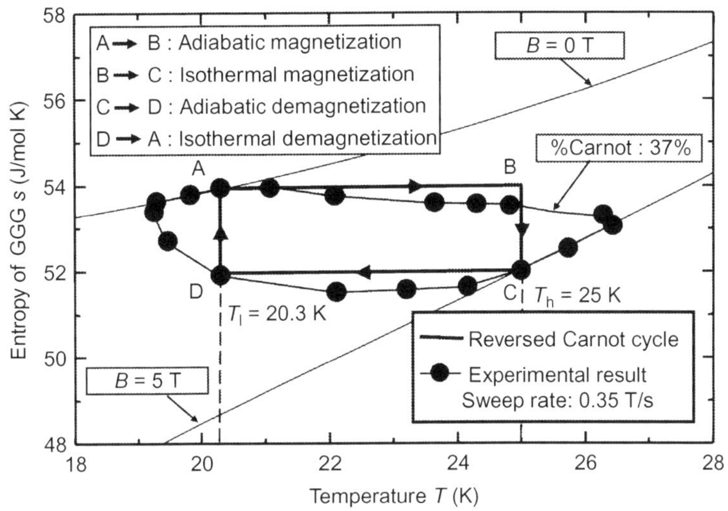

Figure 3.6 Comparison between experimental results and an ideal reversed Carnot cycle of magnetic refrigeration.

liquefaction efficiency, in contrast to the compressed-gas refrigeration method shown in Figure 3.5.

The maximum efficiency is expected to be about 50% in terms of %Carnot efficiency (percent of Carnot efficiency) (Waynert et al., 1989), compared to that of around 38% for a large-scale hydrogen liquefier (liquefaction capacity: 60 ton/day) using the compressed-gas method (Strobridge, 1974). The definition of %Carnot efficiency is the ratio of the Carnot work to actual work per unit mass liquefied. Also, given the use of solid magnetic material, which has much greater entropy density than a gas, the liquefier can be made more compact (less massive).

Figure 3.6 is an entropy–temperature diagram for magnetic material when hydrogen gas that has vaporized from a liquid hydrogen vessel is then liquefied by magnetic refrigeration (Ohira et al., 2000). Here, gadolinium gallium garnet (GGG: $Gd_3Ga_5O_{12}$) was selected as the magnetic material, and a superconductive pulse magnet (maximum field of 5 T, maximum magnetization speed of 0.35 T/s) was used to generate the magnetic field. A Gifford–McMahon refrigerator is used as a heat rejection source at the high-temperature T_h of 25 K.

The reasons an ideal reversed Carnot cycle could not be achieved were the insufficient heat transfer performance of the heat absorption and heat rejection switches, and the influence of uncondensed hydrogen gas in the vicinity of the magnetic material. Nevertheless, the achievement of 37% for %Carnot value in a small-scale (3.55 g/h, or 50 cc/h) liquefaction experiment demonstrates the high efficiency of this method.

For hydrogen liquefaction, a multistage magnetic refrigerator from room temperature to liquid hydrogen temperature has been proposed, and research and development work is proceeding (Barclay and Steyert, 1982; Zhang et al., 1993; Ohira,

2003a). A method has also been proposed for producing slush hydrogen from liquid hydrogen, using magnetic refrigeration to produce temperatures below 13.8 K (Waynert et al., 1989).

3.4.2 Slush hydrogen production

Various methods for slush hydrogen production have been implemented, most notably the freeze–thaw method for the National Aero-Space Plane project in the United States. The use of slush hydrogen with a solid fraction of 0.50 is expected to reduce the total takeoff weight of the space plane by 13–26% (DeWitt et al., 1990).

The major production methods are spray, freeze–thaw, and auger. In the spray method, a nozzle is used for the adiabatic expansion of liquid hydrogen to produce solid particles. However, the particles formed in this manner tend to melt when mixed into liquid hydrogen, making this method unsuitable for the efficient mass production of slush.

Freeze–thaw is a relatively straightforward method, the steps of which are described below, and is applicable at the experimental level or for small- to medium-scale production.

First, a vacuum pump is used for pressure reduction of liquid hydrogen at the normal boiling point pressure. As the liquid boils and vaporizes in accordance with the vapor–liquid equilibrium curve in Figure 3.3, the latent heat is removed and the temperature of the liquid is reduced. When the liquid reaches the triple-point pressure, a frozen layer is formed on the liquid surface. When evacuation is stopped, part of the frozen layer melts and solid hydrogen sinks into the liquid. A stirrer is then used to break up the solid into small particles having a diameter of several mm. Slush hydrogen is produced by performing the freeze and thaw cycle repeatedly.

For slush hydrogen, particle diameters are reported to range from 0.5 to 7 mm (Sindt and Ludtke, 1970), while for slush nitrogen they are reported as 0.5–2 mm, with an average diameter of 1.36 mm (Ohira, 2011). The difference in this regard between hydrogen and nitrogen is considered to be due to differing vacuum and stirring times during production. The freeze–thaw method requires a large vacuum pump, and since slush hydrogen is produced at the triple-point pressure (0.007 MPa), leak-tightness is required to prevent the penetration of air or other contaminants into the production tank.

The slush hydrogen and slush nitrogen used for the experimental work described in this chapter were produced using the freeze–thaw method.

In the auger method, liquid hydrogen is subjected to heat exchange with lower-temperature helium. The solid hydrogen that forms at the heat transfer surface is scraped off by a rotating auger, forming fine particles of solid hydrogen. Since the heat exchanger and auger are immersed in the liquid hydrogen, the solid hydrogen particles are mixed into the liquid during the production process.

If liquid hydrogen and cryogenic helium are continuously supplied, slush hydrogen can be continuously produced at a higher pressure than the triple-point pressure, making this method suitable for mass production. By adjusting the gap between the auger and the heat transfer surface and/or changing the rotational speed of the auger, the size of the solid particles can be controlled.

Daney et al. (1990) compared the ideal work per unit of refrigeration (W/W) in the cases of the freeze–thaw and auger methods used to produce slush hydrogen with a solid fraction of 0.50. These were found to be 29 W/W for the helium refrigerator operating at 10 K and 28 W/W for the freeze–thaw method, i.e., comparable.

Also, numerical analysis using heat transfer models of the auger and heat exchanger showed that the refrigerant temperature of the supercritical helium and the rotational speed of the auger had the greatest influence on the slush production rate. At a refrigerant flow rate of 25 g/s and rotation of 60 rpm, the production rate of slush hydrogen with a solid fraction of 0.50 predicted was 14 g/s (620 l/h) at 10 K, but 30 g/s (1330 l/h) at 6 K.

At present, large-scale production experiments have yet to be carried out. Figure 3.1 is predicated upon a combination of hydrogen liquefier using the helium Brayton cycle and the auger method, resulting in the continuous mass production of slush hydrogen.

Figure 3.7 shows a cross-sectional view of the heat exchanger and auger used in a small-scale production experiment (Ohira et al., 1994). Cryogenic helium is introduced from the top of the heat exchanger and is then directed upward from the bottom, with heat exchanged via copper heat transfer fins. The thin layer of solid hydrogen that forms on the reverse side of the heat transfer surface is scraped off by the auger, forming solid particles.

Figure 3.8 presents the solid hydrogen production rate at differing auger speeds, together with the amount of refrigeration supplied in a small-scale production experiment (Ohira et al., 1994; Ohira, 2004b). At auger speeds of 30–80 rpm, solid hydrogen production increases with speed, and the particle size becomes smaller. The

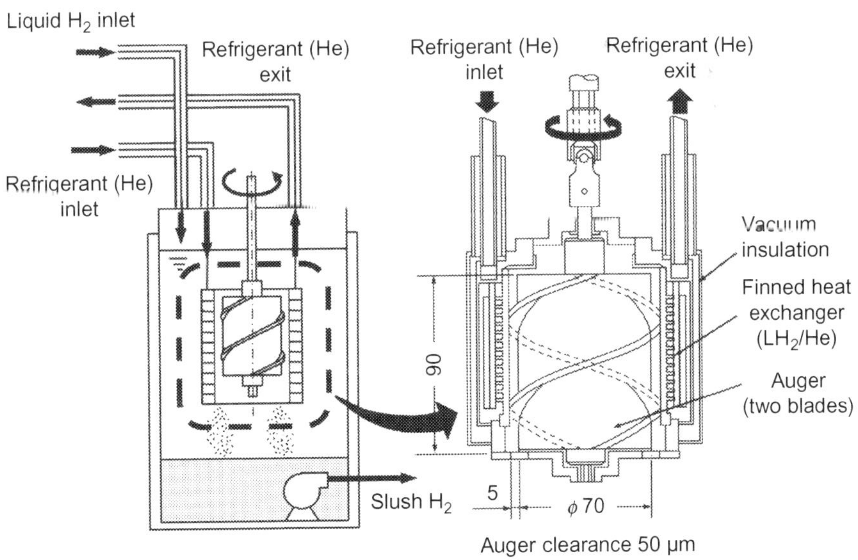

Figure 3.7 Cross-sectional view of the heat exchanger and auger (Ohira et al., 1994).

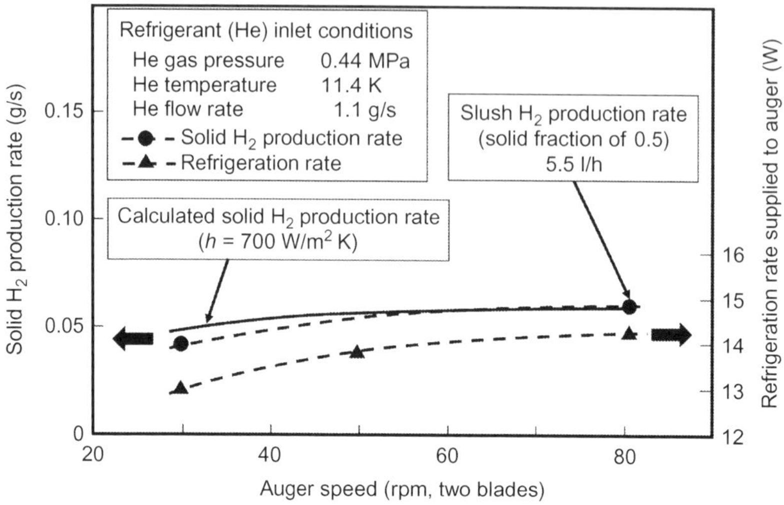

Figure 3.8 Solid hydrogen production rate and refrigeration rate supplied to auger due to auger speed (Ohira et al., 1994).

maximum solid production rate of 0.062 g/s is reached at an auger speed of 80 rpm (two blades), equivalent to 5.5 l/h of slush hydrogen with a solid fraction of 0.50.

The gas pressure, temperature, and mass flow rate of supercritical helium at the heat exchanger inlet are, respectively, 0.44 MPa, 11.4 K, and 1.1 g/s. The measured amount of refrigeration supplied is 13.0–14.3 W. However, estimating from the solid production rate, the refrigeration actually devoted to solid production is 2.5–3.6 W, with approximately 10 W being consumed due to heat inleak.

The reason why the solid production rate and refrigeration supplied increase with auger speed is that the solid layer being scraped off by the auger is thinner. The thermal conductivity of solid hydrogen is about 1/1000 compared to the copper used for the heat transfer surface, such that the heat transfer rate is reduced when the thickness of the solid layer on the heat transfer surface increases.

From the analytical results based on the heat transfer model of Daney et al. (1990), the heat transfer coefficient h of helium to the finned heat transfer surface is approximately 700 W/m^2 K, as shown in Figure 3.8.

In order to achieve practical application of the auger method for high-efficiency mass production, improved performance of the heat exchanger and long-term reliability of the auger mechanism rotating at cryogenic temperatures are necessary.

3.5 Density and mass flow meters for slush hydrogen

By measuring the fluid temperature, pressure, and volumetric flow, it is easy to determine the density or mass flow rate of single-phase fluids such as liquid hydrogen. However, for slush hydrogen, because density varies in accordance with the solid fraction, it is necessary to measure the solid fraction directly.

Slush hydrogen density meters can use gamma ray attenuation, changes in capacitance, or microwave propagation properties as the means of measurement, with ±0.5% accuracy for each of these techniques (Weitzel et al., 1971). Mass flow meters typically combine a waveguide and horn antennas, determining density and flow velocity from changes in amplitude and frequency (the Doppler effect) during microwave propagation, with error around 10% (Ellerbruch, 1971).

Regardless of the particular method being considered, the influence of solid particles on density and flow velocity measurements must be clarified to improve measurement accuracies. Given the objective of higher accuracy in density and mass flow rate measurements, new density meter and mass flow meter structures are discussed in the following. These approaches are based on changes in the specific dielectric constant or changes in the microwave propagation properties, due to variation in the solid fraction.

3.5.1 Density meter

In the case of solid particles mixed in a liquid, such as with slush hydrogen, it is important to consider the behavior of the solid particles in terms of electrode configuration if capacitance densimeter accuracy is to be improved. Figure 3.9 illustrates the principles associated with the parallel flat-plate type for single-phase liquid, and the flat-plate and cylinder type for slush.

A combination design featuring a flat-plate together with two cylinders allows solid particles to easily enter between the electrodes, with a simple construction that offers high accuracy. The relationship between the capacitance C of the capacitance-type densimeter and the specific dielectric constant ε is expressed as:

$$C = C_0\varepsilon + C_d \tag{3.1}$$

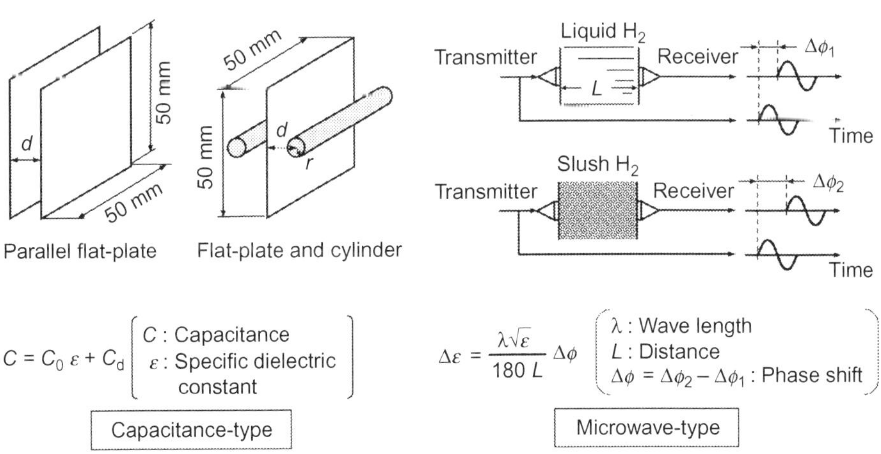

Figure 3.9 Principles of capacitance-type and microwave-type density meters for slush hydrogen (Ohira, 2004c).

Here, C_0 is a fixed constant determined by the size and configuration of the electrode, representing sensitivity to capacitance with respect to ε. C_d is invalid capacitance generated by the nonelectrode structures of the density meter, a value specific to each density meter and a constant, regardless of change in ε.

When C_0 and C_d are measured in advance in liquid hydrogen of differing temperatures (differing specific dielectric constants), the measurement of C for the density meter placed in slush hydrogen allows determination of ε. The volumetric solid fraction can be calculated from the specific dielectric constants ε_s and ε_l of solid and liquid hydrogen, while the density of the slush hydrogen can be determined from densities of the solid and liquid.

For example, considering the change in ε at the triple point (13.8 K), where (as shown in Table 3.2) liquid ($\varepsilon_l = 1.252$) changes to solid ($\varepsilon_s = 1.286$), the increase is a mere 2.7%. Thus, for the sake of improved densimeter accuracy, it is desirable to have a high sensitivity to capacitance (C_0) and a structure that allows easy entry of the solid particles between the electrodes.

Figure 3.10 presents the solid fraction measurement results for two varieties of flat-plate and cylinder type electrodes (with the minimum gap between the flat-plate and the cylinder $d - r = 3$ or 5 mm), and for a parallel flat-plate type (with the gap between the plates $d = 10$ mm). For both designs, the size of the flat-plates was 50 mm × 50 mm. The calculated value used for comparison with the measured value is the average solid fraction, calculated from the reduction in the amount of liquid

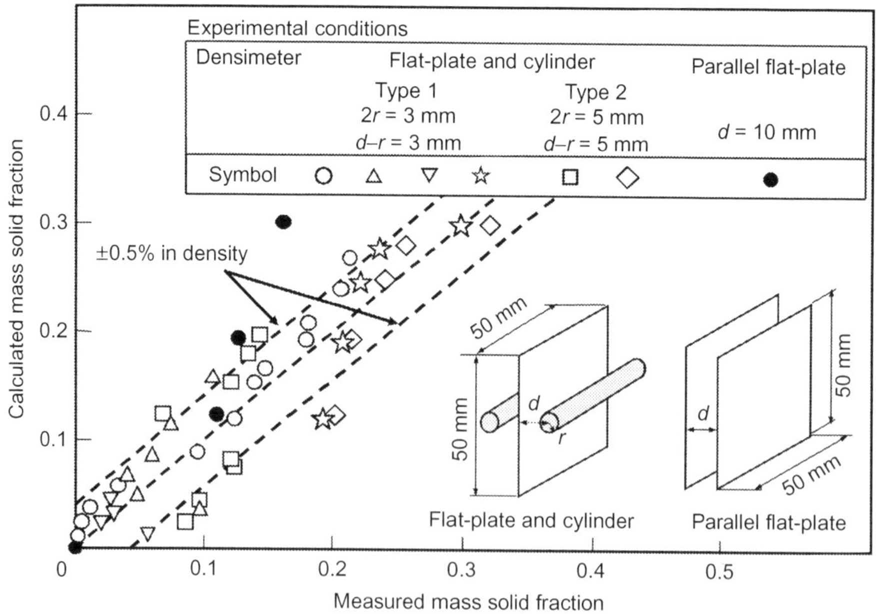

Figure 3.10 Comparison between measured and calculated mass solid fractions (Ohira, 2004c).

during the slush hydrogen production using the freeze–thaw method (Ohira and Nakamichi, 2000; Ohira, 2004c).

In the case of the parallel flat-plate type, as the solid fraction increased, the solid particles were unable to sufficiently enter within the electrode structure, and the solid fraction obtained was lower than the actual density. In contrast, almost all of the measured results for the flat-plate and cylinder type were in good agreement with the calculated value within ±0.5% in density.

The microwave-type density meter, as shown in Figure 3.9, consists of a microwave transmitter and receiver antennas, employing a network analyzer to measure the phase shift in the microwaves propagated between the antennas. In the conventional type, composed of a waveguide and a horn antenna, a disadvantage was that solid particles entered within the horn antenna, and the density measurement accuracy of ±2% was obtained (Ellerbruch, 1971). However, as shown in Figure 3.11, a new type has been developed, incorporating cryogenic coaxial cables and patch antennas, enabling improved measurement accuracy (Ohira et al., 2003b; Ohira, 2004c).

As can be seen in Figure 3.9, measurement is done with respect to the phase shifts $\Delta\phi_1$, $\Delta\phi_2$ in the microwave propagated between the antennas in liquid and slush hydrogen, respectively. When the distance L between the transmitter and receiver antennas is the same, the relationship between the phase-shift difference $\Delta\phi = \Delta\phi_2 - \Delta\phi_1$ on the one hand, and the change in the specific dielectric constant $\Delta\varepsilon$ for liquid and slush hydrogen on the other, is described by:

$$\Delta\varepsilon = \frac{\lambda\sqrt{\varepsilon}}{180L}\Delta\phi \tag{3.2}$$

Here, λ is the microwave wavelength. The amount of change from liquid to solid in ε is very small, such that when ε is regarded as a constant, the coefficient of $\Delta\varepsilon$ with

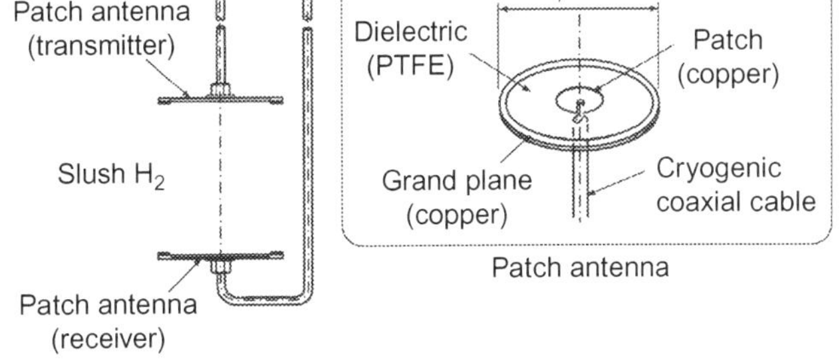

Figure 3.11 Microwave-type density meter for slush hydrogen (Ohira et al., 2003b).

respect to $\Delta\phi$ becomes a constant value specific to the density meter. Accordingly, the relationship between $\Delta\phi$ and $\Delta\varepsilon$ can be represented by:

$$\Delta\varepsilon = F\Delta\phi \tag{3.3}$$

As is the case with the capacitance-type density meter, when a fixed constant value F is measured in advance in liquid hydrogen, the density of slush hydrogen can be determined from the measurement of $\Delta\phi$. In experimental results, density measurement accuracy was obtained within $\pm 0.5\%$.

3.5.2 Mass flow meter

Figure 3.12 shows the new measurement principles of the capacitance-type and microwave-type flow meters in pipe flow (Ohira et al., 2003a, 2005; Ohira, 2004c). These allow the determination of mass flow rate from the measurement of density and flow velocity of slush hydrogen. A noteworthy aspect of flow velocity measurement is the slight fluctuation in the distribution of solid particles, i.e., slush hydrogen density fluctuation.

First, capacitance or microwave (waveguide) type densimeters are placed at two locations along the pipe flow, and the density is measured using the LCR meter or waveguide. Then, the flow velocity u is calculated from the delay time t where the cross-correlation function for the density measurement signals from the two locations

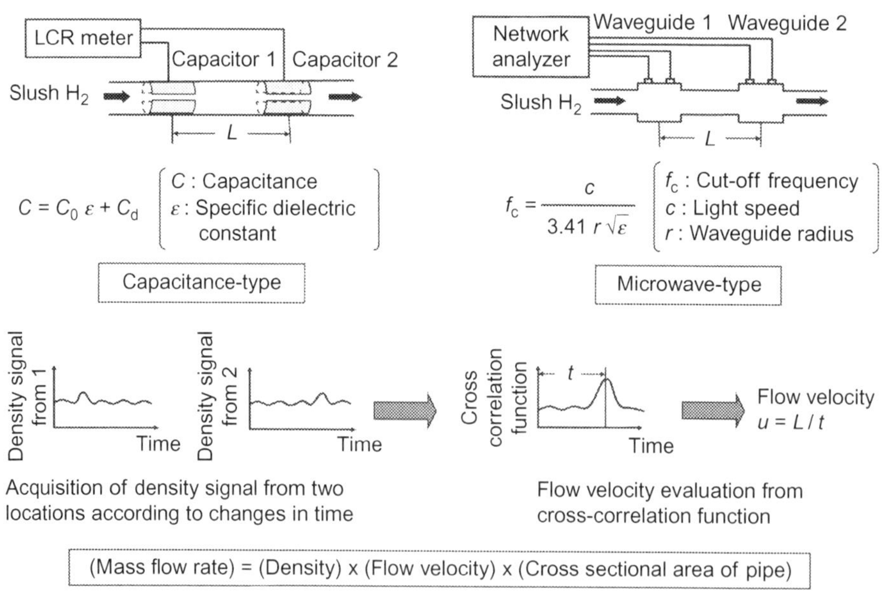

Figure 3.12 Principles of capacitance-type and microwave-type mass flow meters for slush hydrogen (Ohira, 2004c).

is at a maximum, together with the distance L between the densimeters, such that the mass flow rate can be determined.

Since these approaches do not utilize any moving parts (unlike turbine flow meters), there is no need to install precooling piping and valves. Since the difference between the capacitance and the microwave-type mass flow meters is in the density measurement method, only the microwave-type is described below.

Figure 3.13 illustrates the circular waveguide configuration, installed at upstream and downstream locations along the pipe. The relationship shown by Equation (3.4) is for the specific dielectric constant ε of the slush hydrogen within the waveguide, and the cut-off frequency f_c of the microwaves propagating the pipe

$$f_c = \frac{c}{3.41 r \sqrt{\varepsilon}} \tag{3.4}$$

Here, c is the speed of light in a vacuum, and r is the waveguide tube radius.

Given that f_c changes with ε of the slush hydrogen within the waveguide, the density of the slush hydrogen can be measured. As in the case of the capacitance-type densimeter, Equation (3.4) is approximated using Equation (3.5), ε is determined from the measured f_c, and the slush hydrogen density is obtained

$$f_c = C_1 \varepsilon + C_2 \tag{3.5}$$

Here, C_1 and C_2 are constant values unique to the waveguide.

With respect to flow velocity, assuming that the gain between the waveguide transmitter and the receiver signals corresponds to changes in density, the gain measured at the upstream and downstream locations can be used to calculate the cross-correlation function. The flow velocity can thus be determined.

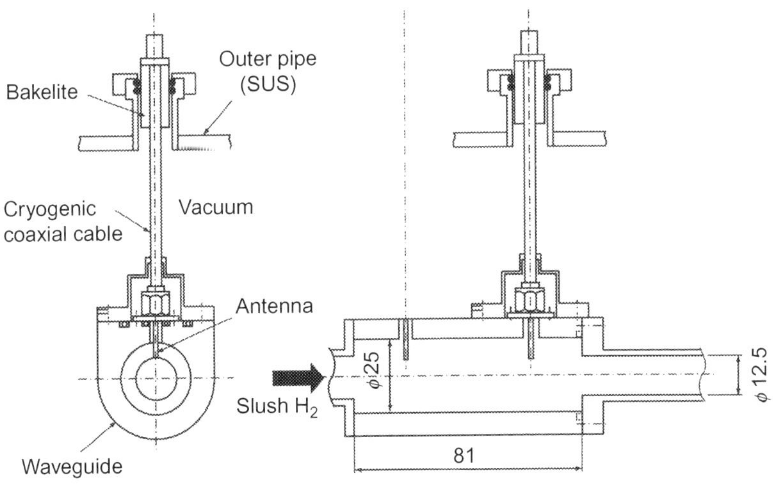

Figure 3.13 Configuration of circular waveguide (density meter for slush hydrogen) (Ohira et al., 2005).

Experimental work was performed using the capacitance and microwave-type flow meters up to solid fractions of 0.14 and 0.08, respectively, and density measurement accuracy of ±0.5% was obtained. In velocity measurement, both flow meters were employed up to flow velocities of 0.5 and 1.1 m/s, respectively. Since the accuracy of the flow velocimeter used for calibration is ±5%, the exact measurement accuracy of this method could not be immediately confirmed.

3.6 Advantages and disadvantages of transporting slush hydrogen via pipeline

In the transportation of slush hydrogen via pipeline, it is important to consider pressure drop due to straight pipes, as well as flow restrictions such as valves, converging–diverging pipes, and orifices. Such pipelines also often make use of corrugated pipes to absorb the cryogenic contraction (stainless-steel pipe contracts by approximately 3 mm per 1 m). Increased pressure drop during slush hydrogen transport requires greater pump power, thus reducing the overall performance of the transport system.

As indicated in Figure 3.1, if slush hydrogen is to be used as the refrigerant for superconductive power transmission, heat generation can be anticipated due to superconductor quench, making it important to also consider pressure loss of slush hydrogen during heat transfer, together with forced convection heat transfer to slush hydrogen.

3.6.1 Transfer pump for slush hydrogen

Slush hydrogen performance tests have been carried out using a liquid hydrogen centrifugal pump having an impeller diameter of 70 mm, under rotational speed and average solid fraction conditions of 8000–19,000 rpm and 0.33 (Vaniman et al., 1969). No difference is apparent in terms of operating characteristics and pump performance among normal boiling point liquid, triple-point liquid, and slush hydrogen.

3.6.2 Pressure drop and heat transfer in pipe flow

Since the solid–liquid density ratio is 1.12 (1.18 for slush nitrogen), the flow pattern, as in the case of slurry, depends on the solid fraction and flow velocity. These regimes can be classified as saltation, heterogeneous, and pseudo-homogeneous flows when slush hydrogen flows in a horizontal and straight pipe. There are two essential types of saltation flow: moving bed, where solid particles accumulate at and flow along the bottom, and static bed, where the solid particles do not move.

In cases of saltation flow where the solid fraction is high and the flow velocity is low, sufficient attention should be directed to blockages (plugging) caused by solid particles in narrow flow channels such as valves. In heterogeneous flow, the solid particles are nonuniformly dispersed perpendicular to the direction of flow, flowing at a velocity less than that of the liquid. In the case of pseudo-homogeneous flow, the solid particles are uniformly dispersed within the liquid.

Little detailed experimental flow data are available for slush hydrogen. In the ranges of 0.18–0.50 for solid fraction and 0.46–12 m/s for flow velocity, pressure drop of slush hydrogen flow in a 16.6 mm (inner diameter) horizontal pipe was reported as described in the following (Sindt and Ludtke, 1970).

At a low Reynolds number of 6×10^4 and a solid fraction of 0.50, the pressure drop for slush hydrogen was around twice that for liquid hydrogen at the triple-point temperature. At a Reynolds number of 7×10^5 and a solid fraction of 0.50, the pressure drop was about the same as for liquid.

For Reynolds numbers of 3×10^5 and above, and for solid fractions of 0.40 and under, pressure drop for slush hydrogen was a maximum of 8% less than for liquid, with pressure drop reduction appearing similar to a non-Newtonian fluid. Also, for an identical solid fraction, the pressure drop for aged slush hydrogen was 4–10% greater than for fresh slush hydrogen. The critical flow velocity at which solid particles no longer flow with the liquid is approximately 0.46 m/s, equivalent to a Reynolds number of 2.3×10^4.

In Figure 3.14, at solid fractions of 0.20 and 0.30, the broken and chain lines illustrate the relationship between the Reynolds number Re and the pipe friction factor λ. The value of the pipe friction factor λ is calculated by Equation (3.6), using the pressure drop ΔP

$$\lambda = \frac{2\Delta P\, d}{\rho u^2 L} \tag{3.6}$$

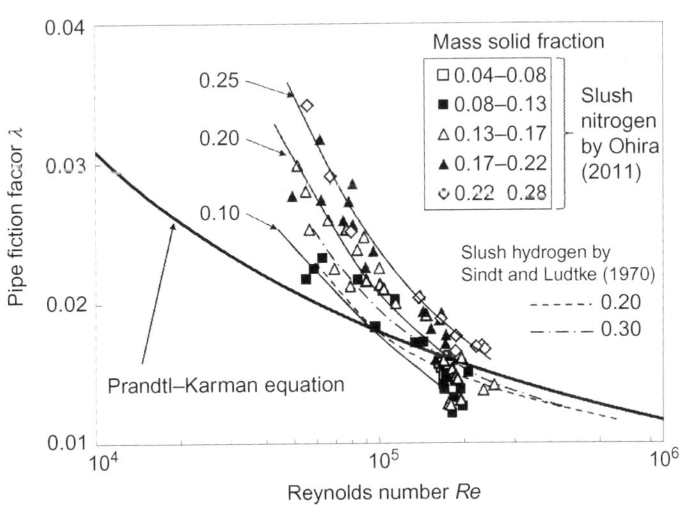

Figure 3.14 Pipe friction factor versus the Reynolds number for slush hydrogen and slush nitrogen.

Here, d is the inner diameter, ρ is the slush hydrogen density, u is the flow velocity, and L is the pipe length. The thick solid line in the figure corresponds to the Prandtl--Karman Equation (3.7) for a smooth circular pipe

$$\frac{1}{\sqrt{\lambda}} = 2\log_{10}\left(Re\sqrt{\lambda}\right) - 0.8 \tag{3.7}$$

Next, experimental results are cited for slush and liquid nitrogen at the triple-point temperature, obtained using a horizontal pipe with an inner diameter of 15 mm (Ohira, 2011). By comparing slush hydrogen and slush nitrogen, similarities and differences in terms of flow characteristics can be found. Fresh slush nitrogen was used in the experimental work, and the particle image velocimetry (PIV) method was employed to measure the solid particle velocity.

The experimental results for slush nitrogen are presented in Figure 3.14. The experimental range for flow velocity is 1.1–5.9 m/s ($4.95 \times 10^4 < Re < 2.54 \times 10^5$), while that for the solid fraction is 0.04–0.28. Based on these results, the pressure drop trends are represented by the solid line for solid fractions of 0.10, 0.20, and 0.25.

At flow velocities of 1.1–3.6 m/s ($Re < 1.60 \times 10^5$), for all solid fractions, the pressure drop for slush is greater than that calculated for liquid nitrogen using Equation (3.7). Furthermore, as the solid fraction increases, the difference between the pipe friction factors for slush and liquid nitrogen becomes greater. In contrast, pressure drop reduction is observed at flow velocities of 3.6–5.9 m/s and solid fractions of 0.04–0.17, such that the pressure drop for slush nitrogen is less than that for liquid nitrogen.

Compared to Equation (3.7), the maximum pressure drop reduction is 23%, at flow velocity of 4.2 m/s and with a solid fraction of 0.12. During low-velocity flow, due to the effect of gravity, the solid particle velocity distribution becomes vertically asymmetric (heterogeneous), such that interference among particles at the bottom of the pipe and interference between the particles and the pipe walls are more prominent than during high-velocity flow. As the solid fraction becomes greater, this tendency becomes even more notable.

During high-velocity flow, the solid particle velocity distribution becomes nearly symmetrical (pseudo-homogeneous). In the case of low solid fractions, the reduction effect becomes prominent, and the pressure drop reduction emerges. On the other hand, at high solid fractions, interference among the solid particles and increased turbulent energy result in pressure drop that supersedes the reduction effect. The total pressure drop is about the same as for liquid nitrogen.

However, even at higher flow velocities, pressure drop approaches around 1.15 times the Prandtl–Karman equation, showing that the reduction effect continues to play a significant role in the pressure drop.

From flow observation results, together with numerical analysis results (SLUSH-3D) (Ohira et al., 2012a), it can be seen that the flow regime changes from heterogeneous to pseudo-homogeneous in the vicinity of 3 m/s, and that the pressure drop reduction effect is manifested. That is, a phenomenon occurs in pseudo-homogeneous flow, in which solid particles migrate toward the center axis of the pipe with fewer particles in

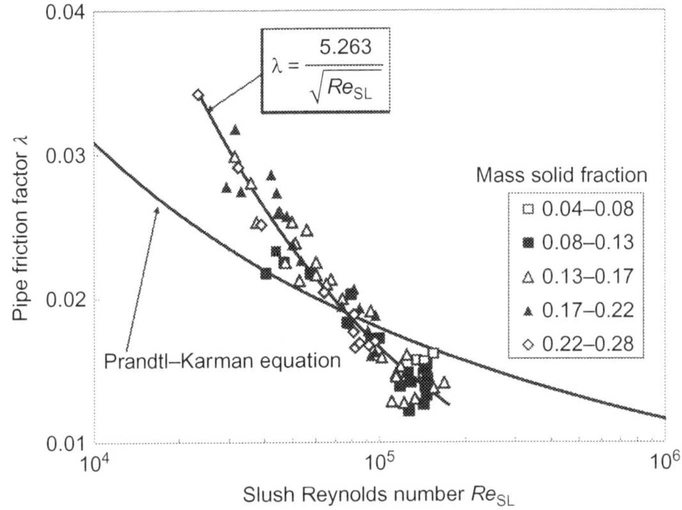

Figure 3.15 Pipe friction factor versus the slush Reynolds number for slush nitrogen (Ohira, 2011).

the vicinity of the pipe wall. A group of particles forming at the pipe center that serves to suppress the development of turbulence induces pressure drop reduction.

Compared with ordinary solid–liquid two-phase fluids, pressure drop is lower even in the case of a high solid fraction at a high flow velocity, thus allowing a reduction in required pump power.

Figure 3.15 shows the relationship between the slush Reynolds number Re_{SL} and the pipe friction factor λ, using Equation (3.8) (Barron, 1999; Orr and Dallavalle, 1954) for the apparent viscosity μ_{SL} of slush nitrogen. Here, x is the volumetric solid fraction and μ_l is the liquid viscosity

$$\mu_{SL} - \mu_l \left[1 - \left(\frac{x}{0.6}\right)\right]^{-1.8} \tag{3.8}$$

Using Equation (3.9) obtained from the least-squares approximation curve for λ and Re_{SL} indicated in Figure 3.15, the pressure drop (required pump power) can be estimated to within ±20% from the solid fraction and flow velocity

$$\lambda = \frac{5.263}{\sqrt{Re_{SL}}} \quad \left(2.35 \times 10^4 < Re_{SL} < 1.69 \times 10^5\right) \tag{3.9}$$

The result also indicates that the apparent viscosity of slush nitrogen can be approximately expressed by Equation (3.8).

Even in a horizontal pipe having an inner diameter of 10 mm, pressure drop reduction has been observed for slush nitrogen (Ohira et al., 2011), manifested at velocities of 2.0–3.5 m/s and solid fractions of within 0.14. In comparison with pressure drop for liquid nitrogen, a reduction of up to 25% is seen in the case of slush nitrogen. At high velocities of 3.5 m/s and over, pressure drop approaches around 1.1 times that for liquid nitrogen.

The velocity at which pressure drop reduction initiated with respect to slush hydrogen in a horizontal pipe with an inner diameter of 16.6 mm was lower than that in the case of slush nitrogen flow in a 15 mm diameter pipe (nearly the same diameter) at identical solid fractions. That is, as indicated in Figure 3.14, for a solid fraction of 0.20, the Reynolds number at which the reduction initiates is 1.0×10^5 for slush hydrogen and 1.5×10^5 for slush nitrogen. Since the kinetic viscosities are about the same for both, the reduction initiates at lower flow velocity in the case of slush hydrogen.

With numerical analysis results for a pipe inner diameter of 15 mm, flow velocities of 2 and 5 m/s, an average solid fraction of 0.15, and particle diameter of 1.3 mm, Ohira et al. (2012a) showed the cross-sectional distributions of solid fraction within the pipe for slush hydrogen and slush nitrogen. In the case of slush hydrogen, solid particles did not tend to settle at the bottom of the flow channel even at the lower flow velocity, presenting a more uniform distribution.

The solid–liquid density ratio is lower for slush hydrogen (1.12 vs. 1.18 for slush nitrogen), and, as the viscosity of liquid hydrogen is 1/11 that of liquid nitrogen, pseudo-homogeneous flow occurs at lower flow velocity. These results suggest that the pressure drop reduction effect is manifested in pseudo-homogeneous flow.

With respect to the relationship between pressure drop reduction and heat transfer deterioration during pipe flow of slush nitrogen, Ohira et al. (2011) measured the pressure drop and heat transfer coefficient for slush nitrogen in horizontal pipes with inner diameters of 10 and 15 mm, under conditions of 10 and 30 kW/m^2 for heat flux, 0.10–0.40 for solid fraction, and 0.16–5.2 m/s for flow velocity.

The results for the 15 mm pipe with heat flux of 10 kW/m^2 are shown in Figure 3.16. The bold solid lines are the pressure drop and heat transfer coefficient

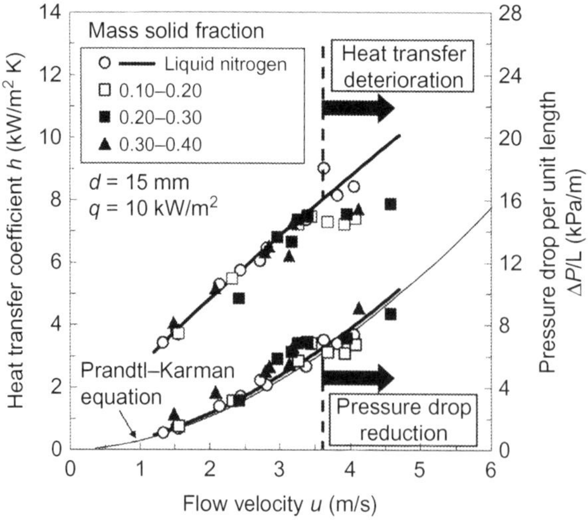

Figure 3.16 Pressure drop and heat transfer coefficient versus flow velocity for slush nitrogen (Ohira et al., 2011).

measured for liquid nitrogen. At a solid fraction in the range of 0.10–0.30, and at a flow velocity of 3.6 m/s or greater, pressure drop for slush nitrogen is lower than that for liquid nitrogen. However, at solid fractions of 0.10–0.40 (and velocity of 3.6 m/s or greater), the heat transfer coefficient deteriorates with respect to that of liquid. Even at solid fractions of 0.30–0.40, where there is no pressure drop reduction, heat transfer deterioration occurs, supporting the continuation of the pressure drop reduction effect even at high solid fractions.

Also, given that the pressure drop reduction region is about the same as for the case of nonheating, the flow structure is considered to be the same as for slush flow during heat transfer. That is, the reason for the deterioration of heat transfer has to do with the aforementioned suppression of turbulence development at the pipe wall, the result of which is to suppress the diffusion of heated liquid in the pipe wall vicinity toward the pipe center.

In Figure 3.17, the results for an inner diameter of 15 mm and heat flux of 10 kW/m^2 are expressed as the pressure drop ratio r and the heat transfer coefficient ratio r_h of slush to liquid nitrogen at the same flow velocity. It is often the case that $r > r_h$ at the same flow velocity.

At higher solid fractions, r tends to become even greater than r_h. That is, although interference among particles becomes greater, such that pressure drop is increased and the amount of reduction is less, heat transfer occurs in the liquid layer in the vicinity of the pipe wall where there are few solid particles, meaning that there is little influence from the solid particles. At low solid fractions, where pressure drop reduction is readily apparent, the difference between r and r_h becomes small ($r \approx r_h$).

In heat transfer experiments in a 15 mm pipe, the maximum amounts of reduction in the pressure drop and the heat transfer coefficient at heat flux of 10 kW/m^2 are, respectively, 17% and 20%. These are 39% and 28% at 30 kW/m^2. Similarly, in a 10 mm pipe, these are 23% and 27% at 10 kW/m^2, and 21% and 18% at 30 kW/m^2.

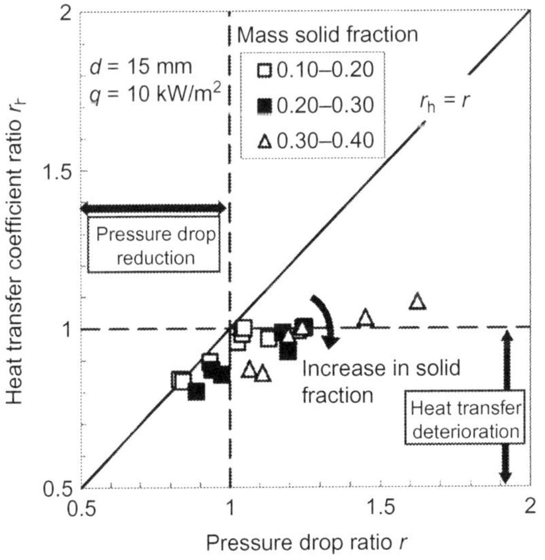

Figure 3.17 Correlation between heat transfer coefficient ratio and pressure drop ratio for slush nitrogen (Ohira et al., 2011).

3.6.3 Pressure drop in flow restrictions

Using slush hydrogen and liquid hydrogen at the triple-point temperature, Sindt and Ludtke (1970) conducted experimental work in a horizontal pipe having an inner diameter of 16.6 mm, and fitted with a globe valve (19.1 mm or 3/4 in.), two types of orifices and a venturi.

In the valve experiment, flow tests were conducted with the valve partially open (6.4 mm) and fully open, using solid fractions in the range of 0.25–0.45, and the relationship expressed by Equation (3.10) between the pressure drop coefficient K and the amount of volumetric flow rate was obtained. It was also confirmed that there was no plugging in any of the slush hydrogen tests

$$K = \frac{2\Delta P}{\rho u^2} \tag{3.10}$$

The coefficient K for slush hydrogen was greater than for liquid hydrogen when the valve was fully open. However, when partially open at solid fractions of 0.35–0.45, the coefficient K for slush hydrogen was up to 23% less than that for liquid. This is of interest in conjunction with pressure drop reduction for slush nitrogen flowing through converging–diverging pipe, as discussed in the following.

The relationship between pressure drop and mass flow rate was also evaluated for pipes having 6.4 and 9.4 mm diameter orifices in the solid fraction range of 0.25–0.50. For these orifice sizes, and regardless of changes in the solid fraction, the slush pressure drop was the same as for liquid at the same mass flow rate, enabling utilization as a volumetric flow meter. Pressure drop became substantial under conditions where cavitation occurred, but the pressure drop was about the same for slush and liquid hydrogen.

Using a venturi with a throat diameter of 9.5 mm and a throat length of 9.5 mm, the relationship between pressure drop and mass flow rate was measured at solid fractions of 0.25–0.45. Aged slush hydrogen showed a somewhat higher value for pressure drop than liquid at the same mass flow rate, while the same values were obtained for fresh slush regardless of changes in the solid fraction, again enabling utilization as a volumetric flow meter.

Next, we turn to the flow properties of slush nitrogen in a horizontal pipe having an inner diameter of 15 mm and fitted with a converging–diverging pipe and an elbow.

Ohira et al. (2012b) used a converging–diverging pipe with throat diameter and length of 10 and 12 mm, respectively. For solid fractions of 0.02–0.31, pressure drop measurement results for flow velocities of 0.9–5.0 m/s ($4.0 \times 10^4 < Re < 2.2 \times 10^5$) are presented in Figure 3.18. Liquid nitrogen measurement results and the Prandtl–Karman Equation (3.7) for an inner diameter of 15 mm are also shown. Pressure drop reduction is clearly apparent at flow velocities of 1.5 m/s and over, and pressure drop for slush nitrogen was about 50% less than for liquid.

Figure 3.19 shows the relationship between the coefficient K expressed in Equation (3.10) and the flow velocity. From the velocity measurement results of solid nitrogen particle using the PIV method, the following conclusions can be drawn concerning a primary factor in pressure drop reduction.

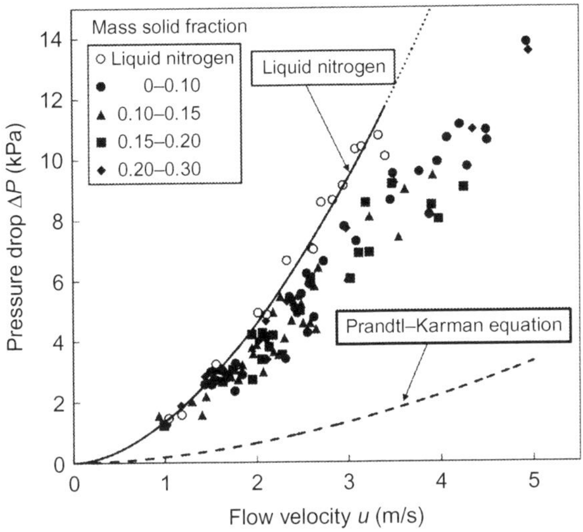

Figure 3.18 Pressure drop versus flow velocity of slush nitrogen for the converging–diverging pipe.

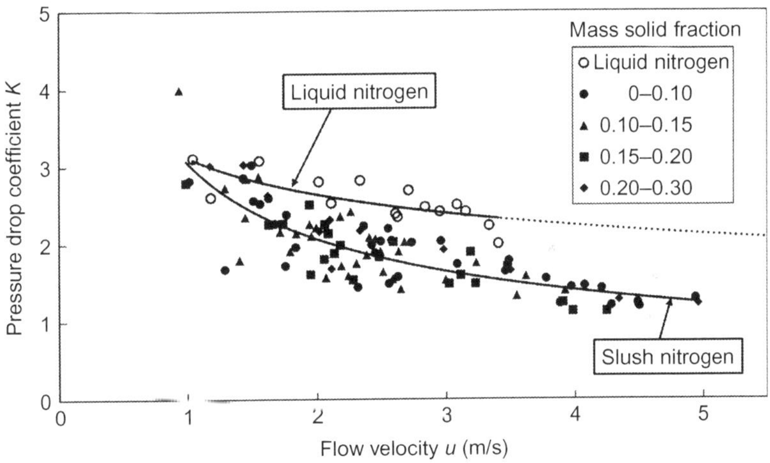

Figure 3.19 Pressure drop coefficient versus flow velocity of slush nitrogen for the converging–diverging pipe.

Solid particles with greater inertial force than liquid become grouped at the pipe center as discussed in Section 3.6.2, serving to suppress the occurrence of liquid separation (recirculation) in the throat. Also, very few of the solid particles are caught up in the separation region of the liquid, and most of them pass through the throat without much loss of kinetic energy. In the throat downstream, the solid particles undergo momentum exchange with liquid, and as a result, increased pressure recovery by the liquid contributes to pressure drop reduction.

Pressure drop in a horizontal elbow section with an inner diameter of 15 mm was measured for solid fractions of 0.04–0.26 at flow velocities of 1.1–4.2 m/s ($4.8 \times 10^4 < Re < 1.8 \times 10^5$) (Nozawa et al., 2009). Here, the maximum value for pressure drop reduction was 10% at a solid fraction of 0.10. As the solid fraction increases, pressure drop becomes greater than for liquid nitrogen. Since pressure loss due to the collision between the solid particles and the pipe wall in the bend section is considerably greater than the pressure drop reduction experienced in a straight section, pressure drop is greater than for liquid except when the solid fraction is low.

3.6.4 Pressure drop in corrugated pipes

Slush nitrogen pressure drop phenomena in two types of corrugated pipes, A and B, were ascertained for solid fractions of 0.10–0.32 at flow velocities of 1.4–4.5 m/s ($6.1 \times 10^4 < Re < 1.75 \times 10^5$) (Ohira et al., 2012b). Types A and B are characterized, respectively, by inner diameter d of 12 and 15 mm, pitch s of 3.3 and 3.0 mm, and corrugation height t of 2.75 and 2.20 mm.

Figure 3.20 presents the relationship between pressure drop and flow velocity for slush and liquid nitrogen at the triple-point temperature in corrugated pipe B. The figure shows the Prandtl–Karman Equation (3.7), as well as the Hawthorne Equation (3.11), which expresses pressure drop for a liquid flowing in a corrugated pipe (Hawthorne and von Helms, 1963; Yeaple, 1995)

$$\Delta P = \lambda \frac{1}{2}\rho u^2 \frac{L}{d} = \frac{d}{s}\left[1 - \left(\frac{d}{d+\gamma s}\right)^2\right]^2 \frac{1}{2}\rho u^2 \frac{L}{d} \tag{3.11}$$

Here, γ (= 0.438) is the experimental constant.

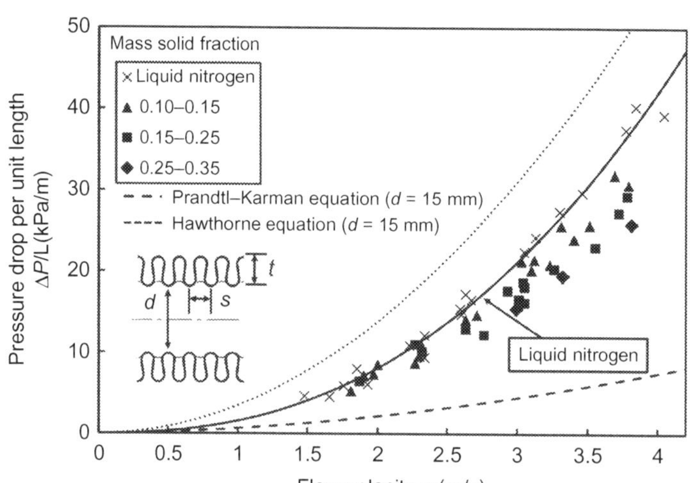

Figure 3.20 Pressure drop versus flow velocity of slush nitrogen for the corrugated pipe B ($d = 15$ mm).

At the same flow velocity, pressure drop values for liquid nitrogen in pipes A and B are, respectively, five to nine and three to six times those obtained using Equation (3.7). Additionally, the Hawthorne equation overestimates liquid nitrogen pressure drop by a factor of around 1.3. For slush nitrogen in the low flow velocity region of within 2 m/s, pressure drop in either type of pipe is about the same as or slightly greater than the value for liquid nitrogen.

Pressure drop reduction appears at velocities of 2 m/s and over, with maximum reductions of 37% (A: solid fraction of 0.30) and 31% (B: 0.27) as compared to liquid nitrogen. With higher solid fractions and/or flow velocities, the amount of pressure drop reduction increases further. Figure 3.21 shows the relationship between the Reynolds number and the pipe friction factor. When the solid fraction is the same, the pipe friction factor for slush shows a nearly constant value regardless of changes in the Reynolds number, while it shows a smaller value for higher solid fractions.

Next, Figure 3.22 shows the relationship between the Froude number Fr, defined in Equation (3.12), and the pressure drop ratio r. Here, g is the gravitational acceleration

$$Fr = \frac{u^2}{gd} \tag{3.12}$$

The least-squares approximation curves for the experimental values obtained in the respective solid fraction ranges are expressed as broken (A) and solid (B) lines.

The pressure drop reduction effect increases in conjunction with a greater solid fraction or Froude number, but there is a tendency for the amount of reduction to be saturated after a certain amount of decrease. This is due to greater interference

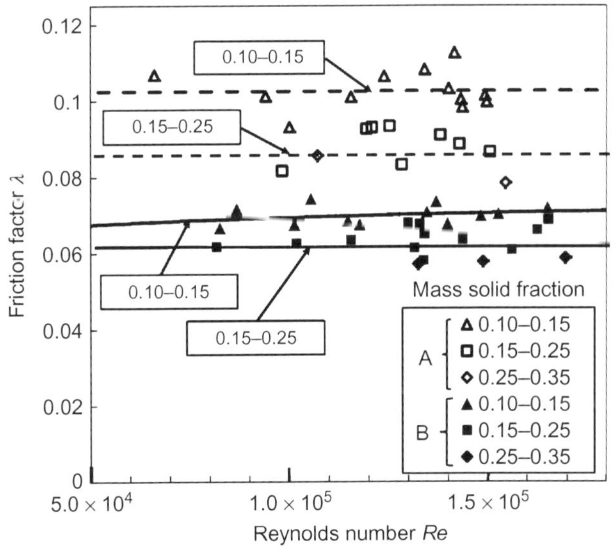

Figure 3.21 Correlation between friction factor and the Reynolds number for the corrugated pipes A ($d = 12$ mm) and B ($d = 15$ mm).

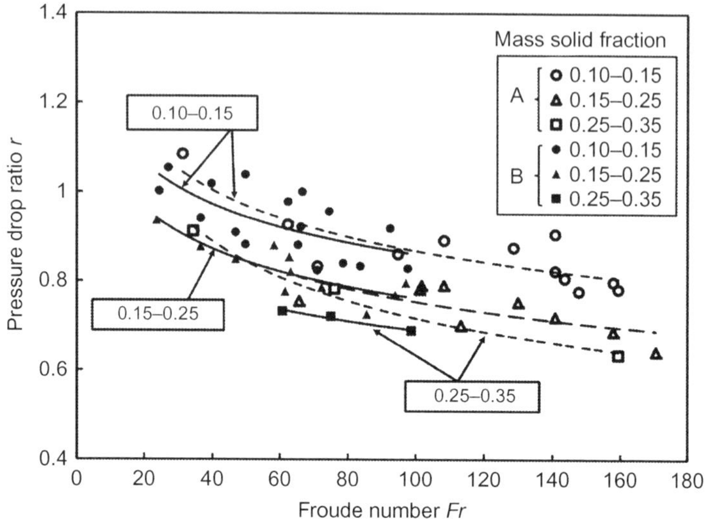

Figure 3.22 Correlation between pressure drop ratio and the Froude number for the corrugated pipes A ($d = 12$ mm) and B ($d = 15$ mm) (Ohira et al., 2012b).

between the liquid and the solid particles, among the solid particles, and between the solid particles and the pipe wall. Since the broken (A) and solid (B) lines in Figure 3.22 are in good agreement, they can be used to estimate the amount of pressure drop reduction in corrugated pipes with different configurations, provided that the Froude number is the same. However, clarification was not obtained, in part because the differences between A and B are slight.

As a result of particle-path measurement for solid particles by using the PIV method, the following conclusions can be drawn concerning a major factor in pressure drop reduction in a corrugated pipe.

As in a smooth cylindrical pipe, solid particles migrate toward the pipe center as the flow velocity increases, such that interference decreases between the corrugation and the solid particles. Furthermore, the group of solid particles flowing at the pipe center serves to suppress the development of turbulence in the liquid near and within the corrugation, as well as suppressing the turbulence diffusion toward the pipe center.

3.7 Uses of stored slush and liquid hydrogen

The advantage of slush hydrogen is the reduction of liquid hydrogen vaporization, since the solid particles' heat of fusion absorbs heat inleak during storage. On the other hand, when slush hydrogen is stored below atmospheric pressure, leak-tightness is required to prevent the penetration of air or other contaminants into the storage tank.

When stored slush or liquid hydrogen is used as refrigerant for SMES, it is important to consider nucleate pool boiling heat transfer properties.

3.7.1 Nucleate pool boiling heat transfer to slush and liquid hydrogen

Sindt (1974) and Ohira (2003b) used a 25.4 mm diameter circular flat-plate made of stainless steel and copper as a heat transfer surface, respectively. The heat transfer surface was placed in differing orientations (horizontal upward facing, vertical, and horizontal downward facing) to investigate nucleate pool boiling heat transfer for liquid hydrogen at the normal boiling point (0.1 MPa, 20.3 K: NBP liquid), liquid hydrogen at the triple-point pressure (0.007 MPa, 13.8 K: TP liquid), and settled slush hydrogen at the triple-point pressure (0.007 MPa, 13.8 K: TP slush).

The experimental work by Sindt (1974) employed slush hydrogen with a solid fraction of 0.45, also measuring heat transfer for pressurized slush hydrogen at 0.1 MPa by helium gas (13.8 K: slush at 0.1 MPa), and determining natural convection heat transfer in addition to nucleate pool boiling heat transfer. In the experimental work by Ohira (2003b), solid fractions of 0.20–0.35 were used, and the critical heat flux (CHF) was verified. Heat transfer was also similarly measured for slush and liquid nitrogen.

In the case of natural convection heat transfer, the classical methods using the Rayleigh and Nusselt numbers (*Ra* and *Nu*) (Jacob, 1949) were able to predict the heat transfer coefficients for NBP liquid with the exception of the horizontal downward facing orientation. For TP liquid, TP slush, and slush at 0.1 MPa, and for all of the heat transfer surface orientations, the heat transfer coefficients could be predicted using Equation (3.13) and the correction for surface orientation given by Jacob. The coefficient A and the exponent n depend on the Rayleigh number range (Incropera and DeWitt, 2002)

$$Nu = A Ra^n \tag{3.13}$$

The natural convection heat transfer coefficients for TP liquid and TP slush showed nearly the same values even when the orientation of the heat transfer surface was changed.

Figure 3.23 shows nucleate pool boiling heat transfer in the case of the horizontal upward facing orientation for NBP liquid, TP liquid, TP slush, and slush at 0.1 MPa (Sindt, 1974; Ohira, 2003b). Heat flux q is represented on the vertical axis, while the temperature difference ΔT (superheat) is represented on the horizontal axis. Also, using the Kutateladze Equation (3.14) (Frost, 1975) applied to the horizontal upward facing orientation, calculated CHF values q^* for liquid hydrogen and liquid helium are given in Figure 3.23 (Brentari et al., 1965; Flynn, 2005)

$$\frac{q^*}{h_{\mathrm{lv}}\rho_{\mathrm{v}}} = K\left[\frac{\sigma g(\rho_{\mathrm{l}} - \rho_{\mathrm{v}})}{\rho_{\mathrm{v}}^2}\right]^{1/4} \tag{3.14}$$

Here, h_{lv}, ρ_{v}, ρ_{l}, and σ are the latent heat of vaporization, vapor density, liquid density, and surface tension, respectively, while the Kutateladze factor $K = 0.16$ is generally used.

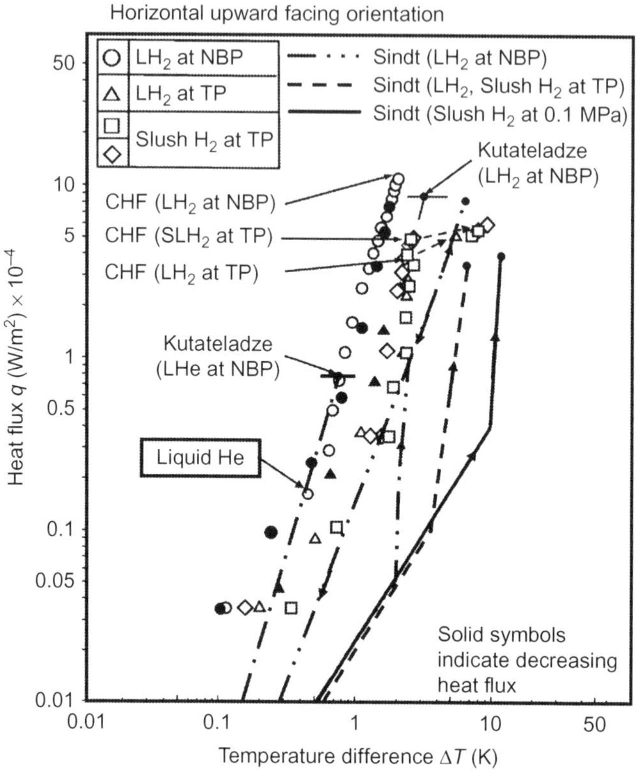

Figure 3.23 Nucleate pool boiling heat transfer to slush and liquid hydrogen (Sindt, 1974; Ohira, 2003b).

Sindt obtained smaller results than Ohira in terms of heat transfer coefficient ($h = q/\Delta T$), and the reason for this is probably due to the differences in the heat transfer surface material and its finish. However, both heat transfer tendencies are about the same for NBP liquid, TP liquid, and TP slush. The heat transfer coefficients obtained by Sindt for TP liquid and TP slush showed nearly the same values as shown in Figure 3.23, and the same results were obtained even when the orientation of the heat transfer surface was changed. The results of Ohira obtained for the copper heat transfer surface are described as follows.

For NBP liquid, the heat transfer coefficient was about that of liquid helium, while the heat transfer coefficient for TP slush in the high heat flux region fell to approximately 0.5 times that of NBP liquid (Figure 3.24). Furthermore, as in the cases of NBP liquid and TP liquid, the heat transfer coefficient for TP slush in the low heat flux region improved when the heat transfer surface angle increased. However, the difference in improvement between the vertical and horizontal downward facing orientations was small.

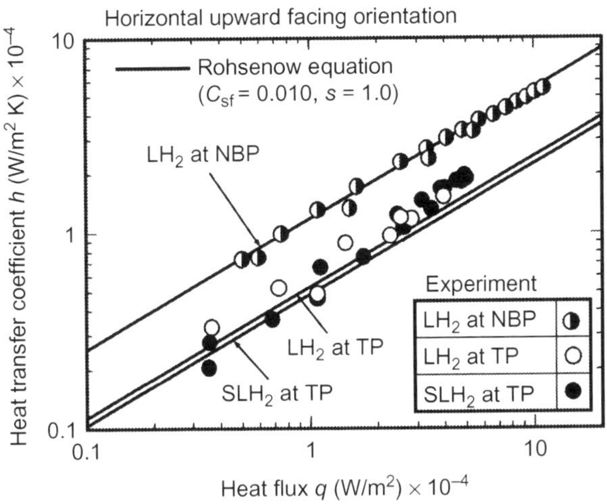

Figure 3.24 Comparison of the Rohsenow equation with heat transfer coefficients of slush and liquid hydrogen.

Using the Rohsenow Equation (3.15) (Frost, 1975) in the case of a horizontal upward facing orientation, the heat transfer coefficients of TP liquid and TP slush were predicted using experimental data for NBP liquid hydrogen and NBP liquid nitrogen. The predictability of the heat transfer coefficients for slush is potentially important in consideration of related experimental difficulties. These results for hydrogen and nitrogen are shown in Figures 3.24 and 3.25 as solid lines, respectively

$$\frac{c_l \Delta T}{h_{lv}} = C_{sf} \left\{ \frac{q}{\mu_l h_{lv}} \left[\frac{\sigma}{g(\rho_l - \rho_v)} \right]^{1/2} \right\}^{0.33} Pr_l^s \tag{3.15}$$

Here, c_l and Pr_l are, respectively, the specific heat and the Prandtl number of the liquid.

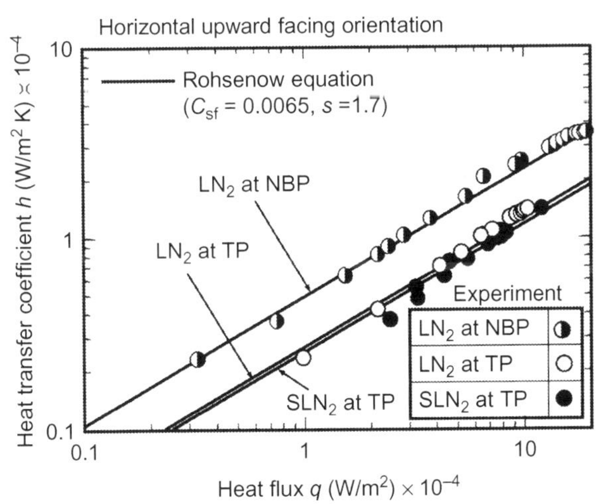

Figure 3.25 Comparison of the Rohsenow equation with heat transfer coefficients of slush and liquid nitrogen.

The factor C_{sf} and the exponent s are values determined by the combination of the heat transfer surface material and the liquid, and the same values can apply to different pressures. The value 1.7 is generally used for the exponent s. In the case of hydrogen, the result of the prediction for slush is slightly lower than that obtained experimentally, but the predicted and experimentally obtained values are in good agreement for slush nitrogen.

In descending order, as shown in Figure 3.23, CHF values for the horizontal upward facing orientation of the heat transfer surface are NBP liquid, TP slush, and TP liquid. Figure 3.26 presents the relationship between the reduced pressure P/P_c and the Kutateladze factor K. The calculations for slush including the heat of fusion of the solid in the latent heat of vaporization are also plotted. In ascending order, the values of K are NBP liquid, TP liquid, and TP slush (for which $K=0.23$). However, the difference is small for NBP liquid and TP liquid ($K=0.20$).

When comparing CHF values $q^*=11$ W/cm^2 ($\Delta T=2$ K) for NBP liquid with $q^*=5$ W/cm^2 ($\Delta T=2.5$ K) for TP slush hydrogen and $q^*=0.8$ W/cm^2 ($\Delta T=0.8$ K) for liquid helium in the case of a horizontal upward facing orientation, NBP liquid hydrogen is characterized by the greatest heat transfer coefficient and the highest CHF.

For transport via pipeline to a remote location, because heat inleak during such transport would melt the solid hydrogen particles, it is predicted that storage would be in the form of liquid hydrogen. In this case, the use of liquid hydrogen as a refrigerant in a SMES system would be advantageous from the standpoint of heat transfer properties. It is also noteworthy that, as shown in Figure 3.25 for liquid nitrogen, CHF $q^*=20$ W/cm^2 ($\Delta T=6$ K) is about twice as large as for liquid hydrogen, while at the same heat flux, the heat transfer coefficient falls to a level of 0.4 times.

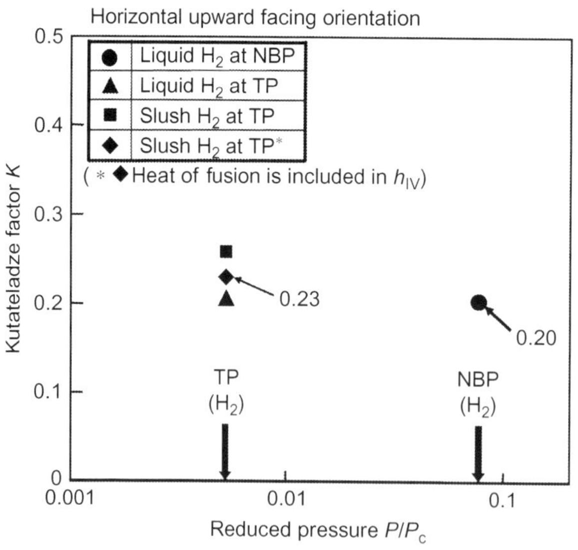

Figure 3.26 Comparison of the Kutateladze factor with experimental results of hydrogen.

3.8 Conclusions

Slush hydrogen is a cryogenic solid–liquid two-phase fluid, wherein solid hydrogen particles having a particle diameter of several mm are contained in liquid hydrogen. Compared to normal boiling liquid hydrogen (20.3 K), slush hydrogen (13.8 K) with a solid fraction of 0.50 features a 15% greater density, and an 18% increase in refrigerant heat capacity (enthalpy). Taking advantage of the solid particles' higher density and heat of fusion, slush hydrogen offers superior characteristics as a functional thermal fluid, and various applications are anticipated for hydrogen transport and storage.

When transporting hydrogen through long-distance pipelines in the form of slush hydrogen, combination with superconducting equipment using MgB_2 enables simultaneous transport and storage of both hydrogen fuel and electrical power, resulting in a synergistic effect. Given the rapid spread of fuel cells and increased demand for electrical power due to information technology requirements, the high-efficiency hydrogen energy system shown in Figure 3.1 has been proposed.

With the objective of practical realization of this system, research and development work associated with slush hydrogen has been conducted on hydrogen liquefaction using the high-efficiency magnetic refrigeration method, large-scale slush hydrogen production using the auger method, high-precision instruments for density and mass flow rate measurements, pressure drop reduction and heat transfer deterioration properties in pipe flow, and heat transfer properties involved in the cooling of magnets used in SMES systems. The current state of this research and development was reviewed in this chapter, along with relevant technology issues.

3.9 Future trends

In order to achieve the practical development of hydrogen energy systems and aerospace propulsion systems using slush hydrogen, further technological development work, for example, mass production, flow and heat transfer, and measurement instrumentation is required as described in Appendices A–C.

3.10 Sources of future information and advice

Efforts are currently underway in Europe, North America, and Asia aimed at the realization of a hydrogen energy society, including household fuel cells, fuel cell vehicles, and the full-scale spread of the infrastructure network of hydrogen stations for fuel cell vehicles (Ohira, 2003a, 2004a).

In Japan, sales of the world's first fuel cell cogeneration system for household use (ENEFARM) began in 2009. General sales of fuel cell vehicles are planned to begin in 2015, with full-scale popularization envisioned for 2030. Fuel cell vehicles are anticipated taking onboard 70 MPa high-pressure hydrogen gas, and the production of 100 MPa gas is anticipated. The actual work per unit mass for gas compression associated with this pressure level is about the same as that per unit mass for hydrogen liquefaction using low-pressure hydrogen gas, due to reduced compression efficiency. Thus, as indicated in Figure 3.1, hydrogen could be in the form of slush or liquid hydrogen for transport and storage, and at the supply point such as a hydrogen fuel station, pumped liquid hydrogen with much less power consumption for pressurization would be vaporized for supply of high-pressure gas hydrogen as needed (Kajikawa et al., 2012).

At the same time, the recent state of development with respect to superconducting technology is encouraging. In the U.S. city of Albany, New York, a superconducting electric power transmission system, incorporating bismuth-based high-temperature superconducting material using liquid nitrogen as the refrigerant, was successfully introduced in 2008, representing the first system connected to a practical-application power grid (Yumura et al., 2008; Ohya et al., 2008; Nakamura et al., 2013; Yang et al., 2013). Furthermore, at a factory in the Japanese city of Kameyama in Mie Prefecture, a commercial SMES system using liquid helium as the refrigerant was introduced in 2003 to protect a liquid crystal panel production line from sudden voltage drops (Ota et al., 2013).

For the practical development of hydrogen energy systems, other proving tests and products closely related to high-temperature superconducting technologies are also being carried out in Europe, North America, and Asia (Watanabe et al., 2013).

Appendix A Production

Development of the freeze–thaw method was vigorously pursued in the United States during the 1970s, with research having been reported on themes such aging effect on solid particles after production and temperature stratification during tank storage (Sindt, 1970; Park, 2010). Since the auger method, potentially suitable for the mass production of slush hydrogen, has been conducted only on a small-scale laboratory basis, establishment of this technology is needed in combination with hydrogen liquefier for large-scale production (Voth, 1978, 1985; Daney et al., 1990; Ohira et al., 1994; Ohira, 2004b).

Looking at the required work per unit mass for the production of slush hydrogen having a solid fraction of 0.50, along with the thermodynamically ideal cycle, it can be seen that the production of slush requires approximately 10% more work than for hydrogen liquefaction. Here, efficiency improvements in the conventional hydrogen liquefier using a compressed-gas method are required to realize reduction in the cost of liquid hydrogen. The use of high-temperature superconducting magnets in the magnetic refrigeration method for hydrogen liquefaction and slush hydrogen production has also been cited as a means of improving efficiency (Waynert et al., 1989; Ohira et al., 2000, Ohira, 2001; Matsumoto and Numazawa, 2011).

Appendix B Flow and heat transfer

While experimental data have been accumulated with respect to both the flow and the heat transfer characteristics of slush nitrogen during pipe flow, only flow characteristics have been reported for slush hydrogen. Thus, more work is needed on heat transfer properties during forced convection, as well as flow and heat transfer properties in slush hydrogen equipment such as heat exchangers.

Numerical analysis methods have been developed to elucidate the pipe flow and heat transfer properties of slush hydrogen and to apply them to engineering design, although analysis models for pressure drop reduction and heat transfer deterioration phenomena are still insufficient (Hardy, 1990; Hardy and Whalen, 1992; Gamma et al., 1998; Crivellari et al., 1999; Ohira et al., 2012a, 2013).

It will thus be necessary in the future to accumulate more experimental data to elucidate the pressure drop reduction and heat transfer deterioration phenomena. Such study will help further refine the numerical analysis method incorporating the reduction phenomenon, with the eventual objective of designing slush hydrogen equipment.

Appendix C Measurement instrumentation

Conventional cryogenic instrumentation can be used for slush hydrogen temperature and pressure measurement. With respect to density meters, measurement accuracy has been achieved that is applicable for practical use, although product development is still required. In the case of mass flow meters, confirmation of measurement accuracy for slush hydrogen is needed, as well as product development.

References

Barclay, J.A., Steyert, W.A., 1982. Materials for magnetic refrigeration between 2 K and 20 K. Cryogenics 22, 73–80.

Barron, R.F., 1985. Cryogenic Systems, second ed. Oxford University Press, New York.

Barron, R.F., 1999. Cryogenic Heat Transfer. Taylor & Francis, Philadelphia.

Brentari, E.G., Giarratano, P.J., Smith, R.V., 1965. Boiling Heat Transfer for Oxygen, Nitrogen, Hydrogen, and Helium, NBS TN-317. National Bureau of Standards, Washington, DC.

Crivellari, P., del Monte, L., Gamma, F., 1999. Computational fluid dynamics of slush hydrogen for aerospace vehicles. In: 35th AIAA/ASME/SAE/ASEE Joint Propulsion Conference and Exhibit, AIAA-99-2243.

Daney, D.E., Arp, V.D., Voth, R.O., 1990. Hydrogen slush production with a large auger. Adv. Cryog. Eng. 35, 1767–1776.

DeWitt, R.L., Hardy, T.L., Whalen, M.V., Paul Richter, G., 1990. Slush hydrogen (SLH2) technology development for application to the National Aerospace Plane (NASP). Adv. Cryog. Eng. 35, 1741–1742.

Ellerbruch, D.A., 1971. Microwave methods for cryogenic liquid and slush instrumentation. Adv. Cryog. Eng. 16, 241–250.

Flynn, T.M., 2005. Cryogenic Engineering. Marcel Dekker, New York.

Frost, W., 1975. Heat Transfer at Low Temperatures. Plenum Press, New York.
Gamma, F., del Monte, L., Liberatore, R., 1998. CFD-analysis of slush hydrogen flows in feed-systems of rocket based combined cycles. In: 34th AIAA/ASME/SAE/ASEE Joint Propulsion Conference and Exhibit, AIAA-98-3226.
Hardy, T.L., 1990. FLUSH: A Tool for the Design of Slush Hydrogen Flow System, NASA TM-102467. National Aeronautics and Space Administration, Washington, DC.
Hardy, T.L., Whalen, M.V., 1992. Technology issues associated with using densified hydrogen for space vehicles. In: 28th AIAA/ASME/SAE/ASEE Joint Propulsion Conference and Exhibit, AIAA-92-3079.
Hawthorne, R.C., von Helms, H.C., 1963. Fluid expansion theory computes flow in corrugated hose. Prod. Eng. 98–100.
Incropera, F.P., DeWitt, D.P., 2002. Introduction to Heat Transfer. John Wiley & Sons, New York.
Jacob, M., 1949. Heat Transfer. John Wiley & Sons, New York.
Johnson, V.J., 1961. Properties of Materials at Low Temperature (Phase 1): A Compendium. Pergamon Press, London.
Kajikawa, K., Kuga, H., Inoue, T., Watanabe, K., Uchida, Y., Nakamura, T., Kobayashi, H., Hongo, M., Kojima, T., Taguchi, H., Naruo, Y., Wakuda, T., Tanaka, K., 2012. Development of a liquid hydrogen transfer pump system with MgB_2 wires. Cryogenics 52, 615–619.
Matsumoto, K., Numazawa, T., 2011. Magnetic refrigeration for hydrogen liquefaction. In: Proc. ICEC 23-ICMC 2010, pp. 359–364.
McCarty, R.D., 1975. Hydrogen Technological Survey-Thermophysical Properties, NASA SP-3089. National Aeronautics and Space Administration, Washington, DC.
Nakamura, N., Shimoda, M., Komagome, T., Ohno, R., Yaguchi, H., Ichikawa, H., Mimura, T., Hara, T., Watanabe, M., Masuda, T., 2013. Recent progress of liquid nitrogen cooling system for Yokohama HTS cable project. In: Proc. ICEC 24-ICMC 2012, pp. 693–696.
Nozawa, M., Ohira, K., Okazaki, N., Ishimoto, J., Kamiya, T., 2009. Flow characteristics of slush nitrogen in various types of pipes. In: Proc. ICEC 22-ICMC 2008, pp. 255–260.
Ohira, K., 2001. Laminar film condensation heat transfer of hydrogen and nitrogen inside a vertical tube. Heat Transfer-Asian Res. 30, 542–560.
Ohira, K., 2003a. Research and development work on liquid hydrogen technologies in Japan's WE-NET project. In: Proc. 19th Int. Cryo Eng. Conf. (ICEC 19), pp. 557–560.
Ohira, K., 2003b. Study of nucleate boiling heat transfer to slush hydrogen and slush nitrogen. Heat Transfer-Asian Res. 32, 13–28.
Ohira, K., 2004a. A summary of liquid hydrogen and cryogenic technologies in Japan's WE-NET project. Adv. Cryog. Eng. 49A, 27–34.
Ohira, K., 2004b. Study of production technology for slush hydrogen. Adv. Cryog. Eng. 49A, 56–63.
Ohira, K., 2004c. Development of density and mass flow rate measurement technologies for slush hydrogen. Cryogenics 44, 59–68. http://dx.doi.org/10.1016/j.cryogenics.2003.08.001.
Ohira, K., 2011. Pressure drop reduction phenomenon of slush nitrogen flow in a horizontal pipe. Cryogenics 51, 389–396. http://dx.doi.org/10.1016/j.cryogenics.2011.04.001.
Ohira, K., Nakamichi, K., 2000. Development of a high-accuracy capacitance-type densimeter for slush hydrogen. JSME Int. J. Ser. B 43, 162–170.
Ohira, K., Matsuo, S., Furumoto, H., 1994. An experimental investigation of production and density measurement of slush hydrogen. Cryogenics 34 (ICEC Suppl.), 397–400.
Ohira, K., Nakamichi, K., Furumoto, H., 2000. Experimental study on magnetic refrigeration for the liquefaction of hydrogen. Adv. Cryog. Eng. 45, 1747–1754.

Ohira, K., Nakamichi, K., Kihara, Y., 2003a. Study on the development of a capacitance-type flowmeter for slush hydrogen. Cryogenics 43, 607–613. http://dx.doi.org/10.1016/S0011-2275(03)00171-1.

Ohira, K., Nakamichi, K., Kihara, Y., 2003b. Development of a microwave-type densimeter for slush hydrogen. Cryogenics 43, 615–620. http://dx.doi.org/10.1016/S0011-2275(03)00172-3.

Ohira, K., Nakamichi, K., Kihara, Y., 2005. Development of a waveguide-type flowmeter using a microwave method for slush hydrogen. JSME Int. J. Ser. B 48, 114–121. http://dx.doi.org/10.1299/jsmeb.48.114.

Ohira, K., Nakagomi, K., Takahashi, N., 2011. Pressure-drop reduction and heat-transfer deterioration of slush nitrogen in horizontal pipe flow. Cryogenics 51, 563–575. http://dx.doi.org/10.1016/j.cryogenics.2011.07.008.

Ohira, K., Ota, A., Mukai, Y., Hosono, T., 2012a. Numerical study of flow and heat-transfer characteristics of cryogenic slush fluid in a horizontal circular pipe (SLUSH-3D). Cryogenics 52, 428–440. http://dx.doi.org/10.1016/j.cryogenics.2012.04.006.

Ohira, K., Okuyama, J., Nakagomi, K., Takahashi, K., 2012b. Pressure drop of slush nitrogen flow in converging–diverging pipes and corrugated pipes. Cryogenics 52, 771–783. http://dx.doi.org/10.1016/j.cryogenics.2012.09.001.

Ohira, K., Ota, A., Aoki, I., 2013. Numerical study of cryogenic slush flow in a horizontal square pipe for a high-efficiency hydrogen energy system (SLUSH-3D). In: Proc. ICEC 24-ICMC 2012, pp. 105–110.

Ohya, M., Yumura, H., Ashibe, Y., Ito, H., Masuda, T., Sato, K., 2008. Design and evaluation of YBCO cable for the Albany HTS cable project. Adv. Cryog. Eng. 53B, 1059–1066.

Orr Jr., C., Dallavalle, J.M., 1954. Heat-transfer properties of liquid–solid suspensions. Chem. Eng. Prog. Symp. Ser. 50, 29–45.

Ota, N., Tanoue, N., Takao, T., Makida, Y., Shintomi, T., Hamajima, T., Tsuda, M., Miyagi, D., Munakata, K., Kajiwara, M., 2013. Design study of SMES system using MgB_2 conductors for compensating fluctuation of renewable energy. In: Proc. ICEC 24-ICMC 2012, pp. 533–536.

Park, Y.M., 2010. Literature research on the production, loading, flow, and heat transfer of slush hydrogen. Int. J. Hydrogen Energy 35, 12993–13003.

Sindt, C.F., 1970. A summary of the characterization study of slush hydrogen. Cryogenics 10, 372–380.

Sindt, C.F., 1974. Heat transfer to slush hydrogen. Adv. Cryog. Eng. 19, 427–436.

Sindt, C.F., Ludtke, P.R., 1970. Slush hydrogen flow characteristics and solid fraction upgrading. Adv. Cryog. Eng. 15, 382–390.

Strobridge, T.R., 1974. Cryogenic Refrigerators—An Updated Survey, NBS TN-655. National Bureau of Standards, Washington, DC.

Vaniman, J.L., Worlund, A.L., Winstead, T.W., 1969. Slush and subcooled propellants for lunar and interplanetary missions. Adv. Cryog. Eng. 14, 20–29.

Voth, R.O., 1978. Producing Liquid–Solid Mixtures of Hydrogen Using an Auger, NBSIR 78-875. National Bureau of Standards, Washington, DC.

Voth, R.O., 1985. Producing liquid–solid mixtures (slushes) of oxygen or hydrogen using an auger. Cryogenics 25, 511–517.

Watanabe, K., Inoue, T., Kajikawa, K., Kobayashi, H., Taguchi, H., Wakuda, T., Tanaka, K., 2013. Operation tests of superconducting level sensors for liquid hydrogen with CuNi-sheathed MgB_2 wires. In: Proc. ICEC 24-ICMC 2012, pp. 755–758.

Waynert, L.A., Barclay, J.A., Claybaker, C., Foster, R.W., Jaeger, S.R., Kral, S., Zimm, C., 1989. Production of slush hydrogen using magnetic refrigeration. In: Proc. ASME 12th Energy-Sources Technology Conf., pp. 9–13.

Weitzel, D.H., Cruz, J.E., Lowe, L.T., Richards, R.J., Mann, D.B., 1971. Instrumentation for storage and transfer of hydrogen slush. Adv. Cryog. Eng. 16, 230–240.

Yang, H.S., Sohn, S.H., Lim, J.H., Namgung, E.S., Oh, S.R., Yim, S.W., Kim, Y.H., Lee, K.T., Jang, H.M., Hwang, S.D., 2013. Long term performance test of cooling system for HTS power cable in the KEPCO power grid. In: Proc. ICEC 24-ICMC 2012, pp. 707–710.

Yeaple, F., 1995. Fluid Power Design Handbook, third ed. Marcel Dekker Inc., New York.

Yumura, H., Masuda, T., Watanabe, M., Takigawa, H., Ashibe, Y., Ito, H., Hirose, M., Sato, K., 2008. Albany HTS cable project long term in-grid operation status update. Adv. Cryog. Eng. 53B, 1051–1058.

Zhang, L., Sherif, S.A., Veziroglu, T.N., Sheffield, J.W., 1993. Second law analysis of active magnetic regenerative hydrogen liquefiers. Cryogenics 33, 667–674.

Underground and pipeline hydrogen storage

4

M. Panfilov
LEMTA - Université de Lorraine/CNRS, Nancy, France

Nomenclature

Ψ	rate of biomass production per single bacterium
ψ	rate of respiration per mole and per bacterium
φ	rate of bacterial death reported to a single bacterium
θ	rate of CO_2 consumption for biomass production per bacterium
$c_w^{H_2}$	mole fraction of H_2 in water in a unit of porous volume (mol/m^3)
$c_w^{CO_2}$	mole fraction of CO_2 in water in a unit of porous volume (mol/m^3)
$c_g^{H_2}$	mole fraction of H_2 in gas in a unit of porous volume (mol/m^3)
$c_g^{CO_2}$	mole fraction of CO_2 in gas in a unit of porous volume (mol/m^3)
n	number of bacteria in a unit of volume (1/m^3)
ρ_w	molar density of water (mol/m^3)
ρ_g	molar density of gas (mol of/m^3)
S	water saturation
t_*	time of the biomass growth
t_d	time of the population decay
ν and n_m	empirical coefficients

4.1 Underground hydrogen storage as an element of energy cycle

4.1.1 Industrial needs in underground hydrogen storage (UHS)

One cubic meter of hydrogen produces 12.7 MJ of energy by combustion, which is a very high energy potential, although it is lower than that of methane (40 MJ). However, hydrogen cannot be considered as an energy source, because the energy released is lower than that consumed for hydrogen production. However, being convertible to electricity or heat, the hydrogen becomes an efficient energy carrier capable of transporting and storing energy. The transport of energy in the form of a gas is with much less loss (<0.1%) than in a power network (8%).

Due to its high energy potential, hydrogen is capable of replacing up to 60% of the natural gas used for nonindustrial activities (Davison et al., 2009).

The storage of hydrogen is thus the storage of energy. The need for such storage is determined by the imbalance between energy production and its consumption. Energy consumption usually fluctuates over time, while energy production is generally constant or randomly intermittent (as in the case of renewable sources). These fluctuations

Compendium of Hydrogen Energy. http://dx.doi.org/10.1016/B978-1-78242-362-1.00004-3

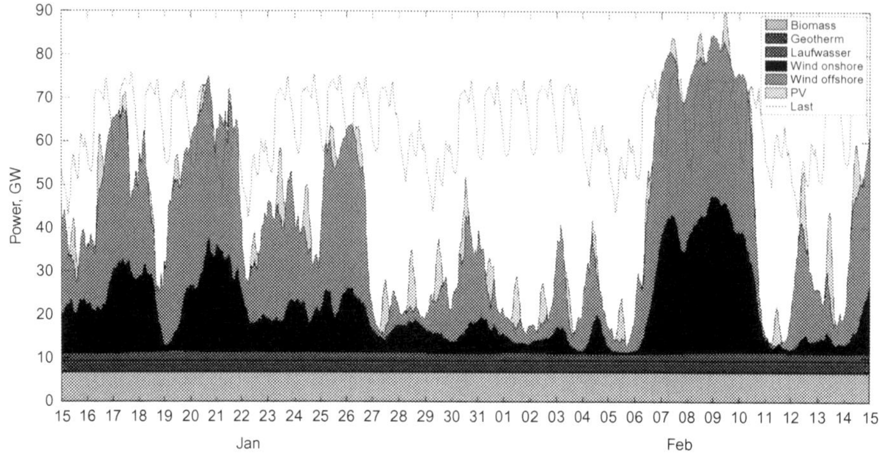

Figure 4.1 Fluctuations of renewable energy production in Germany (German Federal Environment Ministry (BMU) Lead Scenarios).

in consumption are caused by variations in daily and seasonal changes (Kepplinger et al., 2011; Crotogino et al., 2010), but also by emergency situations that lead to sudden and sharp energy losses. To regulate these fluctuations, the excess electricity produced should be temporarily stored and used later when consumption is higher than production. Figure 4.1 shows the feed-in of power produced from renewable sources in Germany and concerns the basic scenario from mid-January to mid-February, 2005 (Meteo: year 2006), according to German Certification of the generation of electricity from renewable sources.

The net average load is 60 GW, which means 15.55×10^{17} J is produced per month. The energy deficit needed to redress the fluctuations is 47%, i.e., 7.26×10^{17} J nonproduced for a month, or 87×10^{17} J nonproduced for a year. This is equivalent to 7.25×10^{10} kg of hydrogen. The hydrogen density is 0.09 kg/m^3 under normal conditions, therefore at least 10^{12} m^3 of hydrogen per year is needed to compensate for the fluctuations in energy production.

Geological reservoirs are the only way to store such large amounts of hydrogen. These exist in the form of salt caverns or porous media, i.e., aquifers or depleted natural gas reservoirs. In addition, they guarantee security due to the absence of contact with atmospheric oxygen (a mixture of hydrogen and oxygen is explosive at practically any concentration). UHS is also economically promising for storing large amounts of hydrogen inexpensively, which would enable an economic threshold to be reached in the medium term (Zittel and Wurster, 1996; Crotogino et al., 2010).

Thus, the main objectives of UHS are:

(i) to regulate the energy supply and demand in cases where the energy produced is in excess of the needs of consumers;
(ii) to regulate energy prices: energy generated at a time when power is inexpensive can be saved and sold later when power is expensive; and
(iii) to offer industry (refineries, etc.) an instantaneous hydrogen backup supply.

4.1.2 Conversion of hydrogen into other forms of energy and vice versa

To release the latent energy from hydrogen, it needs to be fired directly or converted to other forms of energy. The conversion techniques significantly affect the type of hydrogen storage needed.

Electricity conversion to hydrogen:

- low-temperature chemical electrolysis of water $2H_2O = 2H_2 + O_2$ is performed due to the action of electrical current and catalyzers; such reactions occur in standard electrolyzers.
- high-temperature electrolysis of water is performed by electrical current and high temperature (800–1200 °C); such reactions occur in nuclear plants with new generation reactors.

Hydrogen conversion to electricity:

- reaction of water synthesis in fuel cells occurs in the presence of a catalyzer: $2H_2 + O_2 = 2H_2O + 290$ J/mol. In this case, very pure hydrogen should be used, since even minor impurities significantly reduce the energy yield.

In a fuel cell, part of the released energy is electricity, while the other part is heat, which can be used to cogenerate electricity.

Hydrogen conversion to methane:

The reaction of catalytic methanation occurs in the mixture of H_2 with CO_2 or CO:

$$4H_2 + CO_2 = CH_4 + 2H_2O - 134\,\mathrm{kJ/mol}, \tag{4.1}$$

$$3H_2 + CO = CH_4 + H_2O - 206.28\,\mathrm{kJ/mol} \tag{4.2}$$

Process (4.1) is called the Sabatier's reaction.

These reactions can occur at high temperatures (~800 °C) in the presence of a catalyzer (nickel), or at low temperatures (~30–40 °C) in the presence of microorganisms (*methanogenic Archaea*).

4.1.3 Four principle types of UHS

UHS represents an element of the general energy cycle "initial energy production—conversion (or not) to hydrogen—hydrogen storage—reconversion (or not) of hydrogen to other type of energy—energy consumption." The goals and the method of UHS depend heavily on the combination of all these elements. For instance, if the aim is to store the pure hydrogen and use it later in fuel cells, then any chemical transformation of hydrogen during storage should be prohibited. In contrast, if the aim is to use the hydrogen for gas-fired turbines or to inject it into natural gas pipeline, the enrichment of the stored gas by methane or other energy carriers is probably welcome.

Four types of UHS can be distinguished depending on the form of the energy initially produced, the form of the final energy consumed, the methods of energy conversion, and the combination between these elements:

I *Underground storage of pure hydrogen*: the final destination of this hydrogen is for use in fuel cells where the H_2 is converted to electricity and next used in vehicles, which is the case for very pure hydrogen. The most convenient storage sites for ultra-pure hydrogen are salt caverns, which are almost completely hermetic, have a high degree of cleanliness, and are characterized by a very low risk of probable gas contamination by impurities. Pure hydrogen can best be obtained from excess electricity by chemical electrolysis, or from the heat of nuclear plants through thermal electrolysis.

The complete energy cycle consists of plants that produce renewable electricity (e.g., windmills, solar cells). Due to their intermittent mode of functioning, part of the energy is produced in excess and should be stored. This electricity is then converted to hydrogen, stored in a salt cavern, extracted during periods of increased consumption, and reconverted into electricity in fuel cells, as shown in Figure 4.2. Instead of a renewable source and electrolyzer, it can be a nuclear plant that serves as the electricity producer and converter to hydrogen.

II *Underground storage of a mixture with natural gas lean in hydrogen*: pure hydrogen produced by water electrolysis is injected into an underground natural gas storage site. Currently, industry only accepts small amounts of H_2 (6–15%; DVGW, 2011) so that the energy potential of the stored gas is not reduced significantly and the existing infrastructure of gas transport can be used without damaging it (hydrogen makes steel brittle). In this case, hydrogen is used as a fuel. This technology is also known as *power-to-gas*.

The global energy cycle consists of a source of renewable electricity or a nuclear plant, an electrolyzer, and underground storage site of natural gas, from which the gas produced is injected into a gas pipeline (Figure 4.3).

The blended H_2 and CH_4 can also be separated after storage, to obtain pure hydrogen. Various techniques of separation exist: pressure-swing adsorption operating at low hydrogen concentrations (<20%); membrane separation, which is efficient with relatively high concentrations of hydrogen, and can give very highly pure hydrogen; and electrochemical hydrogen separation also known as hydrogen pumping.

III *Underground storage of rich hydrogen mixture with CO, CH_4, and CO_2 (syngas or town gas)*: the mixture of H_2 (20–40%) and CO is called syngas, while the mixture of H_2 (50–60%), CO, and CH_4 is called town gas. In both cases, CO_2 can also be present depending on the production technique. CO is considered as an energy carrier (although it has lower potential than hydrogen).

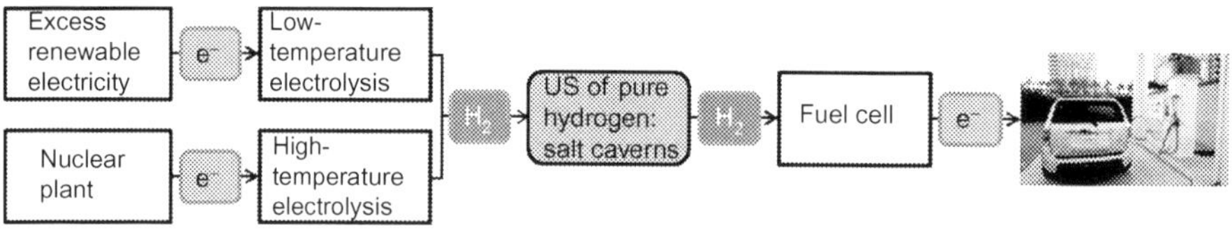

Figure 4.2 The energy cycle including underground storage of pure H_2.

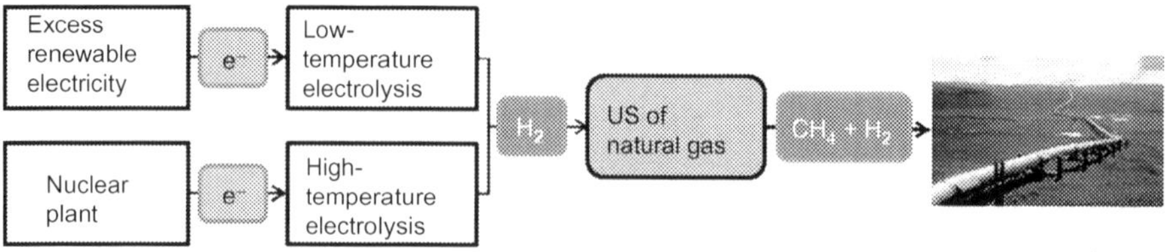

Figure 4.3 The energy cycle including the storage of lean H_2 blended with natural gas.

This type of mixture is produced by superficial or underground coal gasification, which represents coal combustion by injecting vapor at 800 °C along with oxygen. The latest version of this technology allows the mixture to be produced with 70% hydrogen (Kreinin, 1993).

The stored gas can be consumed in two different forms:

- as electricity, through thermo-mechanical conversion in gas turbines;
- as fuel (in the case of town gas) for lighting and heating without any conversion, which is an economically attractive option for operators in several parts of the world, for whom access to natural gas is difficult. For instance, in Honolulu, Hawaii, town gas continues to be delivered with significant blended hydrogen and is used in heating and lighting applications as an economic alternative to natural gas (General Motors, 2010).

The energy cycle consists of a plant producing syngas or town gas by coal gasification, the underground storage of the gas in an aquifer, depleted gas reservoir, or salt cavern, and a gas-fired turbine for conversion of the gas to electricity or for direct use as fuel (Figure 4.4).

IV *Underground methanation reactor (UMR)*: this represents the mixture of hydrogen and CO_2 in an aquifer or depleted gas reservoir, in which methanogenic bacteria initiate the Sabatier's methanation reaction (4.1). The objective of such storage is to enrich the energy potential of the gas by transforming the mixture of H_2 and CO_2 into methane. UMR was first suggested in research by Panfilov (2010a) and Panfilov and Reitenbach (2015). Such a process of underground methanation can occur at low temperature due to bacteria, which can be economically more favorable than the industrial process applied to date at the surface, using high temperatures and expensive catalyzers.

The resulting gas is injected into the grid of natural gas and used as fuel. Consequently, the residual presence of CO_2 should be minimized, since the proportion of injected CO_2 is completely transformed to methane, which is why this type of of storage is different from that of syngas or town gas, where the proportions of the various components are not controlled, and the resulting gas is converted to electricity.

A UMR can be created as follows:

(i) by injecting hydrogen into CO_2 stored underground in an aquifer;
(ii) by injecting CO_2 into underground stocks of radioactive waste, where pure hydrogen is found as the result of container corrosion (Oladyshkin and Panfilov, 2011).

The global energy cycle is shown in Figure 4.5.

Figure 4.4 Energy cycle including the storage of a rich H_2 blend with CO and CO_2.

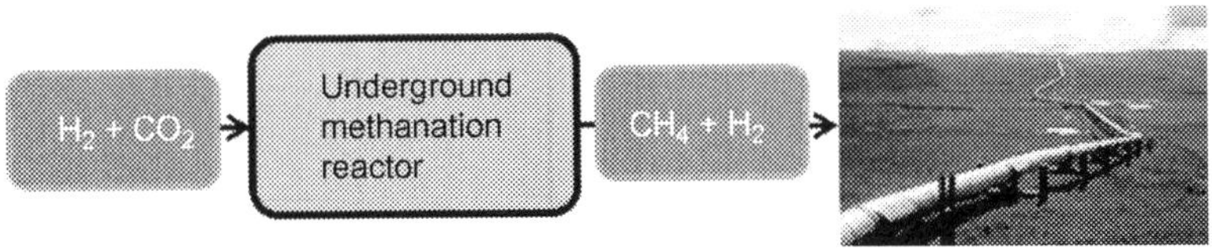

Figure 4.5 Energy cycle including an underground methanation reactor.

4.1.4 Storage in salt caverns and porous media

Salt caverns are frequently used to store natural gas. Hydrogen storage in salt caverns is already used in the United States, Britain, and Germany. The walls of a salt cavern are stable and impervious to gas over the lifespan of the storage facility. Plastic properties of salt protect such caverns against the appearance and spreading of fractures and the loss of impermeability. According to popular opinion, bacteria cannot survive in highly concentrated brine (Kireeva and Berestovskaya, 2012) and thus in salt caverns they do not transform hydrogen into other gases. In contrast, in porous media bacteria actively consume hydrogen. The unit physical volume of salt caverns typically ranges from 100,000 to 1,000,000 m^3, and, depending on the cavern depth, the corresponding working gas volume may range from a few million to 100 million st-m^3. From a practical viewpoint, the salt cavern storage solution is highly flexible in terms of storage volume, and also in terms of modularity, since several caverns can be leached on a single site to adapt the overall storage capacity to changes in demand.

However, salt structures have only limited use due to a restricted cavern volume (compared to an aquifer) and the finite occurrence of salt deposits suitable for salt leached cavern constructions.

Siliciclastic porous deposits, like saline aquifers and depleted gas reservoirs, are another solution of increasing interest for massive storage. They are capable of storing much larger gas volumes than salt caverns (the maximum volume of a Russian aquifer storage, Kasimovskoie, is 18 milliard st-m^3 of gas) and are expected to be technologically and economically feasible because of the long history of storing gases in porous media. However, the physico-chemical behavior of UHS is significantly different than traditional types of gas storage, imposing new scientific and technical problems.

Several examples of hydrogen mixtures and pure hydrogen stored underground are given in Table 4.1.

- A German storage in salt caverns at Bad Lauchstät has now been converted to the storage of natural gas.
- In Germany, at Kiel, manufactured gas with 62% H_2 has been stored in a salt cavern of 32,000 m^3 at 80–100 bar since 1971. This site is now used to store natural gas.

Table 4.1 Underground storage of hydrogen worldwide

	Type	% H_2	P, T	Depth (m)
Bad Lauchstädt, Germany	Salt cavern		150 bar	820
Kiel, Germany	Salt cavern	60–64	80–100 bar	1330
Teesside, UK	Salt cavern	95	50 bar	400
Texas: Air Liquid, USA	Salt cavern	95		
Texas: ConocoPhillips, USA	Salt cavern	95		850
Texas: Praxair, USA	Salt cavern			
Beynes, France	Aquifer	50		430
Ketzin, Germany	Aquifer	62		200–250
Lobodice, Czech	Aquifer	50	90 bar, 34 °C	430
Diadema, Argentina	Natural Gas	10	10 bar, 50 °C	600

- In England, at Teesside in Yorkshire, the British company Imperial Chemical Industries (ICI) stores 1 million N m^3 of pure hydrogen (95% H_2 and 3–4% CO_2) in three salt caverns at about 400 m depth and 50 bar. The hydrogen is ultimately consumed by nearby industrial plants during production of ammonia and methanol.
- In Texas, Air Liquid, a major producer of hydrogen in North America, is finishing the leaching of a large salt cavern on the Gulf Coast. This will allow hydrogen storage with the aim of enhancing flexibility to meet increasing customer demand for hydrogen. The hydrogen network will be extended by 90 miles to southeast Texas.
- In Texas, ConocoPhillips has stored 95% hydrogen in the Clemmons salt dome since the 1980s. The cavern roof is about 850 m underground. The cavern is a cylinder with a diameter of 49 m, a height of 300 m, and a usable hydrogen capacity of 30 million m^3, or 2520 metric tons. This storage is directly connected to the Old Ocean refinery.
- In Texas, Praxair has been operating UHS in a salt cavern for several years, to enable "peak shaving" of its hydrogen production. This facility is connected to the Praxair Gulf Coast hydrogen pipeline network, which serves the petrochemical needs of Texas and Louisiana.
- In France, at Beynes (Yvelines), between 1956 and 1972, Gaz de France (GDF) stored manufactured gas containing 50% of hydrogen in a saline aquifer with 385 million st-m^3 capacity. The porous medium is gray and sandstone. The other parameters are given in Table 4.1. The gas was produced in coking coal and steel in eastern France. The objective was to regulate the fluctuations in gas production/demand. No safety issues or hydrogen losses were reported during the 18 years of operation. However, intensive bacterial activity and consecutive transformation of the gas composition was observed.

  Starting from 1975, the upper stratum was converted to store natural gas coming from Norway. This is also stored in the lower reservoir at a depth of 740 m with 800 million st-m^3 capacity.
- In Germany, in Ketzin, 40 km west of Berlin, manufactured gas was stored in an aquifer at 200–250 m depth in sandstone.

  Since 2008, the lower sandstone reservoir at a depth of 630 m has been the first European onshore pilot site of CO_2 storage. Three wells were drilled. However, the injection of CO_2 was be recently finished (at end of 2013) and, according to the suggestion of GFZ Potsdam, this geological structure can be used further as the pilot project to store hydrogen.
- In the Czech Republic, at Lobodice, town gas with 50% H_2 and 25% of CH_4 is stored in an aquifer.
- In Argentina, Hychico C.A. starts the pilot injection of hydrogen into a sandstone geological structure associated with the storage of natural gas, in Diadema, Patagonia, from the second half of 2015, under a pressure of 10 bar, temperature of 50 °C at a depth of 600–800 m. Diadema has two reservoirs; one serves for the storage of methane, while the second is devoted to the storage of hydrogen produced by a wind plant and electrolyzers.
- In Russia, pure hydrogen has been stored in the ground at 90 bar in large iron tanks, to meet the needs of the aerospace industry (Ponomarev-Stepnoi, 2004).

4.2 Scientific problems related to UHS

4.2.1 State of the art

Estimations of possible hydrogen storage in underground reservoirs have appeared in the literature since the 1970s. In 1979, the Gas Technology Institute in the United

States published a study concerning the mass underground storage of gaseous hydrogen (Foh et al., 1979), confirming its economic and technical feasibility. Since then, economical estimations have appeared regularly in the literature. According to Taylor et al. (1986) underground storage is the cheapest method for storing large quantities of gaseous hydrogen.

However, few articles have been published on the scientific aspects of hydrogen behavior in geological structures. While some reports have been written on hydrodynamic aspects of storing pure hydrogen (Bulatov, 1979; Carden and Paterson, 1979; Lindblom, 1985; Paterson, 1983), they all tend to conclude that hydrogen storage poses no new major problems compared to the storage of natural gas. The only difference that was underlined, for instance, in Simbeck (2004) consists of the high diffusivity of hydrogen, which can cause its leakage through cap rocks in aquifer or through salt walls in caverns. At the same time, existing UHS have never been disturbed by diffusive losses. This can be explained by the fact that hydrogen has very low solubility in water. As cap rocks, as well as salt walls in caverns, are water saturated, there is little risk of diffusive leakage.

In 1990, the first *in situ* data were published on the unusual behavior of hydrogen in underground storage, concerning the storage of town gas in Lobodice in an aquifer (Smigai et al., 1990; Buzek et al., 1994). The data clearly illustrated significant variations in the composition of the stored gas over an injection cycle. During 7 months of storage, the methane content increased two fold while that of hydrogen and acid gases (CO_2+CO) significantly decreased. These observations can be explained by *in situ* generation of methane from hydrogen and CO_2 or CO, according to the Sabatier's reaction (4.1). The reservoir pressure also decreased with respect to the pressure calculated by mass balance, which is explained by the same reaction: the appearance of water as the reaction product means that part of the initial gas was compressed, and the remaining part expanded such that its pressure decreased.

However, Sabatier's reaction can only occur at very high temperature (several hundred degrees of Celsius), or by the action of methanogenic microorganisms capable of destroying the hydrogen bends in H_2 molecules splitting a free electron. The active role of bacteria has been confirmed by isotopic analysis of the extracted methane (Buzek et al., 1994), which showed an isotopic signature for part of it as being typical of methane produced by methanogenic Archaea.

Other observations from the same authors revealed even more unusual effects, such as the appearance of spatial mobile zones preferentially enriched in methane or hydrogen. This was explained as the result of self-organization in nonlinear reaction–diffusion system, similar to the appearance of colored spots on a leopard (Panfilov, 2010a,b; Toleukhanov et al., 2012).

The same authors developed the physico-chemical theory of these phenomena based on the mathematical theory of dynamic systems.

Gaz de France observed the gas transformations in Baynes, which were similar to those described by reaction (4.1).

Another question concerns the abiotic interactions of hydrogen with rocks, which are restricted to redox reactions.

4.2.2 Recent research throughout the world

The real explosion of interest in UHS occurred in 2011–2012, which was influenced, in particular, by the political decisions of the European Commission to diminish power consumption by 20%, to reduce CO_2 emission by 20%, and to increase renewable energy up to 20% in all energy productions by 2020. Within the framework of this policy, hydrogen attracts particular attention as an environmentally clean energy carrier that does not emit CO_2 in the cycles of its conversion to and from electricity.

In 2012, Europe launched several important UHS projects such as the first R&D HyUnder project, "Assessing the potential, actors and business models of large scale underground hydrogen storage in Europe" (1.9 million €). Coordinator: The Foundation for the Development of New Hydrogen Technologies in Aragon. Partnerships: CENEX (Centre of Excellence of Low Carbon and Fuel Cell Technologies), CEA (Commissariat a l'Energie atomique et aux energies alternatives), DEEP Underground Engineering, ECN (The Energy Research Centre of the Netherlands), E.ON Gas Storage, HINICIO, LBST (Ludwig Bölkow Systemtechnik), KBB Underground Technologies, HYRO (National Hydrogen and Fuel Cell Centre from Romania), and Shell Global Solutions International B.V., Solvay. This project concerns engineering and economic issues.

At the same time, Germany launched three research projects on UHS (more than 2 million € per project), all financed by the German Federal Ministry of Education and Research (R&D program of Energy Storing):

- H2STORE: "Investigation into Geohydraulic, Mineralogical, Geochemical and Biogenic Interactions by Underground Storage of Hydrogen in Converted Gas Reservoirs." Coordinator: FSU Jena. Partnerships: TU Clausthal, EFZN, GFZ Helmholtz-Centre Potsdam, and LEMTA-University of Lorraine, Nancy. This project is focused mainly on hydrogen storage in depleted gas reservoirs, with only secondary attention being paid to aquifers.
- InSpEE: "Informational Systems in Salt Structures." Coordinator: KBB Underground Technologies. Partnerships: BGR (Federal Institute for Geosciences and Natural Resources); Leibniz University of Hannover. This project deals with intelligent computer-assisted control of storage behavior.
- ANGUS+: "Impact of using underground geological structures for the storage of heat, electricity and matter." Coordinators: Christian-Albrechts University of Kiel. Partnerships: Deutsches Helmholtz GeoForschungs Zentrum (GFZ), and Helmholtz Centre for Environmental Research-UFZ; Ruhr-University Bochum. This project concerns the mathematical modeling of storage in the subsurface by using existing models and approaches, risk assessment techniques, and experimental studies to observe chemical and microbiological effects of thermal storage.

Austria started the UNDERGROUND SUN STORAGE project in 2013, coordinated by RAG (Rohöl-Aufsuchungs Aktiengesellschaft). Other members of the consortium are the University of Leoben (the Department for Agro biotechnology), IFA-Tulln of the University of Natural Resources and Applied Life Sciences, Vienna; the Energy Institute at the Johannes Kepler University Linz; Verbund; and Axiom Angewandte Prozesstechnik GmbH.

A French research project launched in 2011 by the University of Lorraine and the Carnot Institute ICEEL should also be mentioned as a rare example of fundamental studies in this domain: "Self-organization phenomena in bioreactive multi-component transport through porous media: application to Underground Storage of Hydrogen."

The European Commission also supported a number of research initiatives on the geological storage of large quantities of hydrogen through its Roads2HyCOM (http://www.roads2hy.com) and HyLights (http://www.hylights.org) programs. Some government agencies in the United States are already funding research on the feasibility of a geological storage of hydrogen. These include, for example, the major technological monitoring carried out by the Sandia National Lab on behalf of the Department of State for Energy in the United States (Lord, 2009).

Activity in Russia remains limited to publicity publications (Basniev et al., 2010).

The activity of the USA, UK, Argentina, and Czech Republic, which have already started using UHS, the launching of HyUnder, three German projects since 2012 devoted to UHS, and recent projects in Austria and France illustrates the explosion of industrial and research interest in this field. This research interest is clearly related to the new energy strategies that forces companies and laboratories to increase the efficiency of renewable energy resources and to find efficient methods of energy storage.

4.3 Biochemical transformations of underground hydrogen

Due to the high energy of electron–proton bonding in the hydrogen atom, hydrogen is chemically inactive at reservoir pressures and temperatures, except in cases where it is used by the bacteria present in the rocks. Strictly speaking, bacteria do not consume hydrogen, but rather they consume the energy produced from a redox reaction initiated by the bacteria between hydrogen and other species.

4.3.1 Respiratory and constructive metabolism of microorganisms

There are two kinds of bacterial metabolism:

- *respiratory metabolism*, or simply *respiration*, which consists of initiating redox reactions in the body of a microorganism to capture the energy released from these reactions;
- *constructive metabolism*, or *nutrition*, which consists of fixing carbon and increasing biomass. This kind of metabolism is also called *biomass production*, or *carbon fixation*.

The principle of respiratory metabolism can be illustrated by the Sabatier's reaction (4.1). First, hydrogen and CO_2 molecules dissolve within the bacterial cell. CO_2 dissociates by creating COH^+ cations—the potential electron acceptors. Bacterial proteins separate an electron from a hydrogen atom. The free electron reacts with cations through a long series of multistep reactions, producing methane, water, and energy. This process is called respiratory metabolism, or simply bacterial respiration.

Simultaneously with hydrogen, bacteria need to use electron acceptors—i.e., cations. CO_2 is not the unique substance used for this; other cations are sulfates (SO^{2-}) present in reservoir water or iron-III (Fe^{3+}) frequently found in reservoir rocks.

As a result of respiratory metabolism, the redox reaction occurs where hydrogen plays the role of the reductant providing its electrons. These reactions initiated by bacteria are very fast and cause the transformation of hydrogen into other chemical substances.

Constructive metabolism is designed to produce the biomass of the bacterial population. It consists of carbon consumption from an external organic or nonorganic matter and its transformation into organic substances that are used to synthesize biomass. Some bacteria only use nonorganic CO_2; they are called autotrophic, since they themselves produce the organic matter. Others use organic matter, and a third class can use both organic and nonorganic substances. The group of processes of transformation of CO_2 into organic matter within the body of a cell is called *carbon fixation*. Most methanogenic, acetogenic, and sulfate-reducing bacteria (SRB) are autotrophic. Iron-reducing bacteria (IRB) use both organic matter and CO_2 to produce biomass. Carbon fixation leads to the increase of bacterial mass and dimensions. Once the mass of a cell exceeds a critical value, it splits into two (dichotomy), which is preceded by DNA replication. Globally, the biomass production is equivalent to the increase in the number of bacteria in the population. Respiration and biomass production are not independent of each other. Indeed, bacteria cannot produce biomass without simultaneous respiration, and cannot respire if they have no carbon for fixation. This is why *the synchronization hypothesis* was suggested in paper (Panfilov and Reitenbach, 2015), which dictates that the rate of biomass production is proportional to that of respiration.

4.3.2 Four kinds of hydrogenotrophic biotic reactions

In respiratory metabolism, various microorganisms use hydrogen as the electron donor and other substances as the electron acceptors, among which four essential types can be distinguished, as schematically represented in Figure 4.6.

The corresponding biotic reactions are:

- methanogenesis from hydrogen and CO_2 (4.1) or hydrogen and CO (4.2) induced by methanogenic Archaea.

  All methanogenic Archaea are anaerobes. The presence of 0.004% gaseous O_2 entirely suppresses their growth. There are three groups, *Methanobacteriales*, *Methanococcales*, and *Methanomicrobiales*, and these groups include more than 40 genera. The dimensions of several of them are lower than 0.45 μm. The optimal pressure and temperature for growth are 90 bar and 30–40 °C (mesofils), but they are capable of growing even at 97 °C (Gusev and Mineeva, 1992). At the optimal pH of water (6.5–7.5), 90% of the CO_2 is used for respiration, and only 10% to produce biomass.

  The CO_2 or COH^+ needed for these reactions in UHS is of triple origin:

  (i) it is frequently present in the injected gas. For instance, CO_2 and CO constitute 10–20% of town or manufactured gas;

  (ii) CO_2 is found in the carbonaceous reservoir rocks: the chemical equilibrium between water and carbonates leads to the appearance of COH^+ cations dissociated in water; and

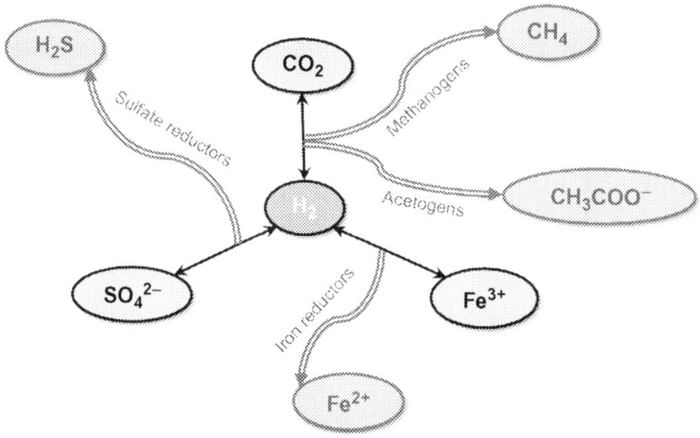

Figure 4.6 Four types of hydrogenotrophic bacteria in UHS.

(iii) acetate produced by hydrogenotrophic bacteria can be decomposed by other bacteria (acetotrophic) into CO_2 and methane: $CH_3COOH = CH_4 + CO_2$.

These reactions can cause nonnegligible losses of stored hydrogen, on the one hand, and gas enrichment by methane, on the other.

- Acetogenic bacteria transform hydrogen and CO_2 into acetate:

$$4H_2 + 2CO_2 = CH_3COOH + 2H_2O \tag{4.3}$$

Acetogenic bacteria include: *Clostridium thermoautotrophicum*, *C. aceticum*, *Butyribacterium methylotrophicum*, *Sporomusa sphaeroides*, *S. ovata*, *Acetogenium kivui*, *Acetobacterium woodii*, and other species.

Acetogenic bacteria have a maximal activity at the same pressure and temperature as methanogens.

- Reaction of sulfate reduction:

The SRB cause the following reaction with hydrogen:

$$5H_2 + SO_4^{2-} = H_2S + 4H_2O \tag{4.4}$$

Sulfate-reducing Archaea were discovered in 1987 and are called *Archaeoglobus*. They are anaerobe and thermophile (prefer high temperatures, up to 92 °C).

The appearance of H_2S is a good indicator of SRB activity.

- Reaction of iron reduction:

IRB are anaerobe and interact directly with reservoir rocks to transform Fe^{3+} into Fe^{2+}:

$$H_2 + Fe_2O_3 = 2FeO + H_2O \tag{4.5}$$

IRB use both CO_2 and organic substances to fix carbon.

The bacteria in question are *Geobacter metallireducens* and *Shewanella putrefaciens*.

Reactions (4.4) and (4.5) produce twice as much water per unit of mass of hydrogen as do the reactions of methanogenesis (4.1), (4.2), or acetogenesis (4.3). The appearance of excessive water could be an indicator of reactions (4.4) and (4.5).

These processes affect all types of underground storage, including salt caverns, due to the presence of residual waste at the bottom of the storage sites, which contains bacteria capable of living in highly concentrated brine.

Other bacteria can implicitly influence the hydrogenotrophy process, even if they do not explicitly use hydrogen. These are the acetotrophic and methanotrophic bacteria. The former use acetate and produce methane and CO_2, while the latter use methane and SO_4^{2-} and produce CO_2 and H_2S. The CO_2 produced can be reused by hydrogenotrophic bacteria. The impact of methanotrophs on hydrogenotrophy in UHS was analyzed numerically in Ebigbo et al. (2013).

4.3.3 Microbial activity in salt caverns

A salt cavern contains a significant amount of water accumulated at the base (up to 1/3 of the cavern volume). This brine contains impurities that appeared during leaching while crossing the strata of other rocks (e.g., dolomite, anhydrite). This sump also contains bacteria that can live in oversaturated brine and create biofilms at the cavern walls. Consequently, the reactions between hydrogen and CO_2 or SO_4^{2-} initiated by bacteria can occur in salt caverns, and these transform the composition of the stored gas.

4.3.4 Hydrogen and bacteria in water

All the reactions mentioned above occur in water, since bacteria can only live in aqueous media. Therefore, the hydrogen injected into UHS is not directly accessible to bacteria, apart from the part that can be dissolved in the reservoir water. The dissolution of hydrogen in water is very low, such that H_2 is frequently considered as the example of complete nonsolubility. At 25 °C, only 0.00002 (mole fraction) of H_2 is dissolved in water at 10 bar, and 0.00018 at 100 bar. In comparison, the dissolved CO_2 under the same conditions is 0.006 and 0.07 mole fractions, which is some 300–400 times higher. Consequently, in a two-phase gas–water system, in which the gas is 99% H_2 and 0.01% CO_2, the water in contact with this gas will contain a higher concentration of CO_2 than H_2.

Concerning the form of existence of the bacterial population in water, the preferable form is a biofilm, either attached to the solid walls, or located at the interface between water and gas. A biofilm consists of polymers and represents the product of bacterial excreta, metabolic wastes, and the cells of dead bacteria (i.e., proteins, lipids, DNA, RNA, etc.). It includes polysaccharides (peptidoglycan, cellulose) or more rarely lipids and proteins and contains a significant proportion of water.

A biofilm protects bacteria against external perturbations (temperature, pH, etc.) and invasion by other bacteria.

Along with biofilms, other forms of bacterial populations exist, such as:

- plankton: freely moving individual bacteria in water;
- neuston: the biofilm or free bacteria attached to the gas–water interface (at the pore-scale).

Due to the exchange between plankton and biofilm, all the bacteria can move and be transported. The main mechanisms of the movement of plankton are

- chaotic movement, described in terms of bacterial diffusion;
- transport at the velocity of water, described as convective transport;
- chemotaxis, which is the oriented motion of bacteria in the direction of nutrients.

4.3.5 Kinetics of reactions and bacterial population growth

Considering the Sabatier's reaction (4.1), let $c_i^{H_2}$ and $c_i^{CO_2}$ be the mole fraction of H_2 and CO_2, respectively, in phase i in a unit porous volume (mol/m^3), while n is the number of bacteria in a unit volume (1/m^3). Let ρ_w and ρ_g be the molar densities of water and gas and S the water saturation.

The mass balance of H_2 and CO_2 is formulated in an asymmetrical way:

$$d_t\left(\rho_w c_w^{H_2} S + \rho_w c_g^{H_2}(1-S)\right) = -4\psi n; \qquad (4.6)$$

$$d_t\left(\rho_w c_w^{CO_2} S + \rho_w c_g^{CO_2}(1-S)\right) = -\psi n - \theta n; \qquad (4.7)$$

$$d_t n = \Psi n - \varphi n \qquad (4.8)$$

where Ψ is the rate of biomass production per single bacterium, ψ is the rate of respiration per mole and per bacterium, φ is the rate of bacterial death reported to a single bacterium, and θ is the rate of CO_2 consumption for biomass production per bacterium.

The population decay is the sum of all effects that reduce the rate of population growth: the reduction of cell activity because of leakage of nutrients; a limited space, which can only encompass a limited number of cells; some other external environmental impacts that destroy cell function; and other microorganisms that use the bacteria as nutrients.

Biomass production occurs in two steps: through carbon fixation by the cell and consecutive cell duplication, which are occur at two different times.

Then, four types of kinetic functions are used to describe the kinetics of bioreactions and population dynamics. All these functions depend on the number of bacteria and the concentrations of CO_2 and H_2 dissolved in water.

At the present stage of knowledge, various simplified kinetic models are used.

In particular, within the project H2STORE the author of the present chapter successfully used the following relationships for ψ for the Sabatier's reaction:

$$\psi = \frac{n}{t_*\left[1+(n/n_m)^2\right]} \frac{c_w^{CO_2} c_w^{H_2}}{\left(1+\nu c_w^{CO_2}\right)\left(1+\nu c_w^{H_2}\right)}, \quad \varphi = \frac{1}{t_d} \qquad (4.9)$$

where t_* is the time of the biomass growth, t_d is the characteristic time of the population decay; and ν and n_m are empirical coefficients. Due to the dependence of the ψ function on n, this model is capable of approximately describing the lag phase, which is a problem for other known models (e.g., Monod's, Moser's, etc.).

4.3.6 Experimental techniques of measuring kinetic functions

Determining the kinetic functions is the first fundamental problem of this theory. Experimental techniques concerning the kinetics of nutrient consumption can be performed in an autoclave under fixed pressure, temperature, and a given initial composition of gas, water, and a sample of reservoir rocks containing microorganisms. By extracting gas samples at different times and measuring their chemical composition, e.g., by mass spectroscopy, the variations in H_2 and CO_2 concentration in the reservoir gas caused by the corresponding biotic reaction can be obtained.

However, all the types of experiments mentioned above are insufficient, because they do not provide any information about the number of bacteria that vary with time. Consequently, they should be completed by other experiments on bacterial kinetics.

Data on kinetics and constants for the Archaea can be found in Boone (2000), Delignette-Muller (n.d.), Moosa et al. (2002, 2005), Odencrantz (1991), and Vavilin et al. (2000).

The general technique of measuring bacterial kinetics consists of using several identical samples of the colony placed under identical thermodynamic conditions. To obtain the Ψ function, the nutrient reserves must be permanently restored in each sample, so that they are abundant at any moment. Then, the process in different samples is stopped at different times, by killing instantaneously all bacteria. By measuring the number of bacteria in different samples, the dependence of the number of bacteria on time is obtained. A second series of experiments can be performed with variable reserves of nutrients.

To measure the number of bacteria, various fluorescent methods are applied. The bacterial activity changes not only the concentration of various molecules present in the water, but also their internal isotopic signatures. The FISH staining technique (fluorescence *in situ* hybridization) is used to obtain the relative number of specific bacteria (e.g., methanogenic Archaea vs. Bacteria vs. Desulfovibrio) and to identify spatial arrangements of bacteria (i.e., consortia of bacteria arranged in aggregates). Other specific fluorescent techniques (DAPI or SYBR green staining) are used to count all microorganisms. The qPCR method (quantitative real-time polymerase chain reaction) can be applied to quantify the specific ribosomal DNA sequence of all microorganisms or that of a species or group.

Another class of methods that can be applied is based on isotopic variation, which can serve as an efficient indicator of bacterial activity, and can separate the variation of the gas–water composition caused by bacteria and by other physico-chemical factors (diffusion, adsorption). There are two basic isotopic transformations that concern methanogenic bacteria:

(i) methane generated by bacteria has a different isotopic signature than the natural methane originally present or injected with hydrogen. Therefore, the monitoring of carbon isotopy in methane yields information about the kinetics of methane production by bacteria;

(ii) bacteria are incapable of separating electrons from deuterium. Thus, only the traditional hydrogen isotope (the protium) will be involved in reactions and transformed into methane and water. Consequently, the concentration of deuterium will increase with time, which is the indicator of the kinetics of bacterial activity.

The measurement of an isotopic signature is performed by means of mass-spectrometry. The techniques of measuring are based on the methodology explained in Whiticar (1999).

In the Lobodice storage of town gas, the isotopic analysis of the extracted methane (Buzek et al., 1994) showed that part of it had a different isotopic signature from that of injected methane. These different observations justify the assumption of *in situ* generation of biotic methane from hydrogen.

4.4 Hydrodynamic losses of H_2 in UHS

4.4.1 Lateral spreading of H_2 and instability of water displacement in aquifers

The Saffman–Taylor instability of water displacement by gas represents the major hydrodynamic problem of all gas storage. The nonviscous highly mobile gas prefers to find preferential paths to penetrate through the viscous displaced liquid. Consequently, any small perturbation of the initial plane interface between gas and liquid leads to gas penetration into water in the form of fingers, with the consecutive fast development of fingers. In the case of a vertical displacement of water by the injected gas, this instability takes the form of lateral spreading of the gas along the cap rocks, as shown in Figure 4.7, in which two lateral fingers appear. This can lead to noncontrolled leakage of gas beyond the geological trap.

This phenomenon is expected to be amplified in the case of hydrogen, which is much less viscous than methane. Simultaneously with lateral spreading, the gas forms additional fingers in the central zone, which penetrate into the water-saturated zone, macroscopic losses of trapped gas.

Spreading and fingering can be stabilized by gravity forces. Indeed, the two-phase system is gravitationally unstable if the upper fluid (gas) is lighter. Therefore, if the role of the dynamic effects is reduced, the fingering will be suppressed by gravity. Thus, an optimal injection rate exists, below which the lateral spreading becomes

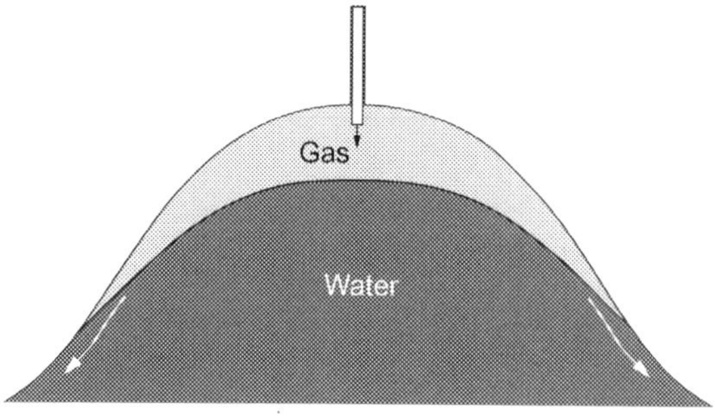

Figure 4.7 Lateral spreading of hydrogen along the cap rocks.

difficult or even impossible. In addition, the displacement can be stabilized by capillary forces, as well as by the reservoir geometry (spreading is reduced for high-amplitude strata). Then the optimal injection rate depends on all these factors.

4.4.2 Selective technology of gas injection/production

A "selective technology" applied to natural gas storage represents the gas injection to the bottom of the reservoir and the gas withdrawal from the top, using two different systems of wells. The objective is to prevent lateral gas spreading along the cap rocks. To decelerate buoyancy-driven rising of gas through water toward the cap rock, the geological stratum should have impermeable or semipermeable horizontal barriers. At low injection velocity, the gas will be fragmented and trapped under the barriers, while at high injection rates the gas will rapidly rise toward the cap rocks and will spread laterally. The problem is finding the optimal injection rate, which at the same time ensures a sufficiently slow gas ascension and avoids gas trapping (Figure 4.8).

4.4.3 Biofilm detachment, transport, and pore clogging in UHS

Hydrogen transport in a porous medium is strongly coupled to bacterial dynamics, which is characterized by bacterial metabolism and chemical impact on gas contents, but also by bacterial movement, which determines the spatial heterogeneities of gas composition (Panfilov, 2010a). A biofilm that is firmly attached to the solid surface is the main form of bacterial existence, and this determines the immobility of bacteria. However, bacteria can be transported by water or gas when the flow velocity is sufficient to detach the biofilm from the surface. The critical velocity of detachment depends on the biofilm thickness. Therefore, in the vicinity of production or injection wells, where the flow velocity is high, the biofilm is detached and transported by gas or water and can create macroscopic aggregates that clog the pores. The objective of these studies is to develop coupled conceptual and numerical model of gas–water

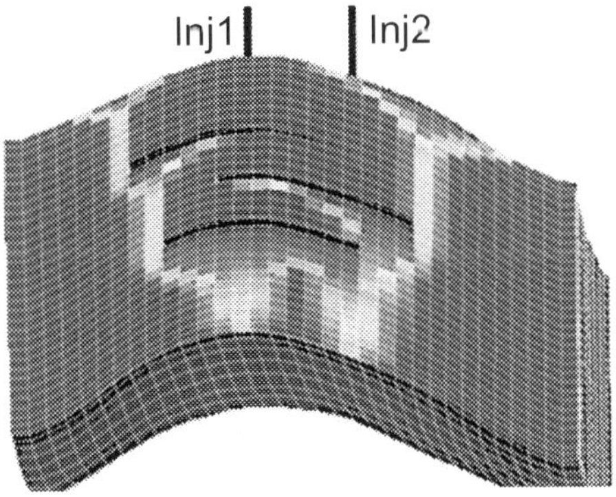

Figure 4.8 Gas shielding by barriers in layered heterogeneous storage.

flow, biofilm detachment and transport, and pore clogging, and the closure parameters for this model.

The study of the dynamics of biofilm formation and detachment by a flow can be analyzed experimentally by using the rotating disk reactor technique developed in Paris et al. (2009), which is capable of measuring the critical force of detachment for a given biofilm, using the balance between the force of fluid friction, adhesion, and biofilm elasticity. This gives the critical velocity as the function of biofilm thickness.

The theoretical study is aimed at developing the model of biofilm growth and detachment that is expected to prove visco-plastic behavior. The effects of pore clogging can be described in terms of the stochastic Smoluchowski model of particle aggregation. This approach was applied by the authors to describe the transport of colloids in porous media (Panfilov et al., 2008; Bedrikovetsky et al., 2012).

4.4.4 Gas evolution and transport in salt caverns

Salt caverns are considered as the best way to ensure hydrogen purity and hermetic storage. Several applications require very pure hydrogen, such as fuel cells. However, high hydrogen diffusivity and bacterial activity can disturb the cavern impermeability and the purity of the stored gas. The mechanisms of gas pollution in salt caverns are different from those in porous media. In salt caverns, bacteria live in the sump at the bottom (the residual water volume that is not evacuated after leaching can reach 30% of the total cavern volume). In the presence of sulfates and carbonates, bacteria consume hydrogen and produce H_2S and/or methane, similar to the bioreactions described in Section 4.3.2. This effect, which leads to gas pollution, is aggravated by water evaporation to gas and by thermal convection in gas that accelerate gaseous mixing. Thermal convection is caused by temperature gradients that are nonnegligible in caverns whose height can be of several hundred meters.

4.4.5 Coupled hydrodynamic, chemical, and bacterial transport: Self-organization phenomena

The dynamics of bacterial population, which comprises the kinetics of its growth and death, the laws of bacterial movement (i.e., diffusion, convection, chemotaxis), and the transformations of various types of bacterial populations (biofilms, plankton), significantly influences the processes in hydrogen storage. The first model of such dynamics was suggested in Panfilov (2010a), for the case of single-phase gas with residual immobile water, a single kind of bacteria, and a single biotic reaction of methanogenesis. A more advanced model of the process was developed by Toleukhanov et al. (2012), Hagemann et al. (2015), and Panfilov and Reitenbach (2015). This model includes a two-phase fluid, several bioreactions, and several types of bacteria. The most advanced kinetic model of bioreactions was developed by Panfilov and Reitenbach (2015).

In several cases, coupling between transport, chemical kinetics, and population growth kinetics leads to the appearance of spatial and time oscillations, which progressively develop and tend to stable periodic limit waves. These waves are called *self-oscillations* or *auto-waves*, and the process is called the *self-organization phenomenon*. In the simplest case, the self-oscillations appear as a result of the nonequilibrium between the rate of nutrient injection (hydrogen) and the rate of nutrient consumption by the bacteria. Indeed, the injection of nutrients leads to an increase in the bacterial population, which leads, in turn, to a decrease in the amount of nutrients; this leads to a decrease in the number of bacteria, and so on. In the theory of dynamic systems, the limiting periodic oscillations correspond to the appearance of a *limit cycle*: a special singular point in the system of differential equations that describe the process.

The existence of self-oscillations (or a limit cycle) corresponds to a fixed range of variation of the governing parameter (the ratio between the rate of hydrogen injection and consumption). The self-oscillations disappear at some critical value of this parameter, which corresponds to the so-called Hopf–Andronov bifurcation. The appearance of the limit cycle and the Hopf–Andronov bifurcation can be analyzed by the methods of nonlinear physics (Murray, 1989; Svirezhev, 1987).

These effects were observed in practice (Buzek et al., 1994) and are explained theoretically in Panfilov (2010a).

Because of these interactions, the existing mathematical and numerical tools of modeling reactive transport in underground reservoirs should either be adjusted, or entirely replaced by new models and algorithms, provided that the population dynamics cannot be introduced in numerical codes as a plugin but changes the mathematical center of the models.

Figure 4.9 shows an example of auto-waves, as calculated by the author of this chapter.

The significance of these phenomena consists of the formation of zones oversaturated with hydrogen and other zones oversaturated with methane. The mathematical analysis shows that in several scenarios these zones remain in steady state over time, while other scenarios lead to permanent fluctuations in space and in time, such that the

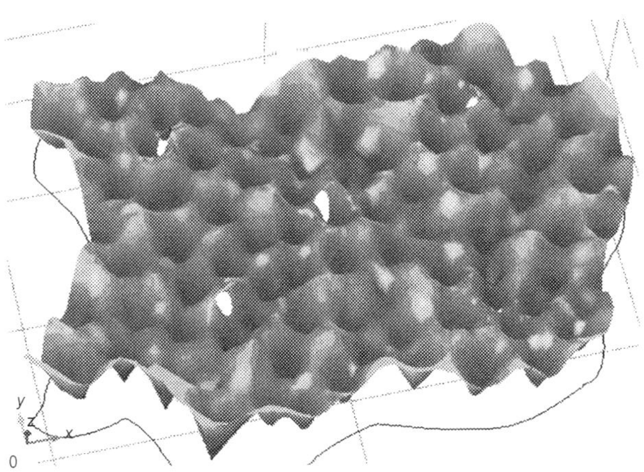

Figure 4.9 Limit nonstationary auto-waves of the concentration of hydrogen in UHS (with five injectors) caused by the coupled effects of bioreactive gas transport and bacterial population growth.

storage behaves as a natural chaotic chemical reactor. The appearance of chaos is mathematically possible. Various scenarios and the bifurcation between them depend significantly on four kinetic functions analyzed in Section 4.3.5: the rate of biomass production Ψ, the rate of respiration ψ, the rate of bacterial death φ, and the rate of CO_2 consumption for biomass production θ. At the present state of knowledge, the information concerning these functions is limited, which unfortunately limits the rapid advancement of mathematical models of gas–water–bacteria transport in UHS.

4.5 Other problems

4.5.1 Abiotic reactions of hydrogen with rocks (pyrite)

The abiotic redox reactions induced by hydrogen are insignificant at temperatures below 100 °C without special catalyzers (Truche, 2009). This is caused by the strong H–H bonding energy (436 kJ/mol), which determines a high energetic barrier that must be overcome to launch electron transfer. For instance, the minimal temperature for the reaction of reduction of pyrite (FeS_2) into pyrrhotite (FeS_{1+x})

$$FeS_2 + (1-x)H_2 = FeS_{1+x} + (1-x)H_2S, \quad 0 < x < 0.125 \tag{4.10}$$

is 90 °C, as demonstrated experimentally in Truche et al. (2010). The reaction rate remains significant even at 50 °C at pressures higher than 30 bar and a pH of water higher than 8.

The experiments modeling such reactions were performed in an abiotic batch reactor in a triphasic water/rock/gas environment under high hydrogen pressure, with chemical and electrochemical monitoring of the gas and liquid phase based on mineral electrode (pyrite electrode) (Truche et al., 2010).

This example shows that for now, heterogeneous abiotic reactions between hydrogen and rocks can be excluded from the consideration as insignificant to UHS. However, other research (Pudlo et al., 2013) reported that a slow reactivity between hydrogen and rocks was observed experimentally.

4.5.2 Hydrogen leakage by diffusion

The frequently repeated argument is that hydrogen may leak from storage, since its molecules are very small and can penetrate through any holes or fissures in cap rocks. This also pertains to salt caverns. However, when the cap rocks are saturated with water they represent a practically impermeable barrier to hydrogen (see Section 4.3.4).

The diffusion coefficient for hydrogen equals 5×10^{-9} m^2/s in free water, and 3×10^{-11} m^2/s in clay rocks at 25 °C saturated with water (Krooss, 2008). The diffusion coefficient of H_2 has been sufficiently well studied in argillite at standard pressure and temperature (Boulin, 2008; Boulin et al., 2008; Didier, 2012) but remains poorly studied at underground storage conditions.

According to Carden and Paterson (1979), during the lifetime of storage only 2% of the stored hydrogen is lost due to diffusion through cap rocks and 2% due to dissolution in reservoir water.

4.6 Pipeline storage of hydrogen

As explained in Section 4.1.3, one of the technologies of rational use of excess electrical power consists of its conversion into hydrogen and its consecutive use as fuel (power-to-gas). In this case hydrogen can be used in different ways:

- injected into a natural gas storage site (UHS of Type II in Section 4.1.3);
- converted into methane through the Sabatier reaction with CO_2 and catalyzers at high temperature and injected into gas pipelines;
- converted into methane through the Sabatier reaction with CO_2 in an aquifer by using underground bacteria as catalyzers (UHS of Type IV in Section 4.1.3); and
- injected directly into the natural gas pipelines, without underground storage.

In the last case, the natural gas network plays the role of permanent storage of a fixed mass of hydrogen (and not only a transporting system), which is why this technique is called pipeline storage of hydrogen. Some advantages and problems of this technique are analyzed in Melaina et al. (2013). An example of the application of this technology is the project developed by Hydrogenics (a fuel cell maker) and Enbridge (a natural gas distributor), which have teamed up to develop such a power-to-gas system in Canada using the existing natural gas infrastructure (Anscombe, 2012).

The following example clearly illustrates the potential of pipeline energy storage. The capacity of the German natural gas network is more than 200,000 GW h, which meets the requirements for several months. In comparison, the capacity of all the German pumped storage power plants only amounts to about 40 GW h.

The main advantage of pipeline storage is that the storage (and transport) of energy through a gas network experiences much less loss (<0.1%) than in a power network (8%).

The main problems of pipeline storage concern system safety, material durability, and hydrogen leakage.

4.6.1 *Safety and material durability*

Safety is essentially related to the risk of ignition and explosion of the blend of methane and hydrogen. According to multiples studies, adding low concentrations of hydrogen (volumes of 20% or less) to existing natural gas pipeline systems results in a minor increase in the risk of ignition.

Material durability of some metal pipes can degrade when they are exposed to hydrogen over long periods, especially with hydrogen at high concentrations and high pressures. Nearly 100% of all onshore and offshore gas transmission pipelines are made of carbon steel, wrapped/coated and cathodically protected against corrosion.

Their diameter is of 5.1–152.4 cm and the operating pressure is 42–84 bar up to 139 bar in some specific cases.

The impact of hydrogen on the properties of steel alloys is known as hydrogen blistering, hydrogen-induced cracking, or hydrogen embrittlement (Kanezaki et al., 2008; Szummer et al., 1999). The still-destroying mechanism consists of the diffusion of small hydrogen atoms into steel microstructures, which is preceded by the dissociation of an H_2 molecule due to chemical adsorption on the still surface. The atoms can accumulate at some locations and create H_2 molecules, which increases the local mechanical stress in the still and leads to the appearance of blistering, fissures, and cracks.

For pipeline storage, the steel of pipes should not be susceptible to hydrogen-induced embrittlement under normal operating conditions, which is the case for metallic pipes in U.S. distribution systems that are primarily made of low-strength steel, typically API 5L A, B, X42, and X46. Consequently, at pressures and stress levels occurring in the natural gas distribution system, hydrogen-induced failure of steel is not a major concern. For other metallic pipes made of ductile iron, cast and wrought iron, and copper, there is no concern of hydrogen damage under general operating conditions. Similarly, there is no concern of the hydrogen impact on polyethylene or polyvinylchloride pipes. Most of the elastomer materials used in natural gas distribution systems are compatible with hydrogen (Melaina et al., 2013). However, hydrogen blends can cause problems (e.g., they can influence the accuracy of existing gas meters).

4.6.2 Leakage

Another problem that might cause some difficulties is related to the leakage of hydrogen through pipe walls. The permeation rate for hydrogen is about four to five times higher than that for methane in the typical polymer pipes used in natural gas distribution systems. Leakage in steel and ductile iron systems occurs mainly through threads or mechanical joints, and the volume leakage rate for hydrogen is about a factor of 3 higher than that for natural gas.

According to Melaina et al. (2013), for a blend of natural gas and 20% hydrogen within the 670 km of PE pipes (in the United States), the gas loss would constitute about 13 million m^3/year, of which 60% is hydrogen (and only 40% natural gas). This volume of leaked gas constitutes 0.0002% of the total amount of natural gas consumed in the United States in 2010. Thus, such leakage is considered economically insignificant. Due to the higher mobility of hydrogen molecules, a hydrogen blend slightly reduces natural gas leakage.

Acknowledgments

The author is grateful to professor Frederic Jorand, from LCPME (Laboratoire de Chimie Physique et Microbiologie pour lEnvironnement, Nancy, France), for his consultations given in the field of microbiology.

References

Anscombe, N., 2012. Energy storage: could hydrogen be the answer? Solar Novus Today, 4 June 2012 (retrieved 03.11.12.).

Basniev, K.S., Omelchenko, R.J., Adzynova, F.A., 2010. Underground hydrogen storage problems in Russia. In: Procs. 18th World Hydrogen Energy Conference 2010—WHEC.

Bedrikovetsky, P., Zeinijahromi, A., Siqueira, F., Furtado, C., de Souza, A., 2012. Particle detachment under velocity alternation during suspension transport in porous media. Transp. Porous Media 91 (1), 173–197.

Boone, D.R., 2000. Biological formation and consumption of methane. In: Khalil, M.A.K. (Ed.), Atmospheric Methane: Its Role in the Global Environment. Springer, New York, pp. 43–59.

Boulin, P., 2008. Expérimentation et Modélisation du Transfert d'hydrogéne á travers des argiles de centre de stockage de déchets radioactifs (Ph.D. thesis). Institut Polytechnique de Grenoble, 241 pp.

Boulin, P., Angulo-Jaramillo, R., Daïan, J.F., Talandier, J., Berne, P., 2008. Experiments to estimate gas intrusion in Callovo-oxfordian argillites. Phys. Chem. Earth A/B/C 33, 225–230.

Bulatov, G.G., 1979. Underground storage of hydrogen (Ph.D. thesis). Moscow Gubkin Oil and Gas University (in Russian).

Buzek, F., Onderka, V., Vancura, P., Wolf, I., 1994. Carbon isotope study of methane production in a town gas storage reservoir. Fuel 73 (5), 747–752.

Carden, P.O., Paterson, L., 1979. Physical, chemical and energy aspects of underground hydrogen storage. Int. J. Hydrog. Energy 4 (6), 559–569.

Crotogino, F., Donadei, S., Bünger, U., Landinger, H., 2010. Large-scale hydrogen underground storage for securing future energy supplies. In: Proc. 18th World Hydrogen Energy Conference, May 16–21, 2010, Essen, Germany.

Davison, J., Arienti, S., Cotone, P., Mansuco, L., 2009. Co-production of hydrogen and electricity with CO_2 capture. Energy Procedia 1, 4063–4070.

Delignette-Muller M.L., n.d. Modélisation du développement microbien dans un aliment et application à l'appréciation quantitative des risques. ENS de Lyon. http://perso.ens-lyon.fr/nathalie.revol/ED/MPetQRA.pdf.

Didier, M., 2012. Etude du transfert ractif de l'hydrogne au sein de l'argilite (Ph.D. thesis). Université de Grenoble.

DVGW, 2011. Arbeitsblatt G262: Nutzung von Gasen aus regenerativen Quellen in der öffentlichen Gasversorgung.

Ebigbo, A., Golfier, F., Quintard, M., 2013. A hybrid two-scale model for methanogenic microbial activity in underground storage of hydrogen. Adv. Water Resour. 21, 74–85.

Foh, S., Novil, M., Rockar, P., Randolph, P., 1979. Underground hydrogen storage. Final report, Institute of Gas Technology, Chicago, 268 pp.

GM, 2011. Looks to Hawaii for Hydrogen Infrastructure Pilot. General Motors News, May 11, 2010. http://media.gm.com/media/us/en/gm/news.detail.html/content/Pages/news/us/en/2010/May/0511hawaii.html (retrieved 12.05.11).

Gusev, M.V., Mineeva, L.A., 1992. Microbiology. Ed. Moscow Lomonosov University, Moscow (in Russian).

Hagemann, B., Rasouldzadeh, M., Panfilov, M., Ganzer, L., Reitenbach, V., 2015. Mathematical modeling of reactive unstable transport in underground hydrogen storage. Comput. Geosci. (accepted for publication).

Kanezaki, T., Narazaki, C., Mine, Y., Matsuoka, S., Murakami, Y., 2008. Effects of hydrogen on fatigue crack growth behavior of austenitic stainless steels. Int. J. Hydrog. Energy 33, 2604–2619.

Kepplinger, J., Crotogino, F., Donadei, S., Wohlers, M., 2011. Present trends in compressed air energy and hydrogen storage in Germany. In: Solution Mining Research Institute SMRI Fall 2011 Conference, York, United Kingdom.

Kireeva, T.A., Berestovskaya, Y.Y., 2012. Microbiological transformations of hydrogenized gases stored in underground reservoirs. Gas Ind. Spec. Issue 684, 51–55.

Kreinin, E.B., 1993. Two-stage underground coal gasification. Coal Chem. Ind. 6 (3), 61–63.

Krooss, B., 2008. Evaluation of database on gas migration through clayey host rocks. Belgian National Agency for Radioactive Waste and Enriched Fissile Material (ONDRAF-NIRAS). RWTH, Aachen.

Lindblom, U.E., 1985. A conceptual design for compressed hydrogen storage in mined caverns. Int. J. Hydrog. Energy 10 (10), 667–675.

Lord, A.S., 2009. Overview of geologic storage of natural gas with emphasis on assessing the feasibility of storing hydrogen. Sandia Report SAND2009-5878, 28 pp.

Melaina, M.W., Antonia, O., Penev, M., 2013. Blending hydrogen into natural gas pipeline networks: a review of key issues. National Renewable Energy Laboratory, Technical Report NREL/TP-5600-51995.

Moosa, M., Nemati, M., Harrison, S.T.L., 2002. A kinetic study on anaerobic reduction of sulphate. Part I. Effect of sulphate concentration. Chem. Eng. Sci. 57, 2773–2780.

Moosa, S., Nemati, M., Harrison, S.T.L., 2005. A kinetic study on anaerobic reduction of sulphate. Part II: Incorporation of temperature effects in the kinetic model. Chem. Eng. Sci. 60, 3517–3524.

Murray, J.D., 1989. Mathematical Biology. Springer, New York.

Odencrantz, J.E., 1991. Modelling the biodegradation kinetics of dissolved organic contaminants in a heterogeneous two-dimensional aquifer (Ph.D. thesis). Civil and Environmental Engineering, IL.

Oladyshkin, S., Panfilov, M., 2011. Hydrogen penetration in water through porous medium: application to a radioactive waste storage site. Environ. Earth Sci. 64 (4), 989–999.

Panfilov, M., 2010a. Underground storage of hydrogen: in situ self-organization and methane generation. Transp. Porous Media 85 (3), 841–865.

Panfilov, M., 2010b. Underground storage of hydrogen: natural methane generation and in-situ self-organisation. Gas Ind. 644, 98–105 (in Russian).

Panfilov, M., Reitenbach, V., 2015. Self-organization and shock waves in underground methanation reactors and hydrogen storages. Comput. Geosci. (under edition).

Panfilov, M., Stepanyants, Y., Panfilova, I., 2008. Mechanisms of particle transport acceleration in porous media. Transp. Porous Media 74, 49–71.

Paris, T., Skali-Lami, S., Block, J.C., 2009. Probing young drinking water biofilms with hard and soft particles. Water Res. 43 (1), 117–126.

Paterson, L., 1983. The implications of fingering in underground hydrogen storage. Int. J. Hydrog. Energy 8 (1), 53–59.

Ponomarev-Stepnoi, N.N., 2004. Nuclear-hydrogen energy. Kurchatov Institute. Available at http://rosatom.ru/vestnik/04/4/5.htm.

Pudlo, D., Ganzer, L., Henkel, S., Liebscher, A., Khn, M., Lucia, M., Panfilov, M., Pilz, P., Reitenbach, V., Albrecht, D., Wrdemann, H., Gaupp, R., 2013. Hydrogen underground storage in siliciclastic reservoirs—intention and topics of the H2STORE project. In: EGU General Assembly Conference Abstracts, Geophysical Research Abstracts, v. 15, EGU2013-4179-3.

Simbeck, D.R., 2004. CO_2 capture and storage—the essential bridge to the hydrogen economy. Energy 29, 1633–1641.
Smigai, P., Greksak, M., Kozankova, J., Buzek, F., Onderka, V., Wolf, I., 1990. Methanogenic bacteria as a key factor involved in changes of town gas in an underground reservoir. FEMS Microbiol. Ecol. 73, 221–224.
Svirezhev, Yu.M., 1987. Nonlinear Waves, Dissipative Structures and Catastrophes in Ecology. Nauka, Moscow (in Russian).
Szummer, A., Jezierska, E., Lublinska, K., 1999. Hydrogen surface effects in ferritic stainless steels. J. Alloys Compd. 293295, 356–360.
Taylor, J.B., Alderson, J.E.A., Kalyanam, K.M., Lyle, A.B., Phillips, L.A., 1986. Technical and economic assessment of methods for the storage of large quantities of hydrogen. Int. J. Hydrog. Energy 11 (1), 5–22.
Toleukhanov, A., Panfilova, I., Panfilov, M., Kaltayev, A., 2012. Bioreactive two-phase transport and population dynamics in underground storage of hydrogen: natural self-organisation, Procs. ECMOR XIII—13th Europ. Conf. Mathematics of Oil Recovery. September 10–13, 2012, Biaritz. Paper B09.
Truche, L., 2009. Transformations minralogiques et gochimiques induites par la prsence d'hydrogne dans un site de stockage de dchets radioactifs. Thse Universit Toulouse III.
Truche, L., Berger, G., Destrigneville, C., Guillaume, D., Giffaut, E., 2010. Kinetics of pyrite to pyrrhotite reduction by hydrogen in calcite buffered solutions between 90 and 180 °C: implications for nuclear waste disposal. Geochim. Cosmochim. Acta 74, 2894–2914.
Vavilin, V.A., Lokshina, L.Ya., Rytov, S.V., Kotsyurbenko, O.R., Nozhevnikova, A.N., 2000. Description of two-step kinetics in methane formation during psychrophilic H_2/CO_2 and mesophilic glucose conversions. Bioresour. Technol. 71, 195–209.
Whiticar, M.J., 1999. Carbon and hydrogen isotope systematics of bacterial formation and oxidation of methane. Chem. Geol. 161, 291–314.
Zittel, W., Wurster, R., 1996. Hydrogen in the Energy Sector, vol. 7. Laudwig-Bolkow Systemtechnik GmbH, Ottobrunn.

Part Two

Physical and chemical storage of hydrogen

Cryo-compressed hydrogen storage

5

R.K. Ahluwalia, J.-K. Peng, T.Q. Hua
Argonne National Laboratory, Argonne, IL, USA

Nomenclature

C	specific heat
h	enthalpy
k	rate constant
m	mass
$\dot{m}$	mass flow rate
P	pressure
$\dot{Q}$	heat transfer rate
t	time
T	temperature
u	internal energy
V	internal volume of tank
ρ	density
X	fraction of ortho hydrogen

Subscripts/superscripts

e	electric (or equilibrium)
g	gas
in	inlet
l	liquid
o	o-H_2
out	outlet
p	p-H_2
r	leakage
s	saturation (or structure)

5.1 Introduction

Cryo-compressed hydrogen storage (CcH_2) refers to the storage of H_2 at cryogenic temperatures in a vessel that can be pressurized (nominally to 250–350 atm), in contrast to current cryogenic vessels that store liquid hydrogen (LH_2) at near-ambient pressures (Aceves et al., 2006, 2010). Cryo-compressed tanks can store liquid hydrogen, supercritical cryogenic hydrogen, or hydrogen in a two-phase region (saturated liquid and vapor).

Compendium of Hydrogen Energy. http://dx.doi.org/10.1016/B978-1-78242-362-1.00005-5

As pointed out by Aceves et al. (2006, 2010), storing liquid hydrogen in insulated pressure vessels overcomes many of the shortcomings of compressed gas (cH_2) or liquid H_2 tanks and may even open new possibilities. Compared to cryogenic H_2 tanks, the dormancy is greatly extended as the allowable pressure inside the vessel increases. The vessel has higher heat receptivity since H_2 contained in the vessel is vented at a higher temperature and this further augments dormancy. A vehicle equipped with an insulated pressure vessel cannot be stranded due to evaporative losses because venting stops when the tank reaches ambient temperature at allowable pressure (>250 atm), at which point the H_2 density is still 30% of the initial liquid density. It may be possible to reduce or eliminate ullage space in insulated pressure vessels since H_2 becomes a supercritical fluid (SCF) at a pressure above 13 atm before there is a need to vent. The storage density may be higher in insulated pressure vessels since liquid H_2 (LH_2) is slightly compressible: at 21 K, the liquid density is 81 g/L at 240 bar compared to 70 g/L at 1 bar. Aceves et al. (2006) suggest that an insulated pressure vessel can be fueled flexibly with LH_2 for extended range and inexpensive cH_2 for normal driving with short trips.

Aceves et al. (2006, 2010) at Lawrence Livermore National Laboratory have demonstrated cryo-compressed storage technology in three generations of cryogenic capable pressure vessels. The first-generation (Gen-1) vessels had 135 L internal volume, could store 9.6 kg LH_2, and were designed to operate at 245 atm peak pressure. One of the vessels was installed on a 1992 Ford Ranger pickup truck to power a hydrogen internal combustion engine. The vehicle was refueled successfully with both LH_2 and cH_2. The second-generation (Gen-2) prototype vessel had larger internal volume (151 L), could store 10.7 kg LH_2, and was designed for higher operating pressure (340 atm). It was installed in an experimental Toyota Prius hydrogen hybrid vehicle, which was test-driven for 1050 km on a single tank filled with LH_2. The third generation (Gen-3) vessel is an improved design of the Gen-2 vessel; it has the same internal volume and hydrogen capacity but the total system weight and volume were reduced by 23%.

In this chapter, we present an integrated thermodynamics, heat transfer, and isomer conversion kinetics model that can be used to evaluate the dynamics of refueling, discharge, dormancy, and storage capacity of cryo-compressed vessels. We also briefly discuss a set of available models that can be used to evaluate the overall efficiency of production, liquefying, delivering, and pumping LH_2 to the vehicle's storage tank. We use the models to assess the performance of the cryo-compressed hydrogen storage option relative to the relevant near-term and ultimate targets for automotive applications (DOE Targets for Onboard Hydrogen Storage Systems for Light-Duty Vehicles, 2009): 5.5 wt% (7.5 wt%) gravimetric capacity, 40 g/L (70 g/L) volumetric capacity, 5.6 kg usable H_2, 1.6 g/s minimum full flow rate of H_2, 1.5 kg/min (2 kg/min) H_2 refueling rate 0.05 g/kg maximum H_2 loss rate, and 60% well-to-tank (WTT) efficiency.

5.2 Thermodynamics and kinetics of cryo-compressed hydrogen storage

Previous work has shown that the important aspects of cryo-compressed hydrogen storage can be captured with a simple model that accounts for thermodynamics, heat transfer, and kinetics of isomer conversion. In this section, the approach developed in

Ahluwalia and Peng (2008) has been generalized to include the effects of para-to-ortho conversion as modeled in Peng and Ahluwalia (2013). As in earlier work, the pressure and temperature in the tank are assumed to be uniform, and H_2 gas and liquid, if present as a mixture, are considered to be in phase equilibrium. The following equations then describe the changes in the mass of H_2 and ortho-H_2 in the storage tank:

$$\frac{dm_{H_2}}{dt} = \dot{m}_{H_2}^{in} - \dot{m}_{H_2}^{out} \tag{5.1}$$

$$\frac{d\left(m_{H_2}^{o}\right)}{dt} = \dot{m}_{H_2}^{po} + \dot{m}_{H_2}^{in}x^{in} - \dot{m}_{H_2}^{out}x \tag{5.2}$$

where x is the mass fraction of o-H_2 and the first term on the right-hand side of Equation (5.2) represents the rate of conversion of o-H_2 to p-H_2. Equation (5.2) can be rearranged to obtain the following equation for o-H_2 fraction:

$$\frac{dx}{dt} = \frac{\dot{m}_{H_2}^{po}}{m_{H_2}} + \frac{\dot{m}_{H_2}^{in}}{m_{H_2}}\left(x^{in} - x\right) \tag{5.3}$$

Following Milenko et al. (1997) we use the following equation to describe the kinetics of para-to-ortho conversion:

$$\frac{\dot{m}_{H_2}^{po}}{m_{H_2}} = -kx\frac{(x - x_e)}{(1 - x_e)} \tag{5.4}$$

where x_e is the equilibrium fraction of o-H_2 (see Figure 5.1a) and k is the rate constant that can be estimated from the following interpolation formula recommended in Milenko et al. (1997):

$$k\left(\rho_{H_2}, T\right) = 2.78 \times 10^{-10} A_0 T^n \rho_{H_2} + 2.2 \times 10^{-13}(C_0 + DT^{-m})\rho_{H_2}^{p} \tag{5.5}$$

where k is in s^{-1}, T is in K, and ρ_{H_2} is in kg m^{-3}; and $A_0 = 18.2 \pm 1.6$, $n = 0.56 \pm 0.02$, $C_0 = 0.77 \pm 0.03$, $D = 921 \pm 91$, $m = 2.5 \pm 0.2$, and $p = 3.6$.

Neglecting the kinetic energy of H_2 flowing in and out of the tank, we write the following equation for the energy balance:

$$\frac{d}{dt}[m_s u_s + m_{H_2} u_{H_2}] = \dot{m}_{H_2}^{in} h_{in} - \dot{m}_{H_2}^{out} h_{H_2} + \dot{Q}_{in}^{r} + \dot{Q}_{in}^{e} \tag{5.6}$$

where $\dot{Q}_{in}^{r}$ is the heat gain from the ambient and $\dot{Q}_{in}^{e}$ is the heat input to the tank. For a constant volume tank, $dV/dt = d\left(m_{H_2}/\rho_{H_2}\right)/dt = 0$, so that

$$\frac{1}{m_{H_2}}\frac{dm_{H_2}}{dt} - \frac{1}{\rho_{H_2}}\frac{d\rho_{H_2}}{dt} = 0. \tag{5.7}$$

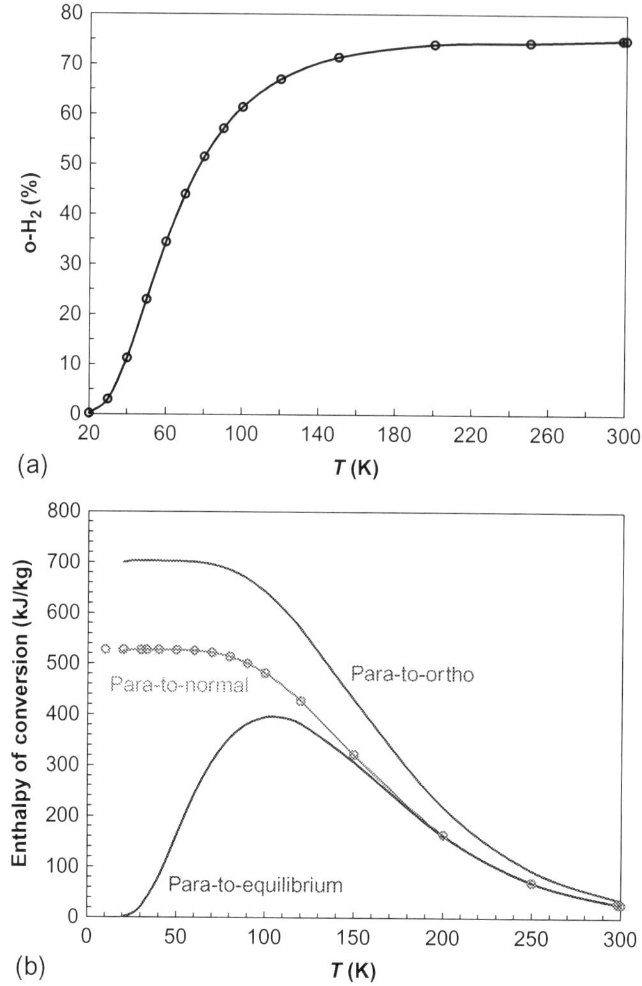

Figure 5.1 Thermodynamics of para-to-ortho conversion: (a) equilibrium composition of o-H_2; (b) enthalpy of conversion. Symbols represent literature data (McCarty et al., 1981; Barron, 1996; Wolf, 2002) and solid lines are model results.

5.2.1 Refueling process

During the refueling process, $\dot{m}_{H_2}^{in}$, $\dot{Q}_{in}^{r}$, and $\dot{Q}_{in}^{e}$ are known and $\dot{m}_{H_2}^{out} = 0$. Consider first the portion of the refueling process in which H_2 is present as a gas–liquid mixture of o-H_2 and p-H_2, so that

$$P = P_s(T)$$

$$m_{H_2} = m_g + m_l$$

$$m_{H_2}u_{H_2} = m_g(h_g - P/\rho_g) + m_l(h_l - P/\rho_l)$$

$$h_g = xh_g^o + (1-x)h_g^p$$

$$h_l = xh_l^o + (1-x)h_l^p$$

$$\rho_g = \left(\frac{x}{\rho_g^o} + \frac{1-x}{\rho_g^p}\right)^{-1}$$

$$\rho_l = \left(\frac{x}{\rho_l^o} + \frac{1-x}{\rho_l^p}\right)^{-1} \tag{5.8}$$

In general, the differences in the densities of p-H_2 and o-H_2 are small, but the thermal properties (enthalpy and thermal conductivity) can show large differences. With the help of Equation (5.8), Equations (5.1), (5.3), (5.6), and (5.7) may be rearranged to obtain the following set of ordinary differential equations (ODEs) for the dependent variables P, T, m_g, m_l, and x:

$$\frac{dP}{dt} = \left(\frac{dP_s}{dT}\right)\frac{dT}{dt}$$

$$A_m\frac{dm_g}{dt} = \frac{-\dot{m}_{H_2}^{in}}{\rho_l} + B_x\frac{dx}{dt} + B_T\frac{dT}{dt} + B_P\frac{dP}{dt}$$

$$\frac{dm_l}{dt} = \dot{m}_{H_2}^{in} - \frac{dm_g}{dt}$$

$$A_T\frac{dT}{dt} = \dot{m}_{H_2}^{in}(h_{in} - h_l) + (h_l - h_g)\frac{dm_g}{dt} - \left[m_g\left(h_g^o - h_g^p\right) + m_l\left(h_l^o - h_l^p\right)\right]\frac{dx}{dt} + \dot{Q}_{in}^r + \dot{Q}_{in}^e$$

$$\frac{dx}{dt} = -kx\frac{(x - x_e)}{1 - x_e} + \frac{\dot{m}_{H_2}^{in}}{m_{H_2}}\left(x^{in} - x\right) \tag{5.9}$$

where

$$A_m = \left(\frac{1}{\rho_g} - \frac{1}{\rho_l}\right)$$

$$A_T = m_sC_s + m_g\left[x\left(\frac{\partial h_g^o}{\partial T}\right)_P + (1-x)\left(\frac{\partial h_g^p}{\partial T}\right)_P\right] + m_l\left[x\left(\frac{\partial h_l^o}{\partial T}\right)_P + (1-x)\left(\frac{\partial h_l^p}{\partial T}\right)_P\right] +$$
$$\left\{m_g\left[x\left(\frac{\partial h_g^o}{\partial P}\right)_T + (1-x)\left(\frac{\partial h_g^p}{\partial P}\right)_T\right] + m_l\left[x\left(\frac{\partial h_l^o}{\partial P}\right)_T + (1-x)\left(\frac{\partial h_l^p}{\partial P}\right)_T\right] - V\right\}\frac{dP_s}{dT}$$

$$B_x = \left[m_g\left(\frac{1}{\rho_g^p}-\frac{1}{\rho_g^o}\right)+m_l\left(\frac{1}{\rho_l^p}-\frac{1}{\rho_l^o}\right)\right]$$

$$B_T = m_g\left[\frac{x}{\left(\rho_g^o\right)^2}\left(\frac{\partial\rho_g^o}{\partial T}\right)_P+\frac{1-x}{\left(\rho_g^p\right)^2}\left(\frac{\partial\rho_g^p}{\partial T}\right)_P\right]+m_l\left[\frac{x}{\left(\rho_l^o\right)^2}\left(\frac{\partial\rho_l^o}{\partial T}\right)_P+\frac{1-x}{\left(\rho_l^p\right)^2}\left(\frac{\partial\rho_l^p}{\partial T}\right)_P\right]$$

$$B_P = m_g\left[\frac{x}{\left(\rho_g^o\right)^2}\left(\frac{\partial\rho_g^o}{\partial P}\right)_T+\frac{1-x}{\left(\rho_g^p\right)^2}\left(\frac{\partial\rho_g^p}{\partial P}\right)_T\right]+m_l\left[\frac{x}{\left(\rho_l^o\right)^2}\left(\frac{\partial\rho_l^o}{\partial P}\right)_T+\frac{1-x}{\left(\rho_l^p\right)^2}\left(\frac{\partial\rho_l^p}{\partial P}\right)_T\right] \tag{5.10}$$

Consider next the portion of the refueling process in which H_2 is present in the tank as a single phase (gas or liquid), so that

$$m_{H_2}u_{H_2} = m_{H_2}\left(h_{H_2} - P/\rho_{H_2}\right)$$

$$h_{H_2} = xh_{H_2}^o + (1-x)h_{H_2}^p$$

$$\rho_{H_2} = \left(\frac{x}{\rho_{H_2}^o}+\frac{1-x}{\rho_{H_2}^p}\right)^{-1} \tag{5.11}$$

Equations (5.1), (5.3), (5.6), and (5.7) may be rearranged with the help of Equation (5.11) to yield the following set of ODEs for the system variables P, T, x, and m_{H_2}.

$$\frac{dm_{H_2}}{dt} = \dot{m}_{H_2}^{in}$$

$$B_{mP}\frac{dP}{dt} = \frac{1}{m_{H_2}}\left(\frac{x}{\rho_{H_2}^o}+\frac{1-x}{\rho_{H_2}^p}\right)\frac{dm_{H_2}}{dt} - \left(\frac{1}{\rho_{H_2}^p}-\frac{1}{\rho_{H_2}^o}\right)\frac{dx}{dt} - B_{mT}\frac{dT}{dt}$$

$$B_{hT}\frac{dT}{dt} = \dot{m}_{H_2}^{in}\left(h_{in}-h_{H_2}\right) - m_{H_2}\left(h_{H_2}^o - h_{H_2}^p\right)\frac{dx}{dt} - B_{hP}\frac{dP}{dt} + \dot{Q}_{in}^r + \dot{Q}_{in}^e$$

$$\frac{dx}{dt} = -kx\frac{(x-x_e)}{1-x_e} + \frac{\dot{m}_{H_2}^{in}}{m_{H_2}}\left(x^{in}-x\right) \tag{5.12}$$

where

$$B_{mT} = \frac{x}{\left(\rho_{H_2}^o\right)^2}\left(\frac{\partial\rho_{H_2}^o}{\partial T}\right)_P + \frac{1-x}{\left(\rho_{H_2}^p\right)^2}\left(\frac{\partial\rho_{H_2}^p}{\partial T}\right)_P$$

$$B_{mP}=\frac{x}{\left(\rho_{\mathrm{H_2}}^{\mathrm{o}}\right)^2}\left(\frac{\partial\rho_{\mathrm{H_2}}^{\mathrm{o}}}{\partial P}\right)_T+\frac{1-x}{\left(\rho_{\mathrm{H_2}}^{\mathrm{p}}\right)^2}\left(\frac{\partial\rho_{\mathrm{H_2}}^{\mathrm{p}}}{\partial P}\right)_T$$

$$B_{hT}=m_{\mathrm{s}}C_{\mathrm{s}}+m_{\mathrm{H_2}}\left[x\left(\frac{\partial h_{\mathrm{H_2}}^{\mathrm{o}}}{\partial T}\right)_P+(1-x)\left(\frac{\partial h_{\mathrm{H_2}}^{\mathrm{p}}}{\partial T}\right)_P\right]$$

$$B_{hP}=m_{\mathrm{H_2}}\left[x\left(\frac{\partial h_{\mathrm{H_2}}^{\mathrm{o}}}{\partial P}\right)_T+(1-x)\left(\frac{\partial h_{\mathrm{H_2}}^{\mathrm{p}}}{\partial P}\right)_T\right]-V \tag{5.13}$$

5.2.2 Discharge process

During the discharge process, $\dot{m}_{\mathrm{H_2}}^{\mathrm{in}}$ is equal to zero. As for refueling process, the following sets of ODEs can be derived for the discharge process.

5.2.2.1 Two-phase region

$$\frac{\mathrm{d}P}{\mathrm{d}t}=\left(\frac{\mathrm{d}P_{\mathrm{s}}}{\mathrm{d}T}\right)\frac{\mathrm{d}T}{\mathrm{d}t}$$

$$A_m\frac{\mathrm{d}m_{\mathrm{g}}}{\mathrm{d}t}=\frac{\dot{m}_{\mathrm{H_2}}^{\mathrm{out}}}{\rho_{\mathrm{l}}}+B_x\frac{\mathrm{d}x}{\mathrm{d}t}+B_T\frac{\mathrm{d}T}{\mathrm{d}t}+B_P\frac{\mathrm{d}P}{\mathrm{d}t}$$

$$\frac{\mathrm{d}m_{\mathrm{l}}}{\mathrm{d}t}=-\dot{m}_{\mathrm{H_2}}^{\mathrm{out}}-\frac{\mathrm{d}m_{\mathrm{g}}}{\mathrm{d}t}$$

$$\begin{aligned}A_T\frac{\mathrm{d}T}{\mathrm{d}t}&=\dot{m}_{\mathrm{H_2}}^{\mathrm{out}}\left(h_{\mathrm{l}}-h_{\mathrm{g}}\right)+\left(h_{\mathrm{l}}-h_{\mathrm{g}}\right)\frac{\mathrm{d}m_{\mathrm{g}}}{\mathrm{d}t}\\&\quad-\left[m_{\mathrm{g}}\left(h_{\mathrm{g}}^{\mathrm{o}}-h_{\mathrm{g}}^{\mathrm{p}}\right)+m_{\mathrm{l}}\left(h_{\mathrm{l}}^{\mathrm{o}}-h_{\mathrm{l}}^{\mathrm{p}}\right)\right]\frac{\mathrm{d}x}{\mathrm{d}t}+\dot{Q}_{\mathrm{in}}^{\mathrm{r}}+\dot{Q}_{\mathrm{in}}^{\mathrm{e}}\end{aligned}$$

$$\frac{\mathrm{d}x}{\mathrm{d}t}=-kx\frac{(x-x_{\mathrm{e}})}{1-x_{\mathrm{e}}} \tag{5.14}$$

5.2.2.2 Single-phase region

$$\frac{\mathrm{d}m_{\mathrm{H_2}}}{\mathrm{d}t}=-\dot{m}_{\mathrm{H_2}}^{\mathrm{out}}$$

$$B_{mP}\frac{\mathrm{d}P}{\mathrm{d}t}=\frac{1}{m_{\mathrm{H_2}}}\left(\frac{x}{\rho_{\mathrm{H_2}}^{\mathrm{o}}}+\frac{1-x}{\rho_{\mathrm{H_2}}^{\mathrm{p}}}\right)\frac{\mathrm{d}m_{\mathrm{H_2}}}{\mathrm{d}t}-\left(\frac{1}{\rho_{\mathrm{H_2}}^{\mathrm{p}}}-\frac{1}{\rho_{\mathrm{H_2}}^{\mathrm{o}}}\right)\frac{\mathrm{d}x}{\mathrm{d}t}-B_{mT}\frac{\mathrm{d}T}{\mathrm{d}t}$$

$$B_{hT}\frac{\mathrm{d}T}{\mathrm{d}t}=-m_{H_2}\left(h_{H_2}^{o}-h_{H_2}^{p}\right)\frac{\mathrm{d}x}{\mathrm{d}t}-B_{hP}\frac{\mathrm{d}P}{\mathrm{d}t}+\dot{Q}_{in}^{r}+\dot{Q}_{in}^{e}$$

$$\frac{\mathrm{d}x}{\mathrm{d}t}=-kx\frac{(x-x_e)}{1-x_e} \tag{5.15}$$

5.2.3 Dormancy process

During dormancy, $\dot{Q}_{in}^{r}$ is known, and $\dot{m}_{H_2}^{in}$ and $\dot{Q}_{in}^{e}$ are both equal to zero. Also, the pressure inside the tank is below the allowable limit so that H_2 does not have to be vented and $\dot{m}_{H_2}^{out}=0$. It can be shown that the thermodynamic state of H_2 stored in the tank during this time period can be determined by solving the following equations.

5.2.3.1 Two-phase region

$$\frac{\mathrm{d}P}{\mathrm{d}t}=\left(\frac{\mathrm{d}P_s}{\mathrm{d}T}\right)\frac{\mathrm{d}T}{\mathrm{d}t}$$

$$A_m\frac{\mathrm{d}m_g}{\mathrm{d}t}=B_x\frac{\mathrm{d}x}{\mathrm{d}t}+B_T\frac{\mathrm{d}T}{\mathrm{d}t}+B_P\frac{\mathrm{d}P}{\mathrm{d}t}$$

$$\frac{\mathrm{d}m_l}{\mathrm{d}t}=-\frac{\mathrm{d}m_g}{\mathrm{d}t}$$

$$A_T\frac{\mathrm{d}T}{\mathrm{d}t}=(h_l-h_g)\frac{\mathrm{d}m_g}{\mathrm{d}t}-\left[m_g\left(h_g^{o}-h_g^{p}\right)+m_l\left(h_l^{o}-h_l^{p}\right)\right]\frac{\mathrm{d}x}{\mathrm{d}t}+\dot{Q}_{in}^{r}$$

$$\frac{\mathrm{d}x}{\mathrm{d}t}=-kx\frac{(x-x_e)}{1-x_e} \tag{5.16}$$

5.2.3.2 Single-phase region

$$\frac{\mathrm{d}m_{H_2}}{\mathrm{d}t}=0$$

$$B_{mP}\frac{\mathrm{d}P}{\mathrm{d}t}=-\left(\frac{1}{\rho_{H_2}^{p}}-\frac{1}{\rho_{H_2}^{o}}\right)\frac{\mathrm{d}x}{\mathrm{d}t}-B_{mT}\frac{\mathrm{d}T}{\mathrm{d}t}$$

$$B_{hT}\frac{\mathrm{d}T}{\mathrm{d}t}=-m_{H_2}\left(h_{H_2}^{o}-h_{H_2}^{p}\right)\frac{\mathrm{d}x}{\mathrm{d}t}-B_{hP}\frac{\mathrm{d}P}{\mathrm{d}t}+\dot{Q}_{in}^{r}$$

$$\frac{\mathrm{d}x}{\mathrm{d}t}=-kx\frac{(x-x_e)}{1-x_e} \tag{5.17}$$

5.2.4 Hydrogen loss

Consider the period in which the pressure inside the tank reaches the set point of the relief valve, which lifts and begins to vent H_2 to maintain $P = P_{max}$, after which $dP/dt = 0$. In this period, $\dot{m}_{H_2}^{out}$ can be determined by solving the following sets of equations.

5.2.4.1 Two-phase region

$$
\begin{aligned}
&\frac{dT}{dt} = \left(\frac{dP_s}{dT}\right)^{-1}\frac{dP}{dt} = 0 \\
&A_m\frac{dm_g}{dt} = \frac{\dot{m}_{H_2}^{out}}{\rho_l} + B_x\frac{dx}{dt} \\
&\frac{dm_l}{dt} = -\frac{dm_g}{dt} - \dot{m}_{H_2}^{out} \\
&\dot{m}_{H_2}^{out} = -\frac{dm_g}{dt} - \frac{h_l}{h_g}\frac{dm_l}{dt} - \left[m_g\frac{\left(h_g^o - h_g^p\right)}{h_g} + m_l\frac{\left(h_l^o - h_l^p\right)}{h_g}\right]\frac{dx}{dt} + \frac{\dot{Q}_{in}^{r}}{h_g} \\
&\frac{dx}{dt} = -kx\frac{(x - x_e)}{1 - x_e}
\end{aligned}
\tag{5.18}
$$

5.2.4.2 Single-phase region

$$
\begin{aligned}
&\frac{dm_{H_2}}{dt} = -\dot{m}_{H_2}^{out} \\
&B_{hT}\frac{dT}{dt} = -m_{H_2}\left(h_{H_2}^o - h_{H_2}^p\right)\frac{dx}{dt} + \dot{Q}_{in}^{r} \\
&\dot{m}_{H_2}^{out} = -\rho_{H_2}m_{H_2}\left\{\left(\frac{1}{\rho_{H_2}^p} - \frac{1}{\rho_{H_2}^o}\right)\frac{dx}{dt} + B_{mT}\frac{dT}{dt}\right\} \\
&\frac{dx}{dt} = -kx\frac{(x - x_e)}{1 - x_e}
\end{aligned}
\tag{5.19}
$$

5.2.5 Method of solution and equation of state

The foregoing sets of ODEs were integrated by using the Adams–Bashford–Molton method as implemented in the Gear package (Gear, 1971). The ODEs are generally not stiff but special care is needed when the thermodynamic state of the stored H_2 changes among the supercritical, subcooled, and two-phase regions.

The Benedict–Webb–Rubin (BWR) equation of state, as implemented in REFPROP (Lemmon et al., 2010), was used to calculate the density, specific heat, and enthalpy of p-H_2, o-H_2, and n-H_2 in liquid and gas phases. Some adjustments had to be made since REFPROP treats saturated ortho, para, or normal liquid at the normal boiling point (20.4 K) as the reference states. REFPROP assigns zero values to enthalpy and entropy of para, ortho, and normal H_2 at the reference state. A common reference state is required to properly calculate the thermodynamics of transitions between p-H_2 and o-H_2. Following McCarty et al. (1981), we selected ideal p-H_2 gas at 0 K as the reference state where the molar internal energy is at its lowest quantized value. Accordingly, the enthalpy and entropy calculated from REFPROP (indicated below as RP) were adjusted to the new reference state by using the following equations:

$$\begin{aligned} h^{\mathrm{i}} &= h^{\mathrm{i}}(\mathrm{RP}) + \Delta h^{\mathrm{i}} \\ s^{\mathrm{i}} &= s^{\mathrm{i}}(\mathrm{RP}) + \Delta s^{\mathrm{i}} \end{aligned} \tag{5.20}$$

where the superscript i denotes para, ortho, or normal H_2, and Δh^{i} and Δs^{i} are −256.06 kJ/kg and 7.987 kJ/kg K for p-H_2, 269.08 kJ/kg and 17.116 kJ/kg K for n-H_2, and 444.695 kJ/kg and 17.066 kJ/kg K for o-H_2. Figure 5.1b shows the heat of conversion for H_2 from para to normal states ($\Delta h^{\mathrm{pn}} = h^{\mathrm{n}} - h^{\mathrm{p}}$) as calculated from REFPROP modified by Equation (5.20) and it compares Δh^{pn} with the data from the standard references (McCarty et al., 1981; Barron, 1996; Wolf, 2002). The complete agreement between the calculations and experimental data validates the use of Equation (5.20) to modify the enthalpies obtained from REFPROP. Figure 5.1b also includes the heat of conversion from para-to-equilibrium states ($\Delta h^{\mathrm{pe}} = h^{\mathrm{e}} - h^{\mathrm{p}}$) and shows a 400 kJ/kg peak at about 100 K. The heat of conversion, Δh^{pe}, is negligible at temperatures below 30 K, because the equilibrium composition consists mostly of p-H_2. Above 160 K, Δh^{pe} approaches Δh^{pn} because of the relatively small difference between the equilibrium and normal compositions in this temperature range.

Another way of validating Equation (5.20) is to calculate the equilibrium composition of H_2 from h^{i} and s^{i}. At equilibrium, p-H_2 and o-H_2 have the same free energies so that the fraction of o-H_2 can be determined from:

$$RT \ln \frac{(1 - x_{\mathrm{e}})}{x_{\mathrm{e}}} = (h^{\mathrm{o}} - h^{\mathrm{p}}) - T(s^{\mathrm{o}} - s^{\mathrm{p}}) \tag{5.21}$$

Figure 5.1a shows that the equilibrium o-H_2 concentration calculated from Equation (5.21), solid curve, is in complete agreement with the experimental data, symbols, from the standard references (McCarty et al., 1981; Barron, 1996; Wolf, 2002).

5.3 Performance of onboard storage system

Figure 5.2 shows a schematic of an onboard hydrogen storage system that mimics the design of the Gen-3 cryo-compressed tank developed by Aceves et al. (2010). The Type 3 tank consists of an Al 6061-T6 alloy liner wrapped with T700S carbon fiber

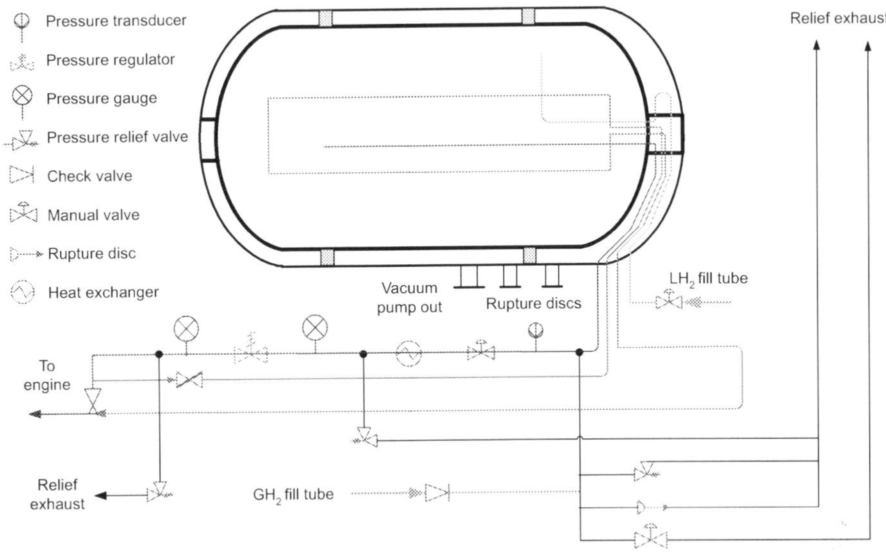

Figure 5.2 Design schematic of the Gen-3 cryo-compressed H_2 storage tank system.

designed to withstand 225% of the nominal storage pressure. Type 3 pressure vessels that are subjected to fluctuating pressure are auto-frettaged to generate residual compressive stresses in the liner, shift tensile loads to the CF composite, and extend the fatigue life of the metal liners (Thesken et al., 2009). The tank is surrounded by a vacuum gap (10^{-5} torr) filled with multilayer vacuum superinsulation. A thin stainless steel or aluminum outer shell, separated from the inner tank by two G-10 space rings, completes the main vessel. There are two in-tank fill tubes connected to the refueling lines; one for liquid hydrogen and another for gaseous hydrogen. The gaseous fill tube also serves as the discharge tube for dispensing hydrogen to the fuel cell. An in-tank exchanger tube is connected to the gas recirculation line, which circulates warm hydrogen through the tank to provide the heat needed for discharge. The rupture discs on the shell provide a path for venting hydrogen in the event of hydrogen leakage into the vacuum space. The main balance-of-plant components include a pressure regulator, pressure gauges and transducers, temperature transducers, heat exchanger, valves, pressure relief devices, and tubings.

Table 5.1 summarizes results from a system model formulated to determine the thickness of the liner, CF composite, and MLSI insulation for prescribed usable H_2 capacity, storage pressure, and heat gain from the ambient. The model was first calibrated against the data for the Gen-3 tank that holds 10.4 kg H_2, and then it was used to scale the tank for 5.6 kg usable H_2 at 272 atm storage pressure. The total system weight and volume for the scaled storage system are 101.4 kg and 133.9 L if the liner thickness is the same as in the larger prototype (9.5 mm), resulting in a system gravimetric capacity of 5.5 wt% and 41.8 g/L, respectively. According to our model, the liner thickness in the scaled tank can be halved if the temperature swing is controlled within a narrow range. In this case, the gravimetric and volumetric capacities increase

Table 5.1 Component and system weights and volumes of the prototype and scaled LLNL Gen-3 cryo-compressed tank system (Ahluwalia et al., 2010)

	Model		Gen-3 Prototype		Model	
	Wt (kg)	**Vol (L)**	**Wt (kg)**	**Vol (L)**	**Wt (kg)**	**Vol (L)**
Stored hydrogen	10.7	151.0	10.7	151.0	5.7	80.8
Usable hydrogen	10.4	151.0			5.6	80.8
Pressure vessel (274 atm)	62.4	29.0	61.0	28.0	39.1	17.7
Aluminum liner	38.8	14.4			12.6	4.7
Boss	22.7	14.1			12.9	8.2
Plug	0.4	0.4			0.4	0.4
In-tank heat exchanger	0.3	0.1			0.3	0.1
	0.3	0.0			0.3	0.0
Insulation and vacuum	52.3	43.7	51.0	45.0	34.6	24.4
shell	1.2	0.7			0.5	0.3
Support rings	2.2	36.8			1.2	20.0
Insulation material	48.9	6.2			34.2	4.3
Vacuum shell (SS 304, 3.2 mm)						
Mounting brackets	6.0	1.0	6.0	1.0	6.0	1.0
Balance-of-plant (BOP)	16.0	10.0	16.0	10.0	16.0	10.0
Computer	0.2	0.5	0.2	0.5	0.2	0.5
Electronic boards	2.2	5.0	2.2	5.0	2.2	5.0
Valves and valve box	6.9	0.8	6.9	0.8	6.9	0.8
Pressure transmitter,	1.1	0.6	1.1	0.6	1.1	0.6
gauge, regulator, and	1.5	1.8	1.5	1.8	1.5	1.8
rupture discs	4.0	1.5	4.0	1.5	4.0	1.5
Heat exchanger						
Miscellaneous tubing, fittings, etc.						
Total	147.4	234.7	144.7	235.0	90.1	129.7
Gravimetric capacity, wt%H_2	7.1	44.5	7.4	45.5	6.2	43.2
Volumetric capacity, g-H_2/L						

to 6.2 wt% and 43.2 g/L, respectively. The gravimetric capacity improves to 9.2 wt% if the shell is made of aluminum instead of 304 stainless steel.

5.3.1 *Refueling dynamics*

Fueling with LH_2 has been analyzed for two different modes of operation of LH_2 storage: cryo-compressed and cryo-supercritical. For either mode, the refueling system uses a single-flow nozzle and a high-pressure LH_2 pump that delivers 1.5 kg/min to the system at a variable pressure (25% above the prevailing pressure in the tank) with an average isentropic efficiency of 80%. The cryo-compressed mode allows the tank to operate mostly in the two-phase region (saturated vapor and liquid) and requires heat to be supplied only during discharge. However, it needs a liquid-level sensor to serve as a fuel gauge. The cryo-supercritical mode does not involve phase change so that a level sensor is not needed. However, it requires heat to be supplied during both refueling and discharge.

The amount of LH_2 that can be charged to the system is a function of the initial tank temperature, as shown in Figure 5.3 for refueling scenarios in which the tank is initially depleted to the 4-atm minimum allowable pressure. The results in Figure 5.3 are for two different modes of refueling, one where the tank is filled to 272 atm regardless of the starting temperature, and the other where the maximum density of the H_2 in the tank is limited to 71 kg/m^3, the density of LH_2 at 1 atm and 20.3 K. In the first mode (top plot in Figure 5.3), the maximum amount of H_2 charged is 6.4 kg corresponding to a stored H_2 density of 81 kg/m^3. In the second mode (bottom plot), the maximum amount of H_2 charged is 5.6 kg, and the final pressure is less than 272 atm if the initial tank temperature is less than 125 K. For both modes of refueling, the maximum amount of H_2 that can be charged into the tank is just slightly greater than 2.1 kg if the initial tank temperature is 300 K.

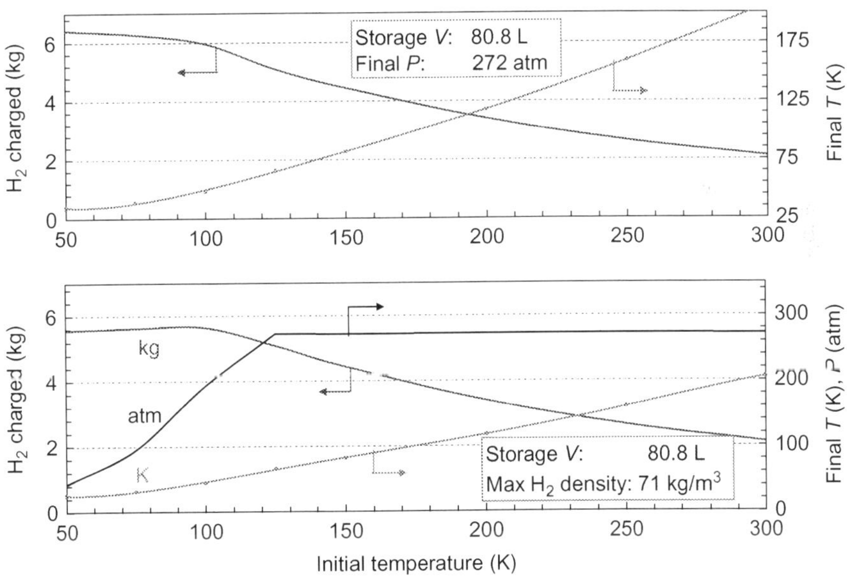

Figure 5.3 Effect of initial tank temperature on the amount of H_2 that can be stored in the scaled Gen-3 system for two different modes of filling with LH_2. Top plot, final pressure is 272 atm, regardless of initial tank temperature. Bottom plot, maximum LH_2 density is limited to that of LH_2 at 1 atm, i.e., 71 kg/m^3.

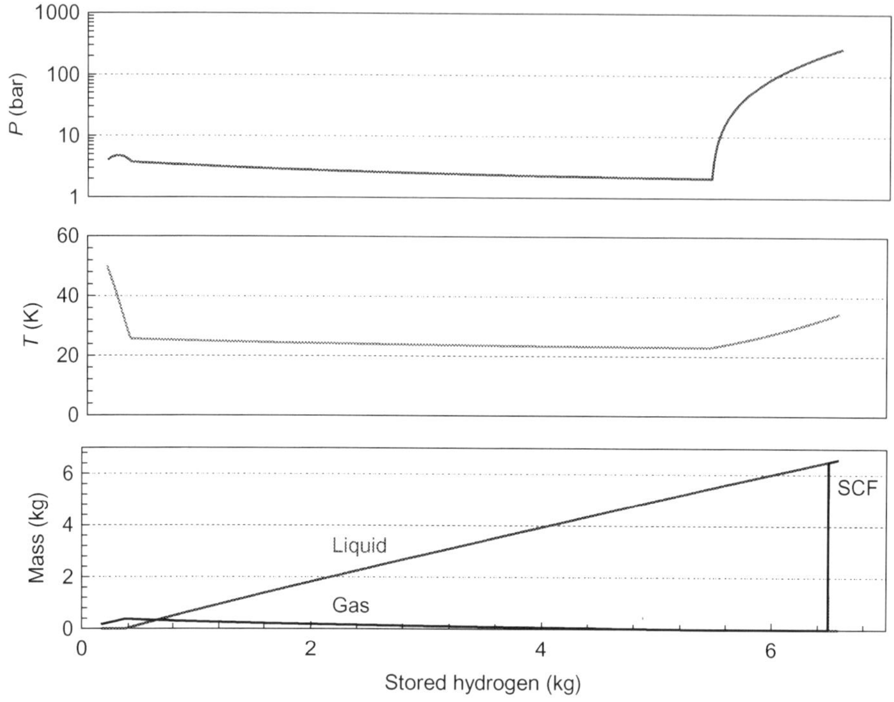

Figure 5.4 System conditions and the mass of liquid and gaseous H_2 in the tank during refueling under the cryo-compressed option. Stored amounts in excess of 6.5 kg result in the H_2 present as a supercritical fluid (SCF).

The system conditions and the mass of liquid and gaseous H_2 in the tank during refueling under the cryo-compressed option are shown in Figure 5.4. These results are based on initial tank conditions of 4 atm and 50 K at the start of the fueling operation. For a stored H_2 mass up to 0.4 kg, all of the H_2 is present as a gas. Initially, the tank temperature decreases toward 22 K and the pressure decreases below 4 atm as the LH_2 fed to the tank cools its contents. As the mass of stored H_2 increases above 0.4 kg, the liquid fraction increases and the gaseous fraction decreases, until at a stored amount of 5.4 kg, all of the hydrogen is present as a saturated liquid in the tank. With continued addition of pumped LH_2, the stored H_2 turns first into a subcooled liquid and then into a SCF when the stored mass exceeds 6.5 kg. The maximum storage capacity of the system is a function of the final pressure, being 5.7 kg at 37.7 atm and 6.6 kg at 272 atm.

5.3.2 Discharge dynamics

Figure 5.5 presents results from discharge simulations in which the initial amount of H_2 stored, pressure, and temperature correspond to the conditions after refueling as determined in Figure 5.4. The top plot in Figure 5.5 shows the amount of usable H_2, the heat input, and the final temperature after discharging a completely full tank at 272 atm down to the final pressure of 4 atm. The lower plot shows similar results for

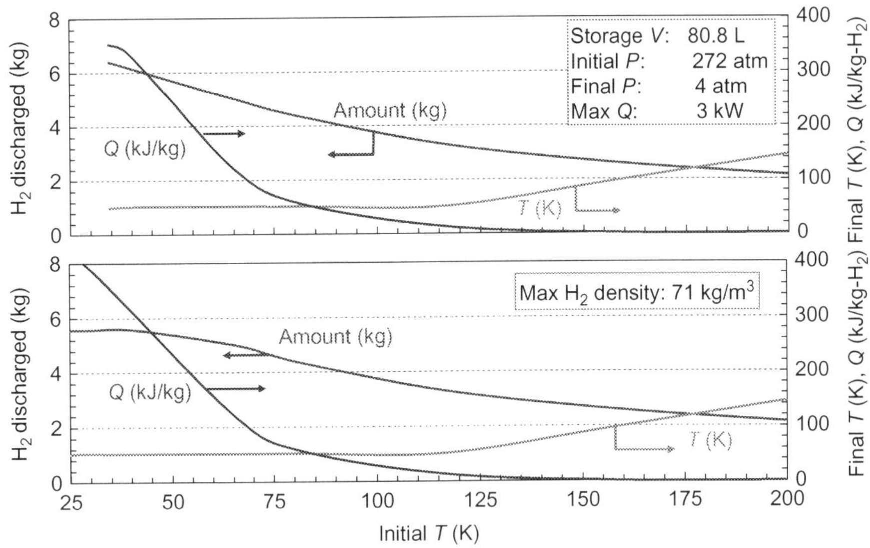

Figure 5.5 Effect of the initial tank temperature at the start of the fueling operation on the maximum amount of usable H_2, the final tank temperature, and the heat input required to maintain the minimum delivery pressure.

the case where the maximum H_2 density is limited to 71 kg/m^3, in which case the maximum amount of usable H_2 is 5.6 kg. In either case, no external heat input is needed if the initial tank temperature is greater than 155 K, for a maximum usable amount of 2.8 kg of H_2. In the first case, the total amount of heat required to discharge the entire 6.4 kg of H_2 is 2.3 MJ at a maximum heat input rate of 3 kW (max Q in Figure 5.6). The total amount of heat input required for the second case to discharge 5.6 kg of H_2 is 2.5 MJ for the same maximum heat input rate.

As the stored H_2 is discharged, the tank pressure, temperature, and the remaining mass of H_2 in the tank all change, as shown in Figure 5.6. This figure also shows the amount of thermal energy that must be provided to the tank (by the recirculation of warmed hydrogen through it) to maintain the 4 atm minimum pressure of the H_2 delivered to the fuel-cell power system. The curves in Figure 5.6 are for an initially full tank at 272 atm and 34.3 K, containing 6.6 kg of H_2.

As shown in Figure 5.6, the tank pressure decreases from 272 atm at the start of discharge to 4 atm when the remaining mass of H_2 decreases to 5.4 kg and the tank temperature drops to 23 K. With continued further withdrawal of H_2 from the tank, maintaining the 4-atm delivery pressure requires the addition of heat to the tank, as shown in the lower-middle plot. The tank temperature and pressure do not change as the H_2 in the tank is maintained in the saturated liquid–vapor form by the addition of ~340 J/g of H_2 withdrawn (~550 W at 1.6 g/s H_2 withdrawal rate), down to a remaining inventory of approximately 0.4 kg. The remaining H_2 exists as a gas, and further withdrawals require increasing heat input to maintain the 4-atm delivery pressure, which thermal energy requirement reaches a maximum of 3 kW.

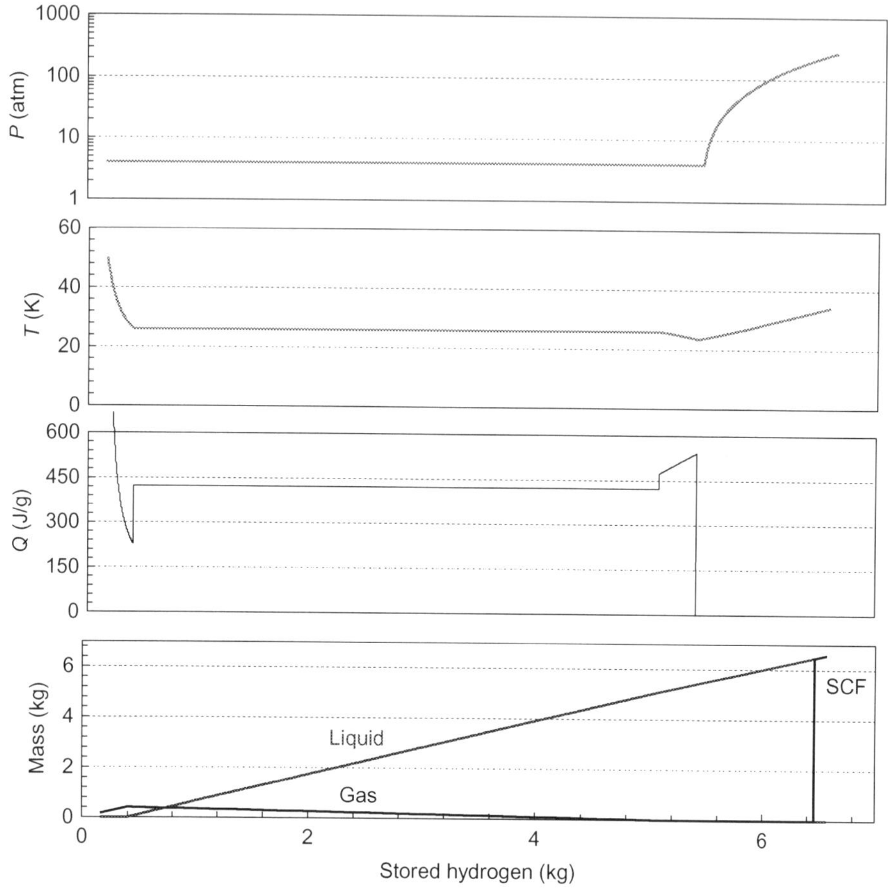

Figure 5.6 Pressure, temperature, and heat input profiles during the discharge of H_2 from an initially full tank at 272 atm, 34.3 K (supercritical fluid, SCF), containing 6.6 kg of H_2. The H_2 is withdrawn continuously at the 1.6 g/s full flow rate until the tank pressure drops below 4 atm.

Of the total inventory of H_2 in the tank, the fraction that is usable has been determined as a function of the total amount of H_2 contained in the tank. This usable fraction varies from a maximum of 97.6% to a minimum of 95.4%, and it is nearly the same whether the H_2 is initially stored as a cryo-compressed two-phase vapor–liquid mixture, or as a single-phase SCF.

5.3.3 Storage capacity

When filled with room temperature cH_2 rather than LH_2, the amount of H_2 that can be charged is a function of the tank temperature at the start of the filling operation. For this mode of refueling, the storage capacity was calculated for adiabatic refueling with compressed H_2 at 300 K and 272 atm, starting from an initial pressure of 4 atm, regardless of the initial temperature. Under these assumptions, Figure 5.7 shows the mass of H_2 that can be charged to the tank, and the final temperature of the tank,

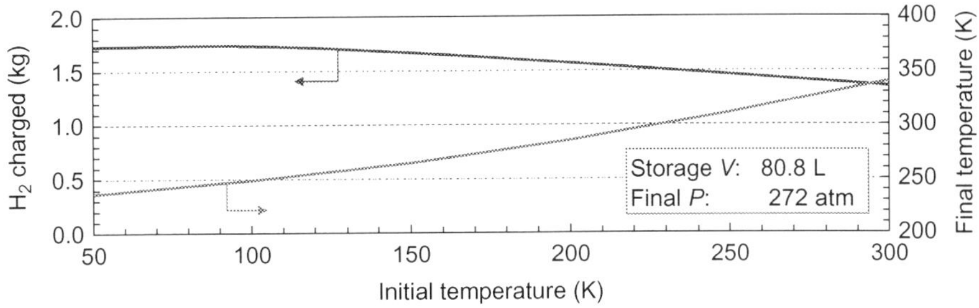

Figure 5.7 Effect of initial tank temperature on the maximum amount of room temperature cH_2 that can be charged into the tank and the final temperature of the tank, with an initial pressure of 4 atm and a final pressure of 272 atm.

as a function of the tank temperature at the start of the fueling operation. For an initial tank temperature of 300 K, ~1.4 kg of cH_2 can be charged into the scaled Gen-3 system, which would then correspond to a gravimetric H_2 storage capacity of 1.3 wt%. The maximum amount of cH_2 that can be charged is 1.7 kg if the initial tank temperature is <90 K, which corresponds to a gravimetric storage capacity of 1.7 wt%. A slightly greater amount of cH_2 can be charged into the tank if the H_2 is precooled to −40 °C (SAE TIR J2601, 2009) and if the tank is filled to a pressure higher than the nominal design maximum operating pressure of 272 atm.

As discussed above, the amount of LH_2 that can be charged to the system is also a function of the initial tank temperature. Figure 5.8 shows that starting from 300 K

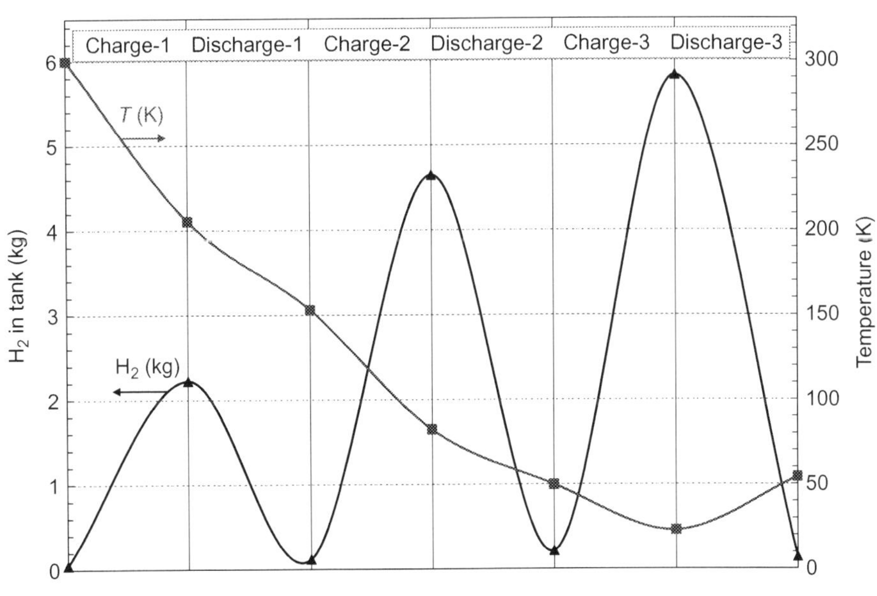

Figure 5.8 Refueling of tank with LH_2 starting from tank and liner at 300 K. Each fill is limited to a 272-atm maximum tank pressure or 5.7 kg stored H_2.

initial temperature, three LH_2 refuelings to rated pressure are needed to reach the full 71 g/L storage density. Over these three cycles, the liner temperature decreases during refueling as the tank is charged with LH_2 and during discharge as the gas undergoes expansion cooling. These results are consistent with the experimental data reported by Kircher et al. (2011) for refueling using a LH_2 cryopump for consecutive single-flow cryo-compressed refuelings at 100 kg/h starting with an empty tank at ambient temperature. He observed that the tank could be filled to its full capacity after five LH_2 refuelings as the H_2 storage density reached ~71 g/L and the liner temperature dropped from ambient to ~50 K.

5.3.4 Dormancy

Dormancy for cryo-compressed hydrogen storage is greatly extended over LH_2 storage at near ambient pressure for several reasons: (1) higher allowable pressure (>200 bar vs. 4 bar), (2) higher heat receptivity since H_2 is vented at higher temperature, (3) higher vessel thermal capacity due to the presence of the added carbon fiber overwrap, and (4) heat absorbed by para-to-ortho hydrogen conversion (Aceves et al., 2006, 2010; Peng and Ahluwalia, 2013). Liquid H_2 is 99.79% para-hydrogen (p-H_2) and 0.21% ortho-hydrogen (o-H_2) while at ambient temperature; the equilibrium mixture is 25% p-H2 and 75% o-H_2. The equilibrium concentration of p-H_2 and o-H_2 as a function of temperature is shown in Figure 5.1a. The conversion from p-H_2 to o-H_2 that occurs as the temperature rises is endothermic. As shown in Figure 5.1b, the heat of conversion is about 702 kJ/kg at 20 K and decreases to about 37 kJ/kg at 300 K (Woolley et al., 1948). For comparison, the heat of vaporization of H_2 is about 446 kJ/kg at its normal boiling point. The natural conversion of p-H_2 to o-H_2 is very slow, and it takes several days for the hydrogen to reach equilibrium (Matthews et al., 2011). Although the conversion time can be reduced to just minutes (Singleton and Lapin, 1965; Nast, 1983) in the presence of an appropriate catalyst, the analysis discussed here was limited to conversion of para-to-ortho hydrogen using natural kinetics.

The dormancy model was validated by using the data from a 35-day long test by Aceves et al. (2013), in which the 151-L tank was mounted onboard a vehicle, filled with liquid H_2 to 95% capacity and let sit dormant, exposed to daily sun. Consistent with an average heat gain of 4.5 W, the tank warmed up and the pressure reached 345 bar in 8.25 days, at which point 1.2 kg of H_2 were intentionally vented to reduce the pressure to 207 bar and cool the remaining H_2 from ~70 to ~60 K. As shown in Figure 5.9, the subsequent test consisted of leaving the tank onboard the vehicle for 27 additional days, with periodic venting of the H_2 to relieve the pressure.

A parametric study was conducted to determine dormancy and H_2 loss in the 151-L tank. Four cases were run in which the tank initially contained 100%, 75%, 50%, and 35% of the rated capacity taken as 10.2 kg H_2, which corresponds to filling 95% of the tank volume with liquid H_2. Thus, the initial densities of H_2 in the tank were 67.5, 50.7, 33.8, and 23.6 kg/m^3. For ease of discussion, the initial gas temperature was fixed at 50 K so that the initial pressures were 206, 91, 50, and 36 bar for the four cases. All simulations were run for an ambient temperature of 300 K and 340 bar as the vent relief pressure.

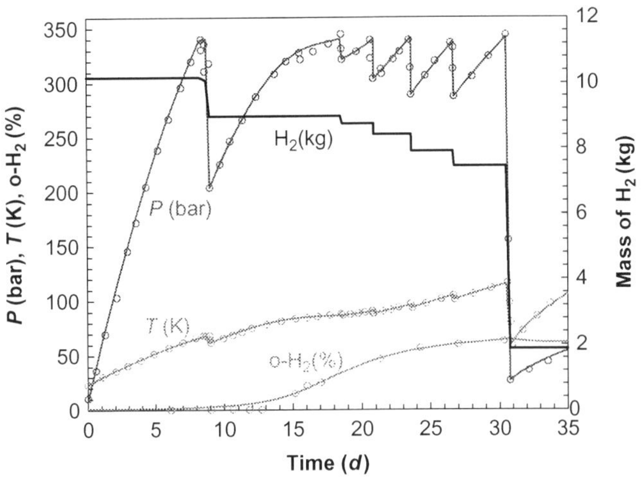

Figure 5.9 Validation of the dormancy model against the long-term exposure data from Aceves et al. (2013). The amount of H_2 vented at discrete times was estimated from the measured temperatures and pressures. The solid lines are the model results and the symbols are the measured data.

Figure 5.10 shows the quantitative effects of endothermic heat absorption on dormancy, defined as the time before any hydrogen needs to be vented to limit the maximum pressure to the 340-bar allowable limit for the 151-L Gen-3 tank. Two sets of results are included in Figure 5.10a: the solid lines represent the dormancy with para-to-ortho conversion and the dashed lines are the results with the conversion kinetics artificially frozen ($k=0$). The dormancy time is about 4.5 days if a 100%-full tank is exposed to 300-K ambient (case 1), and the gas temperature is only 70 K when the tank reaches the 340-bar allowable pressure. In this simulation, the dormancy is not affected by the para-to-ortho conversion, which occurs mainly after the tank has reached the allowable pressure limit. The dormancy time increases to 24 days, with 85% enhancement by the para-to-ortho conversion, if the tank is initially 75% full. The dormancy time further increases to 42 days, with the enhancement by the para-to-ortho conversion dropping to 28%, if the tank is initially 50% full. Further reduction in the initial H_2 in the tank leads to significantly improved dormancy but smaller enhancement by the para-to-ortho conversion.

The role of endothermic para-to-ortho conversion is seen clearly in Figure 5.10b, which shows the dormancy on the basis of the cumulative heat in-leakage (1 W d = 86 kJ) for both para and normal H_2. The difference in dormancy for the two forms of H_2 is the para-to-ortho heat of conversion that depends on the initial amount of H_2 in the tank and on the extent to which the conversion completes prior to reaching the maximum allowable pressure. The results indicate that the extent of conversion (defined as o-H_2 concentration divided by o-H_2 concentration in n-H_2) during the dormancy period is <1%, 84%, 98%, and 100% in the four cases with 100%, 75%, 50%, and 35% full tanks. The maximum amount of heat of conversion is absorbed in the case in which the tank is initially 75% full.

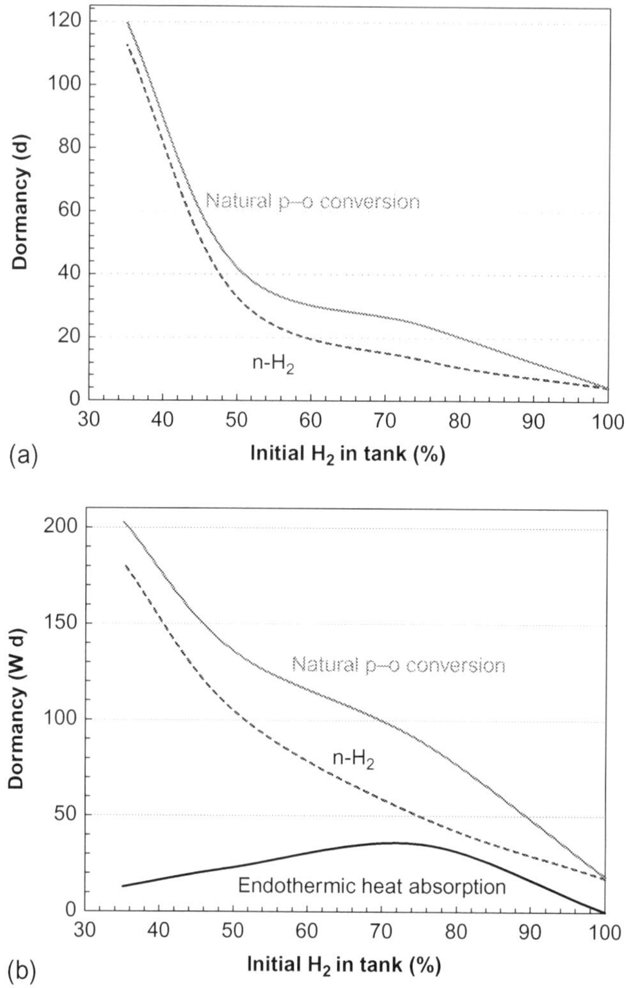

Figure 5.10 Effect of initial amount of H_2 and para-to-ortho conversion on dormancy: (a) Dormancy expressed as time before the tank pressure reaches the allowable limit; (b) dormancy expressed as cumulative heat in-leakage. Included for comparison are results for n-H_2 with frozen gas composition (no isomer transitions).

Figure 5.11 shows the effect of para-to-ortho conversion on the rate of hydrogen loss after the dormancy period. Here, the loss rate is normalized by the H_2 storage capacity of the tank (10.2 kg) and the time is normalized by the length of the dormancy period (t_d). Para-to-ortho conversion has the most pronounced effect on hydrogen loss when the tank is initially 100% full. As discussed above, most of the conversion occurs in this case after the dormancy period. We see an almost 75% reduction in H_2 loss rate because of conversion at $t=t_d$ and a subsequent rapid decline in H_2 loss rate. The

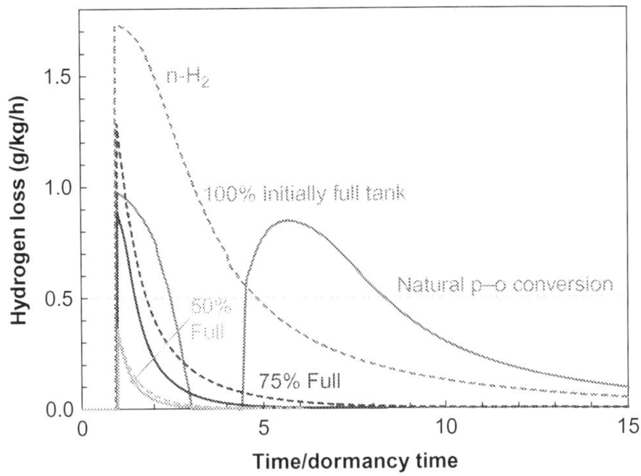

Figure 5.11 Effect of initial amount of H_2 and para-to-ortho conversion on hydrogen loss after the tank pressure reaches the allowable limit. The solid lines represent results assuming that at $t=0$ the gas has composition of liquid H_2 at 20 K. The dashed lines are results for n-H_2 with fixed composition of isomers.

conversion rate peaks at t/t_d of about 3.4 and leads to sufficient heat absorption that the pressure actually decreases for an intermediate period of time and a second dormancy period appears between t/t_d of 2.9 and 4.2. The effect of conversion on H_2 loss rate is less dramatic, although still significant, if the tank is initially 75% full. The effect of conversion on H_2 loss rate is insignificant if the tank is initially <50% full because H_2 has already reached its normal composition during the dormancy period.

5.4 Well-to-tank efficiency

We have evaluated the fuel cycle to support refueling the cryo-compressed H_2 storage system for automotive applications. These off-board assessments make use of existing, publically available models to calculate the performance of the hydrogen fuel cycle on a consistent basis: H2A for H_2 production efficiencies, and Hydrogen Delivery Scenarios Analysis Model (HDSAM) for delivery efficiencies and losses. Details of these models can be found elsewhere (Wang et al., 2007; Ahluwalia et al., 2007; Hydrogen Delivery Scenario Analysis Model (HDSAM), 2009; Hydrogen Production Model (H2A), 2009). The analysis is for 40% H_2 market penetration in a mid-size city—Sacramento, CA. In this scenario, the H_2 demand is about 270,000 kg/day for about 488,000 fuel cell vehicles in the city. To serve this market, a total of 269 refueling stations are needed, where each station has a storage capacity of ~7000 kg and dispenses an average of 1000 kg H_2/day. The vehicles are assumed to have an average fuel economy of 63.4 mpgge (mile per gallon gasoline equivalent), typical for a 2015 mid-sized fuel cell vehicle, and an annual mileage of 12,000 miles. Also in this scenario, H_2

Table 5.2 Assumptions for the well-to-tank (WTT) efficiency calculations

Process/process fuels	Nominal values	Source/comment
Electricity production	32.2% thermal efficiency	EIA projected U.S. grid 2015, inclusive of 8% transmission loss from power plant to user site
North American natural gas production	93.5% efficiency	GREET data
H_2 production by SMR	73% efficiency	H2A
H_2 liquefaction	8.2 kWh/kg	HDSAM, 150 tons/day liquefier
Liquid H_2 delivery by truck	284 km round trip	HDSAM
Truck capacity	4300 kg	HDSAM
Boil-off losses	9.5%	HDSAM liquefaction 0.5%, storage 0.25%/day, loading 0.5%, unloading 2%, cryo-pump 3%
Vehicle refueling with LH_2	2 kg/min. 80% isentropic efficiency	LH_2 pump data
Greenhouse gas emissions	Range	Emission factors from GREET

is produced at a central plant by steam reforming of natural gas without CO_2 sequestration. The LH_2 terminal stores a 10-day reserve to accommodate scheduled and unplanned plant outages. Additional design assumptions and details are given in Table 5.2.

The analysis assumed 93.5% efficiency for delivery of natural gas from the production well to the central plant and 73% efficiency for producing fuel cell quality hydrogen by SMR at the central plant (see Table 5.2 for a summary and bases for all assumptions). The analysis considered that H_2 liquefaction at the central plant consumes 8.2 kWh of electricity per kg of H_2, and that LH_2 is delivered to the refueling stations by 4300-kg tankers (4100-kg refueling capacity). The analysis includes 9.5% H_2 loss from central plant to vehicle including losses during liquefaction, LH_2 storage at the terminal and fueling station, loading of tankers at the terminal, unloading of tankers at the fueling stations, and pumping of LH_2 at the stations.

The pathway assumed that the electricity used in the H_2 production, delivery, and dispensing process is generated using the U.S. Energy Information Administration (EIA) projected 2015 grid mix at 32.2% efficiency, inclusive of 8% transmission losses from the power plant to the central H_2 production and liquefaction plant. Using these assumptions, we estimated that the WTT efficiency for LH_2 refueling of the cryo-compressed systems is 41.1%, based on the lower heating values of the H_2 delivered to the tank and the feedstock natural gas consumed in the process. Figure 5.12 indicates that the WTT efficiency of the cryo-compressed option is lower than the efficiency of the compressed gas storage options but is higher than the efficiencies of several material-based storage options.

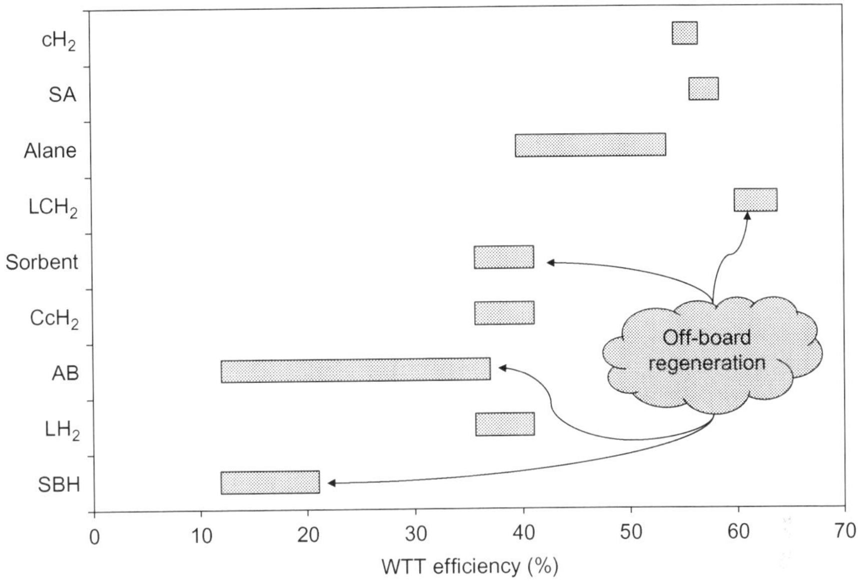

Figure 5.12 Well-to-tank efficiency of various physical- (cH_2, CcH_2, and LH_2) and material-based hydrogen storage options (Ahluwalia et al., 2012): sodium alanate metal hydrides (SA), alane (AlH_3), cryo-sorbents, ammonia borane (AB), and sodium borohydride (SBH).

5.5 Assessment of cryo-compressed hydrogen storage and outlook

Table 5.3 compares some key performance characteristics for compressed gas, liquid, and cryo-compressed H_2 storage options. The key advantages and disadvantages of the cryo-compressed storage are summarized below.

5.5.1 Gravimetric capacity

The Gen-3 cryo-compressed system scaled to 5.6 kg of usable H_2 (using the LH_2 fueling option) has a nominal usable gravimetric capacity of 5.5 wt% at 71 kg/m^3 H_2 density. The theoretical usable capacity is 6.5 wt% if credit is taken for LH_2 compressibility and the tank is refueled to the design pressure of 272 atm and 81 kg/m^3 H_2 density. The nominal capacity increases to 6.2 wt% if the liner thickness can be reduced to 4.3 mm from 9.5 mm in the current design. The nominal capacity further increases to 9.2 wt% if the shell is made of an Al alloy rather than steel. Thus, the cryo-compressed option easily exceeds the near-term target of 5.5 wt% without any changes and can also meet the ultimate target of 7.5 wt% since there is no technical risk in substituting the shell material with a lighter-density alloy that is only required to withstand the vacuum.

Table 5.3 **Summary results for physical storage options**

Performance and cost metric	Units	cH_2 350 bar	cH_2 700 bar	LH_2	CcH_2	2017 targets	Ultimate targets
Tank		1-Tank	1-Tank				
Usable storage capacity (nominal)	kg-H_2	5.6	5.6	5.6	5.6		
Usable storage capacity (maximum)	kg-H_2	5.6	5.6	5.6	6.6		
System gravimetric capacity	wt%	5.4	4.4	5.6	5.5–9.2	5.5	7.5
System volumetric capacity	kg-H_2/m^3	17.8	25.1	23.5	41.8–44.7	40	70
Storage system cost	$/kWh	16	19	TBD	12	12	8
Fuel cost	$/gge	4.2	4.3	TBD	4.80	2–3	2–3
Cycle life (1/4 tank to full)	Cycles	NA	NA	NA	5500	1500	1500
Minimum delivery pressure, FC/ICE	atm	4	4	4	3–4	5/35	3/35
System fill rate	kg-H_2/min	1.5–2	1.5–2	1.5–2	0.5–2	1.5	2.0
Minimum dormancy (full tank)	W d	NA	NA	2	4–30		
H_2 loss rate (maximum)	g/h/kg-H_2	NA	NA	8	0.2–1.6	0.05	0.05
WTT efficiency	%	56.5	54.2	22.3	41.1	60	60
GHG emissions (CO_2 eq)	kg/kg-H_2	14.0	14.8	TBD	19.7		
Ownership cost	$/mile	0.13	0.14	TBD	0.12		

5.5.2 Volumetric capacity

The scaled Gen-3 system has a nominal volumetric capacity of 41.8 g-H_2/L The theoretical volumetric capacity is 47.8 g-H_2/L if credit is taken for LH_2 compressibility and the tank is refueled to the design pressure of 272 atm. The nominal capacity increases to 43.2 g-H_2/L if the liner thickness can be reduced to 4.3 mm from 9.5 mm in the current design. Thus, the scaled Gen-3 system exceeds the near-term target of 40 g-H_2/L without any changes, but cannot satisfy the ultimate target of 70 g-H_2/L even with the credits and modifications considered in this assessment.

5.5.3 Carbon fiber and system cost

A major challenge for the 700-bar compressed hydrogen storage option is the cost, mainly due to the expensive aerospace grade carbon fiber (Toray T700), which accounts for ~75% of the vessel cost and ~63% of the total system cost (Hua et al., 2011; Ahluwalia et al., 2010). In this regard, the cryo-compressed option offers a significant advantage over the 700-bar cH_2 option as the amount of carbon fiber required is reduced by ~85%. The lower cost in carbon fiber is partially offset by the added cost for tank insulation.

5.5.4 Well-to-tank efficiency

Because of the energy consumed in liquefying hydrogen, the WTT efficiency for cryo-compressed storage is only 41.1%, which is more than 13% points lower than that for compressed gas systems. At best, the WTT efficiency may approach 45% if the boil-off losses are greatly reduced or eliminated.

5.5.5 Dormancy and H_2 loss rate

The dormancy of the cryo-compressed tank and H_2 loss rate are strong functions of the initial amount of hydrogen in the tank, initial temperature, and the relief pressure. The dormancy ranges from 52 to 76 W d for a 50%-full tank and 4 to 30 W d for a 100% full tank. The rate of H_2 loss from the system once the dormancy is exceeded has been determined as a function of the amount of H_2 stored in the tank. The maximum H_2 loss rate varies from 0.4 to 2.1 g/h/W, while the average H_2 loss rate ranges from 0.2 to 0.85 g/h/W. There is no venting if the tank contains less than ~1.5 kg of H_2 or if the vehicle is driven daily for a certain minimum distance (Aceves et al., 2006, 2010, 2013; Matthews et al., 2011).

Acknowledgments

This work was supported by the Fuel Cell Technologies Office of the U.S. Department of Energy's (DOE) Office of Energy Efficiency and Renewable Energy. Ms. Grace Ordaz was the DOE technology development manager for this work. Argonne is a DOE, Office of Science Laboratory operated under Contract No. DE-AC02-06CH11357 by UChicago, Argonne, LLC.

References

Aceves, S., Berry, G., Martinez-Frias, J., Espinosa-Loza, F., 2006. Vehicular storage of hydrogen in insulated pressure vessels. Int. J. Hydrog. Energy 31, 2274–2283.

Aceves, S., Espinosa-Loza, F., Ledesma-Orozco, E., Ross, T., Weisberg, A., Brunner, T., Kircher, O., 2010. High-density automotive hydrogen storage with cryogenic capable pressure vessels. Int. J. Hydrog. Energy 35, 1219–1226.

Aceves, S., Petitpas, G., Espinosa-Loza, F., Matthews, M., Ledesma-Orozco, E., 2013. Safe, long range, inexpensive and rapidly refuelable hydrogen vehicles with cryogenic pressure vessels. Int. J. Hydrog. Energy 38, 2480–2489.

Ahluwalia, R., Peng, J., 2008. Dynamics of cryogenic hydrogen storage in insulated pressure vessels for automotive applications. Int. J. Hydrog. Energy 33, 4622–4633.

Ahluwalia, R., Hua, T., Peng, J., 2007. Fuel cycle efficiencies of different automotive onboard hydrogen storage options. Int. J. Hydrog. Energy 32, 3592–3602.

Ahluwalia, R., Hua, T., Peng, J., Lasher, S., McKenney, K., Sinha, J., Gardiner, M., 2010. Technical assessment of cryo-compressed hydrogen storage tank systems for automotive applications. Int. J. Hydrog. Energy 35, 4171–4184.

Ahluwalia, R., Hua, T., Peng, J., 2012. Onboard and off-board performance of hydrogen storage options for light-duty vehicles. Int. J. Hydrog. Energy 37, 2891–2910.

Barron, R., 1996. Cryogenic Systems. McGraw Hill Book Company, New York, NY.

DOE Targets for Onboard Hydrogen Storage Systems for Light-Duty Vehicles, 2009, published on DOE/FCT website: http://www1.eere.energy.gov/hydrogenandfuelcells/storage/pdfs/targets_onboard_hydro_storage.pdf.

Gear, C.W., 1971. Numerical Initial Value Problems in Ordinary Differential Equations. Prentice Hall, Englewood Cliffs, NJ.

Hua, T., Ahluwalia, R., Peng, J., Kromer, M., Lasher, S., McKenney, K., Law, K., Sinha, J., 2011. Technical assessment of compressed hydrogen storage tank systems for automotive applications. Int. J. Hydrog. Energy 36, 3037–3049.

Hydrogen Delivery Scenario Analysis Model (HDSAM), 2009. http://www.hydrogen.energy.gov/h2a_delivery.html.

Hydrogen Production Model (H2A), 2009. http://www.hydrogen.energy.gov/h2a_production.html.

Kircher, O., Greim, G., Burtscher, J., Brunner, T., 2011. Validation of cryo-compressed hydrogen storage (CcH2): a probabilistic approach. In: Proceeding of the Int. Conference on Hydrogen Safety, September 12–14, 2011, San Francisco, USA.

Lemmon E.W., Huber M.L., McLinden M.O., 2010. NIST reference fluid thermodynamic and transport properties database (REFPROP): Version 9.0.

Matthews, M., Petitpas, G., Aceves, M., 2011. A study of spin isomer conversion kinetics in supercritical fluid hydrogen for cryogenic fuel storage technologies. Appl. Phys. Lett. 99, 081906.

McCarty, R.D., Hord, J., Roder, H.M., 1981. Selected Properties of Hydrogen (Engineering Design Data) NBS Monograph 168. U.S. Government Printing Office, Washington, DC.

Milenko, Y., Sibileva, R., Strzhemechny, M., 1997. Natural ortho-para conversion rate in liquid and gaseous hydrogen. J. Low Temp. Phys. 107, 77–92.

Nast, T., 1983. Investigation of a para-ortho hydrogen reactor for application to spacecraft sensor cooling. Lockheed Palo Alto Research Laboratories, Report, LMSC-D877499.

Peng, J., Ahluwalia, R., 2013. Enhanced dormancy due to para-to-ortho hydrogen conversion in insulated cryogenic pressure vessels for automotive applications. Int. J. Hydrog. Energy 38, 13664–13672.

Sae, T.I.R., 2009. J2601 Technical Information Report for Fueling Protocols for Light Duty Gaseous Hydrogen Surface Vehicles. Society of Automotive Engineers, USA.

Singleton, A., Lapin, A., 1965. Design of para-ortho hydrogen catalytic reactors. Adv. Cryog. Eng. 11, 617–630.

Thesken, J., Murthy, P., Phoenix, S., Greene, N., Palko, J., Saulsberry, R., Beeson, H., 2009. A theoretical investigation of composite overwrapped pressure vessels (COPV) mechanics applied to NASA full scale tests. NASA Report NASA/TM-2009-215684.

Wang, M., We, Y., Elgowainy, A., 2007. Operating Manual for GREET: Version 1.7. Argonne National Laboratory Report ANL/ESD/05–3.

Wolf, J., 2002. Liquid hydrogen technology for vehicles. MRS Bull. 27, 684–687.

Woolley, H., Scott, R., Brickwedde, F., 1948. Compilation of thermal properties of hydrogen in its various isotopic and ortho-para modifications. J. Res. Natl. Bur. Stand. (U.S.) 41, 379.

Adsorption of hydrogen on carbon nanostructure

6

P. Bénard, A.-M. Beaulieu, D. Durette, R. Chahine
Institut de recherche sur l'hydrogène, Université du Québec à Trois-Rivières, Trois-Rivières, QC, Canada

Abbreviations

GCMC grand canonical Monte Carlo
MOF metal–organic framework
PIGCMC path integral grand canonical Monte Carlo
SWNT single-wall nanotubes
Wt indicates per weight of adsorbate over the weight of the adsorbent

6.1 Introduction

The use of hydrogen as a source of energy for vehicular and large-scale stationary applications requires high volumetric and gravimetric energy densities. Current commercial storage technologies include storage in tanks as a highly compressed gas (typically 700 bars at ambient temperature) or as a cryogenic liquid. For applications requiring high volumetric efficiency with little gravimetric constraints (such as portable applications), solid-state storage is emerging. Such systems are based in the absorption of hydrogen in a metal (forming a metal hydride), in which atomic hydrogen forms a solid-state solute.

Despite their maturity, compression and liquid storage present challenges in terms of their capacity for improvement in storage density. In order to facilitate the penetration of hydrogen in transportation applications, the U.S. Department of Energy has set stringent short-term targets for (among other things) the gravimetric and volumetric hydrogen storage capacity of light-duty vehicle that would ensure a driving range of at least 500 km without compromising safety, pace, filling time, performance, or cost (Stetson, 2014). Although driving ranges of 500 km have been achieved by hydrogen-powered fuel-cell vehicles, this range, according to the DOE, should be achievable across the full range of vehicle models. The 2017 targets, based on the lower heating value of hydrogen, are (U.S. Department of Energy): 1.8 kWh/kg (5.5 wt%) and 1.3 kWh/L (0.04 kg/L). The ultimate targets are 2.5 kWh/kg (7.5 wt%) and 2.3 kWh/L (0.07 kg/L).

For compression storage, the maximum storage system densities have been achieved at 350 (gravimetric) and 700 bars (volumetric), respectively (Satyapal, 2014), using Type IV pressure vessels. For 700 bars, energy densities of 1.5 kWh/kg and 0.8 kWh/L have been obtained. Lower system weight results in a larger gravimetric energy density of 1.8 kWh/kg for 350 bar, with, however, a smaller volumetric density of 0.6 kWh/L.

Compendium of Hydrogen Energy. http://dx.doi.org/10.1016/B978-1-78242-362-1.00006-7

For liquid hydrogen, densities of 1.7 kWh/kg (gravimetric) and 1.2 kWh/L (volumetric) can currently be achieved (Michel et al., 2006). Liquid hydrogen is ultimately limited by its intrinsic volumetric density of 0.07 kg/L, which prevents it from reaching the ultimate DOE volumetric target when the storage system is taken into account.

Although great strides have been achieved in improving compression and cryogenic storage technologies, meeting the long-term DOE targets for vehicular storage requires going beyond conventional fluid-phase storage technologies. Metal hydrides can achieve volumetric storage densities that surpass the density of liquid hydrogen; however, they generally suffer from low gravimetric storage, which currently limits their use in applications such as portable applications, stationary storage, or vehicles used in underground mining operations where compressed hydrogen cannot be used. Another class of hydrides, called chemical hydrides, also meets several DOE storage targets; however, they cannot be regenerated onboard and the impact of their use on the overall energy efficiency of the storage process can be an issue.

The volumetric storage density of light gases such as hydrogen can be increased through attractive interactions between hydrogen molecules and the surface of an adsorbent. This process (adsorption) occurs within the porous structure of the adsorbent, defined as the volume externally accessible to the adsorbate molecules. For optimal storage, an adsorbent should have a highly microporous pore distribution, with characteristic sizes less than 2 nm, as hydrogen molecules larger mesopores and macropores do not benefit significantly from the local enhancement of their density through the adsorbate–adsorbent interactions. Gas storage applications thus usually require maximizing the microporous volume of an adsorbent. The equation of state of the adsorption process is the adsorption isotherm. The excess adsorption isotherm is usually used to characterize the uptake of hydrogen as a function of pressure and temperature. It is defined as the difference between the amount of hydrogen in the porous structure and the amount of hydrogen that would be present in the same volume in the absence of the adsorbent. Monte Carlo simulations suggest that the adsorption process in highly microporous adsorbents such as MOF 5 and CuBTC may result in average fluid–phase densities above the liquid density of hydrogen (70.97 g/L) in the supercritical state between 50 and 80 K Durette (2012) at high pressure. Although there are indications that some MOF adsorbents can achieve the storage targets on a materials basis (Zacharia et al., 2010), carbon nanostructures have yet to offer both gravimetric and volumetric densities that meet the 2017 targets for small vehicles. They could, however, be used in transportation applications where volumetric considerations are less of an issue but where weight and safety constitute an important factor, such as in public transportation.

Operating conditions for materials-based hydrogen storage systems depend on the characteristic energy binding hydrogen to the substrate. Hydrogen binding in chemical hydride is typically greater than 100 kJ/mol, compared with 50–100 kJ/mol in metal hydrides, and 5–15 kJ/mol for adsorption storage. The characteristic binding energy determines the temperature scale required to store and release hydrogen from storage. Materials with hydrogen binding energies larger than 50 kJ/mol require heating power to release hydrogen from storage. Materials with binding energies smaller than 10 kJ/mol require cryogenic operation to achieve acceptable storage densities.

A target range of 10–50 kJ/mol is generally aimed for to achieve acceptable operating temperatures. The binding energies also set the boil-off properties of materials-based hydrogen storage systems.

In this review, we will focus on the materials-based storage of molecular hydrogen on ordered and disordered carbon nanostructured materials. Following a brief discussion of the general properties of carbon nanostructures from the perspective of storage applications (Section 6.2), we discuss the physiorption of hydrogen on single-wall nanotubes (SWNTs) and C60 fullerenes (Section 6.3), activated carbons (Section 6.4), layered graphene (Section 6.5), and finally zeolite-templated carbons (ZTC) (Section 6.6).

6.2 General considerations for physisorption of hydrogen on carbon nanostructures

Because of the hybridization of the partially filled sp^2 orbitals of the carbon atom, the covalent binding of carbon atoms is possible through single, double, and triple bounds, leading to a diverse family of materials called carbon nanostructures. In addition to catenation, carbon atoms can form pentagonal and heptagonal arrangements of carbon atoms branch out. This flexibility allows the existence of complex two- and three-dimensional nanostructures such as graphene and diamond, which lead to the formation of graphitic nanofibers and caged fullerenes such as C_{60} buckyballs, SWNT, and multiwall nanotubes (MWNT). This flexibility of carbon, in addition to the strong C—C covalent bound, its low atomic weight, and its propensity to bind chemically or physically to other atoms or molecules, makes the resulting carbon nanostructures highly desirable for storage applications. The capacity to bind with three neighboring atoms leads to the formation of graphene layers, in which carbon atoms are arranged hexagonally and separated by 1.41 Å. Graphene planes can then be bound in layers through van der Waals interactions, leading to ordered graphitic structures. The maximum storage capacity by chemisorption in a graphene layer is 8.3%, corresponding to a fully saturated graphene layer with one hydrogen atom per carbon atom (graphane) (Tozzini and Pellegrini, 2013). Graphene can also be shaped into tubular structures called carbon nanotubes, which can be capped by a pentagon, thus forming a cage. The curvature of carbon nanostructures determines the degree of overlap of the Lennard-Jones interaction between individual carbon atoms and an adsorbate molecule, and thus the strength of the overall interaction between the nanostructure and the molecule at a given location close to its surface. For instance, the cylindrical interaction potential (which neglects corrugation effects) used by Stan and Cole (1998a,b) to study the second virial coefficient of the adsorption isotherm of a SWNT shows a significant enhancement of the adsorption inside the SWNT with respect to an equivalent graphene sheet and the external surface of the nanotube, due to the curvature of its surface. This leads to a maximum characteristic interaction energy for hydrogen molecules inside a SWNT of 12.5 kJ/mol for a nanotube radius of 3.43 Å (Chapitre de livre). This radius is close to the size of the smallest freestanding SWNT, which is about 4 Å (Guan et al., 2008), comparable to a C_{60} buckyball, suggesting that

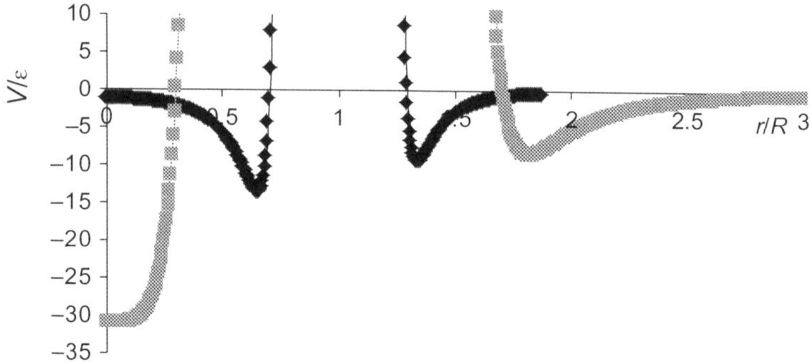

Figure 6.1 Physisorption interaction potential between a SWNT and a helium atom as a function of reduced distance to the center of the nanotube. The distance is normalized by the radius R of the nanotube. The square and the diamond symbols correspond to a nanotube radii of 3.3 and 8 Å, respectively.
Model parameters used are from Stan and Cole (1998a).

the maximum enhancement of the characteristic energy of adsorption due to curvature inside a carbon nanostructure would be of that order. Figure 6.1 shows the effect of the curvature of the surface on the potential depth inside and outside a SWNT.

The situation, however, becomes significantly more complicated when the effect of curvature on carbon bounds are considered. It then appears that convex surfaces lead to increased binding of atomic hydrogen, resulting in increased binding in the external surface of nanotubes compared to flat graphene, acting in the opposite way as it does for physisorption of molecular hydrogen. Convexities seem to stabilize chemisorbed H and slightly destabilize physisorbed H_2 (Tozzini and Pellegrini, 2013).

6.3 Carbon nanotubes and fullerenes

Carbon nanotubes are fullerene nanostructures (Iijima, 1991) that can best be described as a graphene layer folded in such a way as to preserve its periodic structure along its axis. The folding results in a cylindrical carbon allotrope characterized by two integers (n,m) associated with the linear combination of the two unit vectors that project the diameter of the nanotube on a graphene plane. SWNTs are generally classified in three categories: armchair $(n=m)$, zigzag $(m=0)$, and chiral (otherwise). Nanotubes form filaments with extreme width-to-length ratios are typically capped by a carbon fullerene half-shell.

Carbon nanotubes are often found in bundles or ropes of ~100 nanotubes (Figure 6.2) (Mélançon and Bénard, 2004; Stan et al., 2000), arranged in a triangular lattice. A typical SWNT bundle adsorbs preferentially at three locations: the internal volume of the SWNTs, the interstices between the nanotubes forming the lattice, and the grooves on the external surface of the bundle (Figure 6.3).

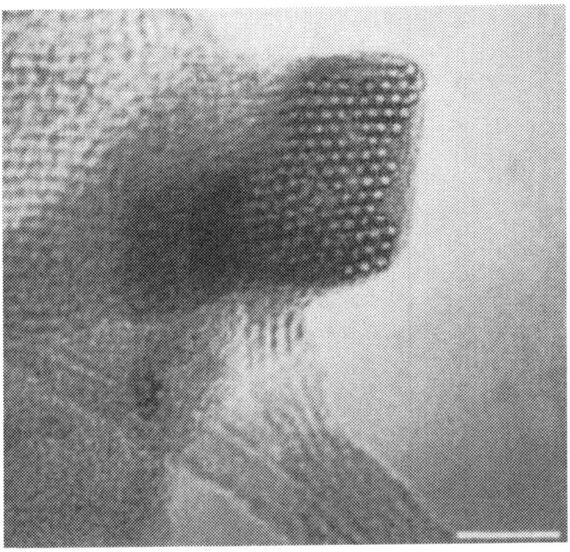

Figure 6.2 Cross-section of a bundle of SWNTs forming a triangular lattice with a lattice parameter of ~17 Å. The reference bar at the bottom left corresponds to 10 nm. Image from Figure 1(c) of Thess et al. (1996).

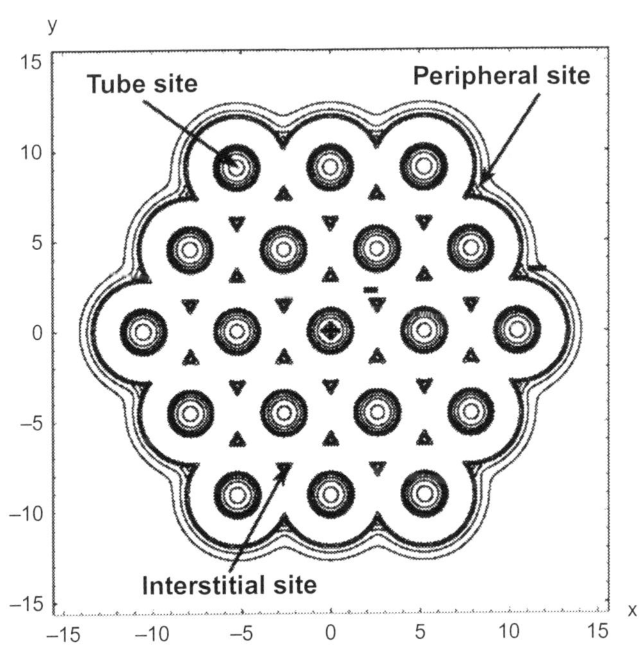

Figure 6.3 Adsorption sites in bundles of single wall nanotubes. Reprinted with permission from Mélançon and Bénard (2004).

The hydrogen adsorption capacity of SWNTs was first investigated by Dillon et al. (1997) using a 1 mg soot sonicated sample containing 0.1–0.2 wt% 12 Å diameter SWNT bundles of 7–14 nanotubes using temperature programmed desorption (TPD). Due to the concave curvature of the nanotubes, an enhancement of the adsorption effect is expected inside the nanotubes and in the interstices between nanotubes forming a bundle, provided the hydrogen molecules can access those sites. Adsorption measurements by TPD were performed over the temperature range 80–500 K and the pressure range 0.03–0.4 bar. The characteristic energy of adsorption was estimated to be 19.6 kJ/mol. This lead to an estimate of 5–10 wt% for the hydrogen storage capacity at room temperature in pure SWNTs. Improved measurement techniques (Poirier et al., 2006) and larger sample sizes later showed that the uptake of high-grade hydrogen did not exhibit a substantial improvement over less costly high-performance activated carbons at higher pressures. Under ambient conditions, it has now been established that pure SWNT adsorb less than 1 wt% (Becher et al., 2003). Under cryogenic conditions (at 77 K and 1 atmosphere), excess adsorbed densities of up to 2.5 wt% have been obtained (Anson et al., 2004; Nishimiya et al., 2003; Poirier et al., 2006; Tarasov et al., 2003). At higher pressures, uptakes of 6 wt% at 77 K and 2 bars (Pradhan et al., 2002) and of 8 wt% at 40 bars and 80 K have been reported (Ye et al., 1999). Because of the structure of SWNTs, synthesis and sample preparation can have a significant impact on their adsorption properties. Access to the internal volume of the nanotubes provides additional surface area and enhanced interactions due to curvature. Chemical etching can increase access to the internal structure of the nanotubes where deep potential wells can be found. Thermal treatments can improve uptake (Kuznetsova et al., 2000) by removing blocks to internal adsorption sites and grooves in nanotube bundles. For example, hydrogen uptake in a SWNT sample could be increased to 4.6 wt% after being processed with fluorhydic acid and a heat treatment (Lafi et al., 2005).

Due to their monoatomic, regular structure, the sorption properties of SWNT bundles can be predicted using statistical physics simulation methods such as classical grand canonical Monte Carlo (GCMC) and quantum path integral grand canonical Monte Carlo (PIGCMC) or molecular dynamics. Classical simulations, at low temperatures ($T < 100$ K), will overestimate the uptake, must be seen as yielding upper limits on hydrogen uptake as a function of pressure and temperature. Classical and quantum simulations study of the optimal configuration of compact-bundled SWNTs show that maximum excess uptake occurs for SWNTs with diameters of 12–13 Å and a van der Waals gap of 5–6 Å (Lafi et al., 2005; Lévesque et al., 2002; Wang and Johnson, 1999a). Figure 6.4 shows the excess uptake isocontours as a function of the van der Waals gap (distance between nanotubes) and the SWNT diameter. The excess adsorbed density and the specific surface areas are calculated at 1 atmosphere and 77 K using classical GCMC as a function of nanotube diameter and bundle lattice parameter. The optimal parameters for hydrogen adsorption correspond to the peak structure in the contour plot, for a diameter of 13.5 Å (which would occur for a (10,10) armchair SWNT) separated by a van der Waals gap of 6 Å.

The corresponding adsorption isotherm would exhibit a maximum excess density of 5% (relative to the mass of the adsorbent) at 25 bars at 77 K when calculated

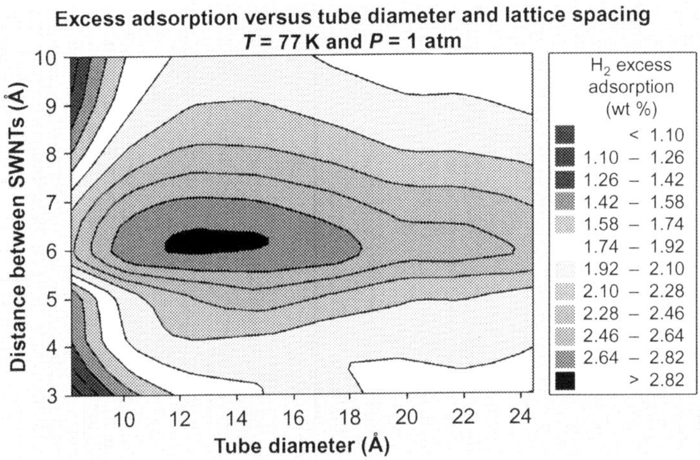

Figure 6.4 Contour plot of excess density adsorbed as a function of SWNT diameter and lattice spacing at 77 K and 1 atm (bundle of seven units), from Lachance and Bénard (2007).

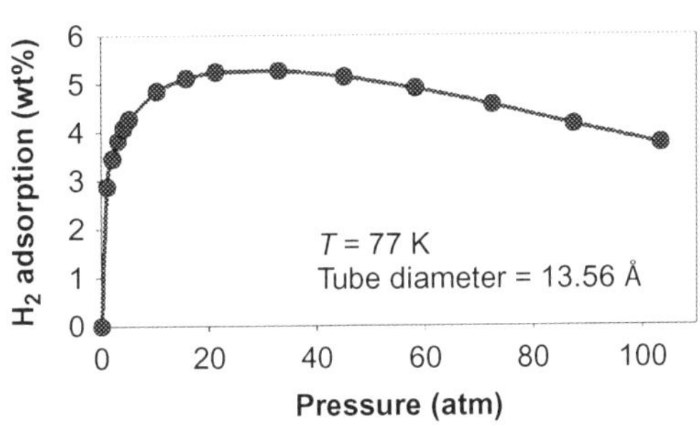

Figure 6.5 Excess adsorption isotherm of hydrogen on an optimal nanotube bundle with a spacing of 6 Å between neighboring nanotubes (Lachance and Bénard, 2007).

classically (Figure 6.5). To show the net gain over compression in a given volume, the excess density shown in Figures 6.4 and 6.5 was calculated by subtracting the amount of hydrogen present in the simulation volume without the adsorbent from the difference between the number of hydrogen molecules in the simulation volume with and without the adsorbent. This yields a smaller value than would have been obtained if the pore volume was used as the reference state for excess adsorption, because the adsorbent volume is not subtracted from the simulation cell. The two can be brought into correspondence by adding the ratio of the density of the bulk gas to the density of the adsorbent in Figures 6.4 and 6.5. This correction is not exact since the pore volume is usually calculated using a reference gas (helium) assumed to be inert in the presence of the adsorbent. The simulation result of 3 wt% at 77 K and 1 bar thus becomes 4–5 wt%.

Due to their curvature and the lack of straightforward access to their internal sites, pure C_{60} Buckminster fullerenes are not expected to exhibit significant hydrogen uptake without modifications. The storage capacity of the internal sites has been investigated numerically by several groups. Pupysheva et al. (2008), using DFT simulations and *ab initio* molecular dynamics, found that C_{60} could hold up to 58 hydrogen atoms, corresponding to a gravimetric uptake of 7.5 wt%. Yoon et al. (2008) examined the storage capacity of calcium-coated C_{60} molecules ($Ca_{32}C_{60}$) numerically. They noted that 92 hydrogen molecules could be stored within the structure, resulting in a maximum gravimetric uptake of 8.4%. Gao et al. (2014) studied the uptake capacity of $Li_8C_{48}B_{12}$ and $Ca_{32}C_{60}$ using DFT simulations and found maximum gravimetric uptakes of 9.2 wt% and 8 wt%, respectively.

6.4 Activated carbons

Activated carbons (Ströbel et al., 2006) are more or less random arrangements of graphitic crystallites exhibiting no long-range order, forming a highly porous three-dimensional structure. As a disordered nanostructure, an activated carbon is characterized by its pore size distribution, which may involve micropores, mesopores, or macropores and its specific surface area. Activated carbons are obtained from the dry distillation of a precursor organic material, which is thermally or chemically processed to tailor its pore size distribution to a specific purpose. The nature of the precursor material and its activation determine the pore size distribution and the specific surface. Activated carbons have been studied as storage materials for low-pressure physisorption storage of natural gas and hydrogen. The energy scales of the adsorption process on activated carbon of both gases differ by a factor of roughly 3 (~15 kJ/mol for methane and ~5 kJ/mol for hydrogen), allowing for room temperature storage by physisorption of natural gas and requiring cryogenic conditions (77 K) for hydrogen.

The storage capacity of activated carbons derived from corncobs can reach 5.8 wt% at 40 bars (Wang et al., 2014) and 296 K. The uptake of activated carbons derived from hydrothermally carbonized organic materials was shown to reach up to 6.4 wt% and yield heat capacities of up to 8.5 kJ/mol. In addition, they were shown to be easily tunable (Sevilla et al., 2011). The superactivated carbon M30, with a specific surface area of 3220 m^2/g exhibited a maximum uptake of 5 wt% of hydrogen at 77 K at pressures ranging from 30 to 50 bars and about 0.7 wt% at 9 bars (Kojima et al., 2006).

The activated carbon AX-21™ (similar to Maxsorb™ MSC-30) has been extensively studied for cryo-adsorption storage of hydrogen due to its strong microporosity, high specific surface area (2800 m^2/g), and chemical and physical resilience. Figure 6.6 shows the adsorption isotherms of hydrogen on the AX-21 carbon obtained using a volumetric method for the temperature range 30–300 K as measured by the Chahine group at UQTR.

As a powdered material, AX-21 has a bulk density of 0.3 g/cm^3 and a micropore volume of 1.06 ml/g (out of a total pore volume of 2.88 ml/g). The adsorption of hydrogen on activated carbons follows Type I fully reversible isotherms. The maximum excess density is 5.4 wt% at 35 bars and 77 K. Under such conditions, the density of hydrogen in the micropores (62 mg/ml) becomes comparable to that of liquid hydrogen (0.71 mg/ml). The fluid-phase density in the pores can range from

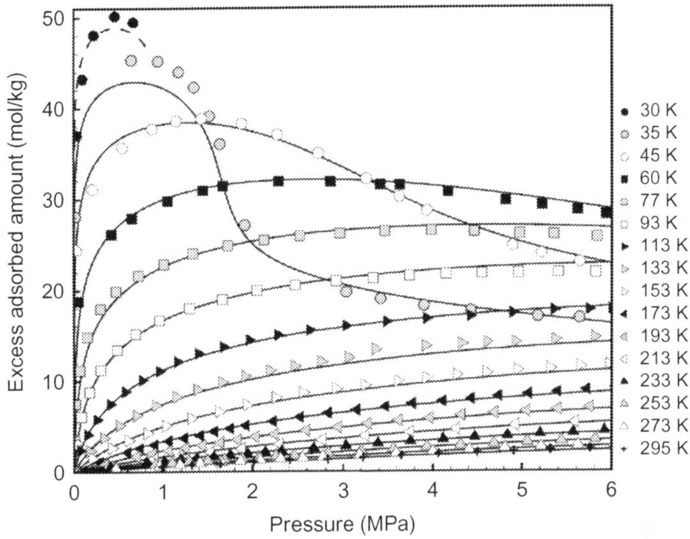

Figure 6.6 Excess adsorption isotherms of hydrogen on the activated carbon AX-21 as a function of pressure for the temperature range 30–295 K. The points represent experimental measurements.
Reprinted with permission from Paggiaro et al. (2010).

38 to 71 mg/ml in activated carbon adsorbents, depending on micropore pore volume and specific surface (Benard and Chahine, 2001).

The excess density of hydrogen adsorbed on activated carbons at 77 K close to the excess maximum has been observed to increase linearly as a function of micropore volume or specific surface at a rate of about 2 vol% per 1000 m^2/g (Chahine and Bose, 1996), representing six molecules per square nanometer (Stadie, 2013). The mass of the storage system notwithstanding, the density of 5.5% required by the 2017 DOE target, would correspond to an activated carbon with a specific surface area of 2750 m^2/g, very close to the theoretical maximum specific surface area (2800 m^2/g) for a graphitic structure in which graphene planes are separated by two layers of hydrogen molecules. Although such values can be achieved with several high-quality activated carbons, these materials are not necessarily consistent with the gravimetric 2017 DOE target, which includes the weight of the storage unit.

6.5 Layered graphene nanostructures

An upper bound to the excess density in activated carbons could in principle be obtained by considering hydrogen adsorption in slit pores. Slit pores are perfectly layered graphene nanostructures. They are often used as idealized versions of activated carbons, the interlayer spacing being representative of the pore size distribution. Simple classical GCMC simulations of the adsorption of hydrogen on slit pores suggest a maximum excess uptake of about 8 wt% at 77 K and 40 bars for an optimal interlayer graphene spacing of about 20 Å (Bénard et al., 2007). Path integral Monte Carlo simulations of hydrogen adsorption on activated carbon of graphitic slit pores were performed by Wang and Johnson (1999b) and Wang et al. (1996). They found that the maximum gravimetric uptake was obtained for 6.67 hydrogen molecule diameters or ~20 Å. Their results were also in relatively good agreement with the data on AX-21 discussed in the previous section.

Patchkovskii et al. (2005) performed a thorough analysis of layered graphene planes and found that improved carbon–H_2 interaction potentials, in addition to a more thorough treatment of the contribution of quantum effects to the free energy and equilibrium constant for hydrogen results in significantly higher adsorbed densities. They predicted an increase in hydrogen binding free energy of up to 10 kJ/mol as a function of interlayer spacing. Favorable adsorption free energies for interlayer separations of 6–7 Å resulted in an effective increase of the pressure inside the layers, which should allow the DOE targets to be met at moderate applied pressures (10 MPa) at room temperature. They conclude that "encapsulating molecular hydrogen in a layered graphitic nanostructure with appropriate spacing should allow the DOE storage targets for hydrogen to be met or even exceeded."

Pillared graphene is a layered nanostructure in which carbon nanotubes are integrated with graphene sheets to form a pillared framework, resulting in a high surface area and highly curved nanostructure with a tuneable pore volume, controlled by the length of the nanotube pillars. Simulation results by Dimitrakakis et al. (2008) show that a pillared graphene based on (6,6) SWNTs, separated by 1.5 nm, linking graphene sheets with a 1.2 nm interlayer distance resulted in less gravimetric uptake than graphene due to the added mass of the nanotubes. However, it resulted in a 25% improvement in the volumetric uptake at ambient temperature. They obtained 7.2 wt% at 77 K, and 1.5 wt% at 300 K and 100 bars, and a volumetric uptake of 50 g H_2/L at 77 K and 9.5 g H_2/L at 300 K. By doping the nanostructure with lithium atoms, the gravimetric uptake was shown to reach 10.5 wt% at 77 K and 7.4 wt% at 300 K at 100 bars. The volumetric uptake increased to 76 g-H_2/L at 77 K and 49 g-H_2/L at 300 K. The gravimetric and volumetric DOE targets could, in principle be achieved, if the nanostructure could be synthesized (Figure 6.7).

6.6 Zeolite-templated carbons

Zeolite templated carbons (ZTC) have generated significant interest recently due to their high uptake capacity of hydrogen and other gases and the capacity to control to some degree the pore size distribution. ZTCs are reported to exceed the storage capacity of Maxsorb MSC-30 by close to 100% at 34 bars. (Stadie, 2013) This level of improvement over activated carbon could be attributed to an extremely narrow pore structure. The heat of adsorption, which would have been expected to be significantly higher than activated carbon, typically remains, somewhat surprisingly, below 8 kJ/mol. (Stadie, 2013).

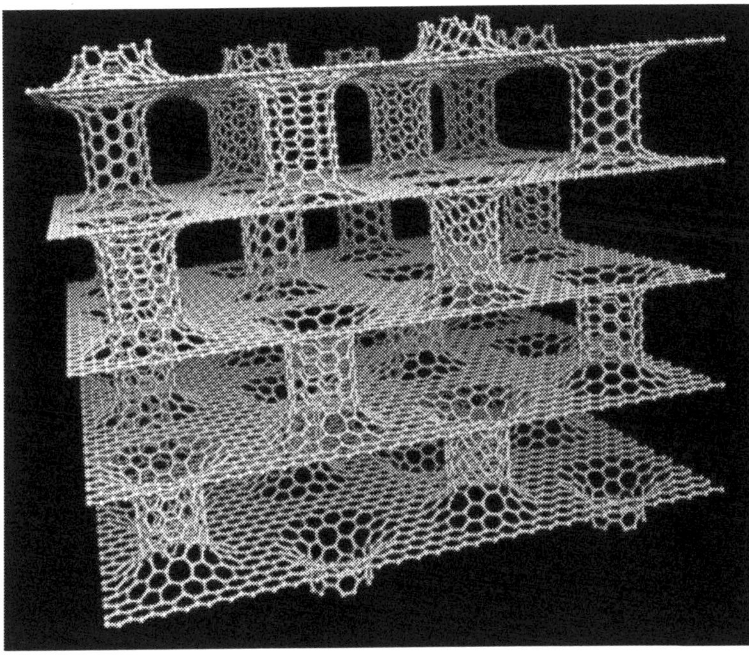

Figure 6.7 Pillared graphene formed of two graphene layers with a spacing of 1.2 nm and (6,6) SWNT pillars separated by 1.5 nm. Figure reprinted with permission from Dimitrakakis et al. (2008), copyright (2008) American Chemical Society.

ZTCs are created when a carbon precursor is polymerized (through, for instance, the polymerization of furfuryl alcohol (Bertarione et al., 2008)) within the pore structure of a zeolite within a zeolite precursor. Zeolites are three-dimensional nanostructured crystalline hydrated aluminosilicate materials. The zeolite structure is generated from tetrahedral units composed of silicon and aluminum atoms connected to four oxygen atoms (Xu et al., 2007) (Figure 6.8a). The oxygen atoms link the tetrahedrons together in a flexible T-O-T structure (Figure 6.8b) which can be used to form one of the 218 different framework types listed by IZA (Database of zeolite structures), determined by the Si/Al ratio and the way they are linked (Sartbaeva et al., 2006; Treacy et al., 2004). Pore sizes range from 3 to 10 Å (Verboekend, 2012).

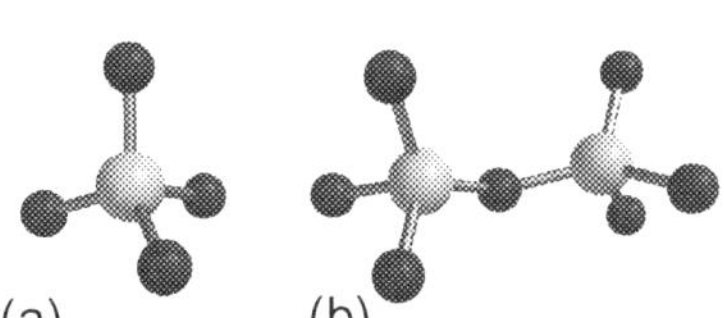

Figure 6.8 (a) Al or Si atoms (white spheres) linked to four oxygen atoms (gray spheres); (b) tetrahedrons are linked by an oxygen atom.

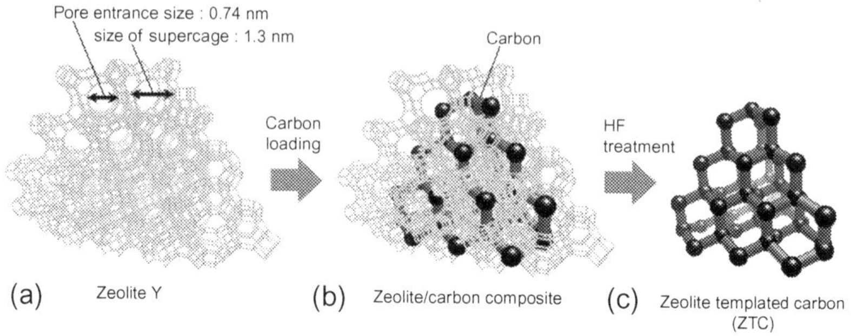

Figure 6.9 (a) Crystal structure of the zeolite Y; (b) carbon polymer within the zeolites pores (zeolite is gray, carbon is black); (c) framework structure of the final ZTC, after removing the zeolite with hydrofluoric acid.
Reprinted with permission from Nishihara et al. (2009a).

The zeolite framework is then removed by dissolution in hydrofluoric acid. The resulting framework is called a ZTC nanostructure. It is the complement of the framework of the zeolite used and inherits its shape and its pore size distribution (Nishihara et al., 2009a), as shown in Figure 6.9.

The carbon precursor can be a liquid polymer infiltrated in the zeolite pores by liquid phase impregnation or a hydrocarbon vapor infiltrated by chemical vapor deposition. The ZTCs have large surface areas (up to 3800 m^2/g) and pore volumes of up to 1.6 cm^3/g. Figure 6.10 shows (a) XRD patterns and (b) pore size distributions for different ZTCs, using furfuryl alcohol and propylene as carbon precursors (Nishihara et al., 2009b). The pore size distribution is strongly microporous and peaked at 1.2 nm.

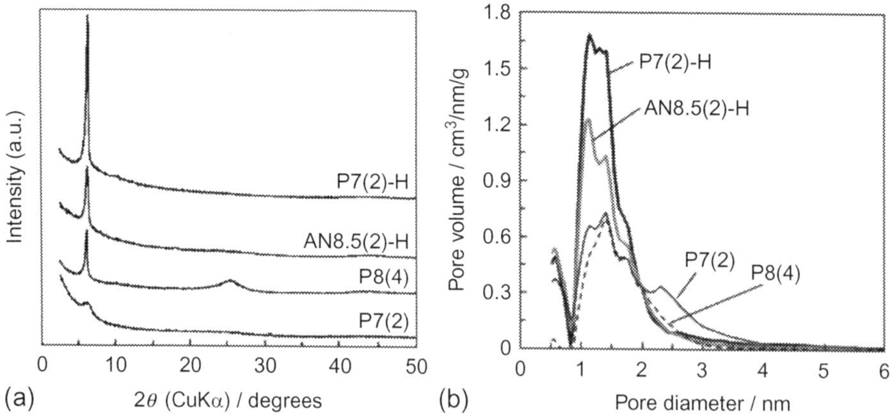

Figure 6.10 (a) XRD patterns and (b) pores size distribution of ZTCs, calculated by DFT method.
Reprinted with permission from Nishihara et al. (2009b).

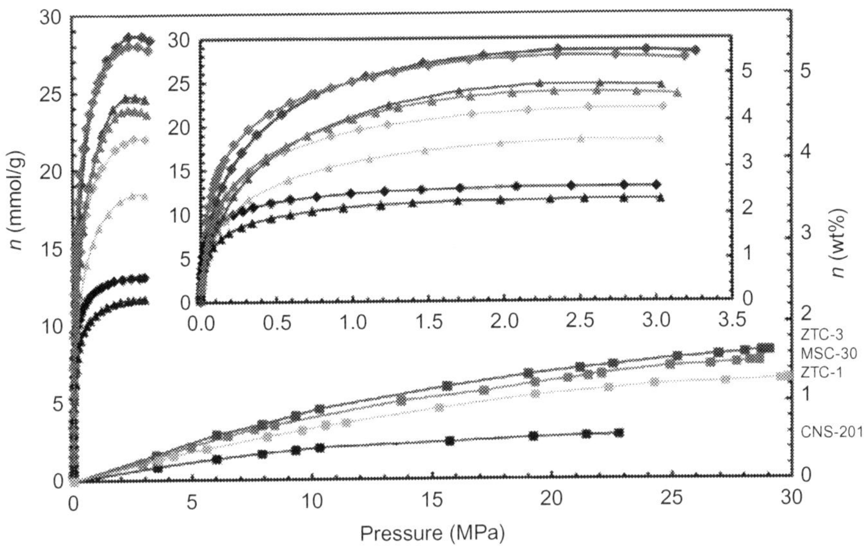

Figure 6.11 Equilibrium adsorption isotherms of hydrogen on two activated carbon (MSC-30 and CNS-201) and two ZTCs (ZTC-2 and ZTC-3) at 77 K (diamond), 87 K (triangle), and 298 K (square) between 0 and 30 MPa.
Reprinted with permission from Stadie et al. (2012).

The adsorption isotherms of hydrogen on two ZTC (ZTC-2 and ZTC-3) and two activated carbons (CNS-201 and MSC-30) are shown in Figure 6.11. The adsorption isotherms are type I reversible isotherms, typical of supercritical physisorption. ZTC-3 exhibits the highest excess uptake at 77 K, with 5.5 wt% at 2.4 MPa, and at room temperature (T = 298 K), with 1.6 wt% at 30 bars (Chapitre de livre; Stadie et al., 2012). For pressures lower than 0.8 MPa, however, Maxsorb MSC-30 exhibits a higher hydrogen capacity than ZTC-3 (Stadie, 2013).

Stadie et al. (2012) also verified that ZTCs are consistent with Chahine's rule, exhibiting 1.8 wt% excess uptake per 1000 m^2/g BET surface area. At 298 K and 30 MPa, he determined that the excess uptake in (wt%) per 1000 m^2/g BET surface area drops to 0.46.

6.7 Conclusion

Meeting the U.S. DOE storage targets for hydrogen has been a challenge ever since they were set. Although some theoretical work suggest that layered graphitic structures may succeed, and several carbon nanostructures exhibit excellent hydrogen storage capacity, there has been no reproducible experimental data to date showing that the full system-based DOE targets for automotive applications could be met with physisorption on pure carbon nanostructures. Note that increasing the binding energies through dipole-induced dipole interactions by embedding partial charges through unsaturated coordinated metal centers may help increase the temperature scale. It is worthwhile, however, to note that considerable progress has been achieved over the last 25 years in understanding the sorption process of hydrogen on carbon nanostructures and in developing carbon-materials with high gravimetric hydrogen uptake. Such materials may find a niche in applications requiring less stringent volumetric constraints and that would benefit from lower operating storage pressures, such as large dedicated vehicles such as buses and stationary applications.

References

Anson, A., Callejas, M.A., Benito, A.M., Maser, W.K., Izquierdo, M.T., Rubio, B., Jagiello, J., Thommes, M., Parra, J.B., Martinez, M.T., 2004. Hydrogen adsorption studies on single wall carbon nanotubes. Carbon 42, 1243–1248.

Becher, M., Haluska, M., Hirscher, M., Quintel, A., Skakalova, V., Dettlaff-Weglikovska, U., Chen, X., Hulman, M., Choi, Y., Roth, S., Meregalli, V., Parrinello, M., Ströbel, R., Jörissen, L., Kappes, M.M., Fink, J., Züttel, A., Stepanek, I., Bernier, P., 2003. Hydrogen storage in carbon tubes. C. R. Phys. 4, 1055–1062.

Benard, P., Chahine, R., 2001. Determination of the adsorption isotherms of hydrogen on activated carbons above the critical temperature of the adsorbate over wide temperature and pressure ranges. Langmuir 17 (6), 1950–1955.

Benard, P., Chahine, R., 2007. Storage of hydrogen by physisorption on carbon and nanostructured materials. Script. Mater. 56, 803–808.

Bénard, P., Chahine, R., Chandonia, P.A., Cossement, D., Dorval-Douville, G., Lafi, L., Lachance, P., Paggiaro, R., Poirier, E., 2007. Comparison of hydrogen adsorption on nanoporous materials. J. Alloys Compd. 446–447, 380–384.

Bénard, P., Chahine, R., 2008. Carbon nanostructures for H_2 storage. In: Walker, G. (Ed.), Solid State Hydrogen Storage—Materials and Chemistry. Woodhead Publishing Ltd., Cambridge, pp. 261–287 (Chapter 10; ISBN: 978-1-84569-270-4).

Bertarione, S., Bonino, F., Cesano, F., Damin, A., Scarano, D., Zecchina, A., 2008. Furfuryl alcohol polymerization in H–Y confined spaces: reaction mechanism and structure of carbocationic intermediates. J. Phys. Chem. B 112, 2580–2589.

Chahine, R., Bose, T.K., 1994. Low-pressure adsorption storage of hydrogen. Int. J. Hydrogen Energy 19, 161–164.

Chahine, R., Bose, T.K., 1996. Characterization and optimization of adsorbents for hydrogen storage. Hydrogen energy progress IX. In: Virzogly, T.N. et al., (Ed.), Proceedings of the 11th World Hydrogen Energy Conference, Stuttgart, Germany 23–28 June. 1259.

Database of zeolite structures. Available at: http://www.iza-structure.org/databases/.

Dillon, A.C., Jones, K.M., Bekkedahi, T.A., Kiang, C.H., Bethune, D.S., Heben, M.J., 1997. Storage of hydrogen in single-walled carbon nanotubes. Nature 396, 377.

Dimitrakakis, G., Tylianakis, E., Froudakis, G., 2008. Pillared graphene: A new 3-D network nanostructure for enhanced hydrogen storage. Nano Lett. 8, 3166–3170.

Durette, D., 2012. Adsorption de l'hydrogène dans les nanotubes de carbone et les composés métallo-organiques: Étude par la méthode de Monte Carlo Grand Canonique Quantique. Master's Thesis 94p.

Gao, Y., Wu, X., Zeng, X., 2014. Designs of fullerene-based frameworks for hydrogen storage. J. Mater. Chem. A 2, 5910–5914.

Guan, L., Suenaga, K., Ijima, S., 2008. Smallest carbon nanotube assigned with atomic resolution accuracy. Nano Lett. 8 (2), 459–462.
Iijima, S., 1991. Helical microtubules of graphitic carbon. Nature 354, 56–58.
Kojima, Y., Kawai, Y.i., Koiwai, A., Suzuki, N., Haga, T., Hioki, T., Tange, K., 2006. Hydrogen adsorption and desorption by carbon materials. J. Alloys Compd. 421, 204–208.
Kuznetsova, A., Mawhinney, D.B., Naumenko, V., Yates, J.T., Liu, J., Smalley, R.E., 2000. Enhancement of adsorption inside of single-walled nanotubes: opening of the entry ports. Chem. Phys. Lett. 321, 292–296.
Lachance, P., Bénard, P., 2007. Specific surface effects on the storage of hydrogen on carbon nanostructures. Int. J. Green Energy 4, 377–384.
Lafi, L., Cossement, D., Chahine, R., 2005. Raman spectroscopy and nitrogen vapour adsorption for the study of structural changes during purification of single-wall carbon nanotubes. Carbon 43, 1347–1357.
Lévesque, D., Gicquel, A., Lamari Darkrim, F., Beyaz Kayiran, S., 2002. Monte Carlo simulations of hydrogen storage in carbon nanotubes. J. Phys. Condens. Matter 14, 9285.
Mélançon, E., Bénard, P., 2004. Theoretical study of the contribution of physisorption to the low-pressure adsorption of hydrogen on carbon nanotubes. Langmuir 20, 7852–7859.
Michel, F., Fieseler, H., Allidières, L., 2006. Liquid hydrogen technologies for mobile use. In: Proceedings of WHEC 16, June 13–16, Lyon, France.
Nishihara, H., Yang, Q.H., Hou, P.X., et al., 2009a. A possible buckybowl-like structure of zeolite templated carbon. Carbon 47, 1220–1230.
Nishihara, H., Hou, P.-X., Li, L.-X., et al., 2009b. High-pressure hydrogen storage in zeolite-templated carbon. J. Phys. Chem. C 113, 3189–3196.
Nishimiya, N., Ishigaki, K., Takikawa, H., Ikeda, M., Hibi, Y., Sakakibara, T., Matsumoto, A., Tsutsumi, K., 2003. Hydrogen sorption by single-walled carbon nanotubes prepared by a torch arc method. J. Alloys Compd. 339, 275–282.
Paggiaro, R., Michl, F., Benard, P., Polifke, W., 2010. Cryo-adsorptive hydrogen storage on activated carbon. II: investigation of the thermal effects during filling at cryogenic temperatures. Int. J. Hydrogen Energy 35, 648–659.
Patchkovskii, S., Tse, J.S., Yurchenko, S.N., Zhechkov, L., Heine, T., Seifer, G., 2005. Graphene nanostructures as tunable storage media for molecular hydrogen. Proc. Natl. Acad. Sci. U.S.A. 102 (30), 10439–10444.
Poirier, E., Chahine, R., Bénard, P., Dorval-Douville, G., Lafi, L., Chandonia, P.A., 2006. Hydrogen adsorption measurements and modeling on metal-organic frameworks and single-walled carbon nanotubes. Langmuir 22, 8784.
Pradhan, B.K., Harutyunyan, A., Stojkovic, D., Zhang, P., Cole, M.W., Crespi, V., Goto, H., Fujiwara, J., Eklund, P.C., 2002. Large cryogenic storage of hydrogen in carbon nanotubes at low pressures. Mat. Res. Soc. Symp. Proc. 706, Z10.3.1–Z10.3.6.
Pupysheva, O., Farajian, A., Yakobson, B., 2008. Fullerene nanocage capacity for hydrogen storage. Nano Lett. 8, 767–774.
Sartbaeva, A., Wells, S.A., Treacy, M.M.J., Thorpe, M.F., 2006. The flexibility window in zeolites. Nat. Mater. 5, 962–965.
Satyapal, S., 2014. U.S. Department of Energy Hydrogen and Fuel Cells Program. In: Annual Merit Review and Peer Evaluation Meeting, June 16–20, Washington. Presentation available at: http://www.hydrogen.energy.gov/pdfs/review14/03_satyapal_plenary_2014_amr.pdf.
Sevilla, M., Fuertes, A.B., Mokaya, R., 2011. High density hydrogen storage in superactivated carbons from hydrothermally carbonized renewable organic materials. Energy Environ. Sci. 4, 1400.

Stadie, N.P., 2013. Synthesis and Thermodynamic Studies of Physisorptive Energy Storage Materials. Ph.D. Thesis, California Institute of Technology Pasadena, California.

Stadie, N.P., Vajo, J.J., Cumberland, R.W., Wilson, A.A., Ahn, C.C., Fultz, B., 2012. Zeolite-templated carbon materials for high-pressure hydrogen storage. Langmuir 28, 10057–10063.

Stan, G., Cole, M.W., 1998a. Low coverage adsorption in cylindrical pores. 395 (2–3), 280–291.

Stan, G., Cole, M.W., 1998b. Hydrogen adsorption in nanotubes. J. Low Temp. Phys. 110, 539.

Stan, G., Bojan, M.J., Curtarolo, S., et al., 2000. Uptake of gases in bundles of carbon nanotubes. Phys. Rev. B 62, 2173.

Stetson, N., 2014. FY 2014 annual progress report—DOE hydrogen and fuel cells program, Section IV.0 hydrogen storage sub-program overview, pp. 3–10.

Ströbel, R., Garche, J., Moseley, P.T., Jörissen, L., Wolf, G., 2006. Hydrogen storage by carbon materials. J. Power Sources 159, 781–801.

Tarasov, B.P., Maehlen, J.P., Lototsky, M.V., Muradyan, V.E., Yartys, V.A., 2003. Hydrogen sorption properties of arc generated single-wall carbon nanotubes. J. Alloys Compd. 356–357, 510–514.

Thess, A., Lee, R., Nikolaev, P., Dai, H., Petit, P., Robert, J., Xu, C., Lee, Y.H., Kim, S.G., Rinzler, A.G., Colbert, D.T., Scuseria, G.E., Tománek, D., Fischer, J.E., Smalley, R.E., 1996. Crystalline ropes of metallic carbon nanotubes. Science 273, 483.

Tozzini, V., Pellegrini, V., 2013. Prospects for hydrogen storage in graphene. Phys. Chem. Chem. Phys. 15, 80.

Treacy, M.M.J., Rivin, I., Balkovsky, E., Randall, K.H., Foster, M.D., 2004. Enumeration of periodic tetrahedral frameworks. II. Polynodal graphs. Microporous and Mesoporous Materials. 74, 121–132.

Technical system targets: onboard hydrogen storage for light-duty fuel cell vehicles, office of energy efficiency and renewable energy, U.S. Department of Energy. Available at: http://energy.gov/eere/fuelcells/downloads/doe-targets-onboard-hydrogen-storage-systems-light-duty-vehicles.

Verboekend, D., 2012. New hierarchical zeolite catalysts by post-synthetic design. Ph.D. Thesis. ETH.

Wang, Q., Johnson, J.L., 1999a. Optimization of carbon nanotube arrays for hydrogen adsorption. J. Phys. Chem. B 103, 4809.

Wang, Q., Johnson, J.K., 1999b. Molecular simulation of hydrogen adsorption in single-walled carbon nanotubes and idealized carbon slit pores. J. Chem. Phys. 110, 577.

Wang, Q., Johnson, J.K., Broughton, J.Q., 1996. Thermodynamic properties and phase equilibrium of fluid hydrogen from path integral simulations. Mol. Phys. 89, 1105.

Wang, D., Geng, Z., Zhang, C., Zhou, X., Liu, X., 2014. Effects of thermal activation conditions on the microstructure regulation of corncob-derived activated carbon for hydrogen storage. J. Energy Chem. 23 (5), 601–608.

Xu, R., Pang, W., Yu, J., Huo, Q., Chen, J., 2007. Structural Chemistry of Microporous Materials. Chemistry of Zeolites and Related Porous Materials. John Wiley & Sons (Asia), Singapour, 19–116.

Ye, Y., Ahn, C.C., Witham, C., Fultz, B., Liu, J., Rinzler, A.G., Colbert, D., Smith, K.A., Smalley, R.E., 1999. Hydrogen adsorption and cohesive energy of single-walled carbon nanotubes. App. Phys. Lett. 74 (16), 2307–2309.

Yoon, M., et al., 2008. Calcium as the superior coating metal in functionalization of carbon fullerenes for high-capacity hydrogen storage. Phys. Rev. Lett. 100, 206806-1–206806-4.

Zacharia, R., Cossement, D., Lafi, L., Chahine, R., 2010. Volumetric hydrogen sorption capacity of monoliths prepared by mechanical densification of MOF-177. J. Mater. Chem. 20, 2145–2151.

Metal–organic frameworks for hydrogen storage

7

H.W. Langmi, J. Ren, N.M. Musyoka
HySA Infrastructure Centre of Competence, Council for Scientific and Industrial Research, Pretoria, South Africa

Abbreviations

AC	activated carbon
BET	Brunauer, Emmett, and Teller
bipy	bipyridine
BTC	benzene-1,3,5-tricarboxylate
diPyNI	*N,N′*-di-(4-pyridyl)-1,4,5,8-naphthalenetetracarboxydiimide
DMA	dimethylammonium
DMF	*N,N′*-dimethylformamide
DOE	Department of Energy
DRIFTS	diffuse reflectance infrared Fourier transform spectroscopy
FTIR	Fourier transform infrared
GCMC	grand canonical Monte Carlo
HBTC	1,3,5-benzenetricarboxylate
HKUST	Hong Kong University of Science and Technology
INS	inelastic neutron scattering
IRMOF	isoreticular metal–organic framework
MIL	Matériaux de l'Institut Lavoisier
MOFs	metal–organic frameworks
NDC	2,6-naphthalenedicarboxylate
NMR	nuclear magnetic resonance
NOTT	Nottingham
NU	Northwestern University
PCN	porous coordination network
SNU	Seoul National University
STEM	scanning transmission electron microscopy
TATB	4,4′,4″-*s*-triazine-2,4,6-triyltribenzoate
TEM	transmission electron microscopy
UiO	University of Oslo
ZIF	zeolitic imidazolate framework

Compendium of Hydrogen Energy. http://dx.doi.org/10.1016/B978-1-78242-362-1.00007-9

7.1 Introduction

Metal–organic frameworks (MOFs) are a class of inorganic–organic hybrid porous crystalline materials consisting of metal ions or metal clusters linked with organic ligands via coordination bonds to form one-, two-, or three-dimensional networks. The flexibility of changing the metal centers and organic ligands allows a wide range of MOFs to be designed, and by careful selection of the constituents, MOFs with desired structures and tailored properties can be produced. MOFs possess several properties that make them particularly attractive for hydrogen storage such as their extraordinarily high surface areas, ultrahigh porosities, tunable pore sizes, and modifiable internal surfaces. Since the first investigation of hydrogen storage in MOFs was reported by Yaghi and co-workers (Rosi et al., 2003), MOFs have been extensively studied for hydrogen storage over the past decade. This chapter provides an overview of MOFs for hydrogen storage. It begins with Section 7.2 discussing synthetic considerations including some of the main methods that have been used to synthesize MOFs. In Section 7.3, hydrogen storage in MOFs at cryogenic temperatures is examined. This section includes studies done at low pressure as well as those conducted at high pressure for various types of MOFs. The factors influencing hydrogen uptake are also presented. Section 7.4 focuses on hydrogen storage at room temperature. Here, some of the efforts to enhance the interactions between hydrogen and MOFs are highlighted. The use of MOFs for nanoconfinement of various chemical hydrides for chemical hydrogen storage is presented in Section 7.5. To conclude, Section 7.6 provides a summary, draws conclusions, and lays out the prospects for MOFs in hydrogen storage.

7.2 Synthetic considerations

MOFs can be constructed by combining metal ions and organic ligands. The characteristics of the ligand play an important part in determining the ensuing framework, which can also be affected by the geometry of the metal ion. In some cases, the metal ion reacts with two different ligands to generate a MOF with mixed ligands. It is also possible to react a ligand with two different metal ions to generate a MOF with mixed metal ions. A principal objective in the synthesis of MOFs is the establishment of synthesis conditions that give defined inorganic building units without decomposing the ligand. Concomitantly, it is required that the crystallization kinetics be adequate for nucleation and growth of the desired MOF phase to occur (Stock and Biswas, 2012). Synthesis of MOFs usually occurs at temperatures from room temperature to about 250 °C in the presence of a solvent. The required energy for the reaction can be supplied from a variety of sources including conventional electric heating, electromagnetic radiation such as microwave, ultrasonic irradiation, mechanical force, and electric potential. The source of energy is linked to several factors (e.g., duration, pressure, energy per molecule that is added to the reaction), which can also significantly affect the resultant MOF (Stock and Biswas, 2012; Bang and Suslick, 2010).

From a practical point of view, MOFs need to be stable in moisture and at elevated temperatures and pressures. Generally, it is quite difficult to completely eliminate water from commercial hydrogen sources. For practical applications, it is not economical to use super high-purity hydrogen because of the additional cost involved in purifying hydrogen. Therefore, water adsorption characteristics and structural stability upon water adsorption are very important issues that need to be addressed in the development of MOFs for practical hydrogen storage applications. For example, the zinc-based MOF (MOF-5) decomposes easily in the presence of water or upon exposure to air, indicating its instability (Panella and Hirscher, 2005). On the other hand, the copper-based MOF (HKUST-1; HKUST, Hong Kong University of Science and Technology) (Chui et al., 1999), zirconium-based MOF (UiO-66; UiO, University of Oslo) (Cavka et al., 2008), and chromium-based MOF (MIL-101; MIL, Matériaux de l'Institut Lavoisier) (Férey et al., 2005) have exceptional thermal and moisture stabilities and can easily be handled in an ambient environment without decomposition. Another point to consider in the synthesis of MOFs for practical applications is that the chosen synthesis route must be up-scalable at a reasonable cost. Rapid reactions would enable a continuous process, which is beneficial for large-scale production. Furthermore, a short reaction time and decreased reaction temperature would be advantageous in terms of energy efficiency and equipment complexity. Use of industrial grade solvents, cheap ligands, low temperature and pressure conditions, and less sophisticated equipment will contribute toward cost reduction. Syntheses of some MOFs have already been scaled-up, and several MOFs are available commercially, e.g., MIL-53(Al), HKUST-1, MOF-5, and ZIF-8 (ZIF, zeolitic imidazolate framework). The commercially available MOFs are yet to meet all the U.S. Department of Energy requirements for onboard hydrogen storage (U.S. DOE, 2009).

To date, several routes such as solvothermal, microwave-assisted, mechanochemical, sonochemical, and electrochemical synthesis have been employed in the synthesis of MOFs (Stock and Biswas, 2012). The most common method used is solvothermal synthesis (also referred to as conventional synthesis), which entails heating a mixture of organic ligands and metal salts in a solvent system at temperatures typically below 250 °C. High boiling point solvents such as water, dimethylformamide, diethylformamide, acetonitrile, and dioxane are often used. Other key variables in the solvothermal synthesis of MOFs include the concentration of reagents, reaction temperature, reaction time, degree of solubility of the reactants, pH of solution, and fractional volume filling of the vessel. The solvothermal method is disadvantageous because of the relatively high temperatures and long reaction times (usually from several hours to days) required as well as the large volumes and high cost of solvents. Microwave-assisted synthesis differs from conventional synthesis in the heating method. The former uses microwave irradiation as the heat source. Microwave-assisted synthesis presents the possibility to homogeneously heat throughout the MOF samples at high heating rates. A wide range of temperatures is accessible and high temperatures can be attained in an extremely short time, leading to increased reaction rate and fast crystallization. Other potential benefits of microwave-assisted synthesis of MOFs are narrow particle size distribution (Kang et al., 1999), phase selectivity (Jhung et al., 2007, 2008), and ease of control of reaction conditions

(Jhung et al., 2004; Hwang et al., 2005). Generally microwave-assisted synthesis enables faster formation of smaller crystals in comparison to solvothermal synthesis employing conventional electric heating. The formation of small-sized crystals that cannot be used to obtain good-quality structural data is the main drawback of the microwave-assisted method.

Mechanochemical synthesis is conducted using a ball mill and it involves the breaking of intramolecular bonds by mechanical forces and the formation of products in a subsequent chemical transformation. This method of synthesis offers several advantages. It is a solvent-free approach and the reaction can be performed at room temperature in a short reaction time (typically 10–60 min), quantitative yields of the MOF product can be achieved, and the synthesis typically generates small-sized particles of the product. In addition, metal oxides can be used in place of metal salt reactants, thereby producing water as the sole by-product. It has been reported that adding minuscule amounts of solvent to the reaction medium to assist in the grinding is beneficial in mechanochemical synthesis in terms of speeding up the reaction and structure-directing (Friic and Fábián, 2009). In sonochemical synthesis, high-energy ultrasound is applied to the reaction mixture. It is anticipated that the sonochemical method applied in the synthesis of MOFs would be advantageous in that it is a room temperature approach, which is not only fast and energy-efficient but is also environmentally benign and produces nanocrystalline MOF particles. These characteristics are of particular relevance for further up-scaling of MOFs for practical applications. Electrochemical synthesis is carried out in an electrochemical cell. In electrochemical synthesis, the anode supplies the metal ions in a continuous manner through dissolution to the reaction medium consisting of organic ligands and a conducting salt. The supply of metal ions in this way implies that anions (such as chlorides and nitrates) from metal salts, which can be problematic in large-scale production processes, are avoided. The electrochemical method also affords faster synthesis at lower temperatures than conventional synthesis. It is clear that the different synthesis routes can produce MOFs that are different in terms of particle sizes and size distributions, and even their morphologies, as a consequence to their hydrogen storage properties. For instance, the diffusion of hydrogen molecules can be affected by different particle sizes of a MOF thereby influencing its hydrogen storage performance.

7.3 Cryo-temperature hydrogen storage at low and high pressures

Hydrogen storage in MOFs at cryogenic temperatures is based on physisorption. Physisorption involves weak interactions, principally van der Waals forces, between the adsorbed hydrogen molecules and the MOF, leading to fast kinetics and complete reversibility of hydrogen storage. The isosteric heat of hydrogen adsorption for most of the reported MOFs are typically in the range 4–12 kJ mol^{-1} (Tan et al., 2011; Chen et al., 2008). Various structural and compositional factors, as discussed in the subsections that follow, affect hydrogen storage in MOFs.

7.3.1 Surface area, pore volume, and pore size

At 77 K, hydrogen uptake in MOFs, especially at high pressure, has been shown to be directly related to the specific surface area (Figure 7.1) and consequently to the pore volume since the pore volume is generally directly proportional to the surface area. The correlation with surface area occurs both for the BET and Langmuir models (Panella et al., 2006; Wong–Foy et al., 2006). Data put together from various studies showed that at 77 K and a low pressure of 1 bar, hydrogen uptake can correlate with BET surface area in the range 100–2000 $m^2\ g^{-1}$. However, above this value the correlation breaks down as the MOF surface cannot be completely covered with hydrogen molecules (Suh et al., 2012). In a study by Wong–Foy et al. (2006), several MOFs were examined including MOF-177, MOF-74, HKUST-1, and the isoreticular (IR) series, IRMOF-1, -6, -11, and -20. The H_2 uptake capacities at saturation were markedly different and ranged from 2.3 wt% (at 26 bar) for MOF-74 to 7.5 wt% (at 70 bar) for MOF-177. Hydrogen uptake values for the MOFs correlated very well with the Langmuir surface area.

For certain framework topologies the surface area (and consequently the pore volume) can be increased by employing elongated ligands. It was shown for a series of NbO-type MOFs (NOTT-101, PCN-46, and NOTT-102, NOTT, Nottingham; PCN, porous coordination network) that high-pressure hydrogen uptake correlated with the length of the ligand, which in turn correlated with the specific surface area and pore

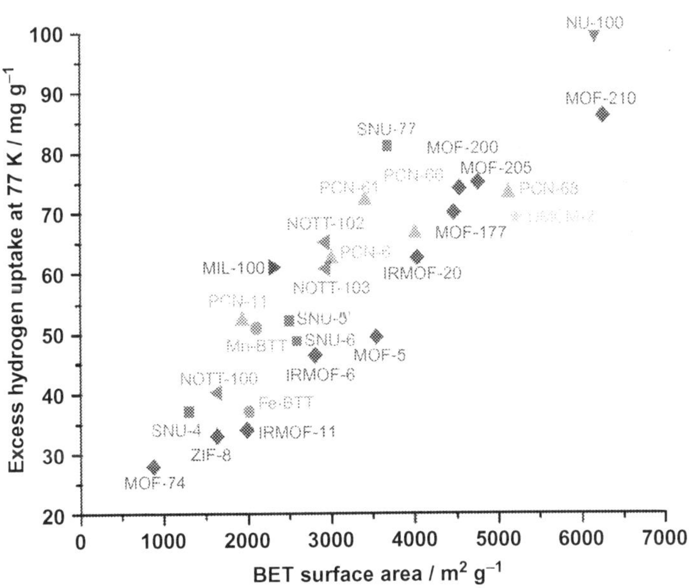

Figure 7.1 Excess high-pressure H_2 uptake capacities at 77 K versus BET surface areas for some highly porous MOFs.

Reprinted with permission from Suh et al. (2012). Copyright (2012) American Chemical Society.

volume (Lin et al., 2009; Zhao et al., 2010). NOTT-101 with the shortest ligand (5.77 Å) displayed the lowest surface area (2316 m^2 g^{-1}), pore volume (0.886 cm^3 g^{-1}), and total H_2 uptake at 77 K and 60 bar (66 mg g^{-1} or 6.2 wt%), while NOTT-102 with the longest ligand (10.098 Å) had the highest surface area (2942 m^2 g^{-1}), pore volume (1.138 cm^3 g^{-1}), and total H_2 uptake (72.0 mg g^{-1} or 6.7 wt%). Fused aromatic rings may also be used to promote high surface area in MOFs. This was illustrated in a study by Wang et al. (2009). An MOF denoted as PCN-20 consisting of a twisted boracite net topology, based on a highly conjugated planar tricarboxylate ligand (triphenylene-2,6,10-tricarboxylate) was evaluated. Its Langmuir surface area (4237 m^2 g^{-1}) was much higher than those of Cu-BTC (BTC, benzene-1,3,5-tricarboxylate) and PCN-6′, implying that fused aromatic rings lead to high surface area. At 77 K and 50 bar, the hydrogen uptake capacities were 6.2, 3.4, and 4.0 wt% for PCN-20, Cu-BTC, and PCN-6′, respectively (Wang et al., 2009; Dybtsev et al., 2004). The higher hydrogen storage capacity of PCN-20 compared to the other MOFs was attributed to the relatively much larger surface area of PCN-20, emanating from the highly conjugated fused triphenylene ring (Wang et al., 2009).

The interaction between hydrogen and MOFs can be enhanced by tailoring the pore size in MOFs such that the potential fields from opposite walls overlap. Pore-size reduction enables the hydrogen molecules to interact with multiple segments of the framework, thereby enhancing hydrogen–framework interactions. It has been reported that the optimal pore size is approximately 6 Å, which is about twice the kinetic diameter of the hydrogen molecule (Wang and Johnson, 1999). This ideal pore size may allow a monolayer of hydrogen molecules to form on opposite pore walls, leading to optimal hydrogen–framework interactions as a result of maximization of the total van der Waals forces acting on the hydrogen molecules (De la Casa-Lillo et al., 2002). In a series of isoreticular MOFs, the maximum H_2 storage capacity was shown to increase in the order PCN-61 $<$ PCN-66 $<$ PCN-68, in accordance with the increase in surface area and pore volume (Yuan et al., 2010). It was also shown that higher pressure is necessary to obtain maximum adsorption for adsorbent materials with higher pore volumes as the maximum adsorption pressure for this series of MOFs also followed the order PCN-61 (33 bar) $<$ PCN-66 (45 bar) $<$ PCN-68 (50 bar). At 77 K and 1 bar, PCN-61 with the smallest pore size, had the highest heat of adsorption and highest H_2 uptake capacity (2.25 wt%). On the other hand, H_2 adsorption capacities of PCN-66 and PCN-68 were 1.79 and 1.87 wt%, respectively, in accordance with their larger pore sizes and respective lower heats of adsorption (Yuan et al., 2010).

Lin et al. (2006) investigated three Cu-based MOFs with the same NbO-type framework topology. The MOFs were constructed from biphenyl, terphenyl, and quater-phenyl tetracarboxylic acid. While the pore size, pore volume, and surface area of the desolvated MOFs increased with the length of the ligand, H_2 adsorption at 1 bar followed the reversed trend. Therefore, a smaller pore size leads to higher hydrogen adsorption affinity as all three MOFs had the same topology as well as chemically very similar internal surfaces, which were made of binuclear Cu-carboxylate nodes and aryl groups. At a higher pressure of 20 bar, the trend in H_2 adsorption also increased with the length of the ligand, which is consistent with the increase in pore volume and surface area. This trend was observed even though the MOF with biphenyl tetracarboxylate ligand had the smallest pore size and highest affinity for H_2 adsorption, demonstrating that at low pressures H_2 uptake is influenced by pore size but at high pressures it is strongly affected by the surface area and pore volume (Lin et al., 2006).

7.3.2 Catenation

Catenation, which is defined as the intergrowth of two or more identical frameworks, can result in reduced pore size for enhanced hydrogen uptake (Ma et al., 2008). When there is maximal displacement of the frameworks from each other the type of catenation is referred to as interpenetration. Whereas when there is minimal displacement between the frameworks, the type of catenation is referred to as interweaving (Batten and Robson, 1998; Chen et al., 2001). Various strategies have been used to control catenation in MOFs, such as rational design of the ligands (Farha et al., 2010a) and alteration of the temperature and concentration in the synthesis reaction (Zhang et al., 2009). The effect of framework catenation on H_2 uptake in the Cu-based MOF ($Cu_3(TATB)_2$ (TATB, 4,4′,4″-*s*-triazine-2,4,6-triyltribenzoate)) isomer pair PCN-6 and PCN-6′ (Figure 7.2) was studied by Ma et al. (2008). It was shown from inelastic neutron scattering (INS) studies that the first sites for H_2 adsorption were the Cu paddle-wheels, which had similar interaction energies. At high H_2 loadings, the hydrogen molecules predominantly adsorbed around the organic linkers, where a stronger interaction was experienced in catenated PCN-6 than in noncatenated PCN-6′. Consequently, H_2 uptake at 77 K and 50 bar for PCN-6 (6.7 wt%) considerably exceeded that for PCN-6′ (4.0 wt%). It was reported that in PCN-6 there were more effective binding sites as a result of catenation, whereas in noncatenated PCN-6′, the overlap of potential fields from opposite walls was not maximized due to the long distance between opposite pore walls (21.4 Å), leading to a reduced number of effective binding sites. Ma et al. (2008) further illustrated that open-metal sites may be blocked due to catenation. The MOFs were first activated at 50 °C. Following this activation, H_2 uptake capacities at 77 K and 1 bar were 1.35 and 1.74 wt% for noncatenated PCN-6′ and catenated PCN-6, respectively. After activation at 150 °C, there were 20% and 10% improvements in H_2 uptake capacity for noncatenated PCN-6′ and catenated PCN-6, respectively. The relatively lower enhancement in H_2 uptake for PCN-6 upon activation at 150 °C was ascribed to blocking of open-metals sites as a result of catenation. Nonetheless, the H_2 uptake capacity for PCN-6

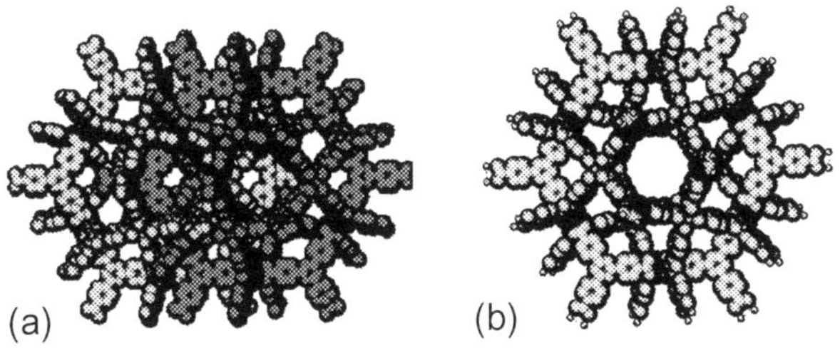

Figure 7.2 (a) Catenated PCN-6 and (b) Noncatenated PCN-6′.
Reprinted with permission from Ma et al. (2008). Copyright (2008) American Chemical Society.

(1.9 wt%) still exceeded that of PCN-6′ (1.62 wt%) under these conditions. Ryan et al. (2008) demonstrated through simulation studies that catenation may offer an advantage at low pressures due to the reduced pore size and increased number of metal sites per unit volume. These characteristics enhance the isosteric heats of adsorption. However, at higher pressures the available free volume becomes more important and consequently, noncatenated MOFs display higher uptake capacities. Frost and Snurr (2007) also performed grand canonical Monte Carlo (GCMC) simulations and showed that for some MOFs catenation leads to lower hydrogen adsorption relative to the noncatenated framework. This is because the increase in isosteric heat of adsorption as a result of catenation does not offset the loss of free volume in the MOFs.

7.3.3 Open-metal sites

Coordinatively, unsaturated metal centers (also known as open-metal sites) provide favorable binding sites for H_2 where the hydrogen molecules bind directly to open-metal sites leading to enhanced isosteric heat of hydrogen adsorption (Chen et al., 2008). The partial positive charges on the metal sites strongly polarize the hydrogen molecules resulting in enhanced H_2 adsorption properties, and high H_2 storage capacities are recorded at low pressures (Dincă et al., 2006; Latroche et al., 2006). Open-metal sites can be created by removal of weakly bound solvent molecules without collapse of the framework (Dincă and Long, 2008; Dincă et al., 2006). Lee et al. (2008) carried out a comparison of the H_2 storage capacities of three isostructural NbO-type MOFs containing Cu or Zn as the metal; namely, SNU-5′ (with open-metal sites; SNU, Seoul National University), SNU-5 (without open-metal sites), and SNU-4 (without open-metal sites). At 77 K and 1 bar, H_2 uptake capacities of the Cu-based MOFs, SNU-5′ and SNU-5 were 2.87 and 1.83 wt%, respectively. Their respective isosteric heats of adsorption at zero coverage were 11.60 and 6.53 kJ mol^{-1}. The Zn-based MOF, SNU-4, had an H_2 uptake capacity and a zero coverage isosteric heat of adsorption of 2.07 wt% and 7.24 kJ mol^{-1}, respectively. The highest H_2 uptake capacity observed for SNU-5′ was ascribed to the stronger interaction of hydrogen molecules with the open-metal centers of the MOF; consequently SNU-5′ also exhibited the highest isostectric heat of adsorption at zero coverage. Yaghi and co-workers also demonstrated remarkably high hydrogen adsorption capacity for MOF-505 with unsaturated Cu metal sites (Chen et al., 2005). The MOF, SNU-15′, in which each Co(II) ion had a vacant coordination site was reported to adsorb only 0.74 wt% hydrogen at 77 K and 15 bar (Cheon and Suh, 2009a). The MOF, however, displayed an extraordinarily high isosteric heat of H_2 adsorption (15.1 kJ mol^{-1}) at zero coverage, which was attributed to the vacant coordination site on each Co(II) ion. Although unsaturated metal sites can lead to enhanced binding between hydrogen and the framework, their effect in hydrogen uptake can be virtually totally masked in the high-pressure range. As mentioned earlier, this is because in the high-pressure range the surface area and pore volume are primary factors affecting hydrogen uptake (Vitillo et al., 2008).

An important aspect to consider is the alignment of the open-metal sites in relation to the H_2 molecules. A MOF designated as NOTT-140 consisting of a 4,8-connected metal–organic polyhedral framework made up of octahedral and cuboctahedral cages, and containing open Cu(II) sites, had a maximum total H_2 adsorption of 6.0 wt% at 20 bar and 77 K (Tan et al., 2011). Compared to other Cu(II)-based metal polyhedral frameworks the isosteric heat at zero surface coverage for NOTT-140 (4.15 kJ mol^{-1}) was low, and this was attributed to the different alignment of open Cu(II) sites. In NOTT-140, the Cu sites were located outside the cages, thus reducing the framework affinity for H_2 molecules. Wang et al. (2008) devised a close-packing strategy aimed at increasing the number of closest neighboring open-metal centers of each void that accommodates hydrogen, and at aligning these sites toward the hydrogen molecules. As such, the open-metal sites can interact directly with the hydrogen molecules inside the void leading to stronger MOF–H_2 interaction. The study reported that PCN-12 in which all the open-metal sites were aligned had much higher H_2 uptake (3.05 wt% at 77 K and 1 bar) than PCN-12′ in which the open-metal sites were misaligned (2.40 wt%).

7.3.4 Dopant cations

Several theoretical studies have predicted that modification of MOFs with metal ions (such as Li^+ ions) could lead to significant improvements in hydrogen uptake arising from the increased strength of binding between hydrogen and the dopant cations (Blomqvist et al., 2007; Han and Goddard, 2007; Mavrandonakis et al., 2008, 2009; Klontzas et al., 2008). A number of experimental studies have been conducted to evaluate the effect of doping MOFs with alkali metals ions. Himsl et al. (2009) prepared a structural analog of MIL-53(Al) containing a pendant hydroxy group from which a lithium alkoxide modified MOF was derived. Hydrogen uptake for the parent MIL-53(Al) was 0.50 wt%, while that for Li-doped MIL-53(Al) was 1.7 wt%. Furthermore, the isosteric heat of hydrogen adsorption for the parent MIL-53(Al) was considerably lower (5.8–4.4 kJ mol^{-1}) than that for Li-doped MIL-53(Al) (11.6–6.4 kJ mol^{-1}).

Independent investigations of doped MOF-5 and a twofold interwoven Zn-based MOF; $Zn_2(NDC)_2$(diPyNI) (NDC, 2,6-naphthalenedicarboxylate; diPyNI, *N,N′*-di-(4-pyridyl)-1,4,5,8-naphthalenetetracarboxydiimide) were carried out (Chu et al., 2012; Mulfort and Hupp, 2008). The latter was first chemically reduced and then doped with Li^+, Na^+, and K^+. There was a considerable increase in hydrogen uptake for the doped MOFs when compared to the parent MOFs. For both types of MOFs, at 77 K and 1 bar, for the same amount of dopant, H_2 uptake followed the trend $Li^+ < Na^+ < K^+$, in agreement with the trend of increasing size of the dopant cation (Figure 7.3). Meanwhile, the H_2 binding strength increased in the order $K^+ < Na^+ < Li^+$. These behaviors were ascribed to structural changes associated with framework reduction such as ligand polarizability and framework displacement, whereby the dopant ions might position themselves between frameworks making accessibility by hydrogen difficult (Chu et al., 2012; Mulfort and Hupp, 2008). It was also suggested that the mechanism of H_2 uptake enhancement was not as a result of generating special metal binding sites. A porous zeolite-like MOF (ZMOF) with *rho* topology was studied (Nouar et al., 2009). The DMA^+ ions (DMA, dimethylammonium) in the parent MOF were exchanged with Li^+ and Mg^{2+} ions. In these MOFs,

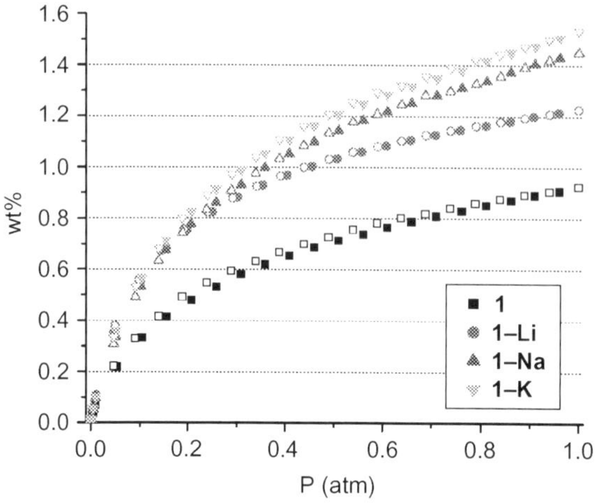

Figure 7.3 77-K H_2 isotherms for **1** and **1 M** (**1** = $Zn_2(NDC)_2(diPyNI)$; **M** = Li, Na, or K). Closed symbols, adsorption; open symbols, desorption. Reprinted with permission from Mulfort and Hupp (2008). Copyright (2008) American Chemical Society.

the presence of an electrostatic field due to the anionic frameworks and charge-balancing cations had an effect on H_2 adsorption as the heats of adsorption were significantly enhanced over those of neutral MOFs. In addition, the charge-balancing cations were fully coordinated by aqua ligands, demonstrating that open-metal binding sites did not contribute to the enhancement of hydrogen binding.

Improved hydrogen adsorption in doped MOFs may arise from the increased surface area and pore volume that occur after exchange of bulky organic cations with alkali metal ions rather than from the direct interaction of hydrogen with metal ion. Yang et al. reported enhancement in H_2 adsorption of an In-based MOF (incorporating bulky $Me_2NH_2^+$) when $Me_2NH_2^+$ was exchanged with Li^+ cations (Yang et al., 2008). The BET surface area, pore size, and pore volume for the parent MOF (NOTT-204) were 820 $m^2\ g^{-1}$, 5.8 Å, and 0.326 $cm^3\ g^{-1}$, respectively. Meanwhile, the corresponding values for the Li^+-exchanged MOF (NOTT-205) were 1024 $m^2\ g^{-1}$, 7.0 Å, and 0.419 $cm^3\ g^{-1}$. At 78 K and 20 bar, H_2 uptake increased from 2.36 wt% in the parent MOF to 2.88 wt% in the Li^+-exchanged MOF. The results illustrated that substituting the large organic cations $Me_2NH_2^+$ by small Li^+ cations afforded a more porous MOF with higher H_2 uptake capacity. The increase in H_2 uptake was attributed to the increase in the accessible pore volume after ion exchange of the large $Me_2NH_2^+$ cations with the small Li^+ cations. The enhancement was not associated with any increase in H_2 adsorption enthalpy; the Li^+-exchanged MOF had an increased pore size, and consequently, its H_2 adsorption enthalpy was lower. Calleja et al. (2010) further demonstrated that the lower H_2 affinity of alkali-exchanged MOFs relative to the parent MOFs containing imidazolium ion could be offset by the inherent increase in the surface area and pore volume of the alkali-exchanged MOFs.

7.3.5 Ligand sites and functionalization

While metal sites are preferential adsorption sites for hydrogen molecules in MOFs it has also been shown that organic ligands are additional favorable sites for hydrogen adsorption. For instance, an INS study on MOF-5 illustrated that there are two distinct binding sites for hydrogen in the MOF; one associated with Zn(II) metal centers and the other with the carboxylate linkers (Rosi et al., 2003). A series of isoreticular MOFs (IRMOF-1, IRMOF-8, IRMOF-11, IRMOF-18) and MOF-177 containing different carboxylate linkers were evaluated (Rowsell et al., 2004). At 1 bar and 77 K, the least number of H_2 molecules per formula unit was 4.2 recorded for IRMOF-18. Meanwhile, the highest value of 9.3 was recorded for IRMOF-11. There was a direct correlation between the number of H_2 molecules per formula unit and the number of organic units (and aromatic rings) per formula unit. The interaction of hydrogen molecules within the framework can be enhanced by increasing the aromaticity of organic linkers. Postsynthetic modification with a series of anhydrides or isocyanates was carried out on a range of MOFs and their hydrogen storage properties were evaluated (Wang et al., 2010). At 77 K and 1 bar, unmodified IRMOF-3 stored 1.51 wt% hydrogen. Hydrogen storage capacities for the modified MOFs, IRMOF-3-AMPh, and IRMOF-3-URPh exceeded that of the unmodified IRMOF-3 reaching 1.73 and 1.54 wt%, respectively. Hydrogen uptake enhancements were also reported for the modified MOFs designated as UMCM-1-AMPh and DMOF-1-AMPh over their respective parent MOFs. Therefore, incorporating aromatic moieties in these MOFs enhanced H_2 binding capacity as the H_2 molecules could specifically interact with the added phenyl groups. Higher isosteric heat of H_2 adsorption was reported for PCN-46 than for NOTT-101 and NOTT-102, due to the stronger interaction between H_2 molecules and the exposed and delocalized *p* electrons in the polyyne unit in PCN-46 as opposed to the phenyl rings in NOTT-101 and NOTT-102 (Lin et al., 2009; Zhao et al., 2010).

Three isostructural MOFs, denoted as NOTT-113, NOTT-114, and NOTT-115, with the same cuboctahedral cage structure but with trimethylphenyl, phenylamine, and triphenylamine moieties, respectively, in the ligands were examined (Yan et al., 2011). At 77 K and 60 bar, the total H_2 uptake for NOTT-115 (7.5 wt%) was significantly higher than those for NOTT-113 (6.7 wt%) and NOTT-114 (6.8 wt%). NOTT-115 displayed the highest heat of adsorption, confirming that functionalization of the ligands with more aromatic rings (i.e., triphenylamine) strengthened the interaction between H_2 molecules and the framework. Conversely, for NOTT-114, amine functionalization of the cage walls resulted in less favorable interaction of H_2 molecules within the framework. Nevertheless, by decorating organic linkers with pendant functional groups, it is possible to enhance interaction of hydrogen within the framework. Introducing —NH_2 functionality in a zeolitic tetrazolate framework led to enhancement in H_2 storage capacity (Panda et al., 2011; Rettig et al., 2000) as the pendant group had the ability to interact with hydrogen. However, introduction of electron withdrawing —Br and —Cl groups in MOF-5 led to destabilization of the coordination bond between the carboxyl group and zinc ions and consequently framework degradation (Yang et al., 2013). This accounted for the significantly lower hydrogen uptake for these mono-substituted MOFs (1.08 and 0.99 wt% for Br-MOF-5 and Cl-MOF-5, respectively) compared to the unsubstituted parent MOF-5 (1.44 wt%). Meanwhile, CH_3-MOF-5 displayed the highest H_2 uptake capacity of 1.47 wt%, most likely because of the narrowing of pores due to bulky CH_3-groups or the presence of an interpenetrated structure.

7.4 Room temperature hydrogen storage at high pressure

A huge challenge facing MOFs is how to reach the DOE hydrogen storage targets at near ambient temperatures. At room temperature, the hydrogen adsorption isotherm usually does not show any saturation represented by a plateau as it does at 77 K. Instead, it shows a linear increase in hydrogen uptake as the pressure is increased, and this relationship is observed in the pressure range typically used in experiments (Lee and Park, 2011; Li and Yang, 2006a). While MOFs exhibit high hydrogen storage capacities at cryogenic temperatures (up to 9 wt% reported experimentally at 56 bar and 77 K (Farha et al., 2010b)), at room temperature only a small amount of hydrogen is stored (typically less than 1 wt%) due to the weak binding (4–12 kJ mol^{-1} (Tan et al., 2011; Chen et al., 2008)) of hydrogen to the MOFs. Furthermore, the isosteric heat of hydrogen adsorption falls considerably as the amount of hydrogen adsorbed increases. Theoretical calculations by Bhatia and Myers (2006) have predicted that the material should have an isosteric heat of adsorption of 15.1 kJ mol^{-1} for ambient temperature hydrogen storage at approximately 30 bar and release at about 1.5 bar. In addition, the isosteric heat of adsorption should not vary with increasing coverage. In another theoretical study, Bae and Snurr (2010) applied GCMC simulations to investigate the influence of heat of adsorption on hydrogen storage and delivery between 1.5 and 120 bar, and revealed that the optimal isosteric heat of hydrogen adsorption is approximately 20 kJ mol^{-1}. It is obviously evident that in order for MOFs to become practical hydrogen storage materials there is a strong need to increase the interactions between hydrogen and the frameworks. As such a number of strategies have been employed to increase hydrogen interactions with MOFs at room temperature.

Although several theoretical calculations (Han and Goddard, 2007; Mavrandonakis et al., 2008; Blomqvist et al., 2007) have predicted that doping MOFs with Li effectively enhances the interaction of hydrogen with MOFs leading to high hydrogen uptake capacity at room temperature, this prediction is yet to be confirmed experimentally. For instance, Han and Goddard (2007) performed GCMC simulations on five Li-doped MOFs and reported that at 300 K the Li-doped MOF denoted as Li-MOF-C30 displayed hydrogen uptake capacities of 3.89 wt% at 20 bar, 4.56 wt% at 50 bar, and 5.16 wt% at 100 bar. Furthermore, for pristine MOF-C30 hydrogen uptake capacities reached only 0.25 and 0.56 wt% at 20 and 50 bar, respectively. The massive enhancement in hydrogen uptake at room temperature for the Li-doped MOFs was suggested to have arisen from the strong interactions between hydrogen and Li^+ ions. For the undoped MOF, the binding energies were relatively lower, i.e., 6.3 and 3.8 kJ mol^{-1} corresponding to the weak binding of hydrogen to the metal oxide clusters and aromatic ligands sites, respectively. On the other hand, for the Li-based MOFs the binding energy was considerably higher, i.e., 16.7 kJ mol^{-1} as a result of a stronger stabilization of hydrogen molecules by the Li^+ sites. Nevertheless, it is very difficult to experimentally attain the ideal metal ion geometry assumed in theoretical studies.

Some experimental studies have reported a link between room temperature hydrogen uptake and pore or channel sizes of the MOFs (Li et al., 2008; Chun et al., 2008).

As mentioned earlier, a reduced pore size potentially enhances the interaction between hydrogen molecules and the framework. Considerably higher hydrogen storage capacities were reported for Ni(HBTC)(4,4′-bipy)·3DMF and Co(HBTC)(4,4′-bipy)·3DMF (HBTC, 1,3,5-benzenetricarboxylate; bipy, bipyridine; DMF, *N*, *N*′-dimethylformamide) (Li et al., 2008) compared to some previously reported IRMOFs (Rowsell and Yaghi, 2006; Dailly et al., 2006). At room temperature and 72 bar, the uptake capacities were 1.20 and 0.96 wt% for Ni(HBTC)(4,4′-bipy)·3DMF and Co(HBTC)(4,4′-bipy)·3DMF, respectively. The relatively higher hydrogen uptake capacities for these two MOFs were attributed to their specific framework channel sizes. When compared to IRMOFs, usually composed of rectangular channels, their frameworks were made up of nonlinear honeycomb channels with only 5 Å at the narrowest and 8 Å at the widest spacing, as well as rectangular channels of dimensions 7 × 6 Å.

Hydrogen spillover, a well-known phenomenon in the area of catalysis, involves the dissociation of hydrogen molecules into atoms on a supported metal surface and subsequent migration of the hydrogen atoms to the support (primary receptor). The hydrogen atoms can further migrate to a secondary receptor, and this migration can be facilitated by a bridge between the receptors (Figure 7.4). In an attempt to advance MOFs research toward practical hydrogen storage, hydrogen spillover has also been widely explored for enhancement of hydrogen storage capacity by inclusion of metal nanoparticles into the pores of the MOFs (Cheon and Suh, 2009b; Li and Yang, 2006a,b, 2008; Liu et al., 2007). These studies indicate that for hydrogen storage spillover will play a significantly larger role at room temperature than at 77 K (Cheon and Suh, 2009b; Li and Yang, 2006a; Liu et al., 2007). Despite reports of promising hydrogen storage capacities for MOFs by spillover at room temperature, the exact mechanism involved in hydrogen storage enhancement from or by spillover is not clear. Hydrogen storage improvement by spillover has been a topic of debate (Hirscher, 2010; Li et al., 2010a; Luzan and Talyzin; 2011).

A typical example of hydrogen enhancement in MOFs by spillover at ambient temperature was reported for MOF-5 and IRMOF-8 by Li and Yang (2006b). MOF-5 and IRMOF 8 were first prepared and then doped with Pt/AC (AC, activated carbon). At 298 K and 100 bar undoped MOF-5 and Pt/C displayed hydrogen storage capacities of 0.4 and 1.0 wt%, respectively. A physical mixture of MOF-5 and Pt/C led to an enhanced hydrogen uptake by a factor of 3.3 over that of MOF-5. Likewise, an enhancement of the hydrogen storage capacity of IRMOF-8 by a factor of 3.1 was obtained for the mixture of Pt/C and IRMOF-8. It was reported that the enhancement of hydrogen uptake in both cases was due to contribution from spillover, with activated carbon acting as the primary receptor for hydrogen spillover and MOF-5 or IRMOF-8 as the secondary spillover receptor. Furthermore, in both cases, hydrogen uptake was completely reversible and the hydrogen uptake plot did not reach a plateau even at a high pressure of 100 bar. Li and Yang (2006a) later illustrated that hydrogen uptake of a physical mixture could be further enhanced by constructing carbon bridges between the primary (Pt/C) and secondary (MOF-5 or IRMOF-8) receptors using carbonization of sucrose. At 100 bar, the bridged MOF-5 mixture displayed a hydrogen uptake capacity of 3 wt%, corresponding to an enhancement factor of 2 compared with

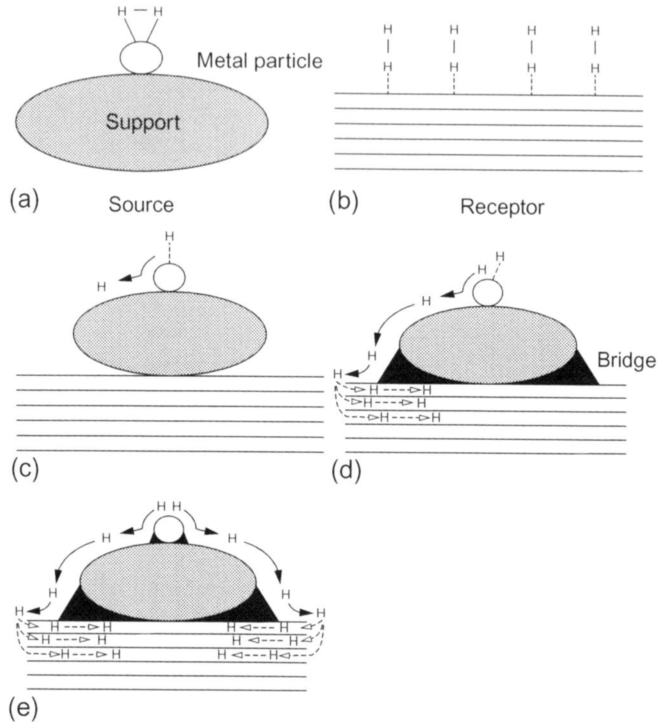

Figure 7.4 Hydrogen spillover in a supported catalyst system: (a) adsorption of hydrogen on a supported metal particle; (b) the low-capacity receptor; (c) primary spillover of atomic hydrogen to the support; (d) secondary spillover to the receptor enhanced by a physical bridge; and (e) primary and secondary spillover enhancement by improved contacts and bridges. Reprinted with permission from Lachawiec et al. (2005). Copyright (2005) American Chemical Society.

unbridged MOF-5 mixture, and a factor of 8 compared with pristine MOF-5. In the same light, enhancement of H_2 uptake by a factor of 8 was also achieved for IRMOF-8 after bridging. At 100 bar, hydrogen uptake for pristine IRMOF-8 was 0.5 wt% and that for the bridged IRMOF-8 mixture was 4 wt%. It was concluded that for both bridged MOFs, this additional improvement in hydrogen uptake was as a result of facilitation of hydrogen uptake by spillover arising from the formation of carbon bridges, which increase the contact between the primary and secondary receptors. The heats of adsorption for the bridged MOFs was estimated to be 20–23 kJ mol^{-1}. These high values were in agreement with the high hydrogen uptakes. Li and Yang (2007; 2008) have also reported hydrogen uptake enhancement by spillover for other MOFs such as MOF-177, HKUST-1, and MIL-101(Cr).

In contrast to the work by Li and Yang, other research groups did not observe any enhancement in hydrogen uptake by spillover for MOF-5 doped with Pt/C (Campesi et al., 2010; Luzan and Talyzin, 2010). Campesi et al. (2010) reported hydrogen uptake for pristine MOF-5 to be 0.32 wt% at 70 bar. While the values for the unbridged and bridged composites were much lower, that is, 0.15 and 0.25 wt%, respectively, in corroboration with the decreased BET surface area and micropore volume of the composites. Similarly, in a study by Luzan and Talyzin (2010) pristine MOF-5 samples adsorbed 0.38 and 0.47 wt% at 120 bar, while the unbridged and bridged mixtures adsorbed 0.43 and 0.27 wt% hydrogen. These hydrogen uptakes were also in close agreement with the BET surface areas and total pore volumes of the materials. The results from both studies contradicts those from the study by Li and Yang (2006a,b), which demonstrated that hydrogen storage was enhanced by factors of 3.3 and 8 for unbridged and bridged MOF-5 mixtures, respectively. It was concluded that hydrogen uptake was due to physisorption and no spillover effect occurred. Nevertheless, other research groups have reported enhancement in hydrogen storage by spillover for MOF-5 (Lee and Park, 2011), MIL-101(Cr) (Liu et al., 2007), and MIL-53(Al) (Liu et al., 2007). For instance, Liu et al. (2007) showed that at 293 K and 50 bar, hydrogen uptake for pristine MIL-101 and pure Pt/C catalyst were 0.37 and 0.54 wt%, respectively. Upon doping MIL-101(Cr) with Pt/C hydrogen uptake increased to 0.75 wt%. After bridging the value further increased to 1.14 wt%. A similar trend was observed for the MIL-53 samples. Improved hydrogen uptake in these MOFs upon doping and bridging was attributed to the spillover effect.

Besides Pt, other noble metals such as Pd have been used to enhance hydrogen uptake at room temperature. The investigations were carried out on SNU-3 and MIL-100 (Al) and their Pd-containing derivatives (Cheon and Suh, 2009b; Zlotea et al., 2010). For MIL-100(Al)/Pd, at low pressure Pd hydride was formed, which was partly responsible for the increased hydrogen uptake, and at higher pressures spillover effect was reported to also account for the increase in hydrogen uptake. Hydride formation has also been reported for MOFs doped with Pt (Proch et al., 2008). Very recently, careful and complex experiments carried out on MIL-101(Cr) did not confirm spillover (Szilágyi et al., 2014). The study showed that Pd nanoparticles could be introduced into the pores of MOFs with good contact, rendering the use of a carbon bridge unnecessary. The addition of Pd nanoparticles increased hydrogen uptake of MIL-101(Cr) as a result of Pd hydride formation and the hydrogen atoms did not chemisorb on the MOF, demonstrating that hydrogen spillover did not occur. It is clear from these studies that caution needs to be taken before associating hydrogen storage in MOFs to spillover from a noble metal.

7.5 Nanoconfinement of chemical hydrides in MOFs

It is widely reported that some of the most attractive bulk forms of solid-state hydrogen storage materials possess unfavorable hydrogen desorption thermodynamics (Bérubé et al., 2007; Fichtner, 2011). For example, materials such as $LiBH_4$ and MgH_2 are known to be too stable whereas others like AlH_3 are too unstable for real practical use in automotive applications. Recent studies by Nielsen et al. (2011) have shown that decreased particle size often leads to significant reduction of desorption temperatures owing to reduced stabilities of these hydrides. There are different ways that have been proposed and applied to reduce particle sizes such as (i) ball milling (Rude et al., 2011), (ii) nanoconfinement (Liu et al., 2011; Stavila et al., 2012), (iii) others, that is, laser ablation, vapor condensation, sputtering, etc. (Bérubé et al., 2007). Among these size reduction options, nanoconfinement, which can be described as a technique for compartmentalization of nanoparticles within porous scaffolds (Rude et al., 2011), is known to inhibit particle growth and agglomeration and can also limit mobility of decomposition products. For example, Zlotea and Latroche (2013) reported that nanoconfined metal particles in porous materials showed no coalescence. Nanostructuring of some complex hydrides (alanates, borohydrides) as well as chemical hydrides such as ammonia borane has been reported to lead to improved reaction kinetics as well as lowering of hydrogen absorption and release temperatures (Gutowska et al., 2005; Bérubé et al., 2007; Fichtner, 2011; Vajo, 2011; Rude et al., 2011). Attractiveness of nanoconfinement is often ascribed to the large atomic population at the particle surface that results in an increase in surface energy. This increased ratio between the surface and bulk atoms has reportedly led to higher surface area and also enhancement of surface reaction kinetics especially those involving diffusion within the particles (Zlotea and Latroche, 2013).

There are many different types of porous scaffolds that have been utilized in nanoconfinement such as carbon-based materials (Baldé et al., 2006; Li et al., 2011; Moussa et al., 2012), mesoporous silica (Gutowska et al., 2005; Zheng et al., 2008), and MOFs (Gadipelli et al., 2011; Sun et al., 2011; Li et al., 2010b), among others. The choice of the scaffolds plays a crucial role since it influences the physical and chemical properties of the confined material (Zlotea and Latroche, 2013). In this case, properties such as pore-size distribution, surface area, and material composition should be evaluated carefully so as to gain full benefits of nanoconfinement. An additional important consideration is related to the thermal and chemical stability of the material especially for inertness toward the confined particles. It is also important that the scaffold is mechanically stable so that it is not weakened easily during subsequent hydrogen loading and unloading cycles. Studies by Nielsen et al. (2009) have highlighted that the pore-size distribution of the scaffold material play a considerable role in desorption kinetics of MgH_2 nanoclusters, hence the need for materials with desirable pore systems.

Recent reviews by Nielsen et al. (2011) and Rude et al. (2011) have presented the three widely used approaches for nanoconfinement. These methods are (i) melt infiltration, (ii) solvent mediated infiltration, and (iii) direct synthesis of nanoconfined materials. For the melt infiltration method (i.e., via capillary suction), there is no need for the use of either solvents or precursors, and posttreatment of the nanoconfined hydride composite is completely avoided. Unfortunately, this technique suffers drawbacks since many metals and their hydrides are very reactive in their molten state (Adelhelm et al., 2010). It is also challenging to identify an inert scaffold material that would withstand the high temperatures involved (Rude et al., 2011). In the case of wet infiltration (i.e., solvent-mediated infiltration), a porous scaffold is often immersed in a hydride-containing solution, which infiltrates the pores of the scaffold "nanomould" and thereafter solidifies upon removal of the solvent (Moussa et al., 2012). This method is attractive since it can be performed under mild conditions but suffers the disadvantage of limited availability of solvents that can dissolve materials to be infiltrated without reacting with either the material or the porous host (Rude et al., 2011). In the case of direct synthesis of nanoconfined metal hydrides, confinement occurs from a one-pot synthesis procedure where the material to be nanoconfined as well as their hosts are present at the same time in the solution. The reader is referred to reviews by Nielsen et al. (2011) and Rude et al. (2011) for more information.

Among the many scaffolds that have been employed in nanoconfinement, it has recently emerged that confinement of metal and chemical hydrides within MOFs often leads to accelerated hydrogen desorption kinetics as well as alteration of reaction thermodynamics (Li et al., 2010b; Gadipelli et al., 2011). Metal–organic frameworks, being porous organic–inorganic hybrid solids possessing zeolite-like structures, have extremely high surface area and tunable pore structures that make them attractive for nanoconfinement. Different researchers have loaded various hydrogen storage materials, e.g., ammonia borane and $NaAlH_4$, into MOFs and have reported and confirmed improved kinetics and thermodynamics (Li et al., 2010b; Ren et al., 2013). A recent short communication by Sun et al. (2011) has shown that when Cu-MOFs was used as the template for loading $LiBH_4$, redox reactions occurring between $LiBH_4$ and Cu—O units enabled dehydrogenation to occur at a much lower temperature. Another report by Wahab et al. (2012) also highlighted that the formation of ammonia in nanoconfined ammonia borane could be prevented if there were metallic sites in the host material such as the MOF framework. A synergistic effect of nanoconfinement in MOF in improving H_2 releasing kinetics, and also in preventing formation of poisonous by-products has been discussed by Ren et al. (2013) and Li et al. (2011). Through these studies and many others reported in the literature, MOFs have demonstrated more advantages as ideal templating material for nanoconfinement compared to other porous scaffolds. A schematic illustration of nanoconfinement of $NaAlH_4$ molecules loaded into the pores of MOF-74(Mg) is shown in Figure 7.5.

Various experimental techniques have been used to probe for evidence of nanoconfinement in MOFs, especially those that are dedicated to modification related to hydrogen storage properties. Electron microscopy techniques such as STEM and TEM have been used to confirm uniform distribution of hydrides throughout the MOF crystals (Zlotea et al., 2010; Campesi et al., 2008). Porosity measurements have shown filling effects and also proven the decrease in pore volume as a result of nanoconfinement (Zlotea et al., 2010; Campesi et al., 2008). Smaller pore sizes that had confined MgH_2 particles were shown to lead to faster desorption rate (Nielsen et al., 2009). FTIR spectroscopy also showed that the MOF framework was stable enough and could not be easily decomposed upon infiltration with $NaAlH_4$ (Moussa et al., 2012). Solid-state NMR spectroscopy was used to confirm the infiltrations of MOF-74 with $NaAlH_4$ nanoclusters as well as prove that desorption and rehydrogenation reactions occurred without formation of an intermediate phase (Stavila et al., 2012). Furthermore, X-ray diffraction has shown success in nanoconfinement in MOFs. For example, no peaks corresponding to $LiBH_4$ were found in the diffraction pattern for the $LiBH_4$ at Cu-MOFs (Sun et al., 2011). This observation could be attributed to the success in loading of $LiBH_4$ into the pores of Cu-MOFs. Calorimetric

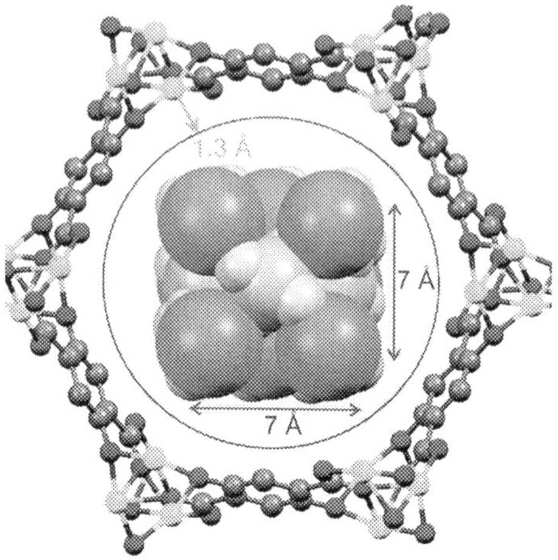

Figure 7.5 Schematic representation of the $(NaAlH_4)_8$ clusters inside the MOF-74(Mg) pores. The space inside the sphere corresponds to the available space outside of magnesium van der Waals radius (1.3 Å).
Reprinted with permission from Stavila et al. (2012). Copyright (2012) American Chemical Society.

techniques have also been employed and were found to generate valuable information from nanoconfinement effects (Li et al., 2011).

7.6 Conclusions and future trends

The last decade has witnessed rapid growth in research on hydrogen storage in MOFs, after the first investigation was reported in 2003. A variety of routes have been taken to synthesize MOFs, with solvothermal synthesis being the most commonly used method. The advantages of each method vary and MOFs with different properties can be synthesized. However, the key aspect is to synthesize MOFs cheaply and efficiently with desirable properties for hydrogen storage applications. At cryogenic temperatures, hydrogen storage capacities of MOFs are quite promising particularly at high pressures; the highest hydrogen storage capacity achieved so far for MOFs is ca. 9 wt% at 77 K and 56 bar for NU-100 (NU, Northwestern University). At this low temperature, hydrogen storage occurs by a physisorption process with fast kinetics and excellent reversibility. As the temperature increases the capacities drop and at room temperature hydrogen storage capacities of MOFs are typically less than 1 wt%. This is due to the weak interactions between hydrogen molecules and the MOFs.

A variety of structural and compositional factors have been shown to affect hydrogen storage in MOFs such as surface area, pore volume, pore size, catenation, unsaturated metal sites, dopant cations, and ligand functionalization. At low temperature and high pressure, hydrogen uptake is dependent on pore volume and surface area. At low temperature and low pressure, hydrogen uptake depends more on the pore size, which determines the isosteric heat of adsorption. Significant enhancement of hydrogen uptake at room temperature has also been achieved by constraining the pore size of MOFs. In some MOFs, catenation is beneficial in reducing the pore size and enhancing hydrogen uptake. However, it is worth noting that catenation may result in blocking of open-metal sites or loss of free volume. Open-metal sites are favorable sites for hydrogen and generating these sites increases the isosteric heat of hydrogen adsorption. Although metal sites are preferential sites for hydrogen adsorption, organic linkers offer additional adsorption sites. By increasing the aromaticity of organic linkers, the hydrogen–framework interactions can be enhanced. Several theoretical studies have predicted that doping MOFs with Li cations can also lead to enhanced hydrogen–framework interactions and consequently an improvement in hydrogen uptake at room temperature. However, it has been difficult to demonstrate these predictions experimentally. Incorporating Pd or Pt nanoparticles in MOFs has been reported to lead to improved hydrogen uptake at room temperature by spillover, though the degree of enhancement varies substantially from one study to another, with some studies reporting no enhancement at all, casting doubts on the role of spillover in hydrogen storage. Nanoconfinement of chemical hydrides in MOFs has been achieved, which may offer benefits such as decreased dehydrogenation temperature and improved hydrogen release kinetics for the hydride, and also inhibition of the formation of unwanted gaseous products.

Despite the significant progress in MOFs research, these materials still do not satisfactorily store hydrogen based on the DOE targets. Many challenges still need to be overcome for MOFs to be used as practical hydrogen storage materials. However, it is expected that MOFs will continue to attract attention for hydrogen storage applications given their numerous attributes such as the huge structural diversity, ability to tune the pore size and surface functionality, exceptional porosity and surface area, and their rich chemistry. A key approach to further development in this field would be to optimize existing MOFs through various modifications and functionalization in order to enhance hydrogen storage. Formation of MOF composites or hybrids is also a promising approach. Alternatively, new MOFs should be synthesized with superior properties to existing MOFs. Future research should be geared toward enhancing the interaction energy between hydrogen molecules and the framework to improve hydrogen storage at room temperature. This is necessary to meet the DOE onboard targets. The ideal MOF will have a combination of an adequate isosteric heat of hydrogen adsorption, large surface area and pore volume, and appropriate pore size. In addition, the MOF should have low cost and high moisture and thermal stability. The development of such a MOF will ultimately position MOFs as a frontrunner in materials-based hydrogen storage.

7.6.1 Sources of further information and advice

For further information on the synthesis of MOFs, the reader is referred to the comprehensive review by Stock and Biswas (2012), which highlights the routes to various MOF topologies, morphologies, and composites. In view of the intense research activities on hydrogen storage in MOFs, there have been several excellent reviews published on the topic. The key recent review articles recommended for further information include Langmi et al. (2014), Suh et al. (2012), Sculley et al. (2011), and Hu and Zhang (2010). It is also recommended to consult the *Handbook of Hydrogen Storage* edited by Michael Hirscher, in which Panella and Hirscher (2010) provide a chapter on physisorption in porous materials. Within the chapter, hydrogen storage in coordination polymers (MOFs) is explicitly discussed. MOFs are increasingly being exploited as scaffolds for nanoconfinement of hydrides. Nanoconfinement is an emerging powerful strategy that can be used to tailor the kinetic and thermodynamic characteristics of hydrides. The reader is referred to the reviews by Nielsen et al. (2011), Rude et al. (2011) and Li and Xu (2013) for more information and further reading on nanoconfinement.

References

Adelhelm, P., Gao, J., Verkuijlen, M.H.W., Rongeat, C., Herrich, M., van Bentum, P.J.M., Gutfeisch, O., Kentgens, A.P.M., de Jong, K.P., de Jongh, P.E., 2010. Comprehensive study of melt infltration for the synthesis of $NaAlH_4$/C Nanocomposites. Chem. Mater. 22, 2233–2238.

Bae, Y.-S., Snurr, R.Q., 2010. Optimal isosteric heat of adsorption for hydrogen storage and delivery using metal-organic frameworks. Microporous Mesoporous Mater. 132, 300–303.

Baldé, C.P., Hereijgers, B.P.C., Bitter, J.H., de Jong, K.P., 2006. Facilitated hydrogen storage in $NaAlH_4$ supported on carbon nanofibers. Angew. Chem. Int. Ed. 45, 3501–3503.

Bang, J.H., Suslick, K.S., 2010. Applications of ultrasound to the synthesis of nanostructured materials. Adv. Mater. 22, 1039–1059.

Batten, S.R., Robson, R., 1998. Interpenetrating nets: ordered, periodic entanglement. Angew. Chem. Int. Ed. 37, 1461–1494.

Bérubé, V., Radtke, G., Dresselhaus, M., Chen, G., 2007. Size effects on the hydrogen storage properties of nanostructured metal hydrides: a review. Int. J. Energy Res. 31, 637–663.

Bhatia, S.K., Myers, A.L., 2006. Optimum conditions for adsorptive storage. Langmuir 22, 1688–1700.

Blomqvist, A., Araújo, C.M., Srepusharawoot, P., Ahuja, R., 2007. Li-decorated metal-organic framework 5: a route to achieving a suitable hydrogen storage medium. Proc. Natl. Acad. Sci. U. S. A. 104, 20173–20176.

Calleja, G., Botas, J.A., Sánchez-Sánchez, M., Orcajo, M.G., 2010. Hydrogen adsorption over zeolite-like MOF materials modified by ion exchange. Int. J. Hydrog. Energy 35, 9916–9923.

Campesi, R., Cuevas, F., Gadiou, R., Leroy, E., Hirscher, M., Vix-Guterl, C., Latroche, M., 2008. Hydrogen storage properties of Pd nanoparticle/carbon template composites. Carbon 46, 206–214.

Campesi, R., Cuevas, F., Latroche, M., Hirscher, M., 2010. Hydrogen spillover measurements of unbridged and bridged metal-organic frameworks—revisited. Phys. Chem. Chem. Phys. 12, 10457–10459.

Cavka, J.H., Jakobsen, S., Olsbye, U., Guillou, N., Lamberti, C., Bordiga, S., Lillerude, K.P., 2008. A new zirconium inorganic building brick forming metal organic frameworks with exceptional stability. J. Am. Chem. Soc. 130, 13850–13851.

Chen, B., Eddaoudi, M., Hyde, S.T., O'Keeffe, M., Yaghi, O.M., 2001. Interwoven metal-organic framework on a periodic minimal surface with extra-large pores. Science 291, 1021–1023.

Chen, B., Ockwig, N.W., Millward, A.R., Contreras, D.S., Yaghi, O.M., 2005. High H2 adsorption in a microporous metal-organic framework with open-metal sites. Angew. Chem. Int. Ed. 44, 4745–4749.

Chen, B., Zhao, X., Putkham, A., Hong, K., Lobkovsky, E.B., Hurtado, E.J., Fletcher, A.J., Thomas, K.M., 2008. Surface interactions and quantum kinetic molecular sieving for H_2 and D_2 adsorption on a mixed metal-organic framework material. J. Am. Chem. Soc. 130, 6411–6423.

Cheon, Y.E., Suh, M.P., 2009a. Selective gas adsorption in a microporous metal-organic-framework constructed of CoII4 clusters. Chem. Commun. 9, 2296–2298.

Cheon, Y.E., Suh, M.P., 2009b. Enhanced hydrogen storage by palladium nanoparticles fabricated in a redox-active metal-organic framework. Angew. Chem. Int. Ed. 48, 2899–2903.

Chu, C.-L., Chen, J.-R., Lee, T.-Y., 2012. Enhancement of hydrogen adsorption by alkali-metal cation doping of metal-organic framework-5. Int. J. Hydrog. Energy 37, 6721–6726.

Chui, S.S.Y., Lo, S.M.F., Charmant, J.P.H., Guy Orpen, A., Williams, I.D., 1999. A chemically functionalizable nanoporous material $[Cu_3(TMA)_2\ (H_2O)_3](_n)$. Science 283, 1148–1150.

Chun, H., Jung, H., Koo, G., Jeong, H., Kim, D.-K., 2008. Efficient hydrogen sorption in 8-connected MOFs based on trinuclear pinwheel motifs. Inorg. Chem. 47, 5355–5359.

Dailly, A., Vajo, J.J., Ahn, C.C., 2006. Saturation of hydrogen sorption in Zn benzenedicarboxylate and Zn naphthalenedicarboxylate. J. Phys. Chem. B 110, 1099–1101.

De la Casa-Lillo, M.A., Lamari-Darkrim, F., Cazorla-Amoro´s, D., Linares-Soano, A., 2002. Hydrogen storage in activated carbons and activated carbon fibers. J. Phys. Chem. B 106, 10930–10934.

Dincă, M., Long, J.R., 2008. Hydrogen storage in microporous metal-organic frameworks with exposed metal sites. Angew. Chem. Int. Ed. Engl. 47, 6766–6779.

Dincă, M., Dailly, A., Liu, Y., Brown, C.M., Neumann, D.A.J., Long, R., 2006. Hydrogen storage in a microporous metal-organic framework with exposed Mn^{2+} coordination sites. J. Am. Chem. Soc. 128, 16876–16883.

Dybtsev, D.N., Chun, H., Yoon, S.H., Kim, D., Kim, K., 2004. Microporous manganese formate: a simple metal-organic porous material with high framework stability and highly selective gas sorption properties. J. Am. Chem. Soc. 126, 32–33.

Farha, O.K., Malliakas, C.D., Kanatzidis, M.G., Hupp, J.T., 2010a. Control over catenation in metal-organic frameworks via rational design of the organic building block. J. Am. Chem. Soc. 132, 950–952.

Farha, O.K., Yazaydin, A.Ö., Eryazici, I., Malliakas, C.D., Hauser, B.G., Kanatzidis, M.G., Nguyen, S.T., Snurr, R.Q., Hupp, J.T., 2010b. De novo synthesis of a metal-organic framework material featuring ultrahigh surface area and gas storage capacities. Nat. Chem. 2, 944–948.

Férey, G., Mellot-Draznieks, C., Serre, C., Millange, F., Dutour, J., Surblé, S., Margiolaki, I., 2005. A chromium terephthalate-based solid with unusually large pore volumes and surface area. Science 309, 2040–2042.

Fichtner, M., 2011. Nanoconfinement effects in energy storage materials. Phys. Chem. Chem. Phys. 13, 21186–21195.

Friic, T., Fábián, L., 2009. Mechanochemical conversion of a metal oxide into coordination polymers and porous frameworks using liquid-assisted grinding (LAG). CrystEngComm 11, 743–745.

Frost, H., Snurr, R.Q., 2007. Design requirements for metal-organic frameworks as hydrogen storage materials. J. Phys. Chem. C 111, 18794–18803.

Gadipelli, S., Ford, J., Zhou, W., Wu, H., Udovic, T.J., Yildirim, T., 2011. Nanoconfinement and catalytic dehydrogenation of ammonia borane by magnesium-metal organic-framework-74. Chem. Eur. J. 17, 6043–6047.

Gutowska, A., Li, L., Shin, Y., Wang, C.M., Li, X.S., Linehan, J.C., Smith, R.S., Kay, B.D., Schmid, B., Shaw, W., Gutowski, M., Autrey, T., 2005. Nanoscaffold mediates hydrogen release and the reactivity of ammonia borane. Angew. Chem. Int. Ed. 44, 3578–3582.

Han, S.S., Goddard, W.A., 2007. Lithium-doped metal-organic frameworks for reversible H_2 storage at ambient temperature. J. Am. Chem. Soc. 129, 8422–8423.

Himsl, D., Wallacher, D., Hartmann, M., 2009. Improving the hydrogen-adsorption properties of a hydroxyl modified MIL-53(Al) structural analogue by lithium doping. Angew. Chem. Int. Ed. 48, 4639–4642.

Hirscher, M., 2010. Remarks about spillover and hydrogen adsorption—comments on the contributions of A.V. Talyzin and R.T. Yang. Microporous Mesoporous Mater. 135, 209–210.

Hu, Y.H., Zhang, L., 2010. Hydrogen storage in metal-organic frameworks. Adv. Mater. 22, E117–E130.

Hwang, Y.K., Chang, J.-S., Park, S.-E., Kim, D.S., Kwon, Y.-U., Jhung, S.H., Hwang, J.-S., Park, M.S., Park, S.-E., 2005. Microwave fabrication of MFI zeolite crystals with a fibrous morphology and their applications. Angew. Chem. Int. Ed. 44, 556–560.

Jhung, S.H., Chang, J.-S., Hwang, Y.K., Park, S.-E., 2004. Crystal morphology control of AFI type molecular sieves with microwave irradiation. J. Mater. Chem. 14, 280–285.

Jhung, S.H., Jin, T., Hwang, J.-S., Chang, J.-S., 2007. Phase transformations and phase-selective syntheses of aluminophosphate molecular sieves. J. Nanosci. Nanotechnol. 7, 2734–2740.

Jhung, S.H., Jin, T., Kim, Y.H., Chang, J.-S., 2008. Phase-selective crystallization of cobalt-incorporated aluminophosphate molecular sieves with large pore by microwave irradiation. Microporous Mesoporous Mater. 109, 58–65.

Kang, K.-K., Park, C.-H., Ahn, W.-S., 1999. Microwave preparation of a titanium-substituted mesoporous molecular sieve. Catal. Lett. 59, 45–49.

Klontzas, E., Mavrandonakis, A., Tylianakis, E., Froudakis, G.E., 2008. Improving hydrogen storage capacity of MOF by functionalization of the organic linker with lithium atoms. Nano Lett. 8, 1572–1576.

Lachawiec Jr., A.J., Qi, G., Yang, R.T., 2005. Hydrogen storage in nanostructured carbons by spillover: bridge-building enhancement. Langmuir 21, 11418–11424.

Langmi, H.W., Ren, J., North, B., Mathe, M., Bessarabov, D., 2014. Hydrogen storage in metal-organic frameworks: a review. Electrochim. Acta 128, 368–392.

Latroche, M., Surble, S., Serre, C., Mellot-Draznieks, C., Llewellyn, P.L., Lee, J.H., Chang, J.S., Jhung, S.H., Ferey, G., 2006. Hydrogen storage in the giant-pore metal-organic frameworks MIL-100 and MIL-101. Angew. Chem. Int. Ed. 45, 8227–8231.

Lee, S.-Y., Park, S.-J., 2011. Effect of platinum doping of activated carbon on hydrogen storage behaviors of metal-organic frameworks-5. Int. J. Hydrog. Energy 36, 8381–8387.

Lee, Y.-G., Moon, H.R., Cheon, Y.E., Suh, M.P., 2008. A comparison of the H_2 sorption capacities of isostructural metal-organic frameworks with and without accessible metal sites: $[\{Zn_2(abtc)(dmf)_2\}_3]$ and $[\{Cu_2(abtc)(dmf)_2\}_3]$ versus$[\{Cu_2(abtc)\}_3]$. Angew. Chem. Int. Ed. 47, 7741–7745.

Li, S.-L., Xu, Q., 2013. Metal-organic frameworks as platforms for clean energy. Energy Environ. Sci. 6, 1656–1683.

Li, Y., Yang, R.T., 2006a. Hydrogen storage in metal-organic frameworks by bridged hydrogen spillover. J. Am. Chem. Soc. 128, 8136–8137.

Li, Y., Yang, R.T., 2006b. Significantly enhanced hydrogen storage in metal-organic frameworks via spillover. J. Am. Chem. Soc. 128, 726–727.

Li, Y., Yang, R.T., 2007. Gas adsorption and storage in metal-organic framework MOF-177. Langmuir 23, 12937–12944.

Li, Y., Yang, R.T., 2008. Hydrogen storage in metal-organic and covalent-organic frameworks by spillover. AICHE J. 54, 269–279.

Li, Y., Xie, L., Liu, Y., Yang, R., Li, X., 2008. Favorable hydrogen storage properties of M (HBTC)(4,4′-bipy) · 3DMF (M = Ni and Co). Inorg. Chem. 47, 10372–10377.

Li, Y.W., Wang, L., Yang, R.T., 2010a. Response to "hydrogen adsorption in Ptcatalyst/MOF-5 materials" by Luzan and Talyzin. Microporous Mesoporous Mater. 135, 206–208.

Li, Z., Zhu, G., Lu, G., Qiu, S., Yao, X., 2010b. Ammonia borane confined by a metal-organic framework for chemical hydrogen storage: enhancing kinetics and eliminating ammonia. J. Am. Chem. Soc. 132, 1490–1491.

Li, C., Peng, P., Zhou, D.W., Wan, L., 2011. Research progress in $LiBH_4$ for hydrogen storage: a review. Int. J. Hydrog. Energy 36, 14512–14526.

Lin, X., Jia, J., Zhao, X., Thomas, K.M., Blake, A.J., Walker, G.S., Champness, N.R., Hubberstey, P., Schröder, M., 2006. High H_2 adsorption by coordination-framework materials. Angew. Chem. Int. Ed. 45, 7358–7364.

Lin, X., Telepeni, I., Blake, A.J., Dailly, A., Brown, C.M., Simmons, J.M., Zoppi, M., Walker, G.S., Thomas, K.M., Mays, T.J., Hubberstey, P., Champness, N.R., Schröder, M., 2009. High capacity hydrogen adsorption in Cu(II) tetracarboxylate framework materials: the role of pore size, ligand functionalization, and exposed metal sites. J. Am. Chem. Soc. 131, 2159–2171.

Liu, Y.-Y., Zeng, J.-L., Zhang, J., Xu, F., Sun, L.X., 2007. Improved hydrogen storage in the modified metal-organic frameworks by hydrogen spillover effect. Int. J. Hydrog. Energy 32, 4005–4010.

Liu, X., Peaslee, D., Jost, C.Z., Bauman, T.F., Mazjoub, E.H., 2011. Systematic pore-size effects of nanoconfinement of $LiBH_4$: elimination of diborane release and tunable behavior for hydrogen storage applications. Chem. Mater. 23, 1331–1336.

Luzan, S.M., Talyzin, A.V., 2010. Hydrogen adsorption in Pt catalyst/MOF-5 materials. Microporous Mesoporous Mater. 135, 201–205.

Luzan, S.M., Talyzin, A.V., 2011. Comment to the "response to "hydrogen adsorptionin Pt catalyst/MOF-5 materials"" by Li et al. Microporous Mesoporous Mater. 139, 216–218.

Ma, S., Eckert, J., Forster, P.M., Ji, W.Y., Young, K.H., Chang, J.-S., Collier, C.D., Parise, J.B., Zhou, H.-C., 2008. Further investigation of the effect of framework catenation on hydrogen uptake in metal-organic frameworks. J. Am. Chem. Soc. 130, 15896–15902.

Mavrandonakis, A., Tylianakis, E., Stubos, A.K., Froudakis, G.E., 2008. Why Li doping in MOFs enhances H_2 storage capacity? A multi-scale theoretical study. J. Phys. Chem. C 112, 7290–7294.

Mavrandonakis, A., Klontzas, E., Tylianakis, E., Froudakis, G.E., 2009. Enhancement of hydrogen adsorption in metal-organic frameworks by the incorporation of the sulfonate group and Li cations. A multiscale computational study. J. Am. Chem. Soc. 131, 13410–13414.

Moussa, G., Bernard, S., Demirci, U.B., Chiriac, R., Miele, P., 2012. Room-temperature hydrogen release from activated carbon-confined ammonia borane. Int. J. Hydrog. Energy 37, 13437–13445.

Mulfort, K.L., Hupp, J.T., 2008. Alkali metal cation effects on hydrogen uptake and binding in metal-organic frameworks. Inorg. Chem. 47, 7936–7938.

Nielsen, T.K., Manickam, K., Hirscher, M., Besenbacher, F., Jensen, T.R., 2009. Confinement of MgH_2 nanoclusters within nanoporous aerogel scaffold materials. ACS Nano 3, 3521–3528.

Nielsen, T.K., Besenbacher, F., Jensen, T.R., 2011. Nanoconfined hydrides for energy storage. Nanoscale 3, 2086–2098.

Nouar, F., Eckert, J., Eubank, J.F., Forster, P., Eddaoudi, M., 2009. Zeolite-like metal-organic frameworks (ZMOFs) as hydrogen storage platform: lithium and magnesium ion-exchange and H-(rho-ZMOF) interaction studies. J. Am. Chem. Soc. 131, 2864–2870.

Panda, T., Pachfule, P., Chen, Y., Jiang, J., Banerjee, R., 2011. Amino functionalized zeolitic tetrazolate framework (ZTF) with high capacity for storage of carbondioxide. Chem. Commun. 47, 2011–2013.

Panella, B., Hirscher, M., 2005. Hydrogen physisorption in metal-organic porous crystals. Adv. Mater. 17, 538–541.

Panella, B., Hirscher, M., 2010. Physisorption in porous materials. In: Hirscher, M. (Ed.), Handbook of Hydrogen Storage: New Materials for Future Energy Storage. Wiley-VCH Verlag GmbH & Co. KGaA, Weinheim. ISBN 978-3-527-32273-2, pp. 39–62.

Panella, B., Hirscher, M., Pütter, H., Müller, U., 2006. Hydrogen adsorption in metal-organic frameworks: Cu-MOFs and Zn-MOFs compared. Adv. Funct. Mater. 16, 520–524.

Proch, S., Herrmannsdörfer, J., Kempe, R., Kern, C., Jess, A., Seyfarth, L., Senker, J., 2008. Pt@MOF-177: synthesis, room-temperature hydrogen storage and oxidation catalysis. Chem. Eur. J. 14, 8204–8212.

Ren, Y., Chia, G.H., Gao, Z., 2013. Metal-organic frameworks in fuel cell technologies. Nano Today 8, 577–597.

Rettig, S.J., Sânchez, V., Storr, A., Thompson, R.C., Trotter, J., 2000. Polybis(4-azabenzimidazolato)-iron(II) and -cobalt(II). 3-D single diamond-like frame-work materials which exhibit spin canting and ferromagnetic ordering at low temperatures. J. Chem. Soc. Dalton Trans. 21, 3931–3937.

Rosi, N.L., Eckert, J., Eddaoudi, M., Vodak, D.T., Kim, J., O'Keeffe, M., Yaghi, O.M., 2003. Hydrogen storage in microporous metal-organic frameworks. Science 300, 1127–1129.

Rowsell, J.L.C., Millward, A.R., Park, K.S., Yaghi, O.M., 2004. Hydrogen sorption in functionalized metal-organic frameworks. J. Am. Chem. Soc. 126, 5666–5667.

Rowsell, J.L.C., Yaghi, O.M., 2006. Effects of functionalization, catenation, and variation of the metal oxide and organic linking units on the low-pressure hydrogen adsorption properties of metal-organic frameworks. J. Am. Chem. Soc. 128, 1304–1315.

Rude, L.H., Nielsen, T.K., Ravnsbaek, D.B., Bosenberg, U., Ley, M.B., Richter, B., Arnbjerg, L.M., Dornheim, M., Filinchuk, Y., Besenbacher, F., Jensen, T.R., 2011. Tailoring properties of borohydrides for hydrogen storage: a review. Phys. Stat. Sol. A 208, 1754–1773.

Ryan, P., Broadbelt, L.J., Snurr, R.Q., 2008. Is catenation beneficial for hydrogen storage in metal-organic frameworks? Chem. Commun. 35, 4132–4134.

Sculley, J., Yuan, D., Zhou, H.-C., 2011. The current status of hydrogen storage in metal-organic frameworks—updated. Energy Environ. Sci. 4, 2721–2735.

Stavila, V., Bhakta, R.K., Alam, T.M., Majzoub, E.H., Allendorf, M.D., 2012. Reversible hydrogen storage by $NaAlH_4$ confined within a titanium-functionalized MOF-74(Mg) nanoreactor. ACS Nano 6, 9807–9817.

Stock, N., Biswas, S., 2012. Synthesis of metal-organic frameworks (MOFs): routes to various MOF topologies, morphologies, and composites. Chem. Rev. 112, 933–969.

Suh, M.P., Park, H.J., Prasad, T.K., Lim, D.-W., 2012. Hydrogen storage in metal-organic frameworks. Chem. Rev. 112, 782–835.

Sun, W., Li, S., Mao, J., Guo, Z., Liu, H., Dou, S., Yu, X., 2011. Nanoconfinement of lithium borohydride in Cu-MOFs towards low temperature dehydrogenation. Dalton Trans. 40, 5673–5676.

Szilágyi, P.Á., Callini, E., Anastasopol, A., Kwakernaak, C., Sachdeva, S., van de Krol, R., Geerlings, H., Borgschulte, A., Züttel, A., Dam, B., 2014. Probing hydrogen spillover in Pd@MIL-101(Cr) with a focus on hydrogen chemisorption. Phys. Chem. Chem. Phys. 16, 5803–5809.

Tan, C., Yang, S., Champness, N.R., Lin, X., Blake, A.J., Lewis, W., Schröder, M., 2011. High capacity gas storage by a 4,8-connected metal-organic polyhedral framework. Chem. Commun. 47, 4487–4489.

U.S. DOE, 2009. Targets for onboard hydrogen storage systems for light duty vehicles. Available from: http://www1.eere.energy.gov/hydrogenandfuelcells/storage/pdfs/targets_onboard_hydro_storage_explanation.pdf [accessed 29.06.14.].

Vajo, J.J., 2011. Influence of nano-confinement on the thermodynamics and dehydrogenation kinetics of metal hydrides. Curr. Opin. Solid State Mater. Sci. 15, 52–61.

Vitillo, J.G., Regli, L., Chavan, S., Ricchiardi, G., Spoto, G., Dietzel, P.D.C., Bordiga, S., Zecchina, A., 2008. Role of exposed metal sites in hydrogen storage in MOFs. J. Am. Chem. Soc. 130, 8386–8396.

Wahab, M.A., Zhao, H., Yao, X.D., 2012. Nano-confined ammonia borane for chemical hydrogen Storage. Front. Chem. Sci. Eng. 6, 27–33.

Wang, Q., Johnson, J.K., 1999. Molecular simulation of hydrogen adsorption in single-walled carbon nanotubes and idealized carbon slit pores. J. Chem. Phys. 110, 577–586.

Wang, X.-S., Ma, S., Forster, P.M., Yuan, D., Eckert, J., López, J.J., Murphy, B.J., Parise, J.B., Zhou, H.-C., 2008. Enhancing H_2 uptake by "close-packing" alignment of open copper sites in metal-organic frameworks. Angew. Chem. Int. Ed. 47, 7263–7266.

Wang, X.-S., Ma, S., Yuan, D., Yoon, J.W., Hwang, Y.K., Chang, J.-S., Wang, X., Jørgensen, M.R., Chen, Y.-S., Zhou, H.-C., 2009. A large-surface-area boracite-network-topology porous MOF constructed from a conjugated ligand exhibiting a high hydrogen uptake capacity. Inorg. Chem. 48, 7519–7521.

Wang, Z., Tanabe, K.K., Cohen, S.M., 2010. Tuning hydrogen sorption properties of metal-organic frameworks by postsynthetic covalent modification. Chem. Eur. J. 16, 212–217.

Wong-Foy, A.G., Matzger, A.J., Yaghi, O.M., 2006. Exceptional H_2 saturation uptake in microporous metal-organic frameworks. J. Am. Chem. Soc. 128, 3494–3495.

Yan, Y., Blake, A.J., Lewis, W., Barnett, S.A., Dailly, A., Champness, N.R., Schröder, M., 2011. Modifying cage structures in metal-organic polyhedral frameworks for H_2 storage. Chem. Eur. J. 17, 11162–11170.

Yang, S., Lin, X., Blake, A.J., Thomas, K.M., Hubberstey, P., Champness, N.R., Schröder, M., 2008. Enhancement of H_2 adsorption in Li^+-exchanged co-ordination framework materials. Chem. Commun. 46, 6108–6110.

Yang, J., Grzech, A., Mulder, F.M., Dingemans, T.J., 2013. The hydrogen storage capacity of mono-substituted MOF-5 derivatives: an experimental and computational approach. Microporous Mesoporous Mater. 171, 65–71.

Yuan, D., Zhao, D., Sun, D., Zhou, H.-C., 2010. An isoreticular series of metal-organic frameworks with dendritic hexacarboxylate ligands and exceptionally high gas-uptake capacity. Angew. Chem. Int. Ed. 49, 5357–5361.

Zhang, J., Wojtas, L., Larsen, R.W., Eddaoudi, M., Zaworotko, M.J., 2009. Temperature and concentration control over interpenetration in a metal-organic material. J. Am. Chem. Soc. 131, 17040–17041.

Zhao, D., Yuan, D., Yakovenko, A., Zhou, H.-C., 2010. A NbO-type metal-organic framework derived from a polyyne-coupled di-isophthalate linker formed in situ. Chem. Commun. 46, 4196–4198.

Zheng, S., Fang, F., Zhou, G., Chen, G., Ouyang, L., Zhu, M., Sun, D., 2008. Hydrogen storage properties of space-confined $NaAlH_4$ nanoparticles in ordered mesoporous silica. Chem. Mater. 20, 3954–3958.

Zlotea, C., Latroche, M., 2013. Role of nanoconfinement on hydrogen sorption properties of metal nanoparticles hybrids. Colloids Surf. A: Physicochem. Eng. Aspects 439, 117–130.

Zlotea, C., Campesi, R., Cuevas, F., Leroy, E., Dibandjo, P., Volkringer, C., Loiseau, T., Férey, G., Latroche, M., 2010. Pd nanoparticles embedded into a metal-organic framework: synthesis, structural characteristics, and hydrogen sorption properties. J. Am. Chem. Soc. 132, 2991–2997.

Other methods for the physical storage of hydrogen

8

N.K. Zhevago
National Research Centre, Kurchatov Institute, Moscow, Russia

Abbreviations

AC	activated carbon
BET	Brunauer–Emmett–Teller equation
CVD	chemical vapor deposition
DOI	digital object identifier at http://dx.doi.org/
DOE	U.S. department of energy
HGM	hollow glass microsphere
IRMOF	isoreticular metal organic framework
MOF	metal–organic framework
PW-HGM	porous-wall hollow glass microsphere
SPV	specific pore volume
SSA	specific surface area
THF	tetrahydrofuran

Nomenclatures

a	interatomic distance
A	specific surface area of an adsorbent
c	length of the preexisting surface crack
E_a	activation energy for hydrogen permeation
E	elastic modulus of glass
G_c	ratio of the hydrogen mass stored to the mass of the storage medium filled with hydrogen (gravimetric capacity of the storage medium)
h	wall thickness of a glass vessel
k	Boltzmann constant
K	hydrogen permeability of glass
K_0	maximum hydrogen permeability
l_0	length parameter of Weibull probability distribution
m	Weibull modulus (shape parameter)
M	number of moles of hydrogen transported through a thin slab of glass
p	hydrogen pressure inside a glass vessel (HGM or capillary)
p_u	ultimate (burst) pressure in a glass vessel
$P_2 - P_1$	pressure differential across a glass slab
r	internal radius of a HGM or a glass capillary
R	universal gas constant
s	surface area over which permeation of hydrogen takes place

Compendium of Hydrogen Energy. http://dx.doi.org/10.1016/B978-1-78242-362-1.00008-0

T	absolute temperature
V_c	ratio of the hydrogen mass to the volume of the storage medium (volumetric capacity of the storage medium)
γ	fracture energy per surface unit
k	concentration of glass network formers
ρ	bulk density of an adsorbent
μ_H	ratio of mass of adsorbed hydrogen to the mass of the adsorbent (hydrogen uptake)
σ	actual tensile strength of glass
σ_t	theoretical tensile strength of glass
σ_s	tensile stress
σ_0	scale parameter (characteristic strength) of Weibull probability distribution
τ	half-life time of hydrogen inside a HGM or a glass capillary due to the leakage through the walls

8.1 Introduction

Onboard storage of hydrogen is one of the key challenges for hydrogen fuel cell vehicles. Compressed hydrogen is the best near-term solution for hydrogen storage on a motor vehicle due to the relative simplicity of producing gaseous hydrogen, rapid refueling capability, and low infrastructure impact. Despite these advantages, onboard high-pressure hydrogen storage must overcome several technical challenges to be viable in the long term. It is well known that hydrogen diffuses more easily through many conventional materials used for vessels, and through gaps that are normally small enough to seal other gases safely. It can cause metal material embrittlement resulting in serious reduction in ductility, cracking, and failures well below the normal yield stresses. Conventionally, hydrogen is stored in high-pressure (up to 70 MPa) and cryogenic (at 21 K) tanks. The currently validated high-pressure tank technology is close to meeting the revised DOE 2017 target 5.5% of system gravimetric capacity and 40 g/l of system volumetric capacity. It utilizes expensive premium carbon reinforcement to meet the challenging structural requirement of supporting over 150 MPa burst pressure as specified in current regulations. However, using tank technology, it is impossible to reach the ultimate DOE targets of 7.5% gravimetric capacity and 70 g/l volumetric capacity of the storage system. Moreover, the problem of safety exists and evidently grows with increasing pressure. Regular tank shapes (cylindrical or spherical) are the easiest for tank fabrication. However, available spaces inside a vehicle are typically not cylindrical or spherical. As a result, hydrogen tanks are typically stored in the trunk, occupying precious cargo space. A better utilization of available space in the vehicle is one of the keys to achieving the hydrogen storage targets. There is the need for more safe, compact, and cost-effective storage of hydrogen. Microscopic voids inside a lightweight medium could probably be the solution to further enhancing both the volumetric capacity and the safety of hydrogen storage system. From this perspective, we will consider hollow glass microspheres (HGMs), glass capillary arrays, flexible glass capillaries, various microporous materials, and hydrogen hydrate clathrates as hydrogen storage candidates for onboard applications. In particular, we will discuss the theoretical background, experimental results, advantages, and drawbacks of the various storage candidates, compared to each other and conventional tanks.

8.2 Storage of compressed hydrogen in glass microcontainers

8.2.1 *Intrinsic and actual strength of glass*

Physical properties of glass are determined by glass composition. Fused quartz glass is pure silicon dioxide (100% SiO_2) in the noncrystalline state where SiO_4 tetrahedra form a random network that does not exhibit any long-range order. However, the tetrahedra themselves represent a high degree of local ordering, i.e., every silicon atom is coordinated by four oxygen atoms and the nearest neighbor Si–O bond length shows a narrow distribution throughout the structure. Most common kinds of glass contain other *glass network creators* (with greater than 80 kcal/mol bond strength with oxygen, e.g., B_2O_3, GeO_2, Al_2O_3) and oxides that do not participate in forming the network structure and are called *network modifiers* (with less than 60 kcal/mol bond strength with oxygen, e.g., Na_2O, K_2O, MgO, CaO). There are also *intermediates* (e.g., TiO_2, ZnO, PbO) that form intermediate bonds to oxygen. They cannot form glasses on their own, but aid with other oxides to form glasses.

The theoretical (intrinsic) strength of glass is directly related to the stress required to break the chemical bonds between two adjacent atoms in the glass structure. During fracturing the chemical bonds are broken and two new surfaces are produced. The theoretical tensile strength σ_t can be approximated as the work per surface area supplied to produce fracture and can be written as:

$$\sigma_t = \sqrt{\gamma \frac{E}{a}} \tag{8.1}$$

where γ = surface energy (fracture energy per surface unit), E = elastic modulus, and a = interatomic distance. If E = 70 GPa, γ = 3.5 J/m^2, and a = 0.2 nm, then σ_t = 35 GPa. The network creators that tend to form tetrahedra or other coordinated networks with oxygen usually strengthen the glass. The effect of network modifiers (which lie in open space) on glass strength is usually negative.

Actual experimental values of strength are only from 10^{-3} to 10^{-2} of the theoretical strength σ_t of glass. In the search for an explanation for why the actual strength of glass is considerably lower than intrinsic one, it was noticed that various defects (e.g., impurities, bubbles, surface cracks) lead to the substantial reduction of glass strength and surface defects are the most significant. According to Griffith (1921), if there is a nanometer-deep crack on the surface, the stress tends to concentrate near the defect and the local stress can exceed the rupture limit of glass, even though the average stress stays below the limit. The Griffith equation for the actual tensile strength of glass σ is

$$\sigma = \sqrt{\frac{2E\gamma}{\pi c}}, \tag{8.2}$$

where c is the length of the preexisting crack (also called the Griffith flaw). The Griffith theory was confirmed by the following. Fresh glass fibers are much stronger than those after a few days of storage without preventing them from atmospheric water vapor. Water diffuses into glass, breaks the bond of silica, and creates cracks. The combination of moisture and stress causes microscopic flaws in the glass to propagate, resulting in fiber failure. Surface etching and flame polishing may eliminate deep cracks and increase the tensile strength. As a rule, thinner samples have better strength due to the lower probability of cracks with the critical depth. Decades ago researchers at Rolls Royce showed that the theoretical strength of silica glass could be obtained by simply redrawing glass into thin fibers using a flame torch (Proctor et al., 1967) with care taken to protect the surface from damage before testing. In the early years of optical fiber development, several industry labs demonstrated theoretical strength levels (France et al., 1983; Kurkjian and Gupta, 2001). To slow erosion, moisture-resistant polymeric coatings, such as UV curable silicone, polyimide, or silane compounds, were applied. The polymer also connects both sides of the microscopic surface cracks, preventing them from growing. At present, quartz optical fibers have tensile strength close to 5000 MPa at 10 km length, 150 μm diameter, and several microns of polymer coating.

Magnesium aluminosilicate glasses (S-glass family) are often the first choice for many structural applications because of their mechanical properties and temperature capabilities. However, their higher melt temperatures require more process energy, making these fibers more expensive than alumino-borosilicate E-glass. The pristine strength of glass fibers decreases as the fibers are exposed to increasing temperature (Hartman et al., n.d.). Glass fiber strength as a function of temperature is presented in Figure 8.1 and Table 8.1. For example, at liquid nitrogen temperature −196°C (77 K) the measured tensile strength of S-2 glass (SiO_2 64 − 66%, Al_2O_3 24 − 26%, MgO 9 − 11%) fibers is 8275 MPa compared to 4890 MPa at room temperature. At 538 °C (1000 °F) all three glass compositions exhibit tensile strengths that are roughly 50% of the room temperature measured value. On the other hand, they have increased strength at liquid nitrogen temperature.

Since it is impossible to locate the deepest crack in a sample, the actual strength of glass can vary from sample to sample. For example, the probability of failure $F(\sigma, l)$ of glass fiber with length l at applied stress σ can be described by the Weibull distribution:

$$F(\sigma, l) = 1 - \exp\left[-\frac{l}{l_0}\left(\frac{\sigma}{\sigma_0}\right)^m\right] \tag{8.3}$$

where l_0, σ_0, and m are the parameters to be derived after many rupture tests of similar samples. The scale parameter σ_0 is often called the characteristic strength, corresponding to the fracture stress with a failure probability of 63.2%. The Weibull modulus m is

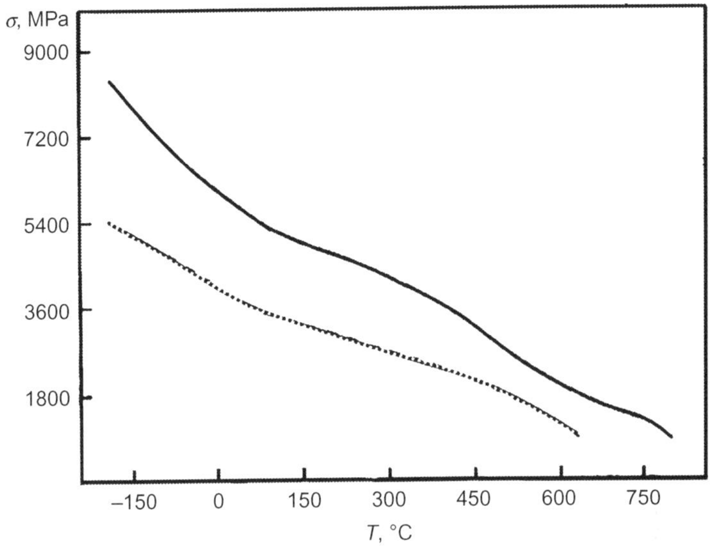

Figure 8.1 Glass fiber strength as a function of temperature. Solid curve—S-2 glass, dotted—E-glass.

Table 8.1 Tensile strength of various types of glass at liquid nitrogen, normal, and elevated temperatures

	E-Glass	S-1 Glass™	S-2 Glass®
Density, g/cm^3	2.58	2.54	2.46
Tensile strength, MPa			
− 196°C	5310	–	8275
23°C	3445	4135	4890
371°C	2620	2930	4445
538°C	1725	2140	2415

also called the shape parameter since it represents the scatter in the fracture strength. Both the Weibull modulus and the characteristic strength have to be positive values. The parameters have no physical meaning but they determine the estimated degree of scatter and the average fiber strength of the distribution. For a relatively small spread of the distribution, the parameter m is large and the Weibull plot has a rapid ascent in the vicinity of the ultimate value of σ. Consequently, parameter m determines the variations of the actual tensile strength. The reliability of the estimated Weibull parameters depends on the number of tested specimens. In general, it is agreed that a minimum number of 30 specimens are required for a good characterization of the strength of a brittle material such as glass.

8.2.2 Hydrogen penetration through glass

Generally, hydrogen permeates through materials in the form of hydrogen molecules, hydrogen atoms, and/or hydrogen ions. In any case, the hydrogen dissolved in a material affects many of its properties, e.g., it causes embrittlement of steel. Most of the hydrogen permeation through glass takes place in the form of hydrogen molecules passing through the large "holes'" induced in the structure with random arrangements of SiO_4 tetrahedra (Kurita et al., 2002). The permeation through a thin wall of a glass capsule is proportional to the product of hydrogen solubility and diffusion coefficient. Based on Henry's law of solubility, the concentration of hydrogen in glass is directly proportional to the applied gas pressure. Assuming diffusion and solubility coefficients of hydrogen to be independent of concentration, the amount M (number of moles) of hydrogen transported through a thin slab of glass in unit time can be expressed as:

$$\frac{\mathrm{d}M}{\mathrm{d}t} = K \cdot \frac{s(P_2 - P_1)}{h} \tag{8.4}$$

It is:

- proportional to the surface area s over which permeation takes place;
- proportional to the difference $P_2 - P_1$ in partial pressures of hydrogen across the slab; and
- inversely proportional to the thickness of the wall h. The temperature dependence of the hydrogen permeability K follows an Arrhenius type law: $K = K_0 \exp(-E_a/kT)$, where K_0 is the maximum permeability (formally at infinite temperature), k is Boltzmann constant and E_a is the so-called activation energy for permeation. For the variety of glasses, the hydrogen permeability K can be described by the empirical formula (Tsugawa et al., 1976) as a function of the absolute temperature T and the concentration κ of glass network creators:

$$K = 8.10 \times 10^{-14} \exp[(-1/T)(4550 + 127.80 \cdot \kappa)] \left[\frac{\mathrm{mol}}{\mathrm{m} \times \mathrm{s} \times \mathrm{Pa}}\right] \tag{8.5}$$

where κ is expressed in %. Fused quartz ($\kappa = 0$) has the highest permeability among all kinds of glass. For S-glass κ varies from 14 to 17. The permeability is strongly dependent on glass temperature T.

8.2.3 Hollow glass microspheres (HGMs)

HGMs are produced from a glass powder with particle size from 20 to 40 μm, mixed with or contains as part of the composition a blowing agent (e.g., sodium sulfate, sodium selenite, urea), which decomposes at high-temperature gas-air flame. As the temperature is quickly raised, the blowing agent decomposes and the resulting gas expands from within, thus forming HGMs. The HGMs then are quenched with a water spray and carried with the quench water and collected by flotation. The important primary formation parameters include the powder feed rate, air-to-gas ratio, flame velocity, and the length of the flame. HGMs have a variety of application: fillers for composite plastics, paints, varnishes etc.

Because they are hollow, they act as excellent insulators against heat, electricity, and even noise. Filled with deuterium–tritium mixture, they were even considered as targets for laser nuclear synthesis. According to Equation (8.5), hydrogen penetrates the walls of HGMs rapidly at elevated temperatures and sufficient pressure differences. At ambient temperature, the penetration rate is so small that they can safely be used as compressed hydrogen containers. Of course, the diffusion through the glass wall does not stop totally at ambient temperature, so that losses will occur continuously.

We define the *gravimetric capacity of the storage medium* G_c as the ratio of the hydrogen mass stored to the mass of the storage medium *filled with hydrogen* and the *volumetric capacity of the storage medium* V_c as the ratio of the hydrogen mass to the volume of the storage medium. These are useful benchmarks for comparing different storage methods. Obviously, both characteristics are important and various storage systems often have good performance in one and poor performance in the other. Comparing their values with DOE targets, we need to distinguish between the storage capacity of the medium and the storage capacity of a storage system. The weight of auxiliaries (e.g., valves for reducing the pressure, pipelines, sensors to control pressure, temperature and tightness, etc.) must be included in the calculation of the gravimetric capacity of a storage system. Operational characteristics such as the temperature, min/max delivery pressure, refilling time, cycle life, and fuel purity are also crucial to the performance of the storage system.

If the HGM wall thickness h is small compared to the HGM internal radius r, then we can consider the tension as uniformly distributed across the wall thickness and the tensile stress σ_s is determined by the following expressions (Timoshenko, 1956):

$$\sigma_s = \frac{pr}{2h} \tag{8.6}$$

Assuming σ_s to be equal to the actual tensile strength σ of glass, we find the ultimate (burst) pressure in HGMs:

$$p_u = \sigma \frac{2h}{r} \tag{8.7}$$

The amount of hydrogen that can be stored in HGMs at pressure p can be calculated using the equation of hydrogen state. A large part of the experimental thermodynamic data on hydrogen is reproduced by the empirical equations (Lemmon et al., 2010) within 0.1% accuracy and practically all data within 0.5%.

In early experiments (Akunets et al., 1994), small number of carefully selected HGMs made of various glasses and polymers were chosen to determine the ultimate pressure p_u. Laser interferometry was used to select HGMs with perfect spherical form. Table 8.2 represents the calculated gravimetric G_c and volumetric V_c, capacity of the imaginary hydrogen storage medium consisting of closely packed magnesium aluminosilicate HGMs tested in (Akunets et al., 1994).

According to the results of the experiment, the HGM medium could have more than 30% gravimetric capacity and relatively high volumetric capacity. On the other hand, recent experiments (Kohli et al., 2007) with HGMs made of boron-containing glasses have shown less than 30 MPa burst pressure resulting in much lower volumetric

Table 8.2 **Storage capacity of the strongest HGMs tested in (Akunets et al., 1994)**

h/r	p_u, atm	V_c, g/l	G_c, %	V_{cw}, g/l	G_{cw}, %
0.024	1790	68.76	26.5	42.9	12.8
0.0117	718	40.15	30.6	26.7	15.7

capacity of 15 g/l and gravimetric capacity 17%. It should be taken into account that V_c, and G_c, listed in Table 8.2, correspond to the burst pressure p_u, while the working pressure, according to the current regulations, must be at least 2.25 times lower. The recalculated values, V_{cw} and G_{cw}, corresponding to the working pressure, are substantially lower than V_c and G_c.

HGM size, wall thickness, and glass type determine not only the ultimate pressure, but the permeation rate that must be acceptable at high temperature and very low at ambient temperature. Using Equation (8.4), the half-life time τ of hydrogen inside HGMs due to the leakage through the walls can be estimated as:

$$\tau \sim \frac{rh}{K \cdot RT}, \tag{8.8}$$

where $R = 83.143$ J/(mol K) denotes the universal gas constant. At fixed pressure p_u, it is proportional to r^2. As the processes for permeation into and out of HGMs is fully reversible, a compromise with respect to glass composition, HGM radius, and the wall thickness have to be found. Assuming HGMs to be heated to 300 °C to release hydrogen at acceptable rate, activation energy for permeability through glass must exceed 57 kJ/mol and the microsphere radius must be smaller than 40 μm. Since common borosilicate glasses have only around 40 kJ/mol activation energy, they must be heated to even higher temperature 600°C, which is too high a temperature for the following reasons. When hydrogen pressure inside HGMs significantly (from 2 to 3 times) increases, the strength of glass significantly decreases. Thus, all HGMs will probably be broken.

8.2.4 Advantages and limitations of HGMs

The tensile strength of glass can be much higher than that of steel. The technology of HGMs production is inexpensive, safe, and requires relatively low energy consumption. A large installation would be possible that loads the HGMs followed by handling the filled HGMs at ambient pressure and temperature. Irregularly shaped containers for the loaded HGMs provide the flexibility needed for mobile application of hydrogen. Filled with highly pressurized hydrogen HGMs are relatively safe compared to conventional steel or composite tanks. Indeed, since both volume and the probability of simultaneous destruction of all HGMs are small, the amount of accidental release of hydrogen is also small.

However, HGMs have the following disadvantages. Their strength depends upon the ideal spherical shape, but it is very difficult to control the diameter and the form of microspheres during the process of manufacturing. Usually, eccentricity and diameter spread of HGMs lead to the collapse of the spheres under the high external pressure applied during the hydrogen filling process. The most substantial disadvantage is that hydrogen can penetrate inside and outside HGMs at an acceptable rate exclusively due to the diffusion through the walls. This rapid diffusion requires the elevated temperature at which hydrogen pressure inside the microspheres can increase above the breakage limit. Furthermore, energy consumption is needed for considerable heating, while the thermal conductivity of HGMs is poor. Due to the above reasons HGMs are not suitable for onboard hydrogen storage until more effective hydrogen retrieving methods will be found.

8.2.5 Enhanced hydrogen retrieving from HGMs

To solve the problem of slow hydrogen retrieving from HGMs, Shelby and co-workers (Alfred University) examined (Halvorson and Shelby, 1998; Rapp and Shelby, 2004; Shelby et al., n.d.; Shelby et al., 2007) the effects of applying various types of radiation (e.g., microwave, infrared, ultraviolet, ultrasonic) in the hope to enhance and control hydrogen permeation through glass. The noticeable effect was discovered with infrared radiation. Particularly, the authors observed (Rapp and Shelby, 2004) that hydrogen permeation through a slab of borosilicate glass doped with specific metal oxides (F_3O_4; CoO; NiO; V_2O_5 or Cr_2O_3) is accelerated when exposed to an incandescent heating lamp, compared with heating in a furnace at 400°C. The outgassing response shown in Figure 8.2 can be considered an indicator of hydrogen release rate. The authors suggested that infrared radiation is contributing the activation energy necessary for hydrogen diffusion. It has been clearly demonstrated that radiation in the range of 1.5–2.2 μm is responsible for the so-called photo-induced hydrogen diffusion effect. Photo-induced hydrogen diffusion has been applied then to the problem of storage of high-pressure hydrogen in HGMs. Samples of these HGMs have been successfully filled to 1500 psi (10.3 MPa) and 5000 psi (34.5 MPa with minimal loss from crushing of the microspheres. The hydrogen release rates obtained by exposing selectively doped glasses to lamp radiation are superior to those obtained from furnace heating. Undoped glasses exhibit a poorer hydrogen outgassing response using lamp exposure compared to furnace heating. The amount of hydrogen released increases exponentially with applied voltage, i.e., infrared radiation intensity. The rate of hydrogen release increases with increasing concentration of the metal oxide and with increasing hydrogen fill pressure within the microspheres. The spectrum and source of the infrared radiation has been found to be extremely important.

The mechanism for photo-induced outgassing is still unknown. Heating of the sample during photo-induced outgassing does occur over extended time of exposure to infrared radiation, nevertheless, measurement of the temperature of the HGMs at equilibrium indicates that the HGMs reach a temperature of approximately 120°C very slowly compared to the almost immediate hydrogen release at switching on the lamp. Thus, photo-induced outgassing cannot be a result of a purely thermal effect, but may

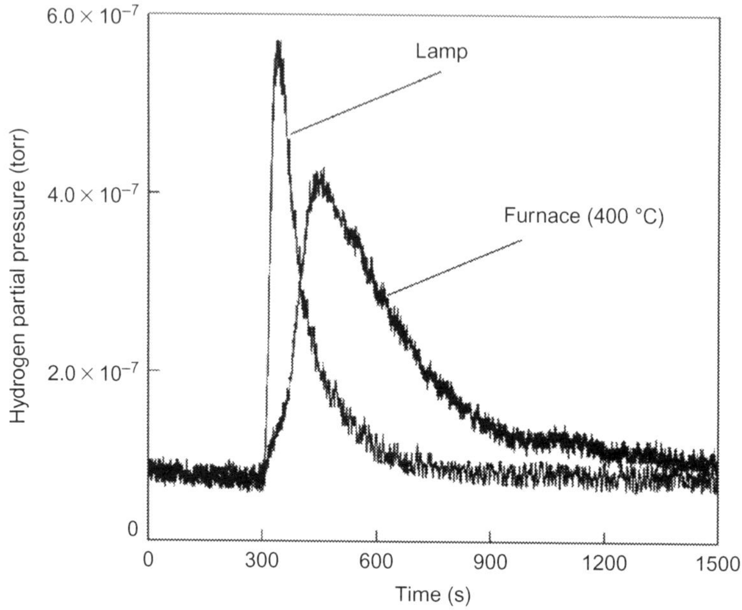

Figure 8.2 Effect of heating method on the hydrogen outgassing response in 0.5 wt% Fe_3O_4 doped Corning Glass Works commercial glass 7070.

be related to the excitation of a metal—hydrogen bond vibrations caused by infrared radiation of specific wavelength. Expansion of the bonds probably opens the doorways between interstices, allowing dramatically increased gas mobility. On the other hand, Kitamura and Pilon (2009) showed that in some cases photo-induced outgassing can be due exclusively to more effective heating with an infrared lamp than with a furnace. The point is that for furnace heating, the silica tube (where borosilicate glass samples were placed) absorbs a large fraction of the radiation emitted by the furnace around wavelength 4.3 μm. This results in a delay in the temperature rise and a reduction in the sample temperature and the hydrogen release rate. On the contrary, the radiation emitted by a heating lamp has a peak emission between 1 and 2 μm and reaches the samples since the silica tube is nearly transparent at wavelengths up to 3.5 μm. However, between 0.8 and 3.2 μm borosilicate glass does not absorb significantly and needs to be doped with metal ions that can significantly increase the absorption coefficient in the wavelength of silica tube transparency. Thus, doped samples heat up much faster than undoped ones when exposed to infrared heating lamps.

In any explanation of photo-induced outgassing, it is not clear how infrared radiation can help to increase the hydrogen-retrieving rate from the bulk of HGMs. Indeed, due to the transmission loss in the doped glass, infrared radiation cannot penetrate at sufficient depth into the bulk. Microwave radiation with longer wavelength could be better than infrared radiation since microwaves can penetrate deeper. In this case, the glass dopants must have the high absorption coefficient in the microwave radiation frequency range. Carbonaceous materials, e.g., fullerene C_{60}, are suitable candidates (Qin and Brosseau, 2012).

The Savannah River National Lab has developed porous-wall hollow glass microspheres (PW-HGMs) (Wicks et al., 2008; Heung et al., n.d.). Borosilicate HGMs were acid leached in hydrochloric acid at 580°C for 15.5 h to preferentially remove a boron oxide leachable phase, thus leaving behind a silica rich phase with interconnected porosity of the walls that can be produced on a scale of 100–300 nm. Some spheres breakage was noted following acid leaching due to mechanical weakening of the walls. The ultimate concentration of SiO_2 in the resulting PW-HGMs can be up to 95% (Qi et al., 2012). Unfortunately, PW-HGMs cannot be used for the storage of highly compressed hydrogen due to their reduced strength and excessive hydrogen permeability of the walls. However, the wall porosity can be used to generate new nanostructures of hydrogen absorbents and then hydrogen can enter the microspheres through the pores and be stored on absorbents inside. Complex hydrides encapsulation in PW-HGMs microspheres was proposed and researched (Mohtadi et al., 2011); in particular, encapsulation of sodium alanate ($NaAlH_4$) within the PW-HGMs by diffusion through wall porosity was illustrated.

8.2.6 Glass capillary arrays

It is well known that capillary arrays made of glass or other materials are commonly used in X-ray optics (MacDonald, 2010; Gao and Janssens, 2004), photonics (http://www.tegs.ru/en/glass/caps.shtml, n.d.; http://www.xos.com/products/x-ray-optics-excitation-systems-x-beam/optics/polycapillary-optics/x-tra-polycapillary-focusing-optics/, n.d.), and are the choice for chromatography columns. The array shown in Figure 8.3 can contain millions of capillaries with a diameter down to one micron or even less and a wall

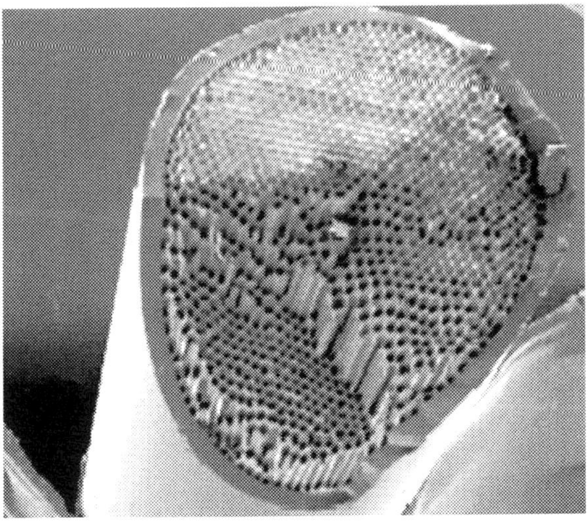

Figure 8.3 SEM image of the capillary array.

thickness-to-radius ratio less than 5%. Furthermore, capillary arrays can be tapered by any tapering envelope. The process of manufacturing of capillary arrays is usually divided into three stages: drawing capillaries with relatively large initial diameter from a preform, redrawing them into a bundle of capillaries with smaller diameter, and sintering capillaries into the array (http://www.tegs.ru/en/glass/caps.shtml, n.d.).

Zhevago and co-workers (Zhevago, 2006; Gnedenko et al., n.d.; Zhevago and Glebov, 2007) proposed using capillary arrays for the mobile storage of hydrogen. Similar to microspheres but in contrast to the tanks, the amount of hydrogen in each individual capillary is very small, preventing the possibility of explosions by improper handling or during accidents. The strength of the capillary arrays and safety of hydrogen storage appears to be comparable to these of HGMs with similar wall thickness-to-radius ratio. Compared to the tanks the capillary storage system can have any desirable dimension and form. There are also substantial advantages of capillary arrays over HGMs, including:

- Capillary diameter and shape can be controlled with good precision during the manufacturing of capillary arrays;
- Better packing ratio than the hollow spheres (less unused space between the capillaries);
- No need for slow diffusion procedure because loading and releasing of hydrogen can be consequently performed through the open ends of the capillaries;
- Hydrogen can be stored and retrieved at cryogenic temperature;
- Capillary arrays can be filled with any kind of pressurized gas, e.g., hydrogen–methane mixtures that bridge the gap between conventional fossil fuels and the clean future of a hydrogen economy; and
- Similar to PW-HGMs they can be used to generate new nanostructures of hydrogen absorbents inside.

The ultimate (burst) pressure in a single cylinder capillary is determined by (Fryer and Harvey, 1998):

$$p_u = \sigma \frac{(r+h)^2 - r^2}{\left[r^2 + (r+h)^2\right]} \tag{8.9}$$

If the wall thickness h is small compared to the internal capillary radius r, then $p_u = \sigma h/r$. The calculated gravimetric (dash-dot curves) and volumetric capacity (solid curves) of closely packed cylinder capillaries made of S-2 glass are presented in Figures 8.4 and 8.5 for normal and liquid nitrogen temperature, respectively, on the assumption that capillaries with 10 μm or less wall thickness can have the same tensile strength as the fibers have (Table 8.1). The packing ratio of capillaries is $\pi/2\sqrt{3} \simeq 0.907$, the working pressure is 2.25 times less than p_u. The horizontal lines indicate the DOE 2017 and ultimate targets.

It follows from the calculations that in the wide range of h/r the gravimetric capacity of the capillary arrays stays above the ultimate DOE target 7.5%. The gravimetric capacity and the volumetric capacity have different tendencies with growing ratio h/r: the gravimetric capacity falls down gradually, but the volumetric capacity grows until

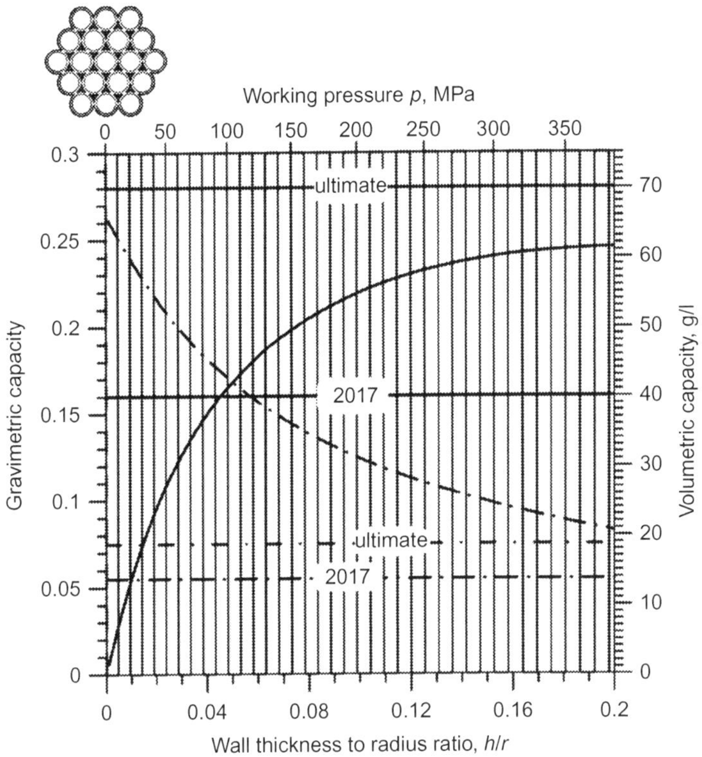

Figure 8.4 Volumetric (solid curves and right *y*-axis) and gravimetric capacity (dash-dotted, left) of the S-2 glass capillary array at room temperature versus the ratio of the capillary wall thickness to radius (or the working hydrogen pressure). DOE target values are shown by the corresponding horizontal lines.

the negative influence of the wall thickness on the storage volume begins to prevail over the increasing hydrogen density. At normal temperature, the DOE 2017 target for the volumetric capacity can be reached at pressure 94 MPa and $h/r = 0.046$, but it is beyond the ultimate target of 70 g/l at any pressure. It should be kept in mind that the actual energy needed for the compression of hydrogen to 94 MPa is about 16% of its calorific value. Further compression needs even more energy, but does not lead to the proportional rise of the volumetric capacity. If the temperature is 77 K, it becomes possible to reach the DOE 2017 targets at quite moderate pressure 15 MPa. Furthermore, at 65 MPa the ultimate DOE targets can be reached. The theoretical maximum of the volumetric density of the S2-glass capillary array is around 90 g/l, which can be reached at 77 K and pressure above 220 MPa. However, it should also be kept in mind that the *volumetric capacity of the storage system* may be still less than the ultimate DOE target because of the increased volume required for the cooling system.

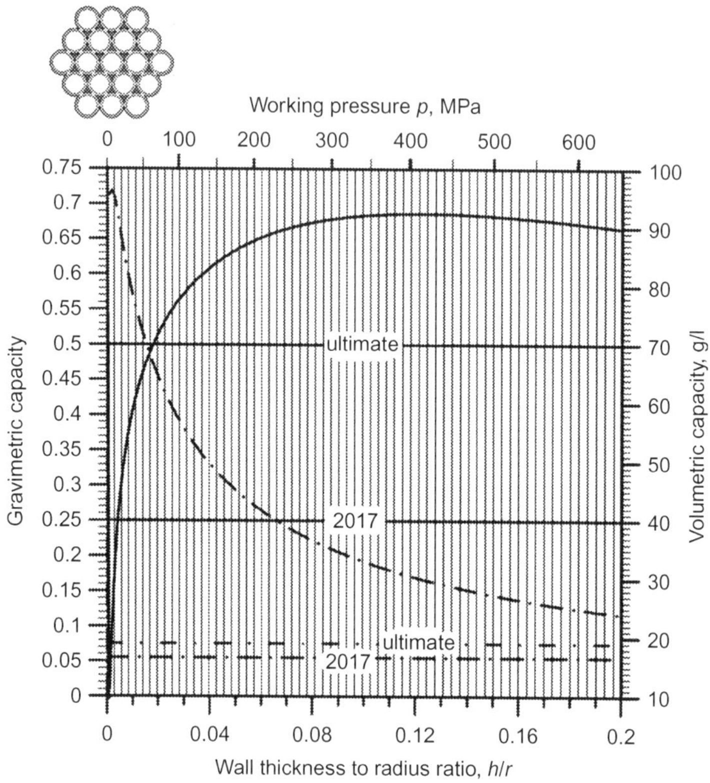

Figure 8.5 Similar to Figure 8.4, but for temperature of 77 K.

8.2.7 Methods of hydrogen encapsulation and retrieving

Various methods of hydrogen encapsulation and retrieving have been developed (Gnedenko et al., n.d.; Zhevago et al., n.d.; Zhevago and Denisov, n.d.) that do not need hydrogen diffusion through the capillary walls. In any case, the array has to be sealed as shown in Figure 8.6 or closed with plugs at one end. The opposite end of the array may be closed with a diffuser plate (Gnedenko et al., n.d.) that includes a layer impermeable to hydrogen material perforated with a plurality of small holes plugged with stoppers that seal the holes. The density of the holes is such that at least one sealed opening is formed for each capillary of the array. If the coefficient of thermal expansion of the layer is higher than the coefficient of thermal expansion of the stoppers, then the holes with stoppers can operate as thermal microvalves.

Another method is closing capillaries with metal stoppers directly inside the autoclave (Zhevago and Denisov, n.d.; Zhevago et al., 2010). A capillary array is placed in the autoclave together with a crucible containing a solid metal alloy. Injected into the autoclave at pressure p_0, hydrogen freely flows into the capillaries through their open ends. Afterward, the autoclave is heated above the melting point T_m of the alloy and the

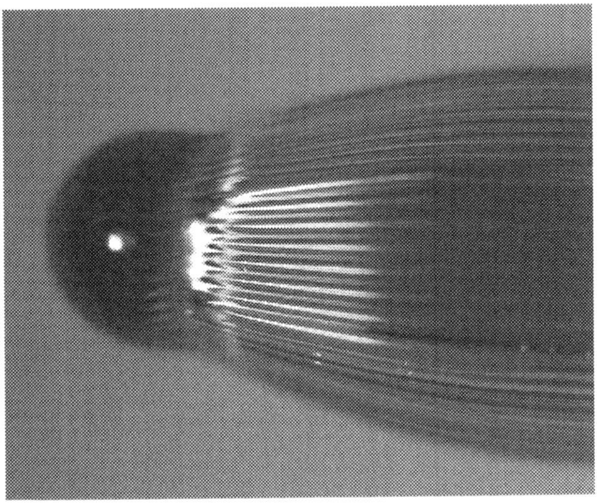

Figure 8.6 Glass capillary array with a sealed end.

open ends of the capillaries are immersed in the melt. Due to heating of the autoclave, the pressure of hydrogen increased from p_0 up to p_T. To force the melt penetrate inside the capillaries at definite length l hydrogen pressure must be elevated to $p_i = p_T + \Delta p$, where the value of Δp usually varies from 0.05 p_T to 0.1 p_T. Then the capillary array with liquid stoppers is removed out of the melt and cooled down below the solidification temperature of the alloy. As a result, a solid stopper is formed inside every capillary of the array. The stoppers prevent hydrogen from escaping out of the capillaries when the pressure in the autoclave is finally set to atmospheric and the capillary array, filled with pressurized hydrogen, can be removed from the autoclave. The alloy for the stoppers must have low enough melting temperature, good adhesion to glass, and should not stress glass too much due to the difference in the linear expansion coefficients of glass and the alloy. Indium and In52Sn alloy are good candidates for the stopper material.

The third method (Zhevago et al., n.d.) is just gluing capillary arrays into a clutch that can be connected with a compressor pipeline. To ease the procedure, relatively thick arrays should be tapered as shown in Figure 8.7.

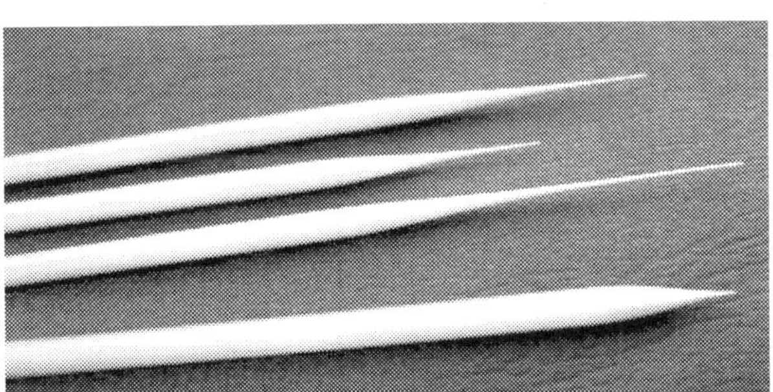

Figure 8.7 Tapered capillary arrays.

8.2.8 Flexible glass capillaries

Flexible capillaries and multicapillaries have some advantages over rigid capillary arrays. They can be produced using the precise optical fiber technology. A glass fiber is usually drawn out of the preform of 10–25 mm in diameter and up to 1 m in length. The preform first passes through a furnace, where it is heated to the softening point of glass. As the fiber is pulled from the preform at speed around 10 m/s, measuring devices monitor its diameter and its concentricity, while another device applies a protective coating. The fiber then passes through a curing furnace and another measuring device that monitors diameter, before being wound on a spool. The structure of the fiber repeats the structure of the preform. For example, if the preform is a capillary array, the fiber will be a flexible multicapillary. Compared to the rigid capillary arrays, they have a relatively small diameter enabling one to attach them more easily to a compressor pipeline (Zhevago et al., 2013). They can be also used for the transportation of highly pressurized hydrogen or other gases. In this case, they should be merged in a cable similar to the cables for optical fiber communications. Since the weight of the cable per unit length is not as important as it is for the mobile hydrogen storage, the cable may contain additional shields of the cable from shocks and flame. Capillary cables are flexible, have relatively small diameter, and therefore can be rapidly deployed in any place. Working pressure of the capillary cables can considerably surpass that of steel pipes. Therefore, there is no the problem of steel embrittlement and gradually loss of pipe strength.

8.2.9 Experimental results and prototypes of capillary vessels

First experiments with hydrogen encapsulation into the capillary arrays were performed by researchers from the National Research Centre "Kurchatov Institute" (Zhevago et al., 2010). After being pressurized up to 250 MPa hydrogen was generated in the hydride compressor due to the disintegration of vanadium hydride inside the closed vessel when it was heated. Capillaries were made from Heraeus HLQ 210 quartz and had near 480 μm external diameter and 25 μm wall thickness. They were sealed at one end, covered by a thin layer of Desotech DSM 950–076 epoxy resin, and 19 of them were merged together via epoxy polymerization under ultraviolet radiation. Another kind of a capillary array was made from borosilicate glass. Closely packed hexahedron capillaries were placed inside a cylinder glass shell and merged together during the process of redrawing. The geometry of the array is shown in Figure 8.8. The measured diameter of the redrawn cylinder shell was $D = 1032$ μm and the wall thickness $S = 30$ μm. There was a honeycomb lattice inside the shell with space period $d = 1032$ μm, and the thickness of the wall separating adjacent cells $s = 0.7$ μm. Hydrogen encapsulation was done using stopper formation inside the autoclave. The highest-pressure values were obtained with the quartz capillary arrays reinforced with the epoxy resin. They withstood up to 171 MPa pressure of hydrogen stored at room temperature resulting in 48.3 g/l volumetric and 10.25% gravimetric capacity of the storage medium. The corresponding results for the borosilicate honeycomb arrays were lower: $p_u = 27$ MPa, $V_c = 15.6$ g/l, and $G_c = 3.2\%$.

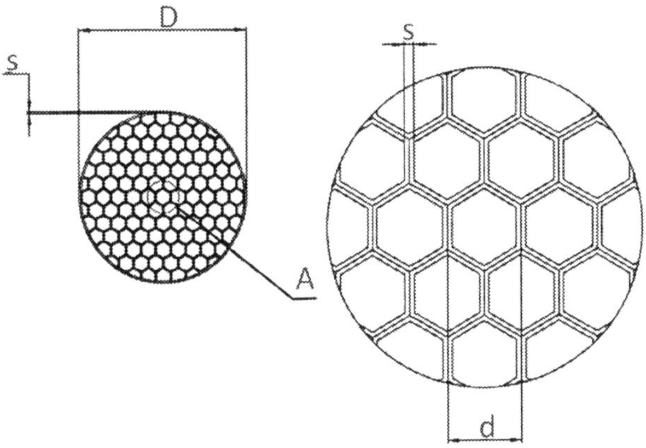

Figure 8.8 Schematic view of the cross section of the honeycomb capillary array from borosilicate glass used in the experiments (Zhevago et al., 2010).

C. En. Ltd. (Switzerland) and Bundesanstalt für Materialforschung und -prüfung (Germany) performed detailed investigations of single capillaries (Holtappels et al., 2009) to find the optimal parameters for future capillary arrays. The pressure resistance of the capillaries was determined in dependence of glass composition, radius, and wall thickness. Capillaries were glued into small 1.16 in stainless steel pipes to connect them to a set-up consisting mainly of a hydrogen supply, a compressor, and a buffering vessel. The main results are summarized in Table 8.3. The considerable spread of the measured burst pressure (from p_u^{min} *to* p_u^{max}) indicates that the tensile strength of the capillaries was determined rather by surface cracks and bulk defects than the glass composition. The highest values of the burst pressure, from 73.7 to 124.2 MPa, were measured with borosilicate samples. Calculations show that if the storage medium could be made of closely packed capillaries of the kind, then the volumetric capacity of the medium would vary from 33.6 to 46.2 g/l and the gravimetric capacity from 7.05% to 9.46%. Based on the patented technology (Zhevago and Glebov, 2007; Zhevago et al., n.d.; Zhevago and Denisov, n.d.) the prototype of the capillary storage system was developed and demonstrated (Holtappels et al., 2012), which showed $35 \times 10 \times 3$ cm^3 volume occupied by 23 capillary arrays with a total number of capillaries 530,000 (Figure 8.9). Unfortunately, designed to operate at pressure below 20 MPa and ambient temperature, it could not ensure high enough volumetric and gravimetric capacity of the storage system. According to Figure 8.4, the theoretical limit for its volumetric capacity is only 14 g/l. However, Incom (http://www.incomusa.com/solutions-for-industry/hydrogen-storage/, n.d.), in conjunction with C.En. Ltd. and BAM, has teamed up to develop a storage system achieving a pressure resistance of a minimum 150 MPa (Figure 8.10).

Experiments with flexible capillaries were performed at room and cryogenic temperatures (Zhevago et al., 2013). Flexible quartz capillaries were produced using fiberoptics technology. They were around 500 m long and had various geometrical parameters, such as the internal radius and wall thickness, and were coated with epoxy resin. The actual weight per unit length of the capillaries was measured with precision

Table 8.3 Burst pressure measured for various capillaries in Holtappels et al. (2009)

Glass composition	Density ρ, $g \times cm^3$	Outside diameter, μm	Inside diameter, μm	Length, mm	p_u^{max}, MPa	p_u^{min}, MPa
Soda-lime	2.52	400	300	100	114.7	25.0
Borosilicate	2.33	400	360	200	124.2	73.7
Aluminosilicate	2.65	340	300	200	62.7	32.6
Quartz	2.2	400	300	200	109.1	39.4

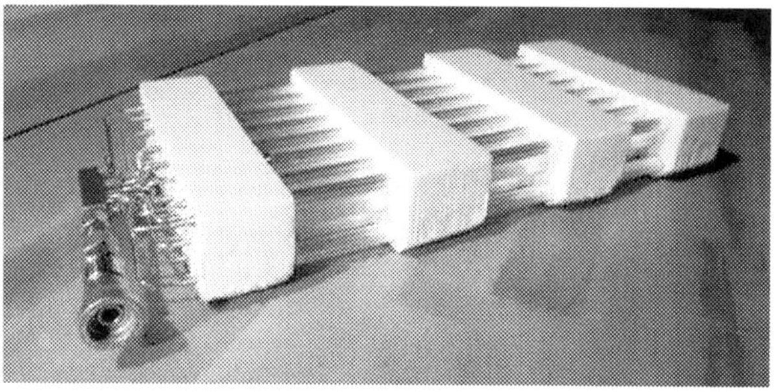

Figure 8.9 Prototype of the capillary storage system (Holtappels et al., 2012) operating at 20 MPa.

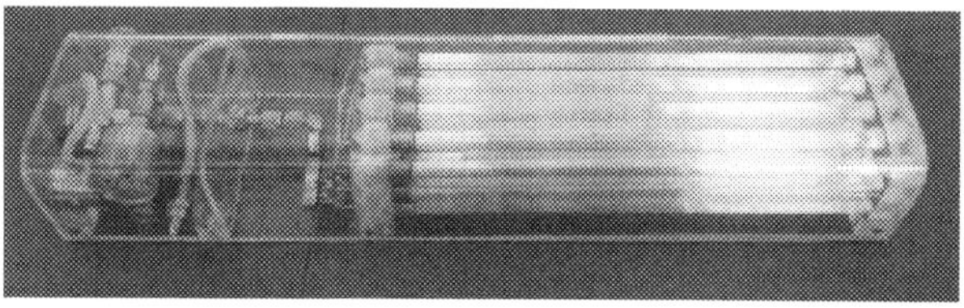

Figure 8.10 Incom's borosilicate glass capillary storage system (http://www.incomusa.com/solutions-for-industry/hydrogen-storage/, n.d.).

0.1%. At first, the tensile strength of short (0.5 m) capillaries was estimated with a stress rupture machine, as the ratio of the breaking tensile force to the cross-section of the quartz core of a capillary. Then 30 cm to 100 m long samples were subjected to the internal hydraulic pressure supplied by the two-stage hydraulic booster.

The capillary near their tips were stripped off the polymer coating using acetone and glued into the pipeline of the hydride compressor. They were cooled down to 77 K in the Dewar vessel with liquid nitrogen. The main results obtained for three different types of quartz capillaries are presented in Table 8.4. The Type 1 and 3 capillaries were made of quartz produced with CVD method while the Type 2 capillaries were made of less expensive fused quartz. The gravimetric and volumetric capacities of the tested capillaries are shown in Figures 8.11 and 8.12 by the solid curves. The marks at the curves correspond to the capillary type and the dots on the curves indicate the experimental values of the pressure applied. Over 9% and 32 g/l volumetric and

Table 8.4 Characteristics of flexible capillaries used in the experiments (Zhevago et al., 2013)

Capillary type #	1	2	3
Mean internal diameter, μm	134	224	200
Mean external diameter, μm	220	272	225
Mean diameter with polymer coating, μm	286	389	310
h/r	0.637	0.214	0.125
Weight per unit length, mg/m	73.1	118	55.5
Estimated maximum tensile strength σ, Mpa	2030	1300	2640
Hydraulic burst pressure, MPa	130–250	100–250	160–230
Hydrogen burst pressure, MPa	184@77 K, 233@293 K	74.5@77 K	85.0@77 K

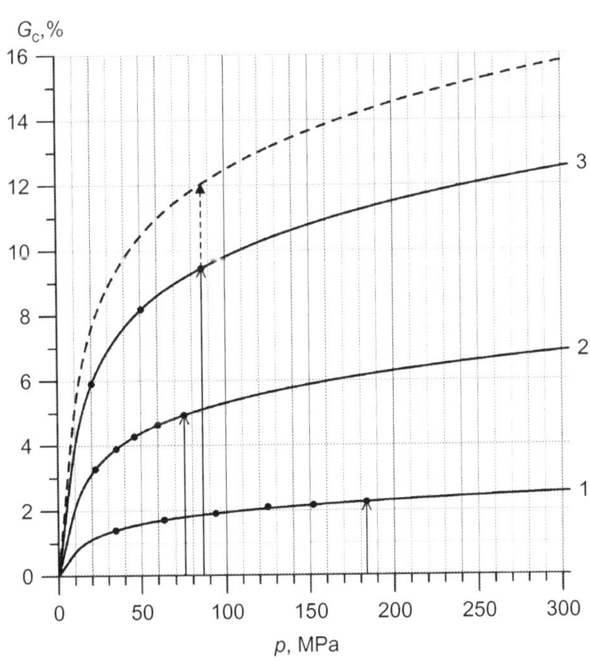

Figure 8.11 Gravimetric capacity of quartz capillaries at 77 K versus hydrogen pressure. The values of the applied pressure are marked by dots on the solid curves. The arrows indicate the maximum values of hydrogen pressure achieved in the experiments. The dashed curve represents the possible result for the capillary number 3 if the coating is reduced to 10 μm.

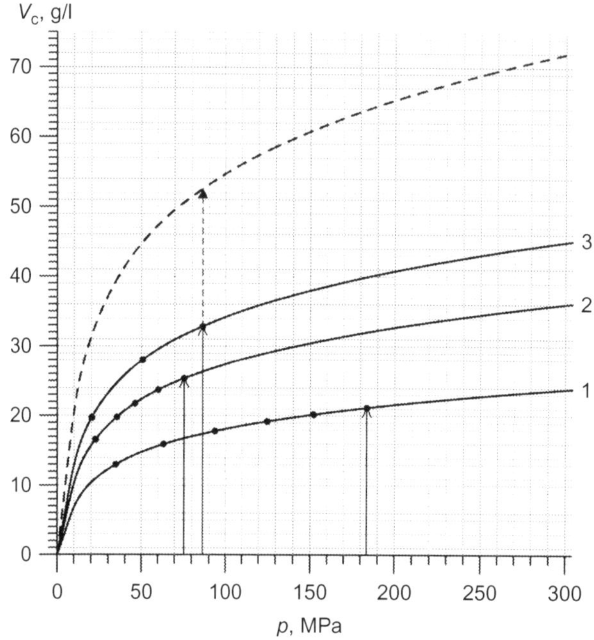

Figure 8.12 Similar to Figure 8.11, but for the volumetric capacity.

gravimetric capacities were reached at 77 K and relatively moderate pressure 70 MPa with Type 3 capillaries. Furthermore, the values could be even higher, if the thickness of the polymer coating of the capillaries could be reduced from 42 to 10 μm. Possible enhancements are illustrated by the dashed curves. It follows from the experimental results that due to the rapidly decreasing compressibility of hydrogen at cryogenic temperature and pressure higher than 40 MPa, it is unreasonable to store hydrogen at pressures above 40 MPa. For example, at 35 MPa and liquid nitrogen temperature $G_c = 7\%$ and $V_c = 25$ g/l were obtained at safe pressure (safety factor 2.25), despite the excessive thickness of the polymer layer. Note that at 184 MPa and 77 K the density of hydrogen in the capillary was 105.8 g/l, i.e., high above the density 70.85 g/l of liquid and 70.6 g/l of solid hydrogen.

The design (Denisov and Zhevago, 2012) of the capillary vessel is schematically presented in Figure 8.13. A number of flexible capillaries with the internal radius r and wall thickness $h \ll r$ are tightly winded around a spool and their ends are glued into the holes in the spool body. Hydrogen flow inside or outside the capillaries can be controlled with the rotating disc with a single hole. It works as a microvalve for many capillaries. The spool is not subjected to high pressures and can be made of thin and light material. If the capillary loops are merged together with polymeric compounds, there is no need for the spool at all. If the capillary vessel is used at low temperature, it should be placed in a Dewar vessel with a coolant circulating between the capillary loops (Zhevago, 2012). The internal space of the spool can be used as a buffer where the pressure should be supported at the level 0.5–1 MPa needed for fuel

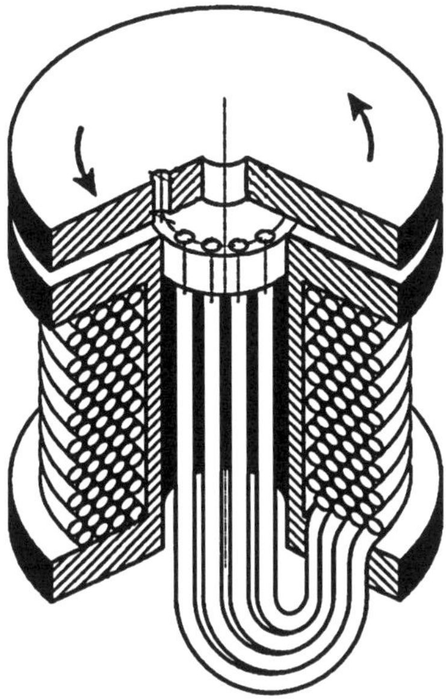

Figure 8.13 Design of the hydrogen storage system with flexible capillaries.

cell functioning. The probability of catastrophic release of total amount of stored hydrogen can be reduced, if a large number of flexible capillaries is used in the vessel. If some part of them fails in an accident, the rest may survive due to the independent outlet control with the rotating valve.

8.3 Hydrogen physisorption in porous materials

Physisorption is the mechanism by which hydrogen is stored in the molecular form on the surface of a solid material due to the weak, compared to the chemical bonding, van der Waals force between molecular hydrogen and solids. Therefore, low heat of adsorption released during physisorption is an advantage for the mobile storage of hydrogen. Moreover, since hydrogen does not chemically react during adsorption, it does not accumulate impurities that can poison the fuel-cell downstream. Physisorption does not involve bulk solid diffusion or chemical dissociation, therefore, it is fast and fully reversible, enabling meeting both the cycle-life and refilling-time demands. On the other hand, due to the weak interaction, the amount of hydrogen stored at room temperature is very low. It increases at a lower temperature and higher pressure, which is why more recent investigations of hydrogen adsorption are mainly focused on liquid nitrogen temperature.

The quantities used to characterize nanoporous materials are the specific surface area (SSA) and the specific pore volume (SPV). The SSA is generally determined from the nitrogen adsorption isotherms at 77 K using the Brunauer–Emmett–Teller

(BET) equation describing adsorption where the adsorbate exceeds a monolayer on the surface of a solid. Since physisorption of hydrogen takes place only on the surface, only porous materials with a high SSA are of practical interest. However, porous materials are not fully characterized by their SSA and SPV. The hydrogen adsorption depends also on the size of the pores. Pores are classified into the following three sizes: micropores (<2 nm), mesopores (2 – 5 nm), and macropores (>5 nm). Micropores are the most valuable for hydrogen physisorption. If pressure continues to rise gradually at a fixed temperature, the adsorption of hydrogen tends to rise with pressure. The absolute amount of hydrogen gas per unit volume of the adsorbent also contains pressurized gas in the pores, which is not adsorbed on the surface. The uptake of hydrogen at high pressure is generally represented as the excess and total adsorption amount. The excess adsorption, which is also known as the Gibbs excess, is the difference in the amount of hydrogen in gas phase that would be present in the equivalent volume of the adsorbed phase in the presence and absence of adsorption. Since the efficiency of gas compressing within the micropores is less than that achieved in a free volume, the excess adsorption will reach a maximum at some pressure (typically 5 – 10 MPa) and then decrease.

Activated carbon adsorbents are attractive for physisorption storage systems due to their simplicity, light weight, and low manufacturing cost, but their cooling by liquid nitrogen or refrigeration is required, in addition to pressurization up to 5 MPa. Carbon adsorbents are mainly prepared by a two-step pyrolysis of carbon-rich precursors with the carbonization process followed by activation at higher temperatures, either by reaction with an oxidizing gas or with inorganic chemicals. The SSA of commercial regular grade activated carbon is 700–1800 $m^2\ g^{-1}$, while that of activated carbon produced by the treatment of potassium hydroxide (KOH) can be above 3000 $m^2\ g^{-1}$. Maxsorb has 3300 $m^2\ g^{-1}$ (based on BET measurements), but it requires a complicated, expensive production process: various kinds of petroleum coke were mixed with an excess amount of KOH and dehydrated at 400°C, followed by activation at 600 – 900°C in an inert atmosphere. The remaining KOH was removed by washing with water after the activation.

Hydrogen uptake μ_H is the ratio of mass of adsorbed hydrogen to the mass of the adsorbent. It is slightly different from the gravimetric capacity G_c defined above as the ratio of mass of hydrogen to the mass of the storage medium *with hydrogen*. The relation between G_c and μ_H is

$$G_c = \mu_H (1 + \mu_H)^{-1} \tag{8.10}$$

The theoretical model of physisorption was proposed by Züttel et al. (2004), where a molecular monolayer possessing the density of liquid hydrogen is ideally adsorbed with a closely packed 2D geometry on both sides of a perfect graphene sheet. It leads to a linear relationship between the hydrogen uptake μ_H, and the SSA A:

$$\mu_H = kA \tag{8.11}$$

If μ_H is measured in % and A in m^2/g, then the theoretical proportionality coefficient k is 2.28×10^{-3} (%g/m^2). At room temperature, maximum hydrogen uptake of 2.7% was found only at extremely high pressures of 50 MPa (Nijkamp et al., 2001). By lowering adsorption temperature to 77 K, the uptake increases and can reach 5.7% in Maxsorb, at moderate pressure of 3 MPa. Most efforts to date have focused on attaining a high gravimetric capacity, which can be at direct odds with achieving a high volumetric capacity $V_c = \mu_H \rho$. This is because materials with a very high surface area also tend to exhibit an increased micropore volume and, consequently, an inherently low bulk density ρ. For example, the gravimetric capacity of Maxsorb is about three times larger than that of regular grade activated carbon. On the other hand, the bulk density of Maxsorb is 0.3 g cm^{-3}, which is considerably smaller than 0.55 g cm^{-3} of commercial activated carbon with a surface area of 1000 m^2 g^{-1}, therefore, the volumetric capacity of Maxsorb is only about 1.6 times larger than that of regular grade activated carbon. It should be taken into account that the volumetric capacity depends significantly on the morphology and shape of the activated carbon, i.e., powder, fiber, or granular, and, consequently, on the packing ratio of adsorbents inside a container.

The measured values of hydrogen uptake for carbon adsorbents with various SSA are presented in Figure 8.14. Panella et al. (2005) found that $k = 1.91 \times 10^{-3}$ (%g/m^2) (dashed-dotted line) for all tested samples including activated carbons and carbon nanotubes at 77 K, smaller than the theoretical value, while Xua et al. (2007) found that $k = 2.35 \times 10^{-3}$ (%g/m^2), very close to the theoretical value. Gao and co-workers (Wang et al., 2009) applied both physical and chemical activation methods to commercial activated carbons, showing that unlike CO_2 activation (physical), KOH activation remarkably altered the pore structures by promoting the development of very narrow

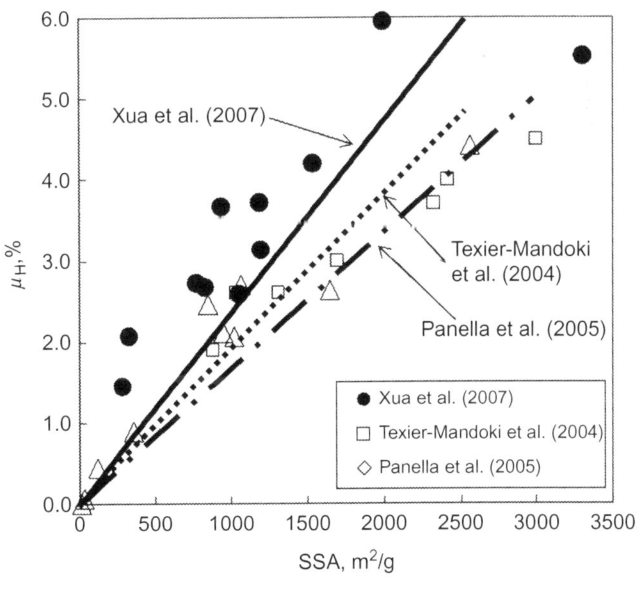

Figure 8.14 Hydrogen uptake in various activated carbons versus their specific surface area.

micropores. The resulting activated carbon AC-K5 had a hydrogen uptake of 7.08% at 77 K and 2 MPa, which was among the largest values reported for porous carbons. Considering that the SSA of AC-K5 was lower than the SSA of Maxsorb, the demonstrated gravimetric capacity indicates that the existence of narrow micropores was critical for the hydrogen storage capacity (Texier-Mandoki et al., 2004). It has also been noted (Nabais et al., 2004) that microwave treatment affected the porosity of the activated carbon, causing a reduction in micropore volume and micropore size. Carbonization of zinc containing metal–organic frameworks (see below) produces porous carbon materials with a linear relationship between the Zn/C ratio of the precursors and the surface area of the resulting carbon materials (Lim et al., 2012).

Zeolites are 3D crystalline aluminosilicate structures built of TO_4 tetrahedrons sharing all four corners, where T typically indicates Si^{4+} and Al^{3+} ions. Zeolites can have a very open microporous structure with different framework types depending on the assembly of the tetrahedral building units. However, since some of the cages in these frameworks are not accessible to gas molecules part of the void volume in the structure does not contribute to hydrogen storage. The extensive experimental survey depicts the gravimetric storage capacity of zeolites to be <2% at cryogenic temperatures and <0.3% at room temperatures and above (Weitkamp et al., 1995). A theoretical gravimetric capacity of 2.8% has been suggested as being an intrinsic geometric constraint of zeolites (Felderhoff et al., 2007). Considering that this calculated value was obtained under extreme physical conditions, i.e., high pressures or very low temperatures, real maximum values can be even lower and zeolites are not typically considered as feasible hydrogen storage materials.

Metal–organic frameworks (MOFs) are a relatively new class of porous polymers. MOFs are crystalline materials consisting of metal ions linked together by organic ligands that generate micropores and channels. Since both the metal centers and the organic ligand can be changed, a huge variety of MOFs with different framework topology, pore size, and surface area can be readily tuned by the selection of molecular building blocks. Many researchers have employed various metal ions such as Zn(II), Cu(II), Mn(II), Cr(III), and lanthanides(III), and diverse types of ligands such as carboxylates, imidazolates, triazolates, and tetrazolates for the construction of the MOFs. Numerous MOFs with relatively small to very large surface areas have been reported. A good example of a MOF structure is provided by MOF-5, Zn_4O (1,4 − *benzenedicarboxylate*)$_3$ (Figure 8.15).

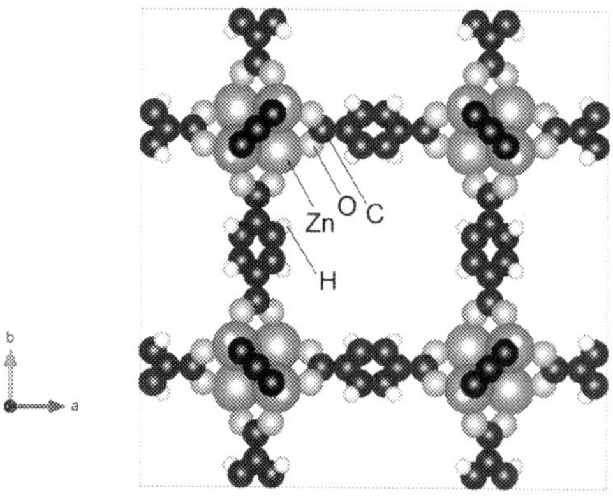

Figure 8.15 Unit cell of MOF-5 displayed along the c-axis.

The 3D structure gives rise to square openings, which are either 13.8 Å or 9.2 Å wide, depending on the orientation of the aromatic rings. As one of the promising hydrogen storage materials, MOFs have been extensively studied for the past decade. Inspired by the performance of MOF-5, researchers have thus far reported hydrogen storage data for hundreds of microporous MOFs (Zhao et al., 2011; Suh et al., 2012). In particular, MOF-177, Zn_4O (1,3,5 − *benzenedicarboxylate*)$_2$, has a BET SSA of 4750 $m^2\ g^{-1}$ and a SPV of 1.59 $cm^3\ g^{-1}$, making it one of the most porous MOFs. Wong-Foy et al. (2006) measured the saturation hydrogen uptake in a series of MOFs at 77 K, in particular, IRMOF-20 (empirical formula: $C_{24}H_6O_{13}S_6Zn_4$) and MOF-177 with a relatively low proportion of metal oxide to organic link. Saturation pressures in the range 2.5–8.0 MPa were examined, and the results revealed that the storage capacity of MOF-177 is 7.5% and 32 $g\ l^{-1}$, and IRMOF-20, 6.7% and 34 $g\ l^{-1}$. The MOF NU-100 was synthesized at Northwestern University (Farha et al., 2010). It has the highest SSA of 6143 $m^2\ g^{-1}$. It has also the highest gravimetric capacity reported so far for MOFs: 9.95% at 5.6 MPa and 77 K, while the volumetric capacity of NU-100 is 28 $g\ l^{-1}$. In most cases, the saturation hydrogen uptake in MOFs at 77 K is mainly determined by the surface area and the pore volume, regardless of chemical composition.

8.4 Hydrogen hydrate clathrates

Gas hydrates are ice-like crystalline materials that belong to the class of clathrates (i.e., inclusion compounds). They are composed of a framework of hydrogen-bonded water molecules that form voids (cages) with specific geometry and size, inside which small molecules can be encaged (enclathrated) (Sloan, 1998). The stability of hydrates is due to the weak van der Waals interactions between the lattice of water molecules and the enclathrated gas. Generally, clathrate hydrate is thermodynamically stable at high pressures and/or low temperatures, and its stability largely depends on guest species. Three types of hydrates have been identified based on their crystal structure, known as structures sI, sII, and sH. For example, the unit cell of the sII hydrate (cubic $Fd3m$ space group), shown in Figure 8.16, consists of 136 water molecules that form two types of cages: the small, a pentagonal dodecahedron (5^{12}), and a large that is formed by 12 pentagons and 4 hexagons ($5^{12}6^4$). There are 16 small and 8 large cages per 1.73 nm unit cell of sII hydrate crystal. If hydrates are to be used as gas-storage materials, the number of guest gas molecules inside the cavities is a crucial factor that will determine the total storage capacity of the hydrate. For a long time, it was believed that each cage of the hydrate structure can accommodate at most one guest molecule. However, it has recently been established that multiple occupancy of a cage can occur. It was also once believed that hydrogen molecule is too small to stabilize the hydrate cages and therefore could not form hydrates by itself. However, this assertion was refused by Dyadin et al. (1999) who synthesized pure hydrogen hydrate that was found later by Mao et al. (2002) to be of the sII type. Pure hydrogen hydrate is stable only at very high pressures or low temperatures, at 220 MPa and 280 K, or at ambient pressure and 145 K (Mao et al., 2002). The cage occupancy for this system has also been the subject of some debate. The calculated gravimetric storage capacity of hydrogen hydrate is 5.0% with quadruple occupancy of the large cavities and double occupancy of the small cavities, while it drops to 3.8% with single occupancy of the small cavities while retaining

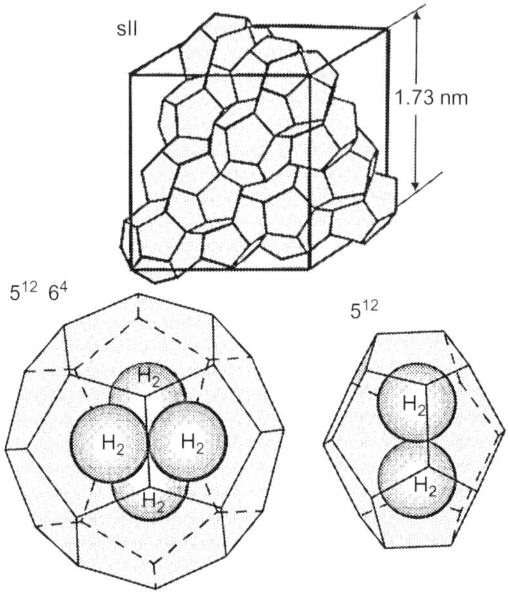

Figure 8.16 Building blocks and unit cell of sII clathrate hydrate.

quadruple occupancy of the large cavities. Unfortunately, the extreme pressure required to stabilize this material makes it impractical for hydrogen storage.

One of the most efficient ways to moderate the thermodynamic stability conditions of clathrate hydrate is addition of relatively large guest species called thermodynamic promoters to form a mixed hydrate of a target gas with a thermodynamic promoter. Florusse et al. (2004) proved that application of a second guest molecule, tetrahydrofuran (THF) made it possible to stabilize the hydrogen clathrate hydrate at pressures as low as 5 MPa and at near ambient temperatures. Unfortunately, in this case the THF molecules can be included only in the large cages of host lattice that significantly reduces amount of stored hydrogen. Moreover, it is already established that THF can be classified as a compound with a low toxicity category and thus can be used only in modest amounts. Nevertheless, Lee et al. (2005) reported that the hydrogen storage capacity in THF-containing binary-clathrate hydrates can be increased to around 4% at 270 K and 12 MPa pressure by tuning their composition to allow the hydrogen guests to enter both the larger and the smaller cages. However, later Strobel et al. (2006), Anderson et al. (2007), and Talyzin (2008) demonstrated that there is no evidence for H_2 entering and stabilizing the large sII cages under the tested conditions. In particular, the hydrogen storage capacity of a THF–hydrogen–water system, directly measured (Talyzin, 2008) with the gravimetric method, was below 0.1%, even after prolonged exposure to hydrogen at 5 MPa. Hence, while methane hydrate clathrates may store up to 40% of all hydrocarbons on Earth, hydrogen hydrate clathrates appear to be impractical for the hydrogen storage.

8.5 Conclusions and outlook

Microporous media, especially MOFs, have relatively good performance at cryogenic temperature. However, the volumetric capacity of MOF storage medium is below 40 g/l. Furthermore, the increased weight and volume, required for the adsorbent container and cooling system, may reduce the performance below the limit of practical utility. The theoretical storage capacity of hydrogen hydrate clathrates at extremely high pressures is below the target as well, while at moderate pressure and temperature it is more than 10 times lower. Among the discussed opportunities for the physical hydrogen storage, glass microvessels seem to be the most practical. However, HGMs do not meet the refilling-time demands at acceptable temperatures, but glass capillaries can, both at ambient and cryogenic temperatures. To reach 40 g/l system volumetric capacity at ambient temperature, hydrogen in capillaries must be pressurized above 100 MPa that need a substantial energy consumption. However, at liquid nitrogen temperature and moderate pressure of 30 MPa, the volumetric capacity of the capillary storage medium can be around 60 g/l, giving the reasonable hope that the nearest DOE target can be reached, regardless of the increased weight and volume for the cooling system. As far as the ultimate targets (7.5% gravimetric capacity and 70 g/l volumetric capacity of the storage system), they can probably be reached at 77 K and hydrogen pressure above 220 MPa, using flexible capillaries of quartz or S-2 glass.

References

Akunets, A.A., Basov, N.G., Bushuev, V.S., et al., 1994. Super-high-strength microballoons for hydrogen storage. Int. J. Hydrog. Energy 19, 697–700. http://dx.doi.org/10.1016/0360-3199(94)90157-0.

Anderson, R., Chapoy, A., Tohidi, B., 2007. Phase relations and binary clathrate hydrate formation in the system H_2—THF—H_2O. Langmuir 23 (6), 3440–3444. http://dx.doi.org/10.1021/la063189m.

Denisov, E.I., Zhevago, N.K., 2012. Multicapillary vessel for gas storage. Russian patent RU 120480 U1; [Pub. 20.09.12]. http://www.fips.ru/cdfi/fips.dll/ru?ty=29&docid=120480&k1=PM.

Dyadin, Y.A., Larionov, E.G., Manakov, A.Y., Zhurko, F.V., Aladko, E.Y., Mikina, T.V., Komarov, V.Y., 1999. Clathrate hydrates of hydrogen and neon. Mendeleev Commun. 9, 209–210. http://dx.doi.org/10.1070/MC1999v009n05ABEH001104.

Farha, O.K., Yazaydn, A.O., Eryazici, I., Malliakas, C.D., Hauser, B.G., Kanatzidis, M.G., Nguyen, S.T., Snurr, R.Q., Hupp, J.T., 2010. De novo synthesis of a metal-organic framework material featuring ultra-high surface area and extraordinary gas storage capacities. Nat. Chem. 2, 944–948. http://dx.doi.org/10.1038/nchem.834.

Felderhoff, M., Weidenthaler, C., von Helmolt, R., Eberle, U., 2007. Hydrogen storage: the remaining scientific and technological challenges. Phys. Chem. Chem. Phys. 9, 2643–2653. http://dx.doi.org/10.1039/b701563c.

Florusse, L.J., Peters, C.J., Schoonman, J., Hester, K.C., Koh, C.A., Dec, S.F., Marsh, K.N., Sloan, E.D., 2004. Stable low-pressure hydrogen clusters stored in a binary clathrate hydrate. Science 306 (5695), 469–471. http://dx.doi.org/10.1126/science.1102076.

France, P.W., Duncan, W.J., Smith, D.J., Beales, K.J., 1983. Strength and fatigue of multicomponent optical glass fibres. J. Mater. Sci. 18, 785–792. http://dx.doi.org/10.1007/BF00745577.

Fryer, D.M., Harvey, J.F., 1998. High Pressure Vessels. Chapman & Hall, New York, NY.

Gao, N., Janssens, K., 2004. Polycapallary X-ray optics. In: Tsuji, K., Injuk, J., Van Grieken, R. (Eds.), X-Ray Spectrometry: Recent Technological Advances. John Wiley & Sons Ltd., New York, pp. 89–110.

Gnedenko, V.G., Goryachev, I.V., Zhevago, N.K., Apparatus for storage of compressed hydrogen gas. U.S. Provisional patent application no. 60/752,379 [Pub. 22.12.05]; Apparatus and cartridge for storage of compressed hydrogen gas and system for filling the cartridge. WO 2007072470 A1 [Pub. 28.06.07]. http://www.google.com/patents/WO2007072470A1.

Griffith, A.A., 1921. The phenomena of rupture and flow in solids. Phil. Trans. R. Soc. Lond. A221, 163–198. http://www.cmse.ed.ac.uk/AdvMat45/Griffith20.pdf.

Halvorson, T., Shelby, J.E., Controlled permeation of hydrogen through glass. Final report prepared for US DOE. March 1998. http://www.osti.gov/scitech/biblio/674639.

Hartman, D., Greenwood, M.E., Miller, M.D., High strength glass fibers. AGY technical paper. http://www.agy.com/wp-content/uploads/2014/03/High_Strength_Glass_Fibers-Technical.pdf.

Heung, L.K., Schumacher, R.F., Wicks, G.G., Hollow porous-wall glass microspheres for hydrogen storage. U.S. patent 7666807 B2 [Pub. 23.02.10]. http://www.google.com/patents/US7666807.

Holtappels, K., Beckmann-Kluge, M., Gebauer, M., Grüneberg, M., Eliezer, D., 2009. Hydrogen storage in glass capillary arrays for portable and mobile systems. In: 3rd International Conference on Hydrogen Safety, Corsica, France, Paper 242. http://conference.ing.unipi.it/ichs2009/images/stories/papers/242.pdf.

Holtappels, K., Krause, A., Dame, K., Eliezer, D., 2012. A new storage technology for compressed hydrogen based on glass structures. Report at WHEC 2012, Toronto (Canada).

http://www.incomusa.com/solutions-for-industry/hydrogen-storage/.

Kitamura, R., Pilon, L., 2009. Radiative heat transfer in enhanced hydrogen outgassing of glass. Int. J. Hydrog. Energy 34 (16), 6690–6704. http://dx.doi.org/10.1016/j.ijhydene.2009.05.113.

Kohli, D.K., Kharderkar, R.K., Singh, R., Gupta, P.K., 2007. Glass microcontainer based hydrogen storage scheme. Int. J. Hydrog. Energy 33, 417–422. http://dx.doi.org/10.1016/j.ijhydene.2007.07.044.

Kurita, N., Fukatsu, N., Otsuka, H., Ohashi, T., 2002. Measurements of hydrogen permeation through fused silica and borosilicate glass by electrochemical pumping using oxide protonic conductor. Solid State Ionics 146, 101–111. http://dx.doi.org/10.1016/S0167-2738(01)00989-4.

Kurkjian, C.R., Gupta, P.K., 2001. Intrinsic strength and the structure of glass. Proc. Int. Congr. Glass 1, 11–18. http://glass-fracture.org/Documents/170%20%20ICG%20paper%202001.pdf.

Lee, H., Lee, J.W., Kim, D.Y., Park, J., Seo, Y.T., Zeng, H., Moudrakovski, I.L., Ratcliffe, C.I., Ripmeester, J.A., 2005. Tuning clathrate hydrates for hydrogen storage. Nature 434 (7034), 743–746. http://dx.doi.org/10.1038/nature03457.

Lemmon, E.W., Huber, M.L., McLinden, M.O., 2013. NIST Standard Reference Database 23: Reference Fluid Thermodynamic and Transport Properties—REFPROP, Version 9.1. National Institute of Standards and Technology, Standard Reference Data Program, Gaithersburg. http://www.nist.gov/srd/nist23.cfm.

Lim, S., Suh, K., Kim, Y., Yoon, M., Park, H., Dybtsev, D.N., Kim, K., 2012. Porous carbon materials with a controllable surface area synthesized from metal-organic frameworks. Chem. Commun. 48, 7447–7449. http://dx.doi.org/10.1039/c2cc33439a.

MacDonald, C.A., 2010. Focusing polycapillary optics and their applications. X-Ray Opt. Instrum. 2010, 17. http://dx.doi.org/10.1155/2010/867049, Article ID 867049.

Mao, W.L., Mao, H.K., Goncharov, A.F., Struzhkin, V.V., Guo, Q., Hu, J., Shu, J., Hemley, R.J., Somayazulu, M., Zhao, Y., 2002. Hydrogen clusters in clathrate hydrate. Science 297, 2247–2249. http://dx.doi.org/10.1126/science.1075394.

Mohtadi, R., Matsunaga, T., Heung, K., Schumacher, R., Wicks, G., 2011. Hollow glass microspheres as micro media for complex metal hydrides hydrogen storage compounds. J. S. C. Acad. Sci. 9 (1), 1–4. http://scholarcommons.sc.edu/cgi/viewcontent.cgi?article=1040&context=jscas.

Nabais, M.V., Carrott, P.J.M., Carrott, M.M.L.R., Menendez, J.A., 2004. Preparation and modification of activated carbon fibres by microwave heating. Carbon 42, 1315–1320. http://dx.doi.org/10.1016/j.carbon.2004.01.033.

Nijkamp, M.G., Raaymakers, J., van Dillen, A.J., de Jong, K.P., 2001. Hydrogen storage using physisorbtion-materials demands. Appl. Phys. A 72, 619–623. http://dx.doi.org/10.1007/s003390100847.

Panella, B., Hirscher, M., Roth, S., 2005. Hydrogen adsorption in different carbon nanostructures. Carbon 43, 2209–2214. http://dx.doi.org/10.1016/j.carbon.2005.03.037.

Proctor, B.A., Whitney, I., Johnson, J.W., 1967. The strength of fused silica. Proc. R. Soc. London A297, 534–557. http://dx.doi.org/10.1098/rspa.1967.0085.

Qi, X., Gao, C., Zhang, Z.W., Chen, S.F., Li, B., Wei, S., 2012. Production and characterization of hollow glass microspheres with high diffusivity for hydrogen storage. Int. J. Hydrog. Energy 37, 1518–1530. http://dx.doi.org/10.1016/j.ijhydene.2011.10.034.

Qin, F., Brosseau, C., 2012. A review and analysis of microwave absorption in polymer composites filled with carbonaceous particles. J. Appl. Phys. 111, 061301. http://dx.doi.org/10.1063/1.3688435.

Rapp, D.B., Shelby, J.E., 2004. Photo-induced hydrogen outgassing of glass. J. NonCryst. Solids 349, 254–259. http://dx.doi.org/10.1016/j.jnoncrysol.2004.08.151.

Shelby, J.E., Hall, M.M., Raszewski, F.C., A radically new method for hydrogen storage in hollow glass microspheres. Final Scientific/Technical Report. August, 2007. http://www.hydrogen.energy.gov/pdfs/progress05/vi_d_8_hall.pdf.

Shelby, J.E., Ordaz, G., Adams, J. Glass microspheres for hydrogen storage. FY 2008 Annual Progress Report: 724. http://www.hydrogen.energy.gov/pdfs/review05/stp_47_hall.pdf.

Sloan, E.D., 1998. Clathrate Hydrates of Natural Gas, second ed. Marcel Dekker, New York.

Strobel, T.A., Taylor, C.J., Hester, K.C., Dec, S.F., Koh, C.A., Miller, K.T., Sloan, E.D., 2006. Molecular hydrogen storage in binary THF-H_2 clathrate hydrates. J. Phys. Chem. B 110 (34), 17121–17125. http://dx.doi.org/10.1021/jp062139n.

Suh, M.P., Park, H.J., Prasad, T.K., Lim, D.-W., 2012. Hydrogen storage in metal organic frameworks. Chem. Rev. 112, 782–835. http://dx.doi.org/10.1021/cr200274s.

Talyzin, A., 2008. Feasibility of H_2—THF—H_2O clathrate hydrates for hydrogen storage applications. Int. J. Hydrog. Energy 33, 111–115. http://dx.doi.org/10.1016/j.ijhydene.2007.09.013.

http://www.tegs.ru/en/glass/caps.shtml.

Texier-Mandoki, N., Dentzer, J., Piquero, T., Saadallah, S., David, P., Vix-Guterl, C., 2004. Hydrogen storage in activated carbon materials: role of the nanoporous texture. Carbon 42, 2744–2747. http://dx.doi.org/10.1016/j.carbon.2004.05.018.

Timoshenko, S., 1956. Strength of Materials. D. van Nostrand Co., Inc., Princeton, NJ. http://www.scribd.com/doc/42152380/Strength-of-Materials-S-timoshenko-2Edition-Part1.

Tsugawa, P.T., Moem, J., Roberts, P.E., Souers, P.G., 1976. Permeation of helium and hydrogen from glass-microsphere laser targets. J. Appl. Phys. 47, 1987–1993. http://dx.doi.org/10.1063/1.322924.

Wang, H., Gao, Q., Hu, J., 2009. High hydrogen storage capacity of porous carbons prepared by using activated carbon. J. Am. Chem. Soc. 131 (20), 7016–7022. http://dx.doi.org/10.1021/ja8083225.

Weitkamp, J., Fritz, M., Ernst, S., 1995. Zeolites as media for hydrogen storage. Int. J. Hydrog. Energy 20, 967–970. http://dx.doi.org/10.1016/0360-3199(95)00058-L.

Wicks, G.G., Heung, L.K., Schumacher, R.F., 2008. Microspheres and microworlds (SRNL'S porous wall, hollow glass balls open new opportunities for hydrogen storage, drug delivery and national defense). Am. Ceram. Soc. Bull. 87, 23–28. http://www.mo-sci.com/uploads/PDF_documents/publications/Jun08_American%20Ceramic%20Society.pdf.

Wong-Foy, A.G., Matzger, A.J., Yaghi, O.M., 2006. Exceptional H_2 saturation uptake in microporous metal-organic frameworks. J. Am. Chem. Soc. 128, 3494–3495. http://dx.doi.org/10.1021/ja058213h.

http://www.xos.com/products/x-ray-optics-excitation-systems-x-beam/optics/polycapillary-optics/x-tra-polycapillary-focusing-optics/.

Xua, W.-C., Takahashia, K., Matsuoa, Y., Hattoria, Y., Kumagaia, M., Ishiyamab, S., Kanekoc, K., Iijima, S., 2007. Investigation of hydrogen storage capacity of various carbon materials. Int. J. Hydrog. Energy 32, 2504–2512. http://dx.doi.org/10.1016/j.ijhydene.2006.11.012.

Zhao, D., Daqiang Yuan, D., Zhou, H.-C., 2011. The current status of hydrogen storage in metal-organic frameworks—updated. Energy Environ. Sci. 4, 2721–2735. http://dx.doi.org/10.1039/C1EE01240A.

Zhevago, N.K. Onboard hydrogen accumulator for vehicles. Report at the International Forum "Hydrogen Technology for Energy Production", Moscow, 6–10 February 2006.

Zhevago, N.K., (2012). Capillary vessel with cooling system. Russian patent RU 123106 U1; [Pub. 20.12.12]. http://www.fips.ru/cdfi/fips.dll/ru?ty=29&docid=123106&ki=PM.

Zhevago, N.K., Denisov, E.I., Apparatus for storage and liberation of compressed hydrogen gas in microcylidrical arrays and system for filling the microcylidrical arrays. European patent application EP 2062850 A2 [Pub. 27.05.09]; U.S. patent US 7870878 B2 [Pub. 18.01.11]. www.google.com/patents/US7870878.

Zhevago, N.K., Glebov, V.I., 2007. Hydrogen storage in capillary arrays. Energy Convers. Manag. 48, 1554–1559. http://dx.doi.org/10.1016/j.enconman.2006.11.017.

Zhevago, N.K., Denisov, E.I., Glebov, V.I., 2010. Experimental investigation of hydrogen storage in capillary arrays. Int. J. Hydrog. Energy 35, 169–175. http://dx.doi.org/10.1016/j.ijhydene.2009.10.011.

Zhevago, N.K., Chabak, A.F., Denisov, E.I., Glebov, V.I., Korobtsev, S.V., 2013a. Storage of cryo-compressed hydrogen in flexible glass capillaries. Int. J. Hydrog. Energy 38, 6694–6703. http://dx.doi.org/10.1016/j.ijhydene.2013.03.107.

Zhevago, N.K., Chabak, A.F., Denisov, E.I., Fateev, V.N., Glebov, V.I., Korobtsev, S.V., 2013b. Safe storage of compressed hydrogen at ambient and cryogenic temperatures in flexible glass capillaries. In: International Conference on Hydrogen Safety ICHS 2013, Brussels, Belgium, Paper 157. http://www.ichs2013.com/images/papers/157.pdf.

Zhevago, N.K., Denisov, E.I., Parnes, Z., Apparatus for gas storage. U.S. patent US 20100059528 A1 [Pub. 11.03.10]. www.google.com/patents/US20100059528; European patent application EP 2163805 A2 [Pub. 17.03.10]. www.google.com/patents/EP2163805A2.

Züttel, A., Sudan, P., Mauron, P., Wenger, P., 2004. Model for the hydrogen adsoption on carbon nanostructures. Appl. Phys. A 78, 941–946. http://dx.doi.org/10.1007/s00339-003-2412-1.

Use of carbohydrates for hydrogen storage

9

J.-E. Kim[1], Y.-H. Percival Zhang[1,2,3]
[1]Virginia Tech, Blacksburg, VA, USA; [2]Cell-Free Bioinnovations Inc., Blacksburg, VA, USA; [3]Institute for Critical Technology and Applied Science (ICTAS), Virginia Tech, Blacksburg, VA, USA

Abbreviations

ATP	adenosine triphosphate
MFC	microbial fuel cell
NADH	nicotinamide adenine dinucleotide
NADPH	nicotinamide adenine dinucleotide phosphate
P_i	phosphate
SFCV	sugar fuel cell vehicle
SyPaB	synthetic pathway biotransformation

Enzymes of cell-free synthetic pathway biotransformation

6PGDH	6-phosphogluconate dehydrogenase
ALD	aldolase
FBP	fructose-1,6-bisphosphatase
G6PDH	glucose 6-phosphate dehydrogenase
GNP	glucan phosphorylase
H_2ase	hydrogenase
PGI	phosphoglucose isomerase
PGM	phosphoglucomutase
PPGK	polyphosphate glucokinase
R5PI	phosphoribose isomerase
Ru5PE	ribulose 5-phosphate epimerase
TAL	transaldolase
TIM	triosephosphate isomerase
TK	transketolase
XI	xylulose isomerase
XK	xylulokinase

Compendium of Hydrogen Energy. http://dx.doi.org/10.1016/B978-1-78242-362-1.00009-2

9.1 Introduction

9.1.1 Carbohydrates

Carbohydrates technically refer to any hydrates of carbons with the empirical formula of $C_m(H_2O)_n$, where m and n may be different. Although some chemical compounds conform to this formula, such as formaldehyde and glycolaldehyde, they are not truly considered as carbohydrates. The most accurate definition of carbohydrates is polyhydroxy aldehydes or polyhydroxy ketones with at least three carbon atoms (Carey, 2003; McNaught, 1996). Carbohydrates, simply called sugar, include all saccharides in different carbon lengths and monomeric units of them. A monosaccharide is the simplest unit of carbohydrates with three or more carbon atoms (McNaught, 1996). Glucose (Figure 9.1a) and xylose (Figure 9.1b) are the two most abundant monosaccharides on the planet because they are the key components of plant-cell walls. An oligosaccharide consists of 2–10 monosaccharides connected by glycosidic linkages. When more than 10 monosaccharides are connected to each other by glycosidic linkages, it is often called a polysaccharide, e.g., starch (Figure 9.1c) and cellulose (Figure 9.1d).

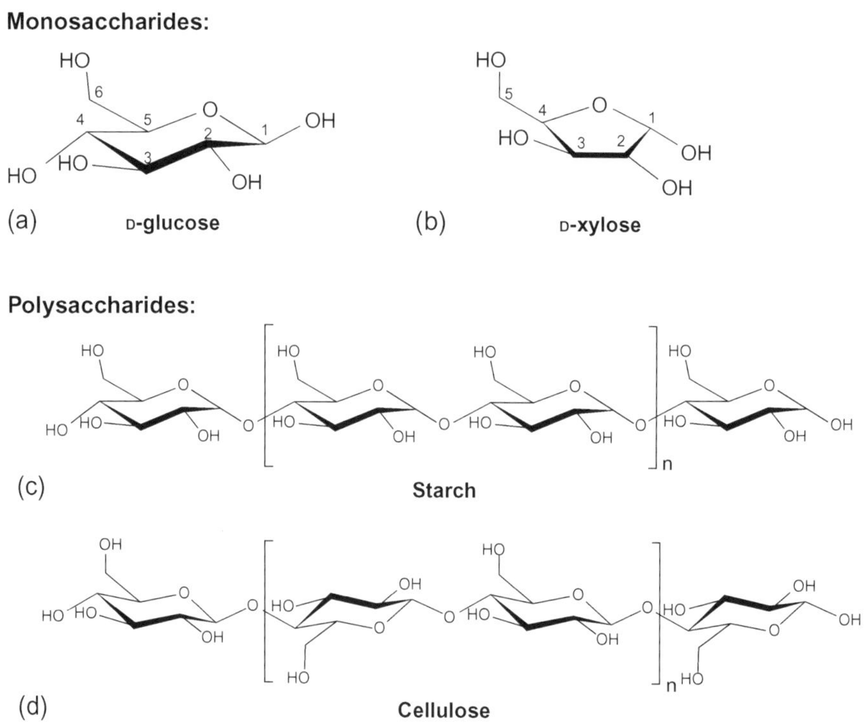

Figure 9.1 Different carbohydrates in their chain length or carbon contents. The most common monosaccharide, D-glucose (a) and the most abundant pentose, five-carbon sugar, D-xylose (b). The two most common polysaccharides are starch (c) and cellulose (d).

Starch and cellulose are different in their glycosidic linkages. Anhydroglucose units of starch are joined by α(1,4)-glycosidic linkages and α(1,6)-glycosidic linkages for branch points, while the same units of cellulose are joined by β(1,4)-glycosidic linkages. Different linkages contribute to their significantly different properties and their biological roles. α(1,4)-Glycosidic linkages in starch result in helical structures, and branched structures of starch make it easy to be hydrolyzed quickly to glucose for the energy metabolism. As the result, starch is a primary source of the energy storage compound for most living organisms (Carey, 2003). β(1,4)-Glycosidic linkages of glucan chains result in linear polysaccharide chains, forming layered sheets tightly held by highly ordered hydrogen bonds, and making cellulose insoluble in water, rigid, inflexible, and low accessible to water and cellulase (Gao et al., 2014; Kirk and Othmer, 2000). Therefore, cellulose serves as the major structural component in plant cell walls.

Carbohydrates produced by plant photosynthesis are the most abundant renewable organic compound on the earth, and its annual production is estimated to be more than 100 billion metric tons (Carey, 2003). Various carbohydrate sources include lignocellulose containing cellulose and hemicellulose, grains (plant seeds) containing starch, and juices of sugarcane and sugar beets containing sucrose. Although cultivated grains and sucrose-producing crops utilize ~30% of arable lands and consume ~70% refresh water withdrawal, their energy content utilized by human beings accounts for approximately 2.5% of the energy stored as lignocellulosic biomass (Zhang, 2013). Only a very small part of lignocellulose (e.g., 5%) is utilized directly or indirectly for meeting human needs, and most of the rest is degraded to CO_2 by the earth biosystem (Naik et al., 2010). Therefore, the utilization of all carbohydrate components in under-utilized lignocellulosic biomass for the production of hydrogen will be important to the bio and hydrogen economy without competing with food sources such as grains and sucrose.

9.1.2 Hydrogen economy and storage

Hydrogen is a colorless, odorless, but very flammable gas. Most hydrogen produced by the reforming of natural gas and coal (Dresselhaus, 2004) is used for petroleum-refining processes and ammonia synthesis (Kirk and Othmer, 2000). The hypothetical hydrogen economy consists of a series of four processes: producing hydrogen, storing hydrogen, distributing hydrogen, and converting the stored hydrogen into electrical energy at the site of end users (Crabtree et al., 2004; Zhang, 2009a). Hydrogen has been suggested as the best alternative fuel to replace fossil fuel-derived fuels primarily used for transportation, mainly due to its clean energy property, high-energy conversion efficiency, and high specific energy density (J/kg) (Crabtree et al., 2004; Dresselhaus, 2004; Zhang, 2009a). Although hydrogen is an attractive energy carrier, the hydrogen economy faces the following technical obstacles: (1) low-cost, scalable, and renewable hydrogen production; (2) a lack of high-density hydrogen storage approaches; (3) high-cost infrastructure for hydrogen distribution; (4) the production of affordable and durable fuel cells; and (5) safety concerns. In order to solve these problems, the use of carbohydrates, such as starch or cellulose, is suggested as an out-of-box solution (Zhang, 2009a; Zhang et al., 2007).

Typical ways to store hydrogen are compressed hydrogen in high-pressure gas cylinders or cryogenic liquid hydrogen (Schlapbach and Zuttel, 2001; Züttel, 2003). However, compressing and liquefying hydrogen processes are too energy-intensive, and their hydrogen storage densities are too low. Because onboard hydrogen storage for a vehicle must be compact, light, safe, and affordable (Schlapbach and Zuttel, 2001), these conventional storage methods cannot fulfill these goals well due to volumetric constrains, high costs, and safety issues (Sakintuna et al., 2007). The goal of novel hydrogen storage approaches for vehicles is to find the materials that can store hydrogen much higher than compressed or cryogenic hydrogen. Intensive research has been conducted to investigate and improve metal hydrides (Chao and Klebanoff, 2012; Imamura et al., 2005; Van Vucht et al., 1970), complex hydrides (Bogdanović et al., 2003; Keaton et al., 2007; Vajo et al., 2005), and carbon materials (Lee and Lee, 2000; Panella et al., 2005). Their relatively high storage capacity of hydrogen and moderate temperature and pressure conditions make them appealing compared to the methods of compressing or cryogenically liquefying hydrogen, but it is not easy to release hydrogen from those materials. The hydrogen storage densities via a number of methods are compared in Figure 9.2, with the hydrogen storage goal proposed by the U.S. Department of Energy (DOE).

Alternatively, some researchers have suggested the production of on-site or on-demand hydrogen via hydrogen carriers such as hydrocarbons (Ahmed and Krumpelt, 2001; Carrette et al., 2001), ethanol (Deluga et al., 2004; Mattos et al., 2012), methanol (Olah, 2005; Pérez-Hernández et al., 2012), ammonia (Klerke et al., 2008; Thomas and Parks, 2006), formic acid (Hull et al., 2012; Schuchmann and Muller, 2013; Yu and Pickup, 2008), and carbohydrates (Ye et al., 2009; Zhang et al., 2007). These different hydrogen carriers are compared based on their costs, typical operational conditions, reactor costs, necessity of product purification, and safety/toxicity (Table 9.1). Hydrocarbons, ethanol, and methanol can be

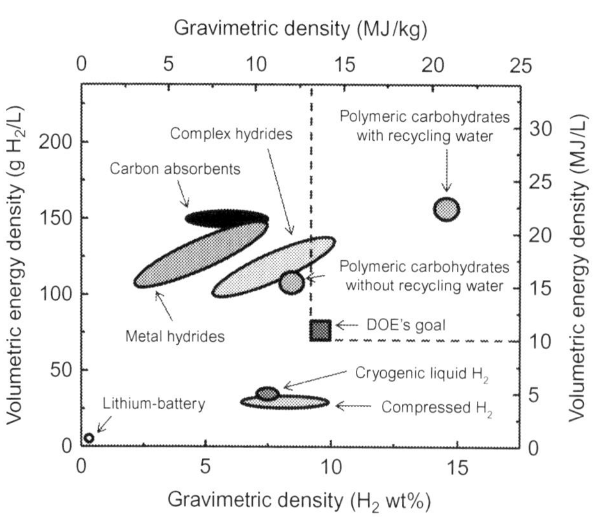

Figure 9.2 The available hydrogen storage means and DOE's goal for hydrogen storage.

Table 9.1 Comparison of hydrogen carriers

Hydrogen carrier	Carrier costs*	Reaction temperature	Reaction pressure	Reactor costs	Purification needed	Safety/ toxicity
Hydrocarbons	$24.9/GJ	≥700 °C	10–40 bar	High	High	Modest
Biodiesel	$39.1/GJ	≥1000 °C	30–80 bar	High	High	Modest
Ethanol	$22.0/GJ	500–700 °C	–	High	Modest	Low
Methanol	$16.8/GJ	200–500 °C	25–50 bar	Modest	Low	Modest
Ammonia	$32.3/GJ	300–700 °C	–	Modest	Modest	Modest
Formic acid	$150/GJ	30–50 °C	~1 bar	Modest	Low	High
Carbohydrates	$12.9/GJ	30–80 °C	~1 bar	Low	No needed	Very low

*These are based on the end users' prices of hydrocarbons (gasoline) for $3.50 per gallon, biodiesel for $5.00 per gallon, ethanol for $2.00 per gallon, methanol for $1.00 per gallon, ammonia for $600 per ton, formic acid for $800 per ton, and delivered carbohydrates for $0.22 per kilogram.

converted to hydrogen and carbon dioxide with carbon monoxide as by-products via several different reforming or oxidation reactions at elevated temperatures. Since a small amount of carbon monoxide is highly toxic to PEM fuel cells (Deluga et al., 2004), extra steps are necessary to remove trace amounts of carbon monoxide, resulting in complicated systems and decreased overall energy efficiencies. Therefore, the concept of hydrocarbons and ethanol as hydrogen carriers for small-size vehicles was abandoned long ago. Relatively low-temperature methanol reforming has been suggested to convert methanol with steam into hydrogen under 20 bars. A recent breakthrough has enabled this reaction to occur under ambient pressure using metal-based catalysts (Nielsen et al., 2013). However, the use of unstable catalysts and low rate of hydrogen production remain challenging. Ammonia is suggested as a hydrogen carrier because it is a carbon-free gas and easily liquefied. However, ammonia is not suitable for PEM fuel cells because trace amounts of ammonia can poison PEM fuel cells (Thomas and Parks, 2006). Recently, formic acid was proposed as another promising hydrogen carrier that can be synthesized from CO_2 and H_2 via either chemical catalysis by iridium metal (Hull et al., 2012) or biocatalysis by enzymes or microorganisms (Schuchmann and Muller, 2013) under mild temperature and pressure conditions. Formic acid can be easily converted back to hydrogen through direct formic acid fuel cells at low temperature (Yu and Pickup, 2008), which offers a great advantage over other hydrogen carriers such as methanol and ethanol requiring high-temperature (>200 °C) reforming processes. However, formic acid has incomparably low hydrogen density (4.3 H_2 wt% for gravimetric density and ~7.5 MJ/L for volumetric energy density) (Yu and Pickup, 2008). In addition to that, highly concentrated formic acid is toxic, very corrosive, and unstable (Reutemann and Kieczka, 1996).

We suggest the use of carbohydrates as a new hydrogen carrier. Carbohydrates have several advantages (Table 9.1). First, carbohydrates are the least costly natural resource and its utilization is a carbon neutral process. Second, the production of highly pure hydrogen is a reality via a new technology called synthetic pathway biotransformation (SyPaB). No product purification steps are required to obtain pure hydrogen for the use with PEM fuel cells. Third, the enzymatic reactions do not require any costly reactors bearing high temperature or pressure because they are conducted under modest reaction conditions (e.g., ~30–80 °C and ~1 bar). Fourth, carbohydrates are nontoxic and nearly inflammable. Therefore, this new biocatalysis breakthrough may make on-site and even onboard vehicle hydrogen production systems feasible.

9.2 Converting carbohydrates to hydrogen by SyPaB

9.2.1 Overview of hydrogen production from carbohydrates

Low-cost renewable carbohydrates are an attractive energy source for green hydrogen production. A number of carbohydrate-to-hydrogen technologies have been under development (Figure 9.3). These technologies can be classified into chemical catalysis, biological transformations, and their hybrids. The theoretical maximum yields of hydrogen production from carbohydrates with water are 12 moles of hydrogen

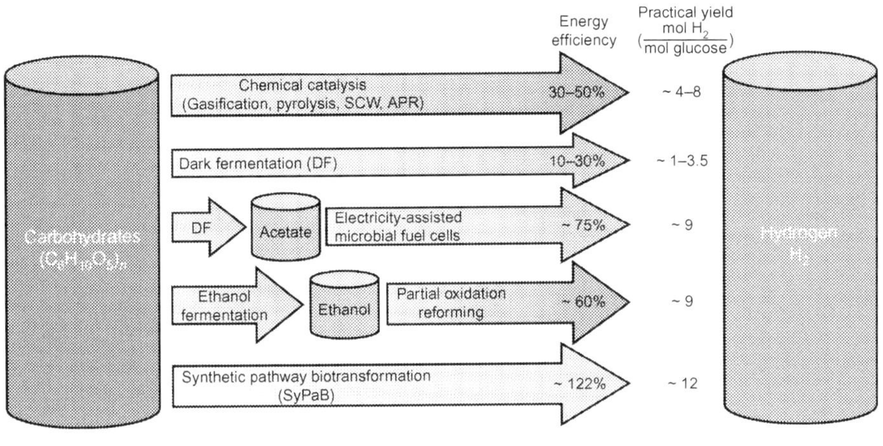

Figure 9.3 Comparison of carbohydrates-to-hydrogen technology.

per mole of hexose and 10 moles of hydrogen per mole of pentose. SyPaB is the only approach that can achieve these theoretical maximum yields. Chemical catalysis features fast reaction rates but harsh reaction conditions (Zhang et al., 2012), while biological transformations (except SyPaB) generally feature modest reaction conditions, easy product separation, and high-purity hydrogen production but low reaction rates and low product yields.

Chemical catalysis includes gasification, pyrolysis, gasification in supercritical water (SCW), and aqueous-phase reforming (APR). Gasification is a process that converts biomass carbohydrates into hydrogen and other products under high temperatures (above 1000 K) and the presence of oxygen and/or water (Navarro et al., 2007). Gasification involves an endothermic reaction of water molecules splitting into H_2 and O by absorbing outside thermal energy or partially combusting carbohydrates. The products of gasification contain some undesired impurities (Rezaiyan and Cheremisinoff, 2005). The regular gasification process has been investigated to produce hydrogen and increase overall yield through the integration with other exothermic processes such as an air-blown bubbling fluidized bed gasifier, a steam reformer, and a water–gas–shift membrane reactor (Ji et al., 2009; Lin et al., 2005). Several reports and reviews about biomass gasification for hydrogen production are also available (Babu, 2002; Cummer and Brown, 2002; Ni et al., 2006; Rezaiyan and Cheremisinoff, 2005; Stevens, 2001). Pyrolysis is a chemical decomposition of biomass at high temperatures (650–800 K) in the absence of oxygen (Rezaiyan and Cheremisinoff, 2005). Unlike gasification process, pyrolysis generally aims to produce liquid bio-oils and solid charcoal plus small amounts of gaseous products (e.g., H_2, CO_2, CO, CH_4) (Ni et al., 2006). The reaction temperature, heating rate, duration, particle size of biomass, and type of catalyst use are important to determine yields of multiple products and their composition (Ni et al., 2006; Probstein and Hicks, 2006; Zanzi et al., 2002). Gaseous products, especially hydrogen, become more favorable under high temperature, high heating rate, and long volatile phase residence time conditions (Demirbaş, 2002). Other gaseous coproducts, such as methane, hydrocarbon vapors, and carbon monoxide, can be converted into more hydrogen through steam reforming or water–gas-shift reaction. In compared to gasification, pyrolysis is a less favorable process for hydrogen production due to much lower hydrogen yields (Ni et al., 2006). One of the major issues for both pyrolysis and gasification is the formation of tar from biomass, which often causes the formation of other undesired products and undergo polymerization, which creates more complex structures, consequentially affecting the production of hydrogen (Ni et al., 2006). Extensive studies on tar reduction have been reported (Corella et al., 1999; Milne et al., 1998; Narvaez et al., 1997; Simell et al., 1997, 1999; Sutton et al., 2001). When biomass contains high moisture content (above 35%), gasification of the biomass can be conducted in the SCW condition (above 647 K and 220 bars). Under this condition, wet biomass can be rapidly converted into gaseous products at a high gasification ratio of 100% and hydrogen volumetric ratio of 50% (Ni et al., 2006). APR is a chemical catalysis occurring in an aqueous phase under relatively low temperature (400–550 K). Unlike vapor-phase reforming processes, APR produces hydrogen without volatilizing water under high pressure (50–70 bars). It is easy to separate hydrogen gas from aqueous water, which means major energy savings. Moderate conditions of APR can minimize undesirable decomposition reactions and provide a favorable condition for the water–gas-shift reaction so that the process generates hydrogen with low amounts of carbon monoxide (Chheda et al., 2007). However, leaching and instability of catalyst components for APR still remain major disadvantages of the process. These thermochemical processes also accompany the production of other impurities (e.g., CO), which require extra steps of separation and purification to obtain pure hydrogen gas.

Biological transformations mediated by microorganisms or isolated enzymes have advantages over chemical catalysis, such as higher selectivity, lower energy input, possibly higher energy efficiency, and less costly bioreactors (Zhang et al., 2012). Biological transformations include dark fermentation, light fermentation, their combination (Argun et al., 2009), electrically assisted microbial fuel cells (MFCs) (Logan and Regan, 2006), microbial ethanol fermentation followed by ethanol partial oxidation reforming (Haryanto et al., 2005), and cell-free SyPaB (Ye et al., 2009; Zhang et al., 2007). In the dark fermentation, natural or engineered microorganisms can produce only 4 moles of hydrogen and 2 moles of acetate per mole of glucose (33% efficiency), which is called the Thauer limit (Thauer et al., 1977). The practical hydrogen yields are much lower than the Thauer limit (Hallenbeck and Benemann, 2002; Kleerebezem and van Loosdrecht, 2007). To increase overall hydrogen yields, the coproduced two moles of acetate can be further converted to 8 moles of hydrogen through the electrically assisted MFCs with extra energy input to overcome thermodynamically unfavorable hydrogen production from acetate (Logan and Regan, 2006). MFCs provide an overall hydrogen yield of approximately 9 moles of hydrogen produced per mole of glucose (75% efficiency). However, high capital investment required for MFCs and slow volumetric productivity are the greatest challenges for its large-scale application (Logan and Regan, 2006). Alternatively, organic acids produced during the dark fermentation can be utilized by photosynthetic bacteria for hydrogen production (Mathews and Wang, 2009; Miyake et al., 1984), but this process is impractical for potential applications because of its very slow rates of hydrogen production and low energy concentration (called a nonpoint energy source, which is difficult for energy collection) (Zhang, 2011a). Ethanol fermentation can be conducted with a nearly theoretical yield, i.e., 2 moles of ethanol produced per mole of glucose by yeasts or bacteria (Lin and Tanaka, 2006). Ethanol can be converted to hydrogen by the partial oxidation-reforming process (Deluga et al., 2004; Haryanto et al., 2005). However, the products of such a reforming process still contain a small amount (though serious enough to damage fuel cells) of CO. The overall process of ethanol fermentation followed by partial oxidation reforming has a practical hydrogen production yield similar to that of MFCs, but lower overall energy efficiency than MFCs because of more energy losses during microbial fermentation, distillation, and chemical reforming (Zhang, 2011a). Overall, these microbe-based biological transformations suffer from low hydrogen yields because the microbial basal metabolism competes with hydrogen production. This competition eventually reduces the overall energy efficiency and product yields.

In vitro SyPaB is a new biomanufacturing platform that assembles a number of purified enzymes and cofactors from different sources *in vitro* for the implementation of complicated biochemical reactions. This biotransformation can achieve the theoretical maximum yield of hydrogen from carbohydrates, 12 moles of hydrogen per mole of hexose (Ye et al., 2009; Zhang et al., 2007). High-yield hydrogen is produced from starch by a reconstituted non-natural catabolic pathway consisting of 13 enzymes in four sub-modules: (1) substrate phosphorylation without adenosine triphosphate (ATP), (2) the oxidative pentose phosphate pathway, (3) hydrogen generation, and (4) partial pathways of glycolysis and gluconeogenesis (Figure 9.4). In the substrate phosphorylation catalyzed by phosphorylases, polysaccharides or oligosaccharides release phosphorylated glucose units without the use of ATP (e.g., glucose 1-phosphate), and they are converted into glucose 6-phosphate, entering in the following reactions. The six-carbon sugar glucose 6-phosphate becomes a five-carbon sugar via the oxidative pentose phosphate pathway after the release of CO_2 and 2 moles of nicotinamide adenine dinucleotide phosphate (NADPH). Two moles of hydrogen are produced from continuously regenerated NADPH. In the carbon rearrangement of the pentose phosphate pathway, phosphate pentoses (e.g., ribulose 5-phosphate) are converted into hexose sugars (e.g., fructose 6-phosphate) and three-carbon sugars (e.g., glyceraldehyde 3-phosphate). The three-carbon sugars are combined to form fructose 6-phosphates by multiple enzymes in the glycolysis and gluconeogenesis pathways. Fructose 6-phosphate is further catalyzed to regenerate glucose 6-phosphate for completing the sugar recycling loop. The overall carbohydrate-to-hydrogen reaction mediated by SyPaB can be summarized as $C_6H_{10}O_5(aq) + 7H_2O(l) \rightarrow 12H_2(g) + 6CO_2(g)$. Most of the reactions are reversible, but the formation of gaseous products makes it easy to separate them from the aqueous phase reaction. Thus, the removal of the gaseous products favors the unidirectional overall reaction and drives the reaction forward to completion. The first proof-of-principle experiment has been conducted and shown to be able to produce nearly 12 moles of hydrogen per glucose consumed at the rate of ~0.4 mmol/h/L (Zhang et al., 2007). The reaction rates have been increased by nearly 10-fold through minor optimization of increasing substrate concentration and rate-limiting enzyme loadings (Ye et al., 2009).

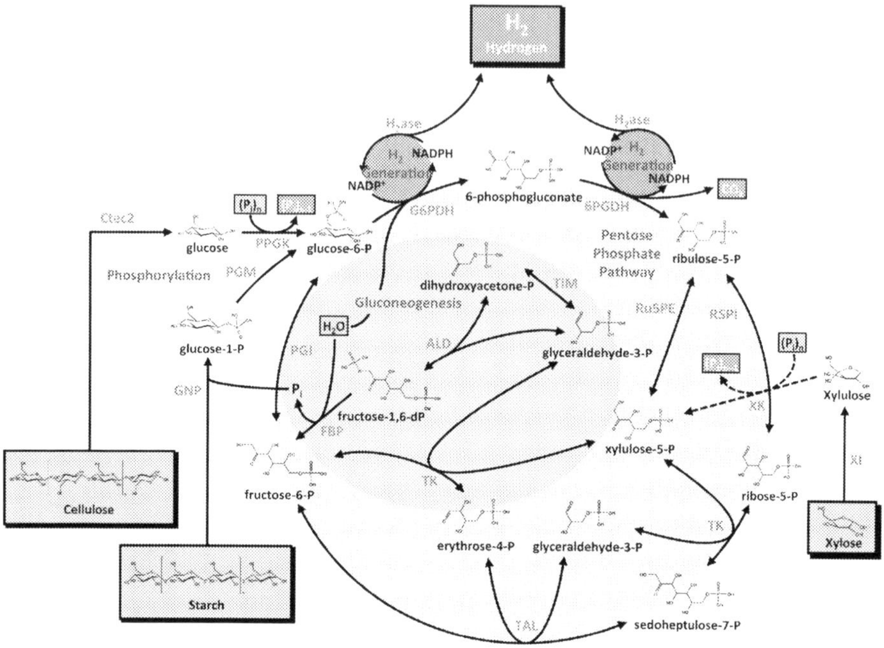

Figure 9.4 Enzymatic synthetic pathway biotransformation (SyPaB) to produce hydrogen from different carbohydrates.

9.2.2 *Advantages of SyPaB*

Generally speaking, biological transformations are advantageous over chemical catalysis in the modest reaction conditions, lower energy, and high initial capital investment required and relatively purer production of hydrogen gas (Zhang, 2009a). Among various biological transformations, SyPaB has several distinguishable advantages: the highest energy efficiency (122%; more output chemical energy than input chemical energy), the highest product yield, the fastest reaction rates, and the simplest products (only hydrogen and carbon dioxide).

SyPaB has the highest energy efficiency in converting carbohydrates to hydrogen. The overall carbohydrate-to-hydrogen reactions are spontaneous ($\Delta G^\circ = \sim -50$ kJ/mol) and endothermic ($\Delta H^\circ = +596$ kJ/mol), meaning that some of ambient thermal energy, which is available from any low-temperature waste heat sources such as refrigerators, air conditioners, and fuel cells, is absorbed and converted to chemical energy used for hydrogen production (Ye et al., 2009; Zhang et al., 2007). This thermodynamic analysis suggests that the overall reactions are driven by entropy gain from the aqueous-to-gaseous phase change, rather than enthalpy losses. These enzymatic reactions for hydrogen production are among rare entropy-driven chemical reactions involving phase changes from highly ordered substrates to poorly ordered products: for example, $N_2O_5(s) \rightarrow 2NO_2(g) + O_2(g)$ and $Ba(OH)_2{\cdot}8H_2O(s) + 2\,NH_4SCN\,(s) \rightarrow Ba(SCN)_2\,(aq) + 2NH_3\,(aq) + 10H_2O\,(l)$, where both reactions involve great phase changes from solid substrates to gaseous, or aqueous and liquid products, respectively. As the result, these entropy-driven reactions can generate more output of chemical energy in the form of hydrogen (122%; based on combustion energy) than input of chemical energy in carbohydrates (Figure 9.3).

SyPaB has higher reaction rates than microbial fermentations because the absence of a cellular membrane and microbial complexity allows faster mass transfer and higher biocatalyst density per volume (Rollin et al., 2013; Zhang and Mielenz, 2011). For example, enzymatic fuel cells can usually generate the higher maximum power than microbial fuel cells due to the absence of cellular membranes limiting mass transfer and high enzyme loading achieved per unit volume or area without unnecessary biomolecules in the reaction solution (Cooney et al., 2008; Osman et al., 2011; Zhu et al., 2014). Today the highest enzymatic hydrogen production rate has been increased to 0.3 g H_2/L/h, compared to the highest microbial hydrogen fermentation on sugars and faster than electrically assisted MFCs.

SyPaB produces H_2 with CO_2 in a ratio of 67% to 33% (Ye et al., 2009; Zhang et al., 2007). The separation cost of hydrogen from carbon dioxide is estimated to be very low; the mixture of hydrogen and carbon dioxide can also be used for PEM fuel cells with only ~1% loss in the fuel cell efficiency because carbon dioxide is an inert gas (Huang and Zhang, 2011a). However, hydrogen production by chemical catalysis contains a small but critical amount of carbon monoxide for damaging fuel cells. Therefore, the high selectivity of SyPaB makes carbohydrates as a possible hydrogen carrier.

9.2.3 Unique advantages of carbohydrates as a hydrogen carrier

The major obstacle of the hydrogen economy is a lack of high-density hydrogen storage approaches at low costs. Clearly, carbohydrates as a hydrogen carrier have several advantages over other hydrogen carriers, such as hydrocarbons (Ahmed and Krumpelt, 2001; Carrette et al., 2001), ethanol (Deluga et al., 2004; Mattos et al., 2012), methanol (Olah, 2005; Pérez-Hernández et al., 2012), ammonia (Klerke et al., 2008; Thomas and Parks, 2006), and formic acid (Hull et al., 2012; Schuchmann and Muller, 2013; Yu and Pickup, 2008). Table 9.1 compares these different hydrogen carriers based on their costs, typical operation conditions, reaction requirements, and safety.

First, the use of carbohydrates is economically beneficial due to its abundant renewability in nature and low costs. Roughly more than 100 billion tons of dry cellulosic material is produced by terrestrial plants annually (Berner, 2003; Falkowski et al., 2000). It is estimated that about 150 billion gallons of gasoline can be replaced with hydrogen energy produced from 700 million tons of biomass via SyPaB (Huang and Zhang, 2011b). Therefore, only 10% utilization of lignocellulose biomass could greatly reduce our energy dependence on fossil fuel. In addition to the economic benefits from renewable carbohydrates as a hydrogen carrier, carbohydrates are nontoxic, nearly inflammable, and well distributed over the world (Zhang, 2013).

Second, carbohydrates have very high hydrogen storage density (Figure 9.2). Polysaccharides have the chemical formula of $(C_6H_{10}O_5)_n$. Gravimetric density of hydrogen in carbohydrates is 14.8 H_2 mass% (gram of hydrogen gas production per gram of polysaccharide consumption = 24 g/162 g) when water molecules are recycled from PEM fuel cells. Volumetric energy density of polysaccharide as hydrogen carrier is much higher than 100 kg/m^3. Clearly, carbohydrates exceed the DOE's hydrogen storage requirement.

Third, carbohydrates are a carbon-neutral energy source in terms of the whole life cycle. The amount of CO_2 released during hydrogen production from carbohydrates via SyPaB would be equal to CO_2 consumed for growing carbohydrates used (Kirk and Othmer, 2000). Thus, the use of carbohydrates as a fuel to produce hydrogen can achieve nearly zero net CO_2 emissions.

Carbohydrates as an energy carrier with these advantages could be the ultimate solution for many energy sustainability challenges such as hydrogen and electricity storage, CO_2 fixation and long-term storage, and transportation fuel production (Zhang and Mielenz, 2011).

9.3 Challenges of carbohydrates as hydrogen storage and respective solutions

Carbohydrates become a possible hydrogen carrier mainly due to SyPaB, which can completely convert carbohydrates with water to highly pure hydrogen under modest reaction conditions. The hydrogen production costs from carbohydrates are strongly related to three major cost factors: carbohydrate costs, enzymes costs, and cofactors costs (Zhang and Mielenz, 2011). The hydrogen generation rates catalyzed by SyPaB determine its potential applications.

9.3.1 Enzyme costs and stability

The costs of enzymes used in SyPaB are influenced greatly by their production costs, i.e., US dollars per kilogram enzyme and their turn-over number (TTN, mole product per mole of enzyme). All enzymes are cytoplasmic enzymes instead of membrane proteins so that all of them can be produced as recombinant proteins in *Escherichia coli*. According to industrial bulk enzyme production knowledge, such bulk enzymes have production costs of ~\$5–100/kg dry protein weight (Zhang, 2011a). Increasing TTN of enzymes exponentially decreases enzyme costs (Zhang, 2010a). Enzyme stability can be improved through enzyme immobilization (Wang and Zhang, 2010), the use of thermostable enzymes (Ye et al., 2010), and protein engineering (Eijsink et al., 2004, 2005; Ye et al., 2012). It is estimated that the enzyme costs would be low enough for industrial-scale use when their TTNs reach more than 10^7–10^8 (Wang et al., 2011; Zhang, 2010a; Zhang et al., 2010). Such high TTNs have been shown to be achievable with the use of thermostable enzymes from thermophilic microbes, such as *Clostridium thermocellum* and *Thermotoga maritima* (Myung et al., 2010, 2011; Wang and Zhang, 2009, 2010). The use of thermostable enzymes produced in mesophilic hosts can also decrease the enzyme purification cost. Enzymes cloned from thermophilic microbes are usually more stable at higher temperature (60–70 °C) (Zhang, 2009b).

Therefore, heat treatment can deactivate *E. coli* endogenous proteins; after centrifugation, only the soluble thermostable proteins remain in the supernatant. This heat precipitation provides a simple and less-costly purification method for thermostable enzymes. In contrast, classic enzyme purification methods in labs, such as chromatographic techniques, are costly and hard to scale up. Immobilization of enzymes not only increases their TTNs but also helps recycle enzymes from the reactants (Zhang, 2009b). Enzyme immobilization is a relatively mature technique, and low-cost enzyme immobilization techniques have been widely used, such as alginate entrapment, cross-linking enzyme aggregate (Sheldon et al., 2007). Furthermore, the cellulose-binding-module-tagged protein immobilization can integrate the purification and immobilization of enzymes in one step (Hong et al., 2008).

Stable enzymes could be stored for long periods, depending on the storage condition and its inherent properties. Proper storage conditions for some enzymes can keep them active up to several years. For example, the shelf life of the proteases used in detergents is several years. Immobilized thermostable glucose isomerase, widely used in the food industry, has a working lifetime of up to 2 years at 55 °C. Immobilized glucose oxidase, used in blood sugar testing strips, can be stored for more than 2 years at ambient temperature.

9.3.2 Cofactor costs and stability

Cofactors are nonprotein chemical compounds required for enzyme functions. Economic analysis suggests that TTN value of cofactors should be higher than 10^6 for the economically viable production of biocommodities (Zhang, 2010a). The issues from cofactor costs and stability can be addressed by the cofactor recycling systems (Liu and Wang, 2007) or the use of low-cost stable biomimetic cofactors (Ansell and Lowe, 1999; Lutz et al., 2004). Efficient regeneration of natural cofactors through recycling systems is economically beneficial to most enzymatic reactions, such as ATP and nicotinamide cofactors (NAD^+ and $NADP^+$) (Liu and Wang, 2007). Among several different $NAD(P)^+$ recycling systems, enzymatic cofactor regeneration is the most promising approach due to its high selectivity and high compatibility with other biological components of the reaction (Weckbecker et al., 2010). Cofactors can be recycled by either reaction-coupled regeneration where a second enzyme regenerates the cofactor by catalyzing a second substrate, or substrate-coupled regeneration where one single enzyme is utilized for both desired product formation and cofactor regeneration. Immobilization of the cofactor has been developed for more efficient cofactor regeneration (Liu and Wang, 2007).

The best solution would be the replacement of native $NAD(P)^+$ cofactors with biomimetic cofactors (mNADs) featuring low cost and high stability. However, most wild-type NAD(P)-dependent redox enzymes cannot work on mNADs, except for horse liver alcohol dehydrogenase (Lo and Fish, 2002), enoate reductase (Paul et al., 2013), and diaphorase (unpublished). Therefore, it is necessary to engineer cofactor preference of Rossmann-motif-based redox enzymes. Cofactor engineering is still in its developing stage, but several successful results have been reported recently (Bastian et al., 2011; Campbell et al., 2010, 2012; Döhr et al., 2001;

Huang et al., 2013; Paul et al., 2013). We also succeeded in converting the cofactor of NADP-dependent glucose 6-phosphate dehydrogenase and 6-phosphogluconate dehydrogenase to the biomimetic cofactor (data unpublished).

9.3.3 Reaction rate

The highest enzymatic hydrogen production rate achieved by us is about 150 mmol/h/L (data unpublished), which was increased by nearly 750 times in the past 10 years (Woodward et al., 2000). Such great enhancements are attributed to elevated reaction temperature, optimized enzyme ratio, high substrate concentration, and high enzyme loading. This reaction rate equals approximately 2 g glucose consumed/L/h. This enzymatic hydrogen production rate is comparative to the ethanol productivity of microbial ethanol fermentation, suggesting the technical feasibility of stationary hydrogen generation systems using SyPaB.

Given that a small-size hydrogen fuel cell vehicle needs to be powered with 1 kg H_2 per hour and the vehicle has a 50-liter on-board bioreactor, the expected hydrogen production rate required for the sugar fuel cell vehicle (SFCV) is estimated to be about 20 g H_2/L/h. This means we have to increase enzymatic hydrogen generation rates by another 60 times.

Table 9.2 shows an analysis of potential rate increases by several orders of magnitude through a combination of known technologies. Increasing reaction temperature from 30 to 80 °C (or even higher) may result in about 32-fold increased reaction rates, according to the Q10 effect where the reaction rate approximately doubles with every 10 °C increase in reaction temperature (Reyes et al., 2008). This trend is supported by

Table 9.2 Potential rate increases for hydrogen production from SyPaB (Zhang, 2011b)

Technology	Potential fold	References	Predicted fold
Increasing reaction temperature from 30 to 80 °C or higher	32	Reyes et al. (2008), Wang and Zhang (2009), and Ye et al. (2009)	4–20
Increasing the rate-limiting step enzyme loading	10	Ye et al. (2009)	2–5
Increasing overall enzyme loading	10	Yoshida et al. (2005)	5
Increasing substrate concentration by 50-fold	10	Ye et al. (2009)	5
Creating substrate channeling among enzymes	2–100	Myung et al. (2010), Srivastava and Bernhard (1986), and Zhang (2011c)	2
Improving catalytic efficiency of enzymes	~10		
Overall increasing rates	640,000–32,000,000		400–5000

the cases of thermostable enzymes (Ma et al., 1994; Wang and Zhang, 2009; Ye et al., 2009). Increasing the loading of enzymes in the rate-limiting steps can practically increase at least twofold of the reaction rate (Ye et al., 2009). Increasing overall enzyme concentration by 10-fold or higher can also achieve another fivefold increase (Yoshida et al., 2005). Increasing substrate concentration by 50-fold or higher is expected to be able to increase the reaction rate by fivefold. Creating substrate channeling among cascade enzymes has been shown to increase the reaction rate because cascade enzymes are closely held by scaffoldins or other reagents (Conrado et al., 2008; Dueber et al., 2009; Wilner et al., 2009; You et al., 2012). Lastly, improving the catalytic efficiency of enzymes would result in a potential increase of about a 10-fold in the reaction rate. Most enzymes have far lower catalytic efficiencies than those of catalytically perfect enzymes (Wolfenden and Snider, 2001) or the same catalytic enzymes with the highest efficiencies reported in the BRENDA database. The overall potential increases in hydrogen production rate can be estimated to be from 640,000 to 32,000,000-fold from the original starting point (Woodward et al., 2000). In previous years, we accomplished about a 750-fold enhancement in carbohydrate-to-hydrogen reaction rates and believe that further reaction enhancement to $\sim$20 g H_2/L/h will be achievable soon. This is because microbial fermentation has already been reported to produce hydrogen with the rate of 23.6 H_2/L/h (Yoshida et al., 2005) and enzymatic reactions usually have faster reaction rates than microbial fermentation (Cooney et al., 2008; Zhang, 2009b; Zhu et al., 2014).

9.4 Future carbohydrate-to-hydrogen systems

The enzymatic carbohydrate-to-hydrogen production opens up several potential applications for the hydrogen economy, where hydrogen is an alternative transportation fuel or a short-term electricity storage carrier. The potential applications include from the most feasible one, i.e., local hydrogen-generating stations (Figure 9.5a), to the most ambitious application, i.e., SFCVs (Zhang, 2009a, 2010b) (Figure 9.5b).

9.4.1 Local hydrogen-generating stations

The near-term application is hydrogen production from local carbohydrate resources in small-size hydrogen generation reactors supplying hydrogen to local users, such as fuel cell vehicles (Figure 9.5a). In contrast with the hydrogen distribution from a centralized hydrogen generation facility to local hydrogen stations, the proposed hydrogen generation and distribution systems are based on local carbohydrates, which would require low capital investment and have a high safety level. Local hydrogen-generating stations system will be supplied with carbohydrates easily delivered from local areas. Each local hydrogen-generating station can have a relatively large size high-pressure hydrogen storage tank under the ground, which is refilled with hydrogen produced from underground bioreactors, like anaerobic digesters (Zhang et al., 2012). Such small-size decentralized hydrogen generation systems could promote hydrogen fuel cell vehicles without costly hydrogen distribution infrastructure.

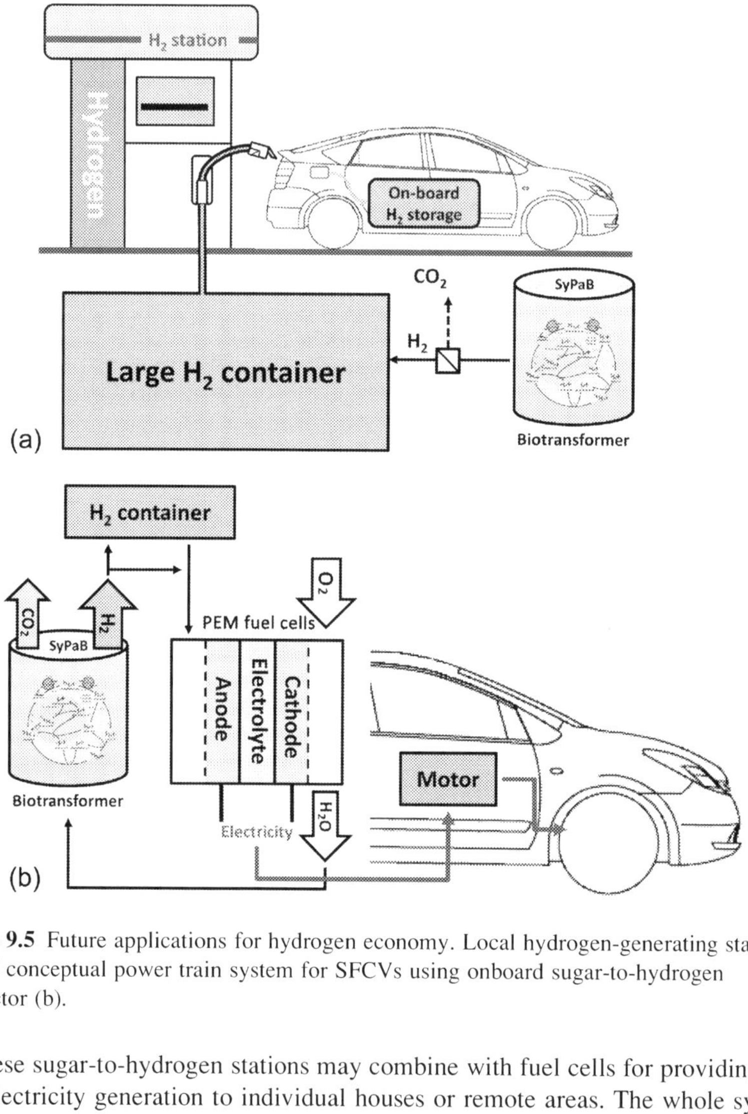

Figure 9.5 Future applications for hydrogen economy. Local hydrogen-generating station (a) and conceptual power train system for SFCVs using onboard sugar-to-hydrogen bioreactor (b).

These sugar-to-hydrogen stations may combine with fuel cells for providing low-cost electricity generation to individual houses or remote areas. The whole systems could have very high energy efficiency because the exothermic reaction of generating electricity from fuels cells is coupled with the endothermic reaction of producing hydrogen from carbohydrates via SyPaB, and the residual heat can cogenerate hot water (Zhang, 2009a).

9.4.2 Sugar fuel cell vehicles

The most ambitious application is SFCVs (Figure 9.5b). These hypothetical SFCVs can be powered by electricity generated from PEM fuel cells, which, in turn, can be supplied hydrogen generated from an onboard SyPaB bioreactor that can produce on-demand hydrogen fast enough. SFCVs can be refilled with solid carbohydrate fuels or a carbohydrate/water slurry within minutes. Highly pure hydrogen generated via SyPaB can greatly decrease the complexity of the whole power train system and increase the operation performance (Zhang, 2009a). Water and heat generated from PEM fuel cells can be recycled to the bioreactor and maintain the reaction temperature of the bioreactor, respectively. Rechargeable batteries will be needed to assist the power train system during the vehicle starting-up or accelerating like typical FCVs (Zhang, 2009a). One important factor of onboard hydrogen storage/production systems is the spatial restriction. At the current hydrogen production rate of 150 mmol/L/h, it requires a reaction volume of 3.33 m^3 to produce 1 kg of hydrogen per hour. This reaction volume seems technically unrealistic, but it is expected to be able to increase the reaction rate by up to another 60-fold (see Section 9.3.3). The advantages of SFCVs are the highest biomass-to-wheel energy conversion efficiency (overall, 55%; carbohydrate-to-hydrogen, 122%; hydrogen-PEM fuel cells, 50%; electricity-motor, 90%), no special infrastructure needed, a high safety level, and the use of carbon-neutral carbohydrates (Zhang, 2009a).

9.5 Conclusions

Hydrogen is the most promising alternative transportation fuel, i.e., high-energy efficiency via PEM fuel cells (Zhang and Mielenz, 2011) and high-energy storage density relative to rechargeable batteries. Among a number of carbohydrate-to-hydrogen technologies (Figure 9.3), SyPaB is the most promising method to producing low-cost hydrogen due to the highest yield, highly pure hydrogen production, modest reaction conditions, competitive reaction rates, and low-cost bioreactor. As compared to other hydrogen carriers, renewable carbohydrates are attractive because of their abundance, renewability, low cost, and even distribution, and they have best hydrogen storage density and safety. However, a few obstacles need to be solved for commercial hydrogen production via SyPaB, such as low-cost and highly stable enzyme production and the use of biomimetic cofactors. It would be a dream that these remaining technical obstacles are solved by a single laboratorial effort. More realistically, it requires international collaborative research to solve those challenges and bring about a new carbon-neutral hydrogen economy.

9.6 Sources of future information and advice

The seminal concept of hydrogen produced from carbohydrates and water via a synthetic enzymatic pathway was first proposed in 2007 (Zhang et al., 2007). Later, high-yield hydrogen was produced from cellulosic materials (Ye et al., 2009) and xylose (Martin del Campo et al., 2013). The recent review covers the advances in SyPaB and remaining technical challenges (Rollin et al., 2013).

The concept of carbohydrates as a hydrogen carrier was proposed first in 2007 (Zhang et al., 2007), and its potential impact on the sustainable revolution was analyzed in 2013 (Zhang, 2013).

References

Ahmed, S., Krumpelt, M., 2001. Hydrogen from hydrocarbon fuels for fuel cells. Int. J. Hydrog. Energy 26 (4), 291–301.

Ansell, R.J., Lowe, C.R., 1999. Artificial redox coenzymes: biomimetic analogues of NAD^+. Appl. Microbiol. Biotechnol. 51 (6), 703–710.

Argun, H., Kargi, F., Kapdan, I.K., 2009. Hydrogen production by combined dark and light fermentation of ground wheat solution. Int. J. Hydrog. Energy 34 (10), 4305–4311.

Babu, S.P., 2002. Biomass Gasification for Hydrogen Production–Process Description and Research Needs. IEA, Gas Technology Institute, IL, USA, Revised draft (29.10.02.).

Bastian, S., Liu, X., Meyerowitz, J.T., Snow, C.D., Chen, M.M., Arnold, F.H., 2011. Engineered ketol-acid reductoisomerase and alcohol dehydrogenase enable anaerobic 2-methylpropan-1-ol production at theoretical yield in *Escherichia coli*. Metab. Eng. 13 (3), 345–352.

Berner, R.A., 2003. The long-term carbon cycle, fossil fuels and atmospheric composition. Nature 426 (6964), 323–326.

Bogdanović, B., Felderhoff, M., Kaskel, S., Pommerin, A., Schlichte, K., Schüth, F., 2003. Improved hydrogen storage properties of Ti-doped sodium alanate using titanium nanoparticles as doping agents. Adv. Mater. 15 (12), 1012–1015.

Campbell, E., Wheeldon, I.R., Banta, S., 2010. Broadening the cofactor specificity of a thermostable alcohol dehydrogenase using rational protein design introduces novel kinetic transient behavior. Biotechnol. Bioeng. 107 (5), 763–774.

Campbell, E., Meredith, M., Minteer, S.D., Banta, S., 2012. Enzymatic biofuel cells utilizing a biomimetic cofactor. Chem. Commun. 48 (13), 1898–1900.

Carey, F.A., 2003. Organic Chemistry. McGraw-Hill, Boston, xxxii, 1191, 131 pp.

Carrette, L., Friedrich, K., Stimming, U., 2001. Fuel cells—fundamentals and applications. Fuel Cells 1 (1), 5–39.

Chao, B., Klebanoff, L., 2012. Hydrogen storage in interstitial metal hydrides. Hydrogen Storage Technol. Mater. Appl. 109.

Chheda, J.N., Huber, G.W., Dumesic, J.A., 2007. Liquid-phase catalytic processing of biomass-derived oxygenated hydrocarbons to fuels and chemicals. Angew. Chem. Int. Ed. 46 (38), 7164–7183.

Conrado, R.J., Varner, J.D., DeLisa, M.P., 2008. Engineering the spatial organization of metabolic enzymes: mimicking nature's synergy. Curr. Opin. Biotechnol. 19 (5), 492–499.

Cooney, M.J., Svoboda, V., Lau, C., Martin, G., Minteer, S.D., 2008. Enzyme catalysed biofuel cells. Energy Environ. Sci. 1 (3), 320–337.

Corella, J., Aznar, M.-P., Gil, J., Caballero, M.A., 1999. Biomass gasification in fluidized bed: where to locate the dolomite to improve gasification? Energy Fuel 13 (6), 1122–1127.

Crabtree, G.W., Dresselhaus, M.S., Buchanan, M.V., 2004. The hydrogen economy. Phys. Today 57, 39.

Cummer, K.R., Brown, R.C., 2002. Ancillary equipment for biomass gasification. Biomass Bioenergy 23 (2), 113–128.

Deluga, G., Salge, J., Schmidt, L., Verykios, X., 2004. Renewable hydrogen from ethanol by autothermal reforming. Science 303 (5660), 993–997.

Demirbaş, A., 2002. Gaseous products from biomass by pyrolysis and gasification: effects of catalyst on hydrogen yield. Energy Convers. Manag. 43 (7), 897–909.

Döhr, O., Paine, M.J., Friedberg, T., Roberts, G.C., Wolf, C.R., 2001. Engineering of a functional human NADH-dependent cytochrome P450 system. Proc. Natl. Acad. Sci. U. S. A. 98 (1), 81–86.

Dresselhaus, M.S., 2004. Basic research needs for the hydrogen economy. Abstr. Pap. Am. Chem. Soc. 227, U1084–U1085.

Dueber, J.E., Wu, G.C., Malmirchegini, G.R., Moon, T.S., Petzold, C.J., Ullal, A.V., Prather, K.L.J., Keasling, J.D., 2009. Synthetic protein scaffolds provide modular control over metabolic flux. Nat. Biotechnol. 27 (8), 753–759.

Eijsink, V.G.H., Bjork, A., Gaseidnes, S., Sirevag, R., Synstad, B., van den Burg, B., Vriend, G., 2004. Rational engineering of enzyme stability. J. Biotechnol. 113 (1–3), 105–120.

Eijsink, V.G.H., Gaseidnes, S., Borchert, T.V., van den Burg, B., 2005. Directed evolution of enzyme stability. Biomol. Eng. 22 (1–3), 21–30.

Falkowski, P., Scholes, R., Boyle, E., Canadell, J., Canfield, D., Elser, J., Gruber, N., Hibbard, K., Högberg, P., Linder, S., 2000. The global carbon cycle: a test of our knowledge of earth as a system. Science 290 (5490), 291–296.

Gao, S., You, C., Renneckar, S., Bao, J., Zhang, Y.-H.P., 2014. New insights into enzymatic hydrolysis of heterogeneous cellulose by using carbohydrate-binding module 3 containing GFP and carbohydrate-binding module 17 containing CFP. Biotechnol. Biofuels 7 (1), 24.

Hallenbeck, P.C., Benemann, J.R., 2002. Biological hydrogen production; fundamentals and limiting processes. Int. J. Hydrog. Energy 27 (11), 1185–1193.

Haryanto, A., Fernando, S., Murali, N., Adhikari, S., 2005. Current status of hydrogen production techniques by steam reforming of ethanol: a review. Energy Fuel 19 (5), 2098–2106.

Hong, J., Wang, Y.R., Ye, X.H., Zhang, Y.-H.P., 2008. Simple protein purification through affinity adsorption on regenerated amorphous cellulose followed by intein self-cleavage. J. Chromatogr. A 1194 (2), 150–154.

Huang, W.-D., Zhang, Y.-H.P., 2011a. Analysis of biofuels production from sugar based on three criteria: thermodynamics, bioenergetics, and product separation. Energy Environ. Sci. 4 (3), 784–792.

Huang, W.-D., Zhang, Y.-H.P., 2011b. Energy efficiency analysis: biomass-to-wheel efficiency related with biofuels production, fuel distribution, and powertrain systems. Plos One 6 (7), e22113.

Huang, Q., Wu, J.W., Xu, H.J., 2013. Biomimetic hydrogenation: a reusable NADH coenzyme model for hydrogenation of alpha, beta-epoxy ketones and 1,2-diketones. Tetrahedron Lett. 54 (29), 3877–3881.

Hull, J.F., Himeda, Y., Wang, W.H., Hashiguchi, B., Periana, R., Szalda, D.J., Muckerman, J.T., Fujita, E., 2012. Reversible hydrogen storage using CO_2 and a proton-switchable iridium catalyst in aqueous media under mild temperatures and pressures. Nat. Chem. 4 (5), 383–388.

Imamura, H., Masanari, K., Kusuhara, M., Katsumoto, H., Sumi, T., Sakata, Y., 2005. High hydrogen storage capacity of nanosized magnesium synthesized by high energy ball-milling. J. Alloys Compd. 386 (1), 211–216.

Ji, P., Feng, W., Chen, B., 2009. Production of ultrapure hydrogen from biomass gasification with air. Chem. Eng. Sci. 64 (3), 582–592.

Keaton, R.J., Blacquiere, J.M., Baker, R.T., 2007. Base metal catalyzed dehydrogenation of ammonia-borane for chemical hydrogen storage. J. Am. Chem. Soc. 129 (7), 1844–1845.

Kirk, R.E., Othmer, D.F., 2000. Kirk-Othmer Encyclopedia of Chemical Technology. Wiley, New York.

Kleerebezem, R., van Loosdrecht, M., 2007. Mixed culture biotechnology for bioenergy production. Curr. Opin. Biotechnol. 18 (3), 207–212.

Klerke, A., Christensen, C.H., Nørskov, J.K., Vegge, T., 2008. Ammonia for hydrogen storage: challenges and opportunities. J. Mater. Chem. 18 (20), 2304–2310.

Lee, S.M., Lee, Y.H., 2000. Hydrogen storage in single-walled carbon nanotubes. Appl. Phys. Lett. 76 (20), 2877–2879.

Lin, Y., Tanaka, S., 2006. Ethanol fermentation from biomass resources: current state and prospects. Appl. Microbiol. Biotechnol. 69 (6), 627–642.

Lin, S., Harada, M., Suzuki, Y., Hatano, H., 2005. Process analysis for hydrogen production by reaction integrated novel gasification (HyPr-RING). Energy Convers. Manag. 46 (6), 869–880.

Liu, W.F., Wang, P., 2007. Cofactor regeneration for sustainable enzymatic biosynthesis. Biotechnol. Adv. 25 (4), 369–384.

Lo, H.C., Fish, R.H., 2002. Biomimetic NAD^+ models for tandem cofactor regeneration, horse liver alcohol dehydrogenase recognition of 1,4-NADH derivatives, and chiral synthesis. Angew. Chem. Int. Ed. 41 (3), 478–481.

Logan, B.E., Regan, J.M., 2006. Microbial fuel cells-challenges and applications. Environ. Sci. Technol. 40 (17), 5172–5180.

Lutz, J., Hollmann, F., Ho, T.V., Schnyder, A., Fish, R.H., Schmid, A., 2004. Bioorganometallic chemistry: biocatalytic oxidation reactions with biomimetic NAD^+/NADH co-factors and $[Cp^*Rh(bpy)H]^+$ for selective organic synthesis. J. Organomet. Chem. 689 (25), 4783–4790.

Ma, K.S., Zhou, Z.H., Adams, M.W.W., 1994. Hydrogen-production from pyruvate by enzymes purified from the hyperthermophilic archaeon, *Pyrococcus furiosus*: a key role for NADPH. Fems Microbiol. Lett. 122 (3), 245–250.

Martin del Campo, J.S., Rollin, J., Myung, S., Chun, Y., Chandrayan, S., Patino, R., Adams, M.W.W., Zhang, Y.-H.P., 2013. High-yield production of dihydrogen from xylose by using a synthetic enzyme cascade in a cell-free system. Angew. Chem. Int. Ed. 52 (17), 4587–4590.

Mathews, J., Wang, G., 2009. Metabolic pathway engineering for enhanced biohydrogen production. Int. J. Hydrog. Energy 34 (17), 7404–7416.

Mattos, L.V., Jacobs, G., Davis, B.H., Noronha, F.b.B., 2012. Production of hydrogen from ethanol: review of reaction mechanism and catalyst deactivation. Chem. Rev. 112 (7), 4094–4123.

McNaught, A.D., 1996. International Union of Pure and Applied Chemistry and International Union of Biochemistry and Molecular Biology—Joint Commission on Biochemical Nomenclature—Nomenclature of carbohydrates—Recommendations 1996. Pure Appl. Chem. 68 (10), 1919–2008.

Milne, T.A., Abatzoglou, N., Evans, R.J., 1998. Biomass Gasifier "tars": Their Nature, Formation, and Conversion. National Renewable Energy Laboratory, Golden, CO.

Miyake, J., Mao, X., Kawamura, S., 1984. Photoproduction of hydrogen from glucose by a co-culture of a photosynthetic bacterium and Clostridium butyricum. J. Ferment. Technol. 62 (6), 531–535.

Myung, S., Wang, Y.R., Zhang, Y.-H.P., 2010. Fructose-1,6-bisphosphatase from a hyperthermophilic bacterium *Thermotoga maritima*: characterization, metabolite stability, and its implications. Process Biochem. 45 (12), 1882–1887.

Myung, S., Zhang, X.Z., Zhang, Y.-H.P., 2011. Ultra-stable phosphoglucose isomerase through immobilization of cellulose-binding module-tagged thermophilic enzyme on low-cost high-capacity cellulosic adsorbent. Biotechnol. Prog. 27 (4), 969–975.

Naik, S.N., Goud, V.V., Rout, P.K., Dalai, A.K., 2010. Production of first and second generation biofuels: a comprehensive review. Renew. Sustain. Energy Rev. 14 (2), 578–597.

Narvaez, I., Corella, J., Orío, A., 1997. Fresh tar (from a biomass gasifier) elimination over a commercial steam-reforming catalyst. Kinetics and effect of different variables of operation. Ind. Eng. Chem. Res. 36 (2), 317–327.

Navarro, R., Pena, M., Fierro, J., 2007. Hydrogen production reactions from carbon feedstocks: fossil fuels and biomass. Chem. Rev. 107 (10), 3952–3991.

Ni, M., Leung, D.Y.C., Leung, M.K.H., Sumathy, K., 2006. An overview of hydrogen production from biomass. Fuel Process. Technol. 87 (5), 461–472.

Nielsen, M., Alberico, E., Baumann, W., Drexler, H.-J., Junge, H., Gladiali, S., Beller, M., 2013. Low-temperature aqueous-phase methanol dehydrogenation to hydrogen and carbon dioxide. Nature 495 (7439), 85–89.

Olah, G.A., 2005. Beyond oil and gas: the methanol economy. Angew. Chem. Int. Ed. 44 (18), 2636–2639.

Osman, M.H., Shah, A.A., Walsh, F.C., 2011. Recent progress and continuing challenges in bio-fuel cells. Part I: enzymatic cells. Biosens. Bioelectron. 26 (7), 3087–3102.

Panella, B., Hirscher, M., Roth, S., 2005. Hydrogen adsorption in different carbon nanostructures. Carbon 43 (10), 2209–2214.

Paul, C.E., Gargiulo, S., Opperman, D.J., Lavandera, I., Gotor-Fernandez, V., Gotor, V., Taglieber, A., Arends, I.W.C.E., Hollmann, F., 2013. Mimicking nature: synthetic nicotinamide cofactors for C=C bioreduction using enoate reductases. Org. Lett. 15 (1), 180–183.

Pérez-Hernández, R., Mendoza-Anaya, D., Martínez, A.G., Gómez-Cortés, A., 2012. Catalytic steam reforming of methanol to produce hydrogen on supported metal catalysts. Hydrogen Energy: Challenges Perspect, 149–256.

Probstein, R.F., Hicks, R.E., 2006. Synthetic Fuels. Courier Dover Publications, New York.

Reutemann, W., Kieczka, H., 1996. Formic acid. Ullmann's Encyclopedia of Industrial Chemistry. Wiley-VCH, Weinheim, Germany.

Reyes, B.A., Pendergast, J.S., Yamazaki, S., 2008. Mammalian peripheral circadian oscillators are temperature compensated. J. Biol. Rhythms 23 (1), 95.

Rezaiyan, J., Cheremisinoff, N.P., 2005. Gasification Technologies: A Primer for Engineers and Scientists. CRC Press, Boca Raton, FL.

Rollin, J.A., Tam, T.K., Zhang, Y.-H.P., 2013. New biotechnology paradigm: cell-free biosystems for biomanufacturing. Green Chem. 15 (7), 1708–1719.

Sakintuna, B., Lamari-Darkrim, F., Hirscher, M., 2007. Metal hydride materials for solid hydrogen storage: a review. Int. J. Hydrog. Energy 32 (9), 1121–1140.

Schlapbach, L., Zuttel, A., 2001. Hydrogen-storage materials for mobile applications. Nature 414 (6861), 353–358.

Schuchmann, K., Muller, V., 2013. Direct and reversible hydrogenation of CO_2 to formate by a bacterial carbon dioxide reductase. Science 342 (6164), 1382–1385.

Sheldon, R.A., Sorgedrager, M., Janssen, M.H.A., 2007. Use of cross-linked enzyme aggregates (CLEAs) for performing biotransformations. Chim. Oggi Chem. Today 25 (1), 48–52.

Simell, P.A., Hakala, N.A., Haario, H.E., Krause, A.O.I., 1997. Catalytic decomposition of gasification gas tar with benzene as the model compound. Ind. Eng. Chem. Res. 36 (1), 42–51.

Simell, P.A., Hirvensalo, E.K., Smolander, V.T., Krause, A.O.I., 1999. Steam reforming of gasification gas tar over dolomite with benzene as a model compound. Ind. Eng. Chem. Res. 38 (4), 1250–1257.

Srivastava, D.K., Bernhard, S.A., 1986. Metabolite transfer via enzyme-enzyme complexes. Science 234 (4780), 1081–1086.

Stevens, D.J., 2001. Hot gas conditioning: recent progress with larger-scale biomass gasification systems. NREL Subcontractor Report (NREL/SR-510-29952).

Sutton, D., Kelleher, B., Ross, J.R., 2001. Review of literature on catalysts for biomass gasification. Fuel Process. Technol. 73 (3), 155–173.

Thauer, R.K., Jungermann, K., Decker, K., 1977. Energy conservation in chemotrophic anaerobic bacteria. Bacteriol. Rev. 41 (1), 100.

Thomas, G., Parks, G., 2006. Potential roles of ammonia in a hydrogen economy: a study of issues related to the use ammonia for on-board vehicular hydrogen storage. US Department of Energy, 23.

Vajo, J.J., Skeith, S.L., Mertens, F., 2005. Reversible storage of hydrogen in destabilized LiBH4. J. Phys. Chem. B 109 (9), 3719–3722.

Van Vucht, J., Kuijpers, F., Bruning, H., 1970. Reversible room-temperature absorption of large quantities of hydrogen by intermetallic compounds. Philips Res. Rep. 25, 133–140.

Wang, Y.R., Zhang, Y.-H.P., 2009. Overexpression and simple purification of the *Thermotoga maritima* 6-phosphogluconate dehydrogenase in *Escherichia coli* and its application for NADPH regeneration. Microb. Cell Fact. 8, 30.

Wang, Y.R., Zhang, Y.-H.P., 2010. A highly active phosphoglucomutase from *Clostridium thermocellum*: cloning, purification, characterization and enhanced thermostability. J. Appl. Microbiol. 108 (1), 39–46.

Wang, Y.R., Huang, W.D., Sathitsuksanoh, N., Zhu, Z.G., Zhang, Y.-H.P., 2011. Biohydrogenation from biomass sugar mediated by in vitro synthetic enzymatic pathways. Chem. Biol. 18 (3), 372–380.

Weckbecker, A., Groger, H., Hummel, W., 2010. Regeneration of nicotinamide coenzymes: principles and applications for the synthesis of chiral compounds. Biosyst. Eng. I: Creating Superior Biocatal. 120, 195–242.

Wilner, O.I., Weizmann, Y., Gill, R., Lioubashevski, O., Freeman, R., Willner, I., 2009. Enzyme cascades activated on topologically programmed DNA scaffolds. Nat. Nanotechnol. 4 (4), 249–254.

Wolfenden, R., Snider, M.J., 2001. The depth of chemical time and the power of enzymes as catalysts. Acc. Chem. Res. 34 (12), 938–945.

Woodward, J., Orr, M., Cordray, K., Greenbaum, E., 2000. Enzymatic production of biohydrogen. Nature 405, 1014–1015.

Ye, X.H., Wang, Y.R., Hopkins, R.C., Adams, M.W.W., Evans, B.R., Mielenz, J.R., Zhang, Y.-H.P., 2009. Spontaneous high-yield production of hydrogen from cellulosic materials and water catalyzed by enzyme cocktails. Chemsuschem 2 (2), 149–152.

Ye, X.H., Rollin, J., Zhang, Y.-H.P., 2010. Thermophilic alpha-glucan phosphorylase from *Clostridium thermocellum*: cloning, characterization and enhanced thermostability. J. Mol. Catal. B Enzymatic 65 (1–4), 110–116.

Ye, X., Zhang, C., Zhang, Y.-H.P., 2012. Engineering a large protein by combined rational and random approaches: stabilizing the *Clostridium thermocellum* cellobiose phosphorylase. Mol. BioSyst. 8, 1815–1823.

Yoshida, A., Nishimura, T., Kawaguchi, H., Inui, M., Yukawa, H., 2005. Enhanced hydrogen production from formic acid by formate hydrogen lyase-overexpressing *Escherichia coli* strains. Appl. Environ. Microbiol. 71 (11), 6762–6768.

You, C., Myung, S., Zhang, Y.-H.P., 2012. Facilitated substrate channeling in a self-assembled trifunctional enzyme complex. Angew. Chem. Int. Ed. 51 (35), 8787–8790.

Yu, X., Pickup, P.G., 2008. Recent advances in direct formic acid fuel cells (DFAFC). J. Power Sources 182 (1), 124–132.

Zanzi, R., Sjöström, K., Björnbom, E., 2002. Rapid pyrolysis of agricultural residues at high temperature. Biomass Bioenergy 23 (5), 357–366.

Zhang, Y.-H.P., 2009a. A sweet out-of-the-box solution to the hydrogen economy: is the sugar-powered car science fiction? Energy Environ. Sci. 2 (3), 272–282.

Zhang, Y.-H.P., 2009b. Using extremophile enzymes to generate hydrogen for electricity. Microbe 4, 560–565.

Zhang, Y.-H.P., 2010a. Production of biocommodities and bioelectricity by cell-free synthetic enzymatic pathway biotransformations: challenges and opportunities. Biotechnol. Bioeng. 105 (4), 663–677.

Zhang, Y.-H.P., 2010b. Renewable carbohydrates are a potential high-density hydrogen carrier. Int. J. Hydrog. Energy 35 (19), 10334–10342.

Zhang, Y.-H.P., 2011a. Hydrogen production from carbohydrates: a mini-review, sustainable production of fuels, chemicals, and fibers from forest biomass. ACS Symp. Series 1067, 203–216.

Zhang, Y.-H.P., 2011b. Simpler is better: high-yield and potential low-cost biofuels production through cell-free synthetic pathway biotransformation (SyPaB). ACS Catal. 1 (9), 998–1009.

Zhang, Y.-H.P., 2011c. Substrate channeling and enzyme complexes for biotechnological applications. Biotechnol. Adv. 29 (6), 715–725.

Zhang, Y.-H.P., 2013. Next generation biorefineries will solve the food, biofuels, and environmental trilemma in the energy-food-water nexus. Energy Sci. Eng. 1, 27–41.

Zhang, Y.-H.P., Mielenz, J.R., 2011. Renewable hydrogen carrier-carbohydrate: constructing the carbon-neutral carbohydrate economy. Energies 4 (2), 254–275.

Zhang, Y.-H.P., Evans, B.R., Mielenz, J.R., Hopkins, R.C., Adams, M.W.W., 2007. High-yield hydrogen production from starch and water by a synthetic enzymatic pathway. Plos One 2 (5), e456.

Zhang, Y.-H.P., Sun, J.B., Zhong, J.J., 2010. Biofuel production by in vitro synthetic enzymatic pathway biotransformation. Curr. Opin. Biotechnol. 21 (5), 663–669.

Zhang, Y.-H.P., Xu, J.H., Zhong, J.J., 2012. A new high-energy density hydrogen carrier—carbohydrate—might be better than methanol. Int. J. Energy Res. 37 (7), 769–779.

Zhu, Z.G., Tam, T.K., Sun, F.F., You, C., Zhang, Y.-H.P., 2014. A high-energy-density sugar biobattery based on a synthetic enzymatic pathway. Nat. Commun. 5, 3026.

Züttel, A., 2003. Materials for hydrogen storage. Mater. Today 6 (9), 24–33.

Conceptual density functional theory (DFT) approach to all-metal aromaticity and hydrogen storage

10

R. Das, A. Chakraborty, S. Pan, P.K. Chattaraj
Indian Institute of Technology, Kharagpur, India

Symbols and abbreviations

A	electron affinity
B3LYP	Becke's three parameter hybrid functional using Lee–Yang–Parr correlation functional
CNFs	carbon nanofibers
CNTs	carbon nanotubes
COFs	covalent organic frameworks
DCD	Dewar–Chatt–Duncanson model
ΔE_{CE}	dissociative chemisorption energy
ΔG	reaction-free energy
ΔH	reaction enthalpy
$\Delta \omega$	reaction electrophilicity
DFT	density functional theory
E_{HOMO}	highest occupied molecular orbital energy
E_{LUMO}	lowest unoccupied molecular orbital energy
η	hardness
f_k^{α}	condensed Fukui function
FF$(f(\bar{r}))$	Fukui function
GE	gain in energy
GOFs	graphene-oxide frameworks
HEDM	high energy density material
I	ionization potential
IE	interaction energy
LAMMPS	large-scale atomic/molecular massively parallel simulator
LDI	London dispersion interactions
MOFs	metal-organic frameworks
μ	chemical potential
M052X	Minnesota 05 global hybrid functional with 52% HF exchange
M06	Minnesota 06 global hybrid functional with 27% HF exchange
MEP	minimum electrophilicity principle
MHP	maximum hardness principle
MP2	second order Møller–Plesset perturbation theory
MPA	Mulliken population analysis
MPP	minimum polarizability principle

Compendium of Hydrogen Energy. http://dx.doi.org/10.1016/B978-1-78242-362-1.00010-9

MPW1K	modified Perdew–Wang 1-parameter model for kinetics
N	number of electrons
NICS	nucleus independent chemical shift
NIMAG	number of imaginary frequency
NPA	natural population analysis
PBE0	Perdew–Burke–Ernzerhof exchange-correlation functional
Q_k	atomic charge
$\rho(\bar{r})$	electron density
SiCNTs	silicon-carbon nanotubes
SWCNTs	single-walled carbon nanotubes
$v(\bar{r})$	external potential
χ	electronegativity
ω	electrophilicity index
ω_k^α	condensed-to-atom local philicity

10.1 Introduction

The utility of hydrogen as a fuel and an energy carrier in automobile applications has been a widely investigated topic of research for both experimentalists and theoreticians since the last 15 years. Worldwide industrialization followed by an incessant rise in the use of automobiles has made a deep impact on the existing reserves of fossil fuels, which have been decreasing at an alarming rate. To combat the curses of the ensuing dearth of fossil fuel reserves and their harmful effects on nature (global warming), the plausible use of various nonconventional matters as alternative sources of energy are investigated. Upon combustion hydrogen produces neither air pollutants nor greenhouse gases as byproducts; only eco-friendly water vapor is produced. Due to its pollution-free nature, nonexclusive access around the globe, and inherent high-energy efficiency, hydrogen is considered to be a potential candidate in solving the upcoming energy crisis caused by the unrestrained use of coal, oil, and other fossil fuels. Hence, the onboard storage of hydrogen is necessary for it to be used as a fuel for mobile applications and as an efficient energy carrier. Hydrogen, as a highly combustible gas at ordinary temperatures, demands adequate safety measures as well as cryogenic conditions to be liquefied and stored securely in manageable quantities in large cylinders. Such provisions are indeed very hard to follow up on a practical day-to-day basis. Moreover, from a practical viewpoint, large-scale manufacturing of such types of cylinders for hydrogen storage with the capability to sustain in such extreme conditions is very expensive and hence not consumer friendly. Therefore, the search for effective molecular templates that are capable of hydrogen storage at ambient conditions is still a challenging task for the scientific community.

With this objective, after much research the US Department of Energy (DOE) has become successful at implementing some standard benchmark criteria that a material should possess to perform as an effective hydrogen storage medium. The demanding criteria of a molecular motif for use as an effective hydrogen storing material for mobile applications are (i) low weight and large stability, (ii) low cost and ease of availability, (iii) large volumetric and gravimetric density of hydrogen, (iv) favorable adsorption and desorption kinetics, and (v) proper thermodynamic parameters. Various groups have successfully modeled several novel molecular motifs that have the potential to store both atomic and molecular hydrogens under optimum conditions. A few papers (e.g., Jena, 2011; Zhou, 2005; Züttel, 2004; Steriotis et al., 2011; Ströbel et al., 2006; Schlapbach and Züttel, 2001; Hoang and Antonelli, 2009; Chakraborty et al., 2012) have highlighted some of these recent achievements and are noteworthy. The conceptual density functional theory (CDFT) (Parr and Yang, 1989; Geerlings et al., 2003; Chattaraj, 2009; Chattaraj and Giri, 2009a; Chakraborty et al., 2011a) also appears to be an important tool for analyzing the hydrogen storage capacity of several molecular motifs. This study attempts to describe the different types of hydrogen storage systems modeled thus far for wider practical applications. In the following sections an overview of different hydrogen storage materials is given.

10.1.1 Molecular clusters designed for hydrogen storage applications

10.1.1.1 Metal hydrides and metal clusters

Metal hydrides represent a class of binary compounds in which different metal atoms coordinate with highly reactive hydrogen molecule under optimum conditions to yield a complex. Similarly, the metal clusters also have a tendency to bind hydrogen. A stable metal-hydrogen linkage is therefore supposed to be the base of the modeling and formation of a huge number of hydrogen-bound complex metal cluster assemblies. Thus, the complex metal hydrides with alkali metals, alkaline earth metals, and the transition metals act as very hydrogen-rich materials. In fact, the metal hydrides contain higher hydrogen density compared to that of gas and liquid forms (Weast et al., 1983) suggesting the metal hydrides as much safe and volume-efficient hydrogen trap. The applicability of the metal hydrides and complex hydrides as rich hydrogen supplying media is well documented elsewhere (Schüth et al., 2004; Sakintuna et al., 2007). The use of Ti-doped alkali metal aluminum hydrides as promising materials for the reversible storage of hydrogen was reported by Bogdanović and Schwickardi (1997). The possibility of hydrogen storage in the slurries of various metal hydrides and organic liquids has also been investigated. Relevant studies show an increase in gravimetric capacity of hydrogen stored as the metal hydrides increases upon dispersion in a polymeric matrix (McClaine et al., 2000; Schmidt, 2001). Sometimes (Alapati et al., 2006) destabilized metal hydrides as hydrogen-trapping materials are found to execute even better than ordinary hydrides. New schemes for the direct synthesis of $LiAlH_4$ and $NaAlH_4$, which are suitable hydrogen supplying materials, have also been reported by Wang et al. (2007). The possibility of using metallo-hydrides such as $NaAlH_4$ and $NaBH_4$ as high-density hydrogen storage materials (Bogdanović et al., 2009) has also been explored. The capability of a diverse variety of metal clusters toward trapping hydrogen was verified by using both experimental and theoretical benchmarks. In this regard, Mg clusters (Wagemans et al., 2005) have turned out to be a very effective candidate among the alkaline earth metals, in part, because its binary hydride also possesses a large amount of hydrogen. However, in spite of the excellent hydrogen storage capacity of metal hydrides, some practical disadvantages such as quite high dehydrogenation temperatures and slow desorption kinetics have also been reported (Bogdanović et al., 2003). McNelles and Naumkin (2009) investigated the stability of the systems by encapsulating one H_2 molecule into the Mg_n ($n=8$–10) cages. Further, titanium centers in various templates are found to be very effective in binding hydrogen. A few studies (Yildirim and Ciraci, 2005; Yildirim et al., 2005; Dag et al., 2005) have demonstrated the effect of titanium doping on C_{60} cage and carbon nanotubes in improving hydrogen storage ability. The mechanism of interaction between H_2 molecules and the Ti-site can be better understood by the Kubas interaction (Kubas, 2001), and the corresponding interaction energy was found somewhat in between those of physisorption and chemisorption. It therefore transpires that a Ti—(C=C) linkage between the Ti-centers and the C—C bonds of the fullerene-type cage was energetically favorable, which further serves as a suitable hydrogen-trapping medium. This concept seems to be applicable to small unsaturated systems such as ethylene (Durgun et al., 2006), where the Kubas concept of binding between the Ti-site of the $(C_2H_4)Ti_n$ complex and the H_2 molecules resembles the well-known Dewar–Chatt–Duncanson (DCD) model of chemical bonding for inorganic complexes. Further study (Zhou et al., 2007) has also been made on the stability of M-C_2H_4 complexes, where M was the other transition metals or light metals such as lithium. Again, the hydrogen storage capability of a transition metal-doped C_5H_5 system (Zhao et al., 2005), C_4H_4–Ti and $(C_4H_4)_2$–Ti complexes (Kiran et al., 2006), transition metal bound organic templates (Weck et al., 2007), titanium metallo-carbohedryne (met-car) clusters (Akman et al., 2006), and multidecker $TiC_3B_2H_5$ complex (Kandalam et al., 2008) have also been tested.

10.1.1.2 Nanomaterials

Nanostructured molecular assemblies are promising as potential hydrogen storage materials due to their unique features such as surface-adsorption, inter- and intragrain boundaries, and bulk absorption (Baburaj et al., 2000; Schulz et al., 1995). A diverse range of carbon-based nanomaterials include the sheet-like 2D and 3D frameworks, cage-like spherical fullerene-based structures including the clathrate hydrate molecules, and the cylindrical tube-like or wire-like systems called nanotubes or nanowires, respectively.

10.1.1.2.1 Sheet-like frameworks

Graphene, an allotrope of carbon, bears a resemblance to a planar sheet-like framework comprising sp^2-hybridized carbon atoms having a one-atom thickness. Graphite, on the other hand is a multilayered polygraphene cluster containing voids in between two graphene sheets. The presence of mobile electrons makes graphene a significant electrical and thermal conductor. Due to some favorable characteristics such as its nontoxic nature, low cost, and very easily synthesizable materials, graphene has obtained much attention from scientists as a plausible hydrogen-storing template. However, it was found that single-layered graphene sheets are not markedly efficient in binding molecular hydrogen. The doping of suitable metal atoms/ions on the graphene sheet enhances the hydrogen storage ability, where metal atoms donate some electrons to the graphene sheet and gain some positive charge. Due to this charge, the metal center can bind hydrogen effectively through an interaction of dipole-induced dipole type. One study (Srinivasu et al., 2008) further established that undoped graphene or carbon nanotube alone is not suitable applicant for storing hydrogen. However, ionic centers generated by doping alkali metal ions on these systems make them effective hydrogen-binding materials. Therefore, a charged surface now behaves as a better template for binding molecular hydrogen. In a similar study, Ataca et al. (2008) also demonstrated that metallized graphene sheets modeled by doping with Li-atoms on both sides of a normal graphene sheet act as better traps for molecular hydrogen in terms of gravimetric means. According to another report (Kim et al., 2009) boron-doped graphene sheets dispersed by alkaline earth metal atoms (e.g., Mg, Ca) is a better template for adsorbing molecular hydrogen. Further, Liu and Zeng (2010) confirmed the increased Li-binding energy of graphene due to the boron doping on it. They have also stated that this is due to the stronger *p-p* hybridization as occurred in Li-dispersed boron-doped graphene instead of relatively weaker *s-p* hybridization in Li-doped graphene. Again, Subrahmanyam et al. (2011) recently showed that few-layer graphene may be utilized as a plausible hydrogen storage material. They showed that Birch reduction of few-layer graphene leads to a hydrogenated sample having up to 5% of hydrogen, which readily decomposes to release free hydrogen upon heating to 500 °C or upon irradiation with UV or laser light.

Multilayered graphene-oxide frameworks (GOFs), formed by the oxidation of graphene sheets, possess a unique constructive specialty that presupposes their plausible use as effective hydrogen-binding templates. In this regard the contribution of Yildirim (Burress et al., 2010) and his group for developing the GOFs as storage materials for hydrogen is worth mentioning. The multidecker GOFs are quite capable of entrapping at least hundred times more hydrogen than the ordinary graphene oxide frameworks. The potential use of a Ti-anchored GOF as a hydrogen storage medium was reported by Wang et al. (2009). Some related theoretical studies by Dimitrakakis et al. (2008) and Han and Jang (2009) illustrate that 3D pillared graphene frameworks perform as effective hydrogen adsorbing material with good gravimetric and volumetric quantities satisfying the goal set by the US Department of Energy (DOE). A first principle calculation by Park et al. (2007) to assess the hydrogen storage capability of 3D covalently bonded graphenes (CBGs) shows that the hydrogen-binding energy can be improved up to 20–15% compared to an isolated graphene sheet. In a recent paper, Chan and Hill (2011) inspected hydrogen storage capacity inside the graphene oxide frameworks through applied mathematical modeling.

The potential of graphyne and graphdiyne as probable hydrogen storage materials has also been recently highlighted. First, principle calculations by Li et al. (2011)

revealed that the binding energy of Ca-graphyne is higher than that of Ca-fullerene with the former having a higher hydrogen storage capacity. *Ab initio* first principle investigations on the hydrogen-binding capability of Li-dispersed graphyne and graphdiyne were also recently reported (Srinivasu and Ghosh, 2012).

10.1.1.2.2 Tubular nanostructures (nanotubes) and cage-like molecules

Several studies have been performed to explore the viability of these cage-like and cylindrical nanomaterials (Bethune et al., 1993) as storage materials for hydrogen in both the endohedral and exohedral fashion. The hydrogen binding was found to increase upon decorating the template surface with a metal ion. In recent years, the cage-like clathrate hydrate molecules, formed by the stacking of water molecules through hydrogen-bonding forces, have also received considerable attention as hydrogen storage materials.

The first experimental report of the utility of these carbon nanotubes (CNTs) as plausible hydrogen storage material is credited to Dillon et al. (1997). The efficiency of carbon-based nanostructures as probable hydrogen storage templates has been investigated (Ding et al., 2011; Hirscher and Becher, 2003; Dillon and Heben, 2001; Hynek et al., 1997). Furthermore, Jiménez et al. (2010) recently investigated the variation of the hydrogen-binding ability of a variety of carbon structures counting activated carbon (AC) and carbon nanofibers (CNFs) with pressure and temperature and found a linear relationship between the capacity of hydrogen physisorbed and specific micropore surface area. In a benchmark study, Ferre-Vilaplana (2005) demonstrated hydrogen physisorption on a single-walled carbon nanotube (SWCNT) and concluded that the physisorption of molecular hydrogen on SWCNTs (both for inside and outside adsorption) depends primarily on the diameter, which is almost independent of the chirality. However, similar conclusions about the effect of curvature and chirality upon hydrogen storage in a SWCNT were drawn by Mpourmpakis et al. (2007); Mpourmpakis et al. (2006a) in their study with the aid of combined *ab initio* and grand canonical Monte Carlo simulations. Kowalczyk et al. (2007) showed theoretically that the most effective structures along with appropriate metal dopants could satisfy the dual norms of DOE. Further, Firlej et al. (2009) in an *ab initio* calculation and Monte Carlo simulation explored the doping or intercalation by other species in an all-carbon skeleton of AC that can improve the heat of hydrogen adsorption. The adsorption energy gets better (up to 10–13.5 kJ/mol) on substitution of 5–10% of carbon atoms in a nanoporous material by boron atoms. Further, Wu et al. (2008) showed that the boron substitutions in carbon nanotubes improve the Li-binding energy, which inhibits the Li-clustering problem and can be a potential material for hydrogen loading. Boron doping in the presence of metals such as metal-diboride nanotubes (Meng et al., 2007) have proven to be promising hydrogen-binding materials. Mpourmpakis et al. (2006b) reported that analogous Si-doping on carbon nanotubes improves the hydrogen-binding energy (almost by 20%) due to improved net charges on the walls of silicon-carbon nanotubes (SiCNTs). Yildirim and Ciraci (2005) doped Ti metal upon CNTs to make them better hydrogen-trapping templates.

Later the effect of charge on the hydrogen-binding energy of fullerenes was explored by Yoon et al. (2007). The hydrogen storage capacity (both exohedral and endohedral) of different-sized fullerene cages have also been studied Yue et al. (2007). A theoretical investigation on the applicability of alkali-metal functionalized fullerenol derivatives ($M^{+}{}_{12}[C_{60}O_{12}]^{12-}$ (M = Li and Na)) as effective hydrogen-binding template has also been carried out (Peng et al., 2009). Recently, Javan et al. (2011) demonstrated the endohedral hydrogen-trapping ability of carbon-nitride heterofullerenes, while Saha and Deng (2011) experimentally showed improved hydrogen adsorption capability of Pd and Ru dispersed C_{60} fullerene compared to the nondispersed one. In a first principle study Sun et al. (2005) revealed that $B_{36}N_{36}$ cage is able to trap up to 18 hydrogen molecules endohedrally at zero temperature, leading to the gravimetric density of 4 wt%. A $B_{36}N_{36}$ cage as effective hydrogen capturing template can be tuned by encapsulating suitable metal ions which, in turn, enhance the hydrogen-binding energy due to the presence of a net charge on metal centers, thereby, binding molecular hydrogen endohedrally through an ion-induced dipole interaction Wen et al. (2008). The applicability of the suitable transition metal-doped B-N systems such as borazine in the hydrogen storage domain has been fruitfully investigated by Shevlin and Guo (2006). A theoretical report regarding the effective usage of metal-doped boron buckyballs (B_{80}) as a probable hydrogen-storing candidate is worth mentioning Wu et al. (2009). Further, Wu et al. (2010a) revealed through their computation that a gravimetric density of 7.43 wt% of hydrogen can be achieved by carbon-doped boron nitride cage. In a comparative study concerning hydrogen-binding ability of $B_{16}N_{16}$ cage with that of smaller BN analogues Cui et al. (2010) concluded that the interaction energy with hydrogen is decreased for BN cages with larger cavity ($B_{16}N_{16}$). The investigation on the usefulness of clathrate compounds as novel hydrogen storage material has recently become a stimulating field of research in both theoretical and experimental spheres. In a review report Struzhkin et al. (2007) explored the utility of different clathrate hydrates as probable hydrogen traps. More comprehensive overviews on the success of clathrate hydrate as plausible hydrogen storage material have also been reported (Sloan and Koh, 2007; Strobel et al., 2007; Lang et al., 2010). Recently, the hydrogen storage capacity of semiclathrate hydrate (Jeffrey, 1984; Shimada et al., 2005; Chapoy et al., 2007; Hashimoto et al., 2010), a special class of clathrate containing ionic guest species, has also been investigated. Reports describing the hydrogen storage capacity in binary clathrate hydrates having *tert*-butylamine as guest molecule should also be mentioned (Prasad et al., 2009; Aladko et al., 2010).

10.1.1.3 Metal-organic frameworks (MOFs)

Metal-organic frameworks (MOFs) constitute highly porous planar and 3D crystalline structures in which a metal center coordinates strongly with an organic ligand to yield a large complex suitable for adsorption. An MOF primarily consists of two parts, a metal ion or a cluster of metal ions, and, a mono- or multivalent organic molecule, i.e., "linker." Application of these extended 3D-MOF structures for hydrogen adsorption has been successful in recent years. The weak interaction, mostly a physisorption process, provides a reversible adsorption–desorption kinetics suitable for practical applications. Further studies (Kuc et al., 2008) reveal that the weak physisorption forces are nonbonded in nature resemble the London dispersion interactions (LDI). Electrostatic interactions can be nullified since the charge separation in the MOFs hardly induces any dipole in the H_2 molecules.

Some critical reviews (James, 2003; Eddaoudi et al., 2001) describe novel strategies to synthesize the MOFs and their further applications in gas storage. Thorough investigations (Britt et al., 2008; Rosi et al., 2003; Zhao et al., 2004, 2008; Murray et al., 2009; Rowsell and Yaghi, 2005; Rowsell et al., 2004; Collins and Zhou, 2007; Hu and Zhang, 2010; Meek et al., 2011; Han et al., 2007; Ma and Zhou, 2010) have also been conducted to access the potentiality of the MOFs to adsorb hydrogen and various gases selectively. A few studies (Dincă and Long, 2008; Dincă et al., 2006; Peterson et al., 2006) highlight the effect of metal doping on the surface of MOFs for improving their hydrogen storage capability in which the metal centers act as the active centers in binding hydrogen.

Klontzas et al. (2010) employed multiscale techniques to design novel 3D covalent organic frameworks (COFs) that can store up to 25 wt% of H_2 at 77 K temperature. A few studies (Tylianakis et al., 2011; Thomas, 2009) have also shown that the hydrogen storage capability of COFs is very comparable or even better than those of the MOFs or other materials. Very similar to the MOFs, in cases of 3D-COFs the hydrogen storage ability has been found to be increased upon metal doping (Klontzas et al., 2008; Wu et al., 2010b). Further study (Suri et al., 2009) has also been conducted to access the hydrogen storage capability of COFs and MOFs by spillover.

The rest of the chapter is divided as follows. Section 10.2 gives a brief introduction to conceptual density functional theory (CDFT) (Parr and Yang, 1989; Geerlings et al., 2003; Chattaraj, 2009; Chattaraj and Giri, 2009a; Chakraborty et al., 2011a) and various reactivity descriptors along with a brief discussion on associated electronic structure principles. At the end of Section 10.2 the applied computational methods are described. Section 10.3 gives a brief description of all-metal aromaticity and its applicability in exploring the stability of different metal and nonmetal clusters. Section 10.4 includes the effect of aromaticity on the stability/reactivity of hydrogen storage material,as well as the structural and bonding aspects of those materials. The application of CDFT (Parr and Yang, 1989; Geerlings et al., 2003; Chattaraj, 2009; Chattaraj and Giri, 2009a; Chakraborty et al., 2011a) and its various allied reactivity descriptors for the intuitive modeling of different molecular assemblies capable of trapping hydrogen is described in Section 10.5. The chapter concludes with a brief discussion on the future of the field of hydrogen storage.

10.2 Background of conceptual DFT

Conceptual density functional theory (Parr and Yang, 1989; Geerlings et al., 2003; Chattaraj, 2009; Chattaraj and Giri, 2009a; Chakraborty et al., 2011a) (CDFT) has strongly influenced the progress of quantum chemistry in the past few decades.

The theory was introduced by Robert G. Parr (Parr and Yang, 1989) to the chemistry community that eventually introduced the CDFT-based reactivity descriptors. The global reactivity descriptors include parameters such as electronegativity (Chattaraj, 1992; Parr et al., 1978 (χ), hardness Parr and Pearson, 1983; Pearson, 1997) (η) and electrophilicity (Parr et al., 1999; Chattaraj et al., 2006a, 2011a; Chattaraj and Roy, 2007a) (ω) that have been established qualitatively to understand molecular stability and reactivity. The condition of stability may be complemented by the concepts of maximization in hardness (Pearson, 1987; Parr and Chattaraj, 1991; Ayers and Parr, 2000) (MHP) and minimization in polarizability (Chattaraj and Sengupta, 1996; Fuentealba et al., 2000) (MPP) and electrophilicity (Chamorro et al., 2003; Parthasarathi et al., 2005) (MEP). The reactivity of a specific atomic site in a molecule can be rationalized by the various local reactivity descriptors such as atomic charges (Mulliken, 1955) (Q_k) and Fukui functions (Parr and Yang, 1984) (f_k) that definitely play an important role in determining the site-selectivity of a chemical species. For an N-electron system, the electronegativity (χ) (Chattaraj, 1992; Parr et al., 1978) and hardness (η) (Parr and Pearson, 1983; Pearson, 1997) can be defined as follows:

$$\chi = -\left(\frac{\partial E}{\partial N}\right)_{v(r)} = -\mu \tag{10.1}$$

$$\eta = \left(\frac{\partial^2 E}{\partial N^2}\right)_{v(r)} \tag{10.2}$$

and

$$\omega = \left(\frac{\chi^2}{2\eta}\right) = \left(\frac{\mu^2}{2\eta}\right) \tag{10.3}$$

where $v(r)$ and μ are the external and chemical potentials, respectively.

With the help of the finite difference method, the electronegativity and hardness can be expressed as follows:

$$\chi = \frac{(I + A)}{2} \quad \text{and} \quad \eta = (I - A) \tag{10.4}$$

where I and A are the ionization potential and electron affinity, respectively, and can be calculated from the energies of the N and $N \pm 1$ electron systems. For an N-electron system with energy $E(N)$, I and A may be expressed as:

$$I \approx E(N-1) - E(N) \tag{10.5}$$

$$A \approx E(N) - E(N+1) \tag{10.6}$$

If E_{HOMO} and E_{LUMO} are the energies of the highest occupied and the lowest unoccupied molecular orbitals, respectively, and then using Koopmans' theorem (Koopmans, 1933) Equation (10.4) can be written as:

$$\chi = -\frac{(E_{\mathrm{HOMO}} + E_{\mathrm{LUMO}})}{2} \quad \text{and} \quad \eta = (E_{\mathrm{LUMO}} - E_{\mathrm{HOMO}}) \tag{10.7}$$

The local reactivity descriptor, the Fukui function (FF) (Parr and Yang, 1984), measures the change in electron density at a particular point when an electron is added to or detached from a system at constant $v(\vec{r})$. It may be written as:

$$f(\vec{r}) = \left(\frac{\partial \rho(\vec{r})}{\partial N}\right)_{v(\vec{r})} = \left(\frac{\delta \mu}{\delta v(\vec{r})}\right)_N \tag{10.8}$$

We can now condense this FF $f(\vec{r})$ (Yang and Mortier, 1986) to an individual atomic site k in a molecule in terms of electron population p_k through the following expressions:

$$f_k^+ = p_k(N+1) - p_k(N) \text{ for nucleophilic attack} \tag{10.9a}$$

$$f_k^- = p_k(N) - p_k(N-1) \text{ for electrophilic attack} \tag{10.9b}$$

$$f_k^0 = [p_k(N+1) - p_k(N-1)]/2 \text{ for radical attack} \tag{10.9c}$$

The site selectivity for a particular atomic site in a molecule can be acquired from their local philicity (ω_k^α) values. The condensed-to-atom local philicity (ω_k^α; $\alpha = +, -$ and 0 representing nucleophilic, electrophilic, and radical attacks, respectively) variants for the kth atomic site in a molecule is expressed as:

$$\omega_k{}^a = \omega \cdot f_k{}^a \tag{10.10}$$

Further, thermodynamic parameters such as interaction energy (IE), gain in energy (GE), reaction enthalpy (ΔH), and dissociative chemisorption energy (ΔE_{CE}) help to determine the manner of hydrogen adsorption on the various molecular clusters that have been designed theoretically. The IE, GE, and ΔE_{CE} are expressed as:

$$n = \text{no. of molecular } H_2 : \mathrm{IE} = E_{nH_2X} - [E_X + nE_{H_2}] \tag{10.11}$$

$$\mathrm{GE} = E_{(n-1)H_2X} + E_{H_2} - E_{nH_2X} \tag{10.12}$$

$$\Delta E_{\mathrm{CE}} = \frac{2}{n}\left[E_X + \frac{n}{2}E_{H_2} - E_{XH_n}\right]; \quad n = \text{no. of H atoms} \tag{10.13}$$

where E_X denotes the energy of the parent moiety.

Additional insights into the stability of these structural motifs can be gained by the criterion of "all-metal aromaticity" (Li et al., 2001a; Schleyer, 2001; Boldyrev and Wang, 2005), which can be mathematically assessed from an evaluation of various aromaticity indices such as nucleus independent chemical shift (Schleyer et al., 1996) (NICS). NICS(0) for a planar ring molecule signifies the amount of absolute magnetic shielding computed at ring center. Subsequently, NICS(1) defines it computed at a distance of 1 Å perpendicular to the ring plane. The NICS rate at a given distance (r) from the ring center is calculated with the following mathematical formula (Stanger, 2006; Oscar et al., 2009; Li et al., 2008; Noorizadeh and Dardaba, 2010):

$$\mathrm{NICS-rate}(r)=\frac{\mathrm{dNICS}}{\mathrm{dr}}\cong \mathrm{Lim}_{\Delta r\to 0}\frac{\mathrm{NICS}(r+\Delta r)-\mathrm{NICS}(r)}{\Delta r} \tag{10.14}$$

10.2.0.1 Computational details

The optimization and frequency calculations are done using GAUSSIAN 03 (GAUSSIAN, 2003) and GAUSSIAN 09 (Frisch et al., 2009) program packages. The computations are performed at various levels of theory (B3LYP, dispersion corrected B3LYP, MP2, MPW1K, M052X, M06, PBE0) using various basis sets (cc-pvdz, 6-31G, 6-31+G(d), 6-311+G(d), 6-311+G(d,p)). Different levels in conjugation with different basis sets have been chosen to study different systems. All the structures reported here are fully optimized and exist at minima on the potential energy surface (PES) due to the absence of any imaginary frequency (NIMAG = 0). The I and A values are computed either by Koopmans' theorem (Koopmans, 1933) or by the ΔSCF technique. The CDFT-based global and local reactivity descriptors are calculated by using Equations (10.3)–(10.10). Mulliken population analysis (MPA) and natural population analysis (NPA) are performed to estimate the atomic charges (Mulliken, 1955) (Q_k) and FF (Parr and Yang, 1984) ($f\left(\vec{r}\right)$). A molecular dynamics simulation is carried out to check the hydrogen storage capability of sI clathrate hydrate. Metastable hydrogen hydrate structures are generated with the help of the LAMMPS software (Plimpton, 1995).

10.3 All-metal aromaticity

The concept of "aromaticity" arose centuries ago in the vocabulary of chemical literature (Hofmann, 1856; Kekulé, 1865, 1866) is a unique phenomenon that describes the unusual stability of a range of benzenoid and allied organic molecules. A more generalized definition by Hückel (1931) rendered the aromaticity criterion a popular qualitative tool to rationalize the "extra stability" of a particular class of organic compounds. However, further advancement of ideas along with the growth of computational chemistry shows that the given term is meaningful from several perspectives and thus can neither be a monopoly of organic chemists, but rather well accounts for the stability of purely inorganic compounds. The extension of this novel aromaticity concept to an all-metal Al_4^{2-} system, a first of its kind by Boldyrev et al.

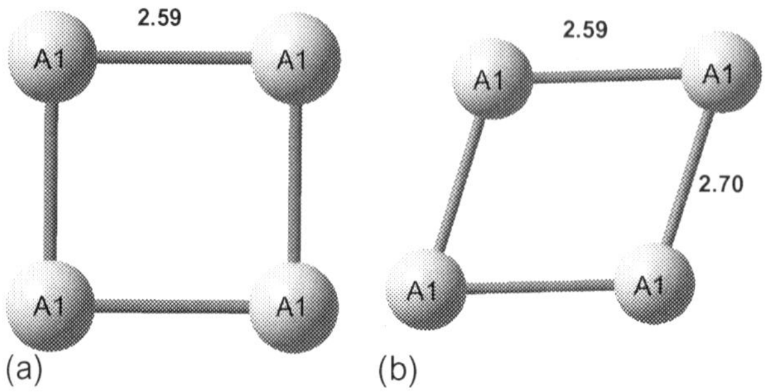

Figure 10.1 Optimized geometries of Al_4^{2-} (a) delocalized (singlet) and (b) localized (triplet) molecules.
This figure was reproduced from Chattaraj et al. (2007) with permission from the American Chemical Society.

(Li et al., 2001) opened new vistas in chemical thinking as he subsequently introduced a seminal term "all-metal aromaticity."

Boldyrev's (Li et al., 2001) novel idea of the existence an "all-metal aromaticity" phenomenon in the Al_4^{2-} system (Figure 10.1) further showed that the given molecule exhibits a prominent delocalization of the σ-electrons in addition to the π-ones.

Therefore, the Al_4^{2-} system is doubly aromatic (having both σ- and π-electron delocalization) in nature. The existence of such a doubly aromatic criterion is well documented (Fowler et al., 2001, 2002; Juselius et al., 2001; Zhan et al., 2002).

The novel concept of aromaticity (Schleyer, 2001; Kekulé, 1865; Minkin et al., 1994) extended to all-metal systems has been further exploited to study the bonding and reactivity in several types of metal-sandwich complexes such as ferrocene (Kealey and Pauson, 1951; Wilkinson et al., 1952; Eiland and Pepinsky, 1952; Coriani et al., 2006; Kudinov and Rybinskaya, 1999; Kudinov et al., 2002; Beck and O'Hare, 2004) and its allied multidecker analogues, which were studied earlier on both experimental (Qian-Shu et al., 1986; Padma Malar, 2005; Chen et al., 2005) and theoretical (Kruszewski and Krygowski, 1972; Krygowski, 1993) grounds. To obtain a quantitative rationale of the so-called qualitative concept of aromaticity a set of aromaticity indices (Schleyer et al., 1996; Chen et al., 2005; Kruszewski and Krygowski, 1972; Krygowski, 1993; Giambiagi et al., 1975, 1990, 2000; Becke and Edgecombe, 1990; Mallion, 1980; Gomes and Mallion, 2001) are introduced to help determine the degree of aromaticity/antiaromaticity of a chemical system.

The concept of "all-metal aromaticity" has been fruitfully used for the modeling of a wide range of molecular clusters, comprising both metals and nonmetals (Boldyrev and Wang, 2005; Li et al., 2001b; Kuznetsov et al., 2001; Mercero and Ugalde, 2004; Tsipis, 2005; Chattaraj et al., 2006b, 2007, 2008; Mallajosyula et al., 2006; Chattaraj and Roy, 2007b; Datta et al., 2007; Chi and Liu, 2007; Roy and Chattaraj, 2008; Chattaraj and Giri, 2008, 2009b; Roy et al., 2008; Duley et al., 2010; Chakraborty et al., 2009, 2011b; Giri et al., 2009, 2012). CDFT has also been applied to assess the stability, reactivity, and aromaticity of various trigonal, all-metal systems (Roy and Chattaraj, 2008; Chattaraj and Giri, 2008; Giri et al., 2009, 2012; Chakraborty et al., 2011b) $[X_3]^{n-}$ (X = Be, Mg, Ca; $n = 1, 2$) along with other systems with organic

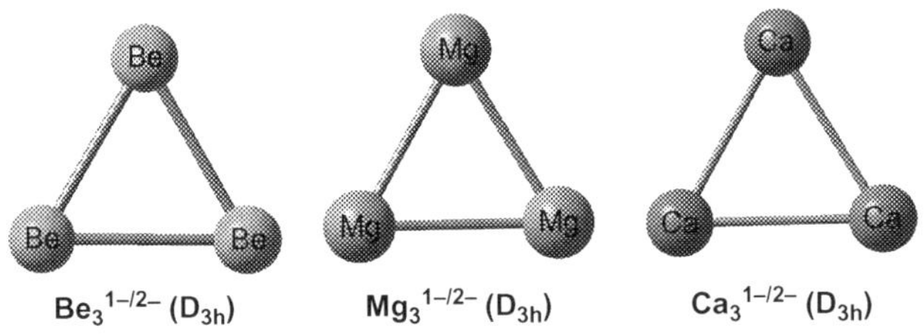

Figure 10.2 Optimized geometries of $(Be_3)^{q-}$, $(Mg_3)^{q-}$ and $(Ca_3)^{q-}$ ($q=1$, 2) molecules. This figure was reproduced from Roy and Chattaraj (2008); Chattaraj and Giri (2008); Giri et al. (2009) with permission from American Chemical Society, Elsevier, John Wiley.

(Boldyrev and Wang, 2005) and inorganic (Chattaraj and Roy, 2007b; Chattaraj et al., 2005; Chattaraj and Giri, 2007; Khatua et al., 2007; Kuznetsov and Boldyrev, 2004) moieties. The trigonal, all-metal $[X_3]^{n-}$ (X = Be, Mg, Ca; $n=1$, 2) systems (Figure 10.2) exhibit prominent π-aromaticity.

The results are quite similar to allied theoretical and experimental findings (Oscar et al., 2009; Kuznetsov and Boldyrev, 2004; Khatua et al., 2008; Middleton and Klein, 1999; Kaplan et al., 2002; Liu et al., 2007). The trigonal, all-metal clusters being anionic in nature contain an excess of electrons that endows the electropositive metal site(s) with a formal negative charge. Such a charged cluster species needs to be coupled with suitable counter-ions to attain further stability. Thus, upon binding with another suitable metal counter-cation the trigonal anionic $[X_3]^{n-}$ (X = Be, Mg, Ca; $n=1$, 2) system forms a stable specimen, which serves as a building block toward designing double (Chattaraj et al., 2008; Roy et al., 2008) and multidecker (Duley et al., 2010) metal clusters. In a recent paper (Chakraborty et al., 2011c), the effect of a counter-cation on a trigonal, anionic, and aromatic all-metal system was discussed.

The formation of a stable metal–metal bond, as stated earlier, attributes to the basis of designing a metal cluster. In their paper, Resa et al. (2004) proved the very existence of such a direct Zn–Zn linkage stabilized upon coordination with a $[C_5Me_5]^-$ group to form a sandwich complex. Prompted by the work of Resa et al. (2004), the trigonal, planar, all-metal aromatic (Roy and Chattaraj, 2008) ($[Be_3]^{2-}$) and ($[Zn_3]^{2-}$) (Yong and Chi, 2007) units have also been employed (Chattaraj et al., 2008; Chakraborty et al., 2009) to stabilize the Zn–Zn bond. Both the $[Be_3]^{2-}$ and $[Zn_3]^{2-}$ units exhibit prominent π- electron delocalization and hence a favorable ring current around the trigonal plane (Figure 10.3).

Gradual substitution of the Zn centers with the Be atoms and the aromatic $[C_5Me_5]^-$ rings with the $[Be_3]^{2-}$ units creates an all-beryllium double-decker $[Be_8]^{2-}$ complex, $[Be_3\text{-}Be\text{-}Be\text{-}Be_3]^{2-}$, which shows the ability of the trigonal, planar all-metal aromatic $[Be_3]^{2-}$ system for stabilizing a direct Be–Be linkage (Chattaraj et al., 2008). The $[Zn_3]^{2-}$ unit has also been conceived as a base and upon coupling with another Zn^{2+} counter-cation produces a pyramidal $[Zn_4]$ unit. The $[Zn_4]$ unit further combines with another $[Zn_3]^{2-}$ moiety to produce an all-zinc anionic sandwich

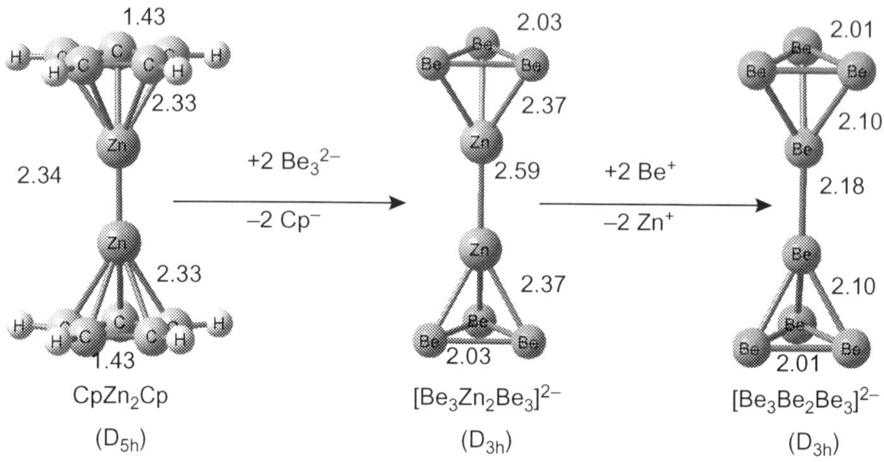

Figure 10.3 Two-step substitution reaction of CpZnZnCp to produce $(Be_3BeBeBe_3)^{2-}$molecules.
This figure is reproduced from Chattaraj et al. (2008) with permission from Elsevier.

complex. The intermediate Zn–Zn chain length can be increased by incorporating more Zn atoms, and a number of anionic "all-zinc" chain clusters are modeled (Figure 10.4) (Chakraborty et al., 2009).

The existence of a favorable aromaticity criterion around the planar $[Zn_3]^{2-}$ units is supposed to account for the stability of such large all-zinc clusters.

The design of unique multidecker (Chattaraj and Giri, 2009b) sandwich complexes using the aromatic $[Be_3]^{2-}$ and $[Mg_3]^{2-}$ as the building units and different plausible substitution reactions of them (Figure 10.5) have also been investigated (Chattaraj and Giri, 2009b).

The stability rationale of such large metal clusters has also been credited to an "all-metal aromaticity" phenomenon of the trigonal rings at the base.

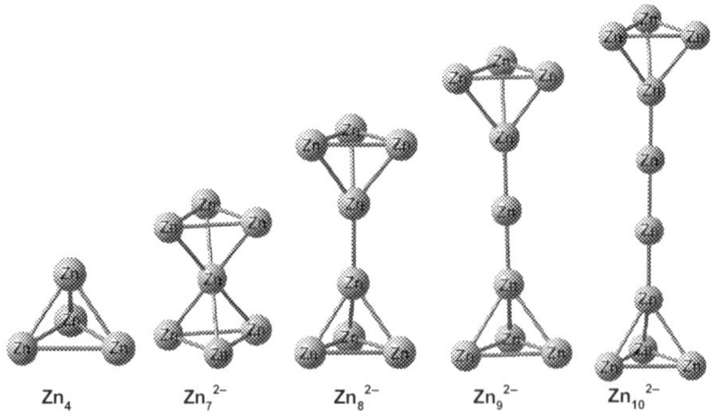

Figure 10.4 Optimized geometries of the $(Zn_n)^{2-}$ clusters molecules.
This figure was reproduced from Chakraborty et al. (2009) with permission from Elsevier.

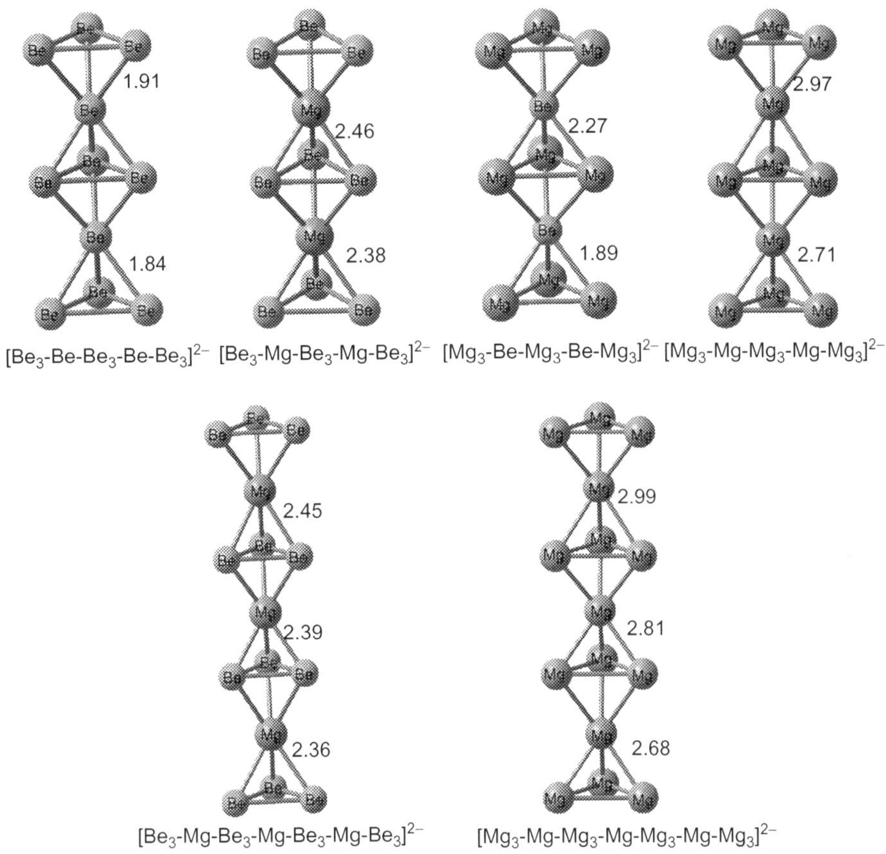

Figure 10.5 Optimized geometries of various "all-metal" multidecker complexes molecules. This figure was reproduced from Chattaraj and Giri (2009b) with permission from Wiley.

10.4 Role of aromaticity in hydrogen storage

Aromaticity is an important concept in the field of organic chemistry, particularly for rationalizing structure and stability of various systems and also for modeling new compounds. Modeling of new materials for hydrogen storage based on the concept of aromaticity is most worthy since aromaticity is associated with thermodynamic stability as well reactivity. Hübner et al. (2004) studied the interaction of molecular hydrogen with different aromatic systems and found that the strength of interaction of molecular hydrogen with the aromatic system is higher for electronically rich aromatic systems. They concluded that larger aromatic systems are more favorable than single aromatic benzene ring for interaction with hydrogen molecules, which helped to set a strategy for modeling new MOFs. Later, Srinivasu et al. (2009) revealed in a theoretical study that the van der Waals interaction between the molecular hydrogen and organic molecular systems is not enough for hydrogen adsorption and alkali metal doping on organic molecular systems improve their hydrogen adsorption capacity. They studied small organic molecular systems and their metal-doped derivative for their potential hydrogen adsorption capacity and found that aromaticity of organic molecules significantly influences the stability of the alkali metal-organic molecule complex, aromatic rings interact more strongly with metal ions. Recently, Bodrenko et al. (2012) considered carbon-based anionic aromatic molecules stabilized by alkali and alkaline earth metal ions for hydrogen adsorption and found that the properties of cations influence the interaction between the molecular hydrogen and organo-metal complexes. The utility of the concept of aromaticity as a primal stability criterion for advocating the formation of a variety of hydrogen loaded all-metal and nonmetal cluster motifs has become a topic of extensive research during the past decade. The presence of a favorable aromaticity phenomenon renders the various cluster-assemblies to act as effective hydrogen storage templates. Based on an earlier study Chakraborty et al. (2010) depicting the noble gas trapping potential of the trigonal aromatic all-metal Li_3^+ and nonmetallic H_3^+ species, Giri et al. (2010) investigated the possibility of trapping hydrogen molecules on small all-metal aromatic clusters such as Li_3^+ and Na_3^+ (Figure 10.6). The use of Mg_n and Ca_n ($n=8$–10) cages (Giri et al., 2010) (Figure 10.7) as a trap for hydrogen has also studied.

Duley et al. (2011) showed that planar, annular N_4^{2-} and N_6^{4-} ring systems exhibit comparable NICS values with that of cyclobutadiene and benzene, respectively. The N_6^{4-} ring is a local minimum and belongs to D_{6h} symmetry and N_4^{2-} ring shows conflicting aromaticity with the simultaneous presence of the σ-antiaromaticity (NICS $(0) > 0$) and the Π- aromaticity [NICS $(1) < 0$]. The electronic stability of both the N_6^{4-} and N_4^{2-} rings may be provided by combining them with suitable counter-ions. In the presence of two Ca^{2+} ions the aromaticity of the planar N_6^{4-} ring is found to increase considerably, whereas the N_4Li_2 system shows conflicting aromaticity, i.e., both σ-antiaromaticity and π-aromaticity are simultaneously observed. The optimized structures of the N_6Ca_2 and N_4Li_2 clusters and their corresponding hydrogen-trapped analogues are depicted in Figure 10.8.

The metal-doped polynitrogen systems, i.e., N_4Li_2 and N_6Ca_2 clusters, are able to trap hydrogen, which can be justified from the negative adsorption energies per hydrogen molecule (ΔE_{ads}). This suggests that N_4Li_2 and N_6Ca_2 are competent as potential hydrogen traps.

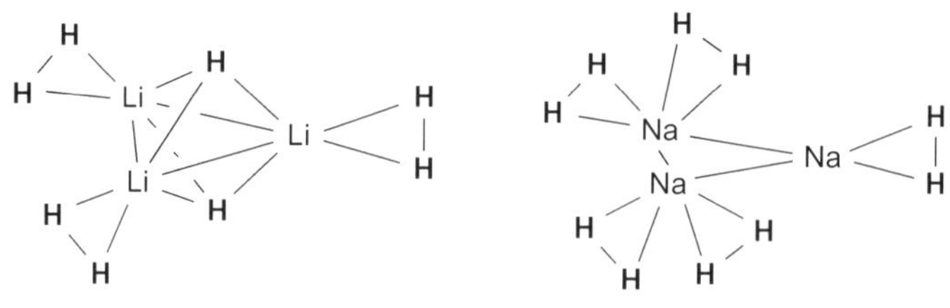

Figure 10.6 Depiction of $H_8Li_3^+$ and $H_{10}Na_3^+$ studied at B3LYP/6-311+G(d) level of theory. This figure was reproduced from Giri et al. (2010) with permission from Springer-Verlag.

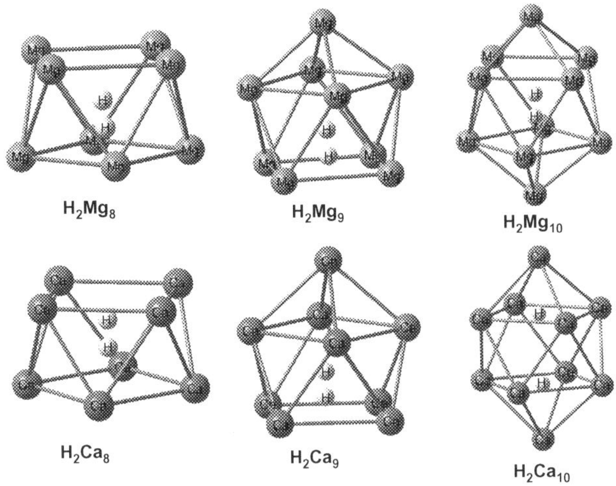

Figure 10.7 Optimized geometries (at B3LYP/6-311+G(d) level) of H_2M_n (where M = Mg, Ca; n = 8, 9, 10) clusters molecules.
This figure was reproduced from Giri et al. (2010) with permission from Springer-Verlag.

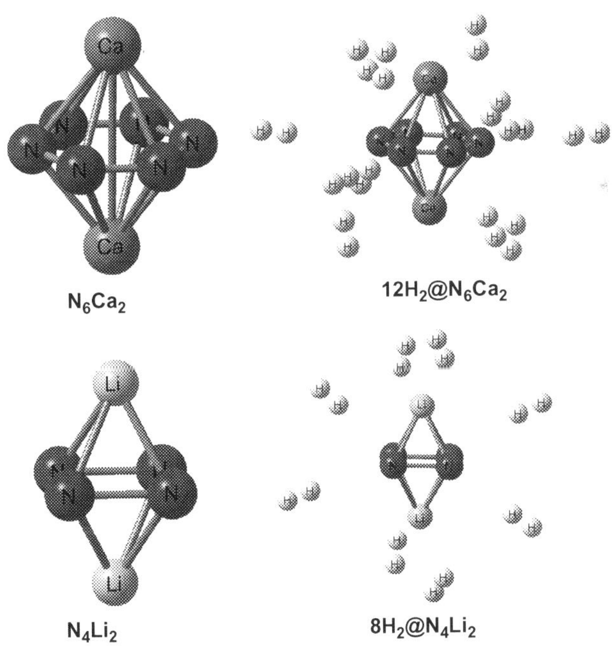

Figure 10.8 Optimized molecular geometries of N_6Ca_2 and N_4Li_2 and their hydrogen trapped complexes.
This figure was reproduced from Duley et al. (2011) with permission from Elsevier.

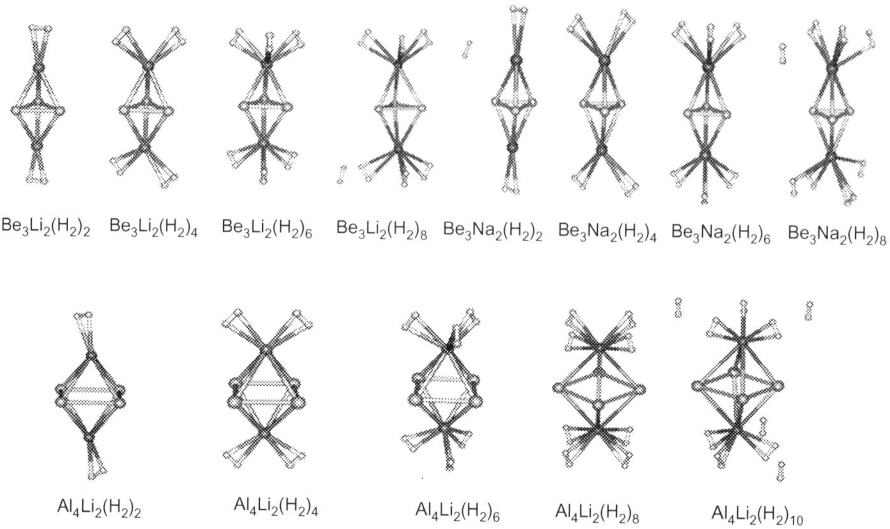

Figure 10.9 Optimized geometrical structures of hydrogen trapped complexes of Be_3Li_2, Be_3Na_2, and Al_4Li_2.
This figure was reproduced from Srinivasu et al. (2012) with permission from Royal Society of Chemistry.

In another work, Srinivasu et al. (2012) explored the efficacy of counter-ion-doped planar all-metal aromatic Be_3^{2-}, Mg_3^{2-}, and Al_4^{2-} systems for binding hydrogen molecules. The planar dianionic systems were coupled with alkali-metal cations to obtain the Be_3M_2, Mg_3M_2, and Al_4M_2 (M = Li, Na and K) complexes. The alkali-metal atom obviously serves as the active binding site and the tendency to bind more hydrogen molecules depends on the effective charge on the associated atom. The geometry of the counter-ion-doped planar all-metal aromatic systems and their hydrogen bound derivatives are shown in Figure 10.9. Thus, the K-site in the respective complex in spite of owning a higher charge cannot attract more H_2 molecules due to its comparatively larger ionic radius and hence a lesser charge density. Mg_3M_2 molecules do not seem to serve as good hydrogen adsorbents compared to the others, probably due to a lesser charge transfer from the alkali-metal center to the Mg_3 moiety. The overall hydrogen adsorption potential is also rationalized from favorable ΔH and $\Delta\omega$ values of the allied hydrogen-binding reactions.

Recently (Pan et al., 2012a), some bond-stretch isomers of $Li_3Al_4^-$ were studied to check their hydrogen-trapping ability. These bond stretch isomers are aromatic in nature and the Li centers are preferable sites for hydrogen binding. The usefulness of these isomers as probable hydrogen storage material are rationalized with the help of favorable interaction energy, reaction enthalpy, reaction electrophilicity, and desorption energy (for the gradual hydrogen-loading processes) values.

Giri et al. (2011) modeled some unique aromatic/antiaromatic "star-like" molecular clusters by substituting each H-center of the well-known planar, mononuclear C_4H_4, $C_5H_5^-$, and C_6H_6 systems with an Li-atom. The resultant parent moieties,

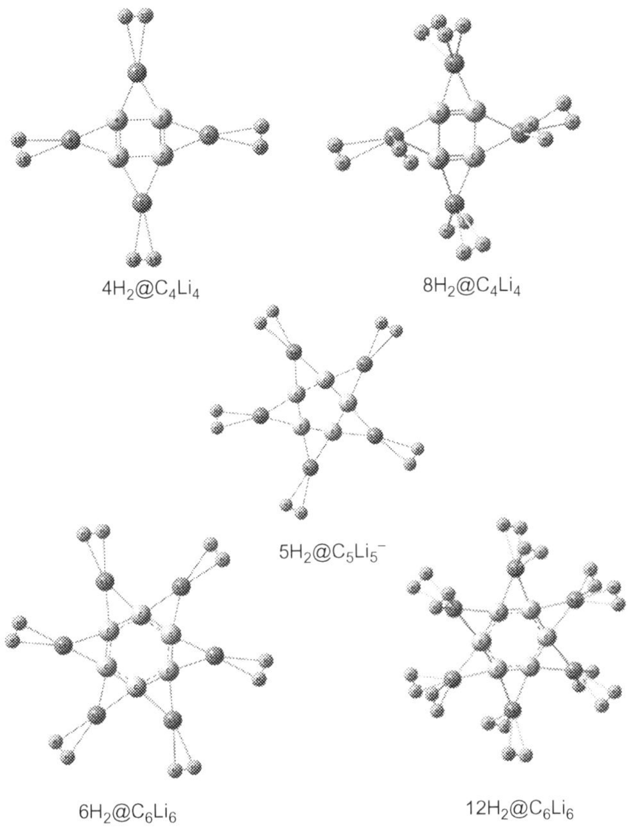

Figure 10.10 Optimized molecular geometries of the Li-decorated "star-like" molecules and their associated hydrogen bound complexes molecules.
This figure was reproduced from Giri et al. (2011) with permission from the PCCP Owner Societies.

C_4Li_4, $C_5Li_5^-$, and C_6Li_6 were then investigated for their capability to bind hydrogen molecules where the Li-center supposedly behaves as the primary H-binding site (Figure 10.10). In addition to that, mononuclear aromatic/antiaromatic systems starting from the smallest C_3-ring to the C_7-analogue (tropylium cation) as well as some polyaromatic hydrocarbons (PAHs) such as naphthalene, anthracene, and phenanthrene were chosen and doped with a suitable counter-ion (Li^+/F^-) (Figure 10.11). The ability of hydrogen adsorption on these counter-ion bound annular complexes was further studied from a CDFT approach. The effect of aromaticity/antiaromaticity of the central ring on the H_2-loading potential of a complex accompanied by some drastic changes if any (conversion from aromatic to antiaromatic or vice versa) has also been carefully scrutinized.

The sustenance of an aromaticity/antiaromaticity criterion in the resulting complexes serves as an important determinant for establishing the stability of the same. The "star-like" complexes are innovative and upon binding with suitable "linkers"

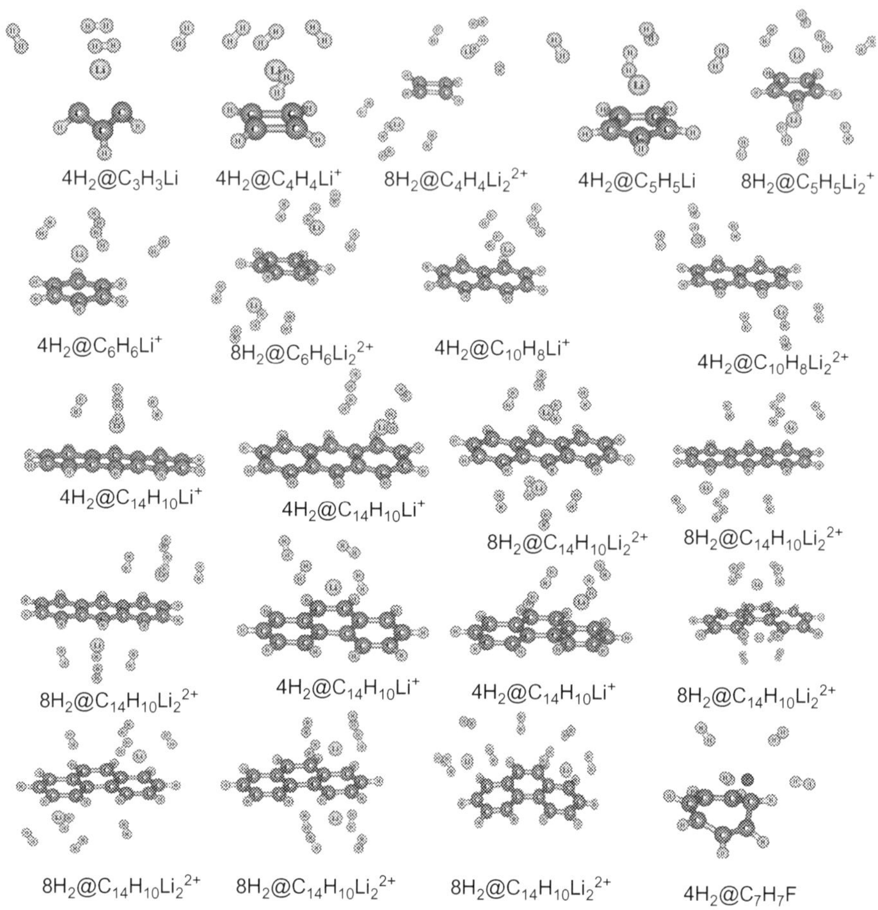

Figure 10.11 Optimized molecular geometries of hydrogen bound complexes of the Li^+/F^--doped aromatic/antiaromatic molecules.
This figure was reproduced from Giri et al. (2011) with permission from the PCCP Owner Societies.

may function as the building blocks for designing larger macromolecular motifs that can store hydrogen in larger gravimetric quantities in practice at ambient temperature and pressure conditions.

A recent study (Pan et al., 2012b) revealed that the aromatic $B_3H_3^{2-}$ unit along with its various Li/Li^+-doped isomers are proficient at trapping hydrogen molecules with high gravimetric wt% (Figure 10.12). The feasibility of the hydrogen-binding process is justified by energy parameters and NICS values.

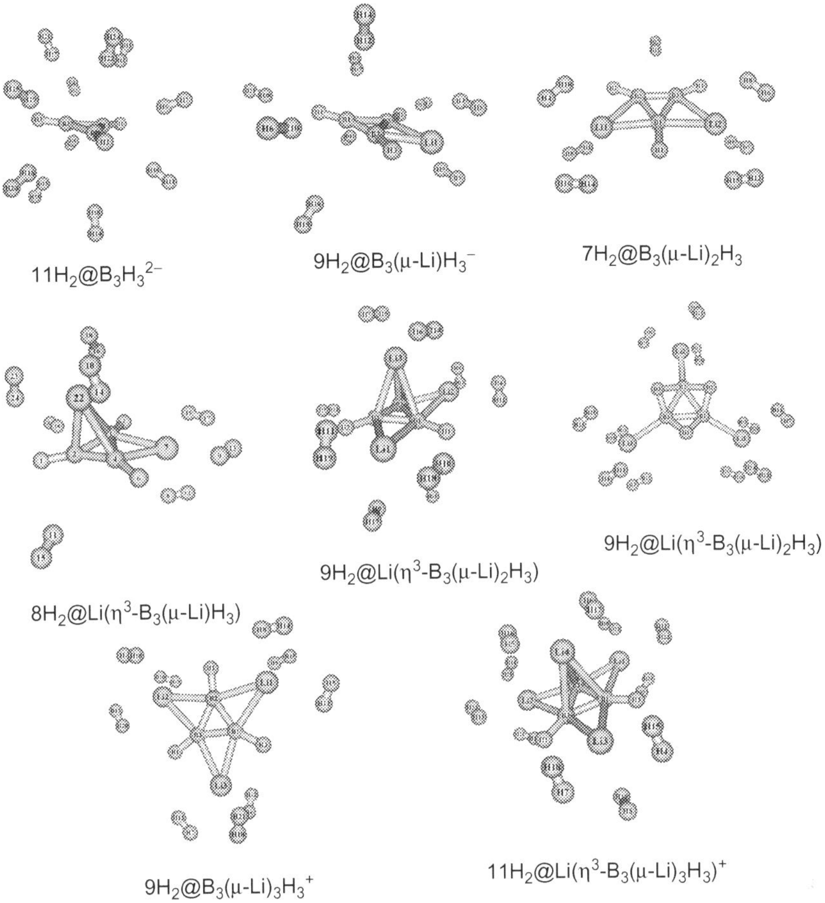

Figure 10.12 The optimized geometries of various hydrogen-loaded clusters studied at B3LYP/6-311+G(d) level.
This figure was reproduced from Pan et al. (2012b) with permission from Wiley.

10.5 Case studies of possible hydrogen-storage materials with the aid of CDFT

10.5.1 $(N_4C_3H)_6Li_6$ and its 3D molecular material

Recently (Das and Chattaraj, 2012) some molecular clusters and 3D functional material were designed and their Li-doped counterparts investigated to access their hydrogen storage capacity. In a CDFT-based calculation, it was explored that the interaction of hydrogen molecules with the Li-doped planar aromatic $(N_4C_3H)_6$ cluster is quite reasonable. It was found that each Li binds with three H_2 molecules (Figure 10.13).

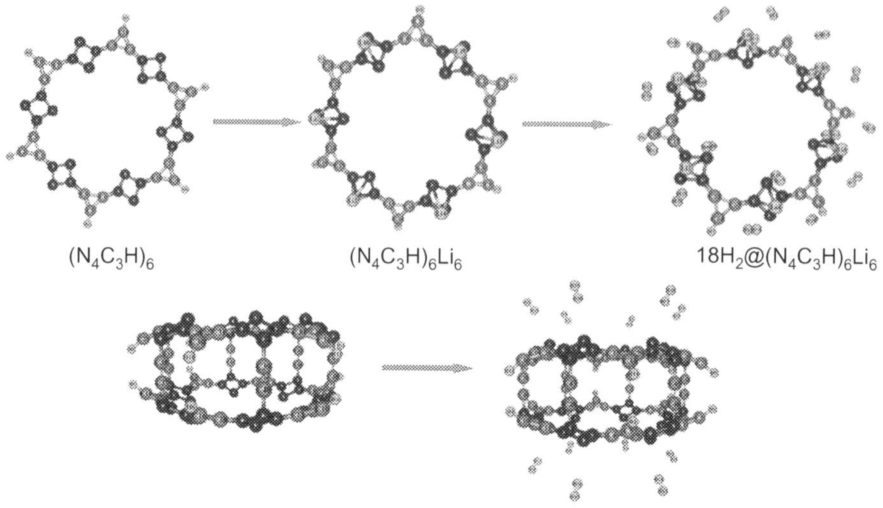

Figure 10.13 Optimized geometries of $(N_4C_3H)_6$, $(N_4C_3H)_6Li_6$, its highest hydrogen-trapped analogue at B3LYP/6-31+G(d) level and 3-D molecular material (both bare and H_2-loaded) at B3LYP/6-31G(d) level.
This figure was reproduced from Das and Chattaraj (2012) with permission from the American Chemical Society.

The obtained thermodynamic quantities revealed that the hydrogen adsorption process on the $(N_4C_3H)_6Li_6$ cluster was spontaneous in a given temperature–pressure zone as exhibited in the associated (T–P) phase diagrams through the negative ΔG region (Figure 10.14). This suggests that the $(N_4C_3H)_6Li_6$ cluster can be used for H_2 storage. Kinetics of adsorption and desorption and the calculated CDFT-based reactivity descriptors also provide additional insight into this process. The H_2 storage in an associated 3D functional material, i.e., $\left((N_4C_3H)_6(-c{\equiv}c-)_6(N_4C_3H)_6\right)$, was also explored. It was found that the 3D functional material can bind up to 12 H_2 molecules (Figure 10.13) having 1.88 gravimetric wt% of H_2. The favorable interaction energy per H_2 molecule suggests that the 3D functional moiety is proficient at trapping H_2.

10.5.2 Microsolvated metal ions as hydrogen storage material

It has already been established that charged systems are efficient at storing H_2 molecules. Based on this a DFT based study (Das et al., 2013) recently explored the H_2 storage capacity of microsolvated alkali (Li^+, Na^+) and alkaline earth (Be^{2+}, Mg^{2+}) metal ions (Figures 10.15 and 10.16). A single molecule of $H_2O/NH_3/HF$ was used as solvent. The effect of microsolvation on the H_2-binding ability of metal ions both in the presence and absence of the counter ions was explored. The study showed that the Be^{2+} ion-doped systems bind H_2 more strongly than the other metal-doped systems. The value of the

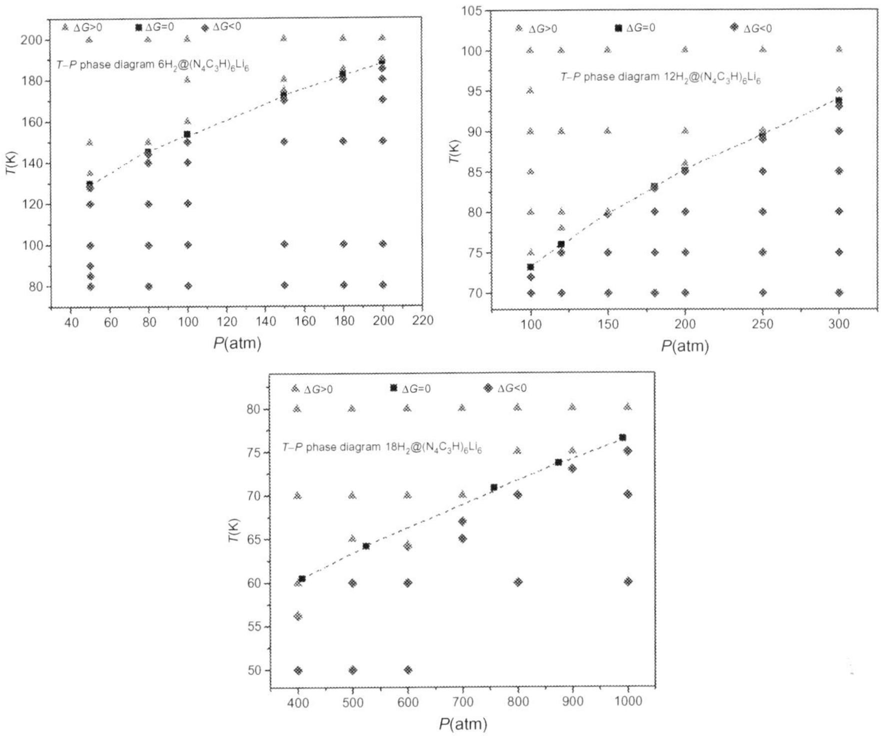

Figure 10.14 The temperature–pressure phase diagram showing the variation of ΔG.
This figure is reproduced from Das and Chattaraj (2012) with permission from the American Chemical Society.

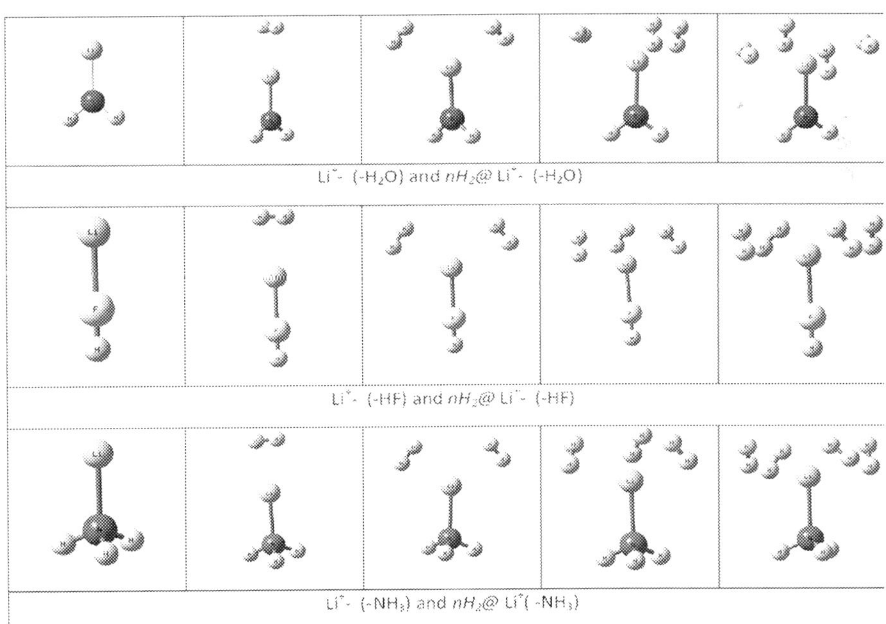

Figure 10.15 Structures of Li^+–(–H_2O, –NH_3, –HF) and nH_2@ Li^+–(–H_2O, –NH_3, –HF); ($n=1$–4).
This figure was reproduced from Das et al. (2013) with permission from Elsevier.

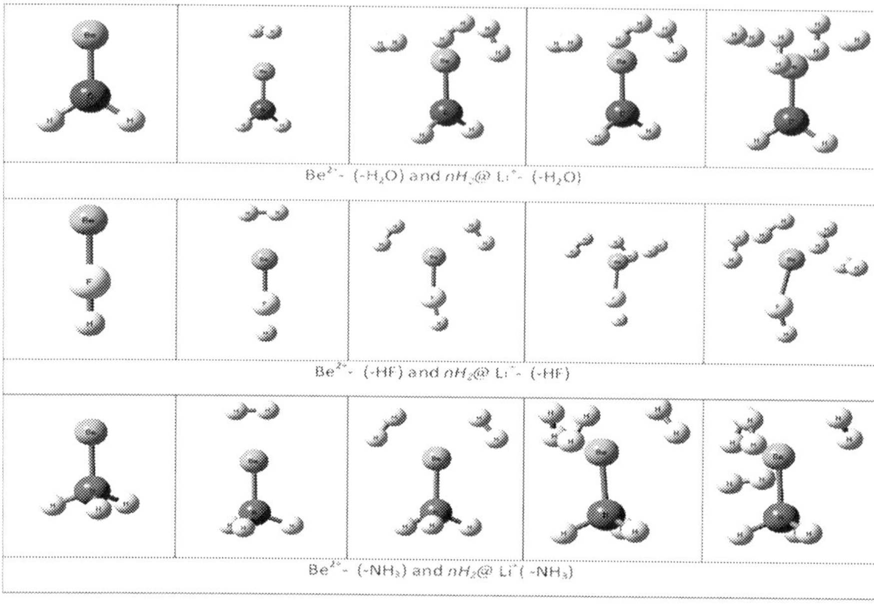

Figure 10.16 Structures of Be^{2+}–(–H_2O, –NH_3, –HF) and $n$$H_2$@ Be^{2+}–(–H_2O, –NH_3, –HF); (n = 1–4).
This figure was reproduced from Das et al. (2013) with permission from Elsevier.

interaction energy was not found to vary much with the change of level of computation. The H_2-binding process on the alkaline earth metal-doped systems even turns out to be exergonic at room temperature (298 K) as suggested by corresponding reaction free energy values. The alkali metal-ion-doped systems are better candidates for binding H_2 than the alkaline earth metal-ion-doped systems in terms of gravimetric wt% of H_2, whereas the reverse is true if we consider the binding energy, reaction enthalpy, and reaction-free energy values associated with the H_2 adsorption.

10.5.3 Exploring the hydrogen-binding capacity and the nature of fundamental interaction between transitional metal (TM) and H_2 in TM doped ethylene complex

Some researchers (Durgun et al., 2006 and Zhou et al., 2007) have explored the usefulness of the ethylene molecules doped with light transition metals such as Ti as a media for H_2 binding. In a recent report Chakraborty et al. (2011d) studied the utility of transition metal–ethylene complex as an H_2-binding template from a CDFT viewpoint (Figure 10.17). The base moiety contained the C_2H_4 molecule and transition metals such as Sc, Ti, Fe, and Ni to yield the complex structure, i.e., M_n–(C_2H_4) (M = Sc, Ti, Fe, Ni; n = 1, 2), where the H_2 binding sites are the metal atoms. Figure 10.17 illustrates H_2-loaded molecular structures containing M_n–(C_2H_4) (M = Sc, Ti, Fe, Ni; n = 1, 2) as the base moiety.

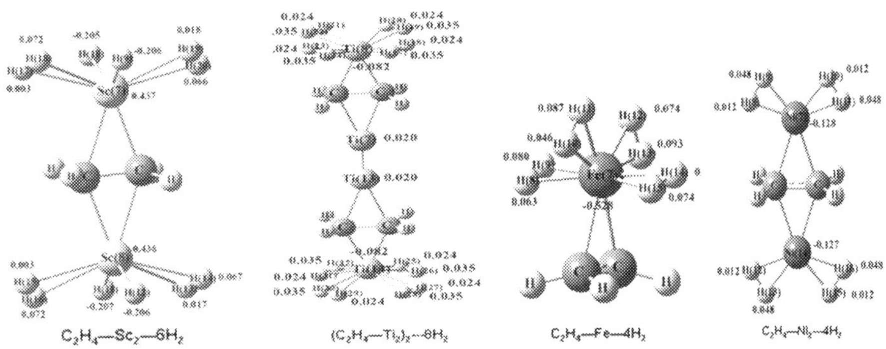

Figure 10.17 M-ethylene (M = Sc, Ti, Fe, Ni) complex and its corresponding H_2-trapped analogues studied at B3LYP/6-311 + G(d,p) level.
This figure was reproduced from Chakraborty et al. (2011d) with permission from Springer-Verlag, Berlin.

The total energy (E) was found to decrease gradually with the increased number of adsorbed H_2 molecules on the M_n–(C_2H_4) (M = Sc, Ti, Fe, Ni; n = 1, 2) clusters. The stability and reactivity of these M_n–(C_2H_4) (M = Sc, Ti, Fe, Ni; n = 1, 2) clusters were studied with the aid of CDFT-based reactivity descriptors. The hardness (η) and electrophilicity (ω) values of the associated H_2-bound complexes corroborated with each other satisfactorily at the B3LYP level, except for the Sc- and Ti-doped ethylene complexes at both B3LYP and MP2 levels. For, Fe, and Ni doped ethylene complexes at B3LYP level the η values increased uniformly with gradual H_2 trapping accompanied by a fall in the associated ω values vis-à-vis the validity of the principles of maximum hardness (Pearson, 1987; Parr and Chattaraj, 1991; Ayers and Parr, 2000) and minimum electrophilicity (Chamorro et al., 2003; Parthasarathi et al., 2005). However, the MP2 calculation did not reproduce the same relationship between η and ω values for the H_2-bound Ni-ethylene complexes. The efficacy of the clusters as an H_2 trap can be justified by the associated interaction energy (IE), reaction electrophilicity ($\Delta\omega$), and the reaction enthalpy (ΔH) values that are found to be quite favorable, indicating the feasibility of the H_2 adsorption processes. The interaction present between bound H_2 and transitional metal may be of two types; namely, the Kubas interaction (Kubas, 2001) and the usual electrostatic interaction. However, in earlier reports (Durgun et al., 2006; Zhou et al., 2007), a Kubas model (Kubas, 2001) of binding between the metal center and the incoming η^2-dihydrogen ligands was explored. In the case of Kubas interaction a simultaneous electron donation from σ orbital of H_2 molecule to metal d-orbital and back-donation from metal d-orbital to σ* orbital of H_2 molecule was observed.

10.5.4 Clathrate hydrate

The H_2 storage capacity of cage-like clathrate hydrates was explored by Chattaraj et al. (2011b). Theoretical investigations showed that only 5^{12}, $5^{12}6^2$, and $5^{12}6^8$ clathrate structures belong to minima on PES. The cage size as well as the structure are the determining factors for the aptitude of storing hydrogens within the clathrate hydrates.

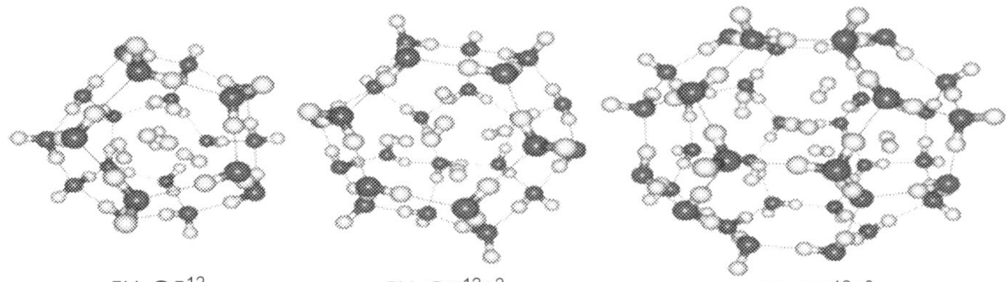

Figure 10.18 Encapsulation of hydrogen trapping within different clathrate cages at B3LYP/6-31G(d) level of theory.
This figure was reproduced from Chattaraj et al. (2011b) with permission from the American Chemical Society.

It was observed that both 5^{12} and $5^{12}6^2$ cages can encapsulate up to five H_2 molecules but the interaction energy turned out to be negative only for the first and the second H_2 adsorption processes. However, the $5^{12}6^8$ cage trapped up to six H_2 molecules endohedrally with the negative interaction energy. It is obvious that the H_2-binding capacity of $5^{12}6^8$ is better than that of 5^{12} and $5^{12}6^2$ cages. Some optimized geometries depicting maximum H_2 encapsulation into the 5^{12}, $5^{12}6^2$ and $5^{12}6^8$ cages are shown in Figure 10.18.

The stability of the H_2-bound clathrate cages was successfully investigated with the help of the CDFT-based reactivity descriptors such as hardness (η) and electrophilicity (ω). The hardness showed an increasing trend whereas the electrophilicity is decreasing with gradual H_2 upload in most of the cases, indicating an increased stability with gradual H_2 loading. Further, in a recent study Mondal et al. (2013a) carried out a molecular dynamics simulation to access the efficacy of sI clathrate hydrate to trap H_2. The study of the radial distribution functions of H_2-loaded hydrates explored that the cage remains intact within the simulated time scale. The interaction energy remained negative up to 50% occupation of the $5^{12}6^2$ channels of sI clathrate hydrate with H_2. This indicates that the designed hydrogen hydrate is metastable and able to act as a hydrogen storage material.

Mondal et al. (2013b) also recently studied the encapsulation of molecular H_2 within HF-doped clathrate hydrates and found that the stability of clathrate hydrates increases due to HF doping as suggested by associated thermodynamic data and interaction energy values. They also studied the dynamics of HF-doped clathrate hydrates as well as its H_2 bound analogues by means of ab initio molecular dynamics simulation.

10.5.5 $C_{12}N_{12}$ cages

In a recent theoretical study (Mondal et al., 2013c) the relative stability of isomeric polycyanogen $C_{12}N_{12}$ cages (A, B, and C types) were evaluated at different levels (Figure 10.19). The applicability of these cages as H_2 trap were also checked theoretically. The calculated values of the gas phase heat of formation indicate the stability as well as usefulness of these $C_{12}N_{12}$ isomers as the high-energy density material

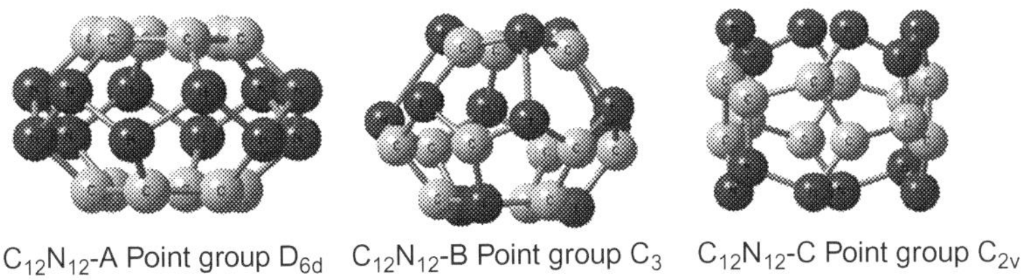

Figure 10.19 The modeled structures of $C_{12}N_{12}$-A, $C_{12}N_{12}$-B, and $C_{12}N_{12}$-C.
This figure is reproduced from Mondal et al. (2013c) with permission from Royal Society of Chemistry.

(HEDM). In all these cages each nitrogen site was found to adsorb one H_2 molecule with a gravimetric density of 7.2 wt%. The optimized geometries of $12H_2@C_{12}N_{12}$-A, $12H_2@C_{12}N_{12}$-B, and $12H_2@C_{12}N_{12}$-C are shown in Figure 10.20.

The isomer $C_{12}N_{12}$-C is the most stable among all the isomers but the $C_{12}N_{12}$-A isomer was found to interact with H_2 more strongly than the other two isomers.

The *T–P* phase diagram gives an overall idea of the optimum temperature-pressure zone for a spontaneous ($\Delta G < 0$) hydrogen adsorption process. It is observed that upon the application of electric field both $C_{12}N_{12}$-A and $C_{12}N_{12}$-C can bind H_2 molecules more strongly and the binding energy shows an increasing trend for adsorption of a single H_2 molecule with an increase in the applied electric field strength. However, in the case of $H_2@C_{12}N_{12}$-B a change in orientation of the adsorbed H_2 molecule is observed at a field strength of 0.03 a.u.

10.5.6 Cucurbiturils

Recently, in a theoretical study Pan et al. (2013) investigated the H_2-binding ability of cucurbit[7]uril. They found that Cucurbit[7]uril is able to interact with H_2 in both a exohedral and endohedral manner. Cucurbit[7]uril binds up to 52 H_2 molecules (Figure 10.21), of which five H_2 molecules are accommodated inside the cavity of the cage with the remaining H_2 molecules prefering to bind exohedrally, giving

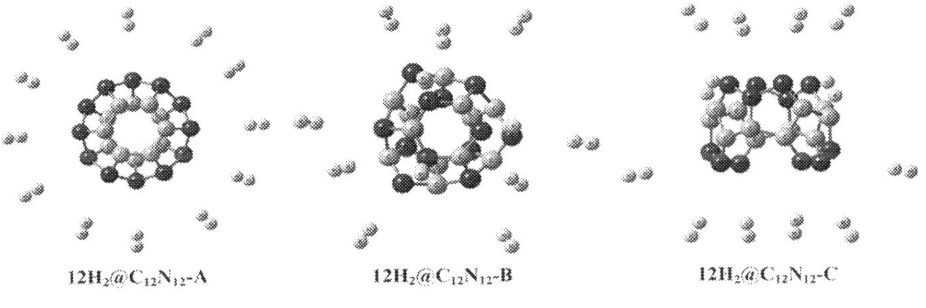

Figure 10.20 Optimized geometries of $12H_2@C_{12}N_{12}$-A, $12H_2@C_{12}N_{12}$-B, and $12H_2@C_{12}N_{12}$-C.
This figure was reproduced from Mondal et al. (2013c) with permission from the Royal Society of Chemistry.

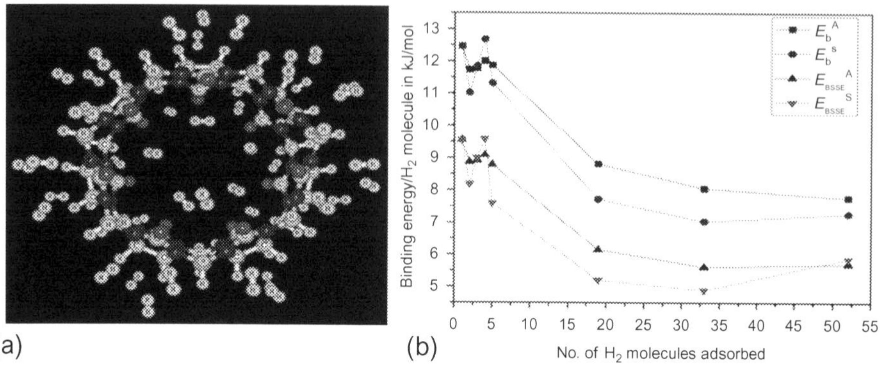

Figure 10.21 (a) Optimized geometries of $52H_2@CB[7]$ at ωB97X-D/6-31G(d,p) level; (b) variation of average binding energy per H_2 molecule (E_b^A, kJ/mol), sequential binding energy per H_2 molecule (E_b^S, kJ/mol), BSSE corrected average binding energy per H_2 molecule (E_{BSSE}^A, kJ/mol), and BSSE corrected sequential binding energy per H_2 molecule (E_{BSSE}^S, kJ/mol) with the number of hydrogen molecules adsorbed at ωB97X-D/6-31G(d,p) level of theory.
This figure is reproduced from Pan et al. (2013) with permission from The Royal Society of Chemistry (RSC) on behalf of the Centre National de la Recherche Scientifique (CNRS) and the RSC.

8.3 wt% of H_2. The N and O centers act as binding sites for H_2 molecules. Each H_2 in $52H_2$@cucurbit[7]uril interacts with cucurbit[7]uril with an average binding energy value of 7.8 kJ /mol, although the binding energy gets reduced a little upon BSSE correction. The H_2-binding processes are exothermic in nature as suggested by the adsorption enthalpy values. Further, the kinetic stability is explained with the help of the HOMO–LUMO gap as they showed that the HOMO–LUMO gap of the H_2-loaded cases are higher compared to the bare ones, suggesting an increased kinetic stability upon H_2 binding. The effect of external electric field on the binding energy and geometry was also analyzed. A gradual improvement in binding energy occurs with the increase in external electric field strength accompanied by a gradual decrease in the O—H_2 bond length and an increase in the H—H bond length.

10.6 Future trends

It can be said that the use of hydrogen as a plausible alternative to combatting an ensuing fuel crisis in an eco-friendly fashion has inspired both the experimentalists and theoreticians to design novel molecular motifs, capable of binding the gas in comparable gravimetric and volumetric amounts. The application of CDFT in combination with the various global and local reactivity descriptors and the associated electronic structure principles such as MHP, MPP, and MEP are useful for designing new molecular motifs having efficient hydrogen storing potential and in explaining their stability and reactivity with hydrogen growth. The situation becomes even better in the presence of aromaticity in the bare as well as their hydrogen-loaded motifs. The active centers of a hydrogen storage material are found to bind with molecular hydrogen through ion-induced dipole type of interaction by virtue of their net atomic charges, thereby having a direct role in the hydrogen storage capability of that material. Some systems exhibit Kubas-type interactions as well. The temperature–pressure regions where the hydrogen adsorption process is thermodynamically spontaneous ($\Delta G < 0$) for the given hydrogen adsorption processes is identified in the corresponding temperature–pressure phase diagram (see Figure 10.14). Again, the interaction energy per H_2 molecule can be improved by applying the external electric field in a proper direction, therefore, the kinetics associated with hydrogen adsorption and desorption processes can be controlled by varying applied electric field.

Since hydrogen is the lightest element and storage of it is quite challenging, there is ample room for searching for effective hydrogen storage material that meets the DOE target. The studies reported here will be helpful in designing new hydrogen storage material with substantial stability and adsorption energy as well a fast enough desorption kinetics, which is also necessary for an ideal hydrogen storage. Aromatic systems, both organic and inorganic, are expected to be ideal building blocks for designing novel hydrogen storage material. Some attention should be paid to developing hydrogen storage material that can adsorb a considerable amount of hydrogen at room temperature. Therefore, we hope this report will help to design efficient hydrogen storage materials that satisfy both the thermodynamics and kinetics requirements for safe and effective hydrogen storage at ambient temperature and pressure.

Acknowledgments

We are grateful to Professor Angelo Basile for inviting us to contribute this chapter. PKC thanks the Indo-EU (HYPOMAP) project for financial assistance. PKC would also like to thank DST, New Delhi for the J. C. Bose National Fellowship. RD thanks UGC and SP thanks CSIR, New Delhi for their fellowships.

References

Akman, N., Durgun, E., Yildirim, T., Ciraci, S., 2006. Hydrogen storage capacity of titanium met-cars. J. Phys. Condens. Mat. 18, 9509.

Aladko, E.Y., Larionov, E.G., Rodionova, T.V., Aladko, L.S., Manakov, A.Y., 2010. Double clathrate hydrates of tetrabutylammonium fluoride + helium, neon, hydrogen and argon at high pressures. J. Incl. Phenom. Macrocycl. Chem. 68, 381.

Alapati, S.V., Johnson, J.K., Sholl, D.S., 2006. Identification of destabilized metal hydrides for hydrogen storage using first principles calculations. J. Phys. Chem. B 110, 8769.

Ataca, C., Aktürk, E., Ciraci, S., Ustunel, H., 2008. High-capacity hydrogen storage by metallized graphene. Appl. Phys. Lett. 93, 043123.

Ayers, P.W., Parr, R.G., 2000. Variational principles for describing chemical reactions: the Fukui function and chemical hardness revisited. J. Am. Chem. Soc. 122, 2010.

Baburaj, E.G., Froes, F.H., Shutthanandan, V., Thevuthasan, S., 2000. Low cost synthesis of nanocrystalline titanium aluminides. Interfacial Chemistry and Engineering Annual Report. Pacific Northwest National Laboratory, Oak Ridge, TN, USA.

Beck, V., O'Hare, D., 2004. Triple-decker transition metal complexes bridged by a single carbocyclic ring. J. Organomet. Chem. 689, 3920.

Becke, A.D., Edgecombe, K.E., 1990. A simple measure of electron localization in atomic and molecular systems. J. Chem. Phys. 92, 5397.

Bethune, D.S., Kiang, C.H., de Vries, M.S., Gorman, G., Savoy, R., Vazquez, J., Beyers, R., 1993. Cobalt-catalysed growth of carbon nanotubes with single-atomic-layer walls. Nature 363, 605.

Bodrenko, I.V., Avdeenkov, A.V., Bessarabov, D.G., Bibikov, A.V., Nikolaev, A.V., Taran, M.D., Tkalya, E.V., 2012. Hydrogen storage in aromatic carbon ring based molecular materials decorated with alkali or alkali-earth metals. J. Phys. Chem. C 116, 25286.

Bogdanović, B., Schwickardi, M., 1997. Ti-doped alkali metal aluminium hydrides as potential novel reversible hydrogen storage materials. J Alloys Compd. 253, 1.

Bogdanović, B., Felderhoff, M., Kaskel, S., Pommerin, A., Schlichte, K., Schüth, F., 2003. Improved hydrogen storage properties of Ti-doped sodium alanate using titanium nanoparticles as doping agents. Adv. Mater. 15, 1012.

Bogdanović, B., Felderhoff, M., Streukens, G., 2009. Hydrogen storage in complex metal hydrides. J. Serb. Chem. Soc. 74, 183.

Boldyrev, A.I., Wang, L.S., 2005. All-metal aromaticity and antiaromaticity. Chem. Rev. 105, 3716, and references therein.

Britt, D., Tranchemontagne, D., Yaghi, O.M., 2008. Metal-organic frameworks with high capacity and selectivity for harmful gases. Proc. Natl. Acad. Sci. 105, 11623.

Burress, J., Simmons, J., Ford, J.,Yildirim, T., Gas Adsorption Properties of Graphene-Oxide-Frameworks and Nanoporous Benzene-BoronicBcid Polymers (Presented at the March meeting of the American Physical Society (APS) in Portland Ore), March 18, 2010.

Chakraborty, A., Giri, S., Chattaraj, P.K., 2009. Structure, bonding, reactivity and aromaticity of some selected Zn-clusters. J. Mol. Struct. (THEOCHEM) 913, 70.

Chakraborty, A., Giri, S., Chattaraj, P.K., 2010. Trapping of noble gases (He-Kr) by the aromatic H_3+ and Li_3+ species: a conceptual DFT approach. New J. Chem. 34, 1936.

Chakraborty, A., Duley, S., Giri, S., Chattaraj, P.K., 2011a. In: Sukumar, N. (Ed.), A Matter of Density: Exploring the Electron Density Concept in the Chemical, Biological, and Materials Sciences. John Wiley and Sons, New York.

Chakraborty, A., Duley, S., Giri, S., Chattaraj, P.K., 2011b. Bond-stretch isomerism and aromaticity in some selected metal clusters. In: Taylor, J.C. (Ed.), Advances in Chemistry Research. Nova publishers, ISBN: 978-1-62081-669-1, pp. 107–141.

Chakraborty, A., Giri, S., Duley, S., Anoop, A., Bultinck, P., Chattaraj, P.K., 2011c. Aromaticity in all-metal annular systems: the counter-ion effect. Phys. Chem. Chem. Phys. 13, 14865.

Chakraborty, A., Giri, S., Chattaraj, P.K., 2011d. Analyzing the efficiency of Mn-(C_2H_4) (M = Sc, Ti, Fe, Ni; n = 1, 2) complexes as effective hydrogen storage materials. Struct. Chem. 22, 823.

Chakraborty, A., Duley, S., Chattaraj, P.K., 2012. Hydrogen storage: an overview with current insights based on a conceptual DFT approach. Ind. J. Chem. 51A, 226.

Chamorro, E., Chattaraj, P.K., Fuentealba, P., 2003. Variation of the electrophilicity index along the reaction path. J. Phys. Chem. A 107, 7068.

Chan, Y., Hill, J.M., 2011. Hydrogen storage inside graphene-oxide frameworks. Nanotechnology 22, 305403.

Chapoy, A., Anderson, R., Tohidi, B., 2007. Low-pressure molecular hydrogen storage in semi-clathrate hydrates of quaternary ammonium compounds. J. Am. Chem. Soc. 129, 746.

Chattaraj, P.K., 1992. Electronegativity and hardness: a density functional treatment. J. Ind. Chem. Soc. 69, 173.

Chattaraj, P.K. (Ed.), 2009. Chemical Reactivity Theory: A Density Functional View. Taylor & Francis/CRC Press, Florida.

Chattaraj, P.K., Giri, S., 2007. Stability, reactivity, and aromaticity of compounds of a multivalent superatom. J. Phys. Chem. A 111, 11116.

Chattaraj, P.K., Giri, S., 2008. Variation in aromaticity and bonding patterns in a reaction cycle involving Be_3^{2-} and Mg_3^{2-} dianions. J. Mol. Struct. (THEOCHEM) 865, 53.

Chattaraj, P.K., Giri, S., 2009a. Electrophilicity index within a conceptual DFT framework. Ann. Rep. Prog. Chem. Sect. C Phys. Chem. 105, 13.

Chattaraj, P.K., Giri, S., 2009b. Multi-decker sandwich complexes using Be_3^{2-} and Mg_3^{2-} dianions. Int. J. Quantum Chem. 109, 2373.

Chattaraj, P.K., Roy, D.R., 2007a. Update 1 of: electrophilicity index. Chem. Rev. 107, PR46.

Chattaraj, P.K., Roy, D.R., 2007b. Popular electronic structure principles in a dynamical context. J. Phys. Chem. A 111, 4684.

Chattaraj, P.K., Sengupta, S., 1996. Aromaticity in polyacene analogues of inorganic ring compounds. J. Phys. Chem. 100, 16126.

Chattaraj, P.K., Roy, D.R., Elango, M., Subramanian, V., 2005. Stability and reactivity of all-metal aromatic and antiaromatic systems in the light of the principles of maximum hardness and minimum polarizability. J. Phys. Chem. A 109, 9590.

Chattaraj, P.K., Sarkar, U., Roy, D.R., 2006a. Electrophilicity index. Chem. Rev. 106, 2065.

Chattaraj, P.K., Roy, D.R., Elango, M., Subramanian, V., 2006b. Chemical reactivity descriptor based aromaticity indices applied to Al_4^{2-} and Al_4^{4-} systems. J. Mol. Struct. (THEOCHEM) 759, 109.

Chattaraj, P.K., Sarkar, U., Roy, D.R., 2007. Electronic structure principles and aromaticity. J. Chem. Edu. 84, 354.

Chattaraj, P.K., Roy, D.R., Duley, S., 2008. Bonding and aromaticity in an all-metal sandwich-like compound, Be_8^{2-}. Chem. Phys. Lett. 460, 382.

Chattaraj, P.K., Giri, S., Duley, S., 2011a. Update 2 of: electrophilicity index. Chem. Rev. 111, PR43.

Chattaraj, P.K., Bandaru, S., Mondal, S., 2011b. Hydrogen storage in clathrate hydrates. J. Phys. Chem. A 115, 187.

Chen, Z., Wannere, C.S., Corminboeuf, C., Puchta, R., Schleyer, P.V.R., 2005. Nucleus-independent chemical shifts (NICS) as an aromaticity criterion. Chem. Rev. 105, 3842.

Chi, X.X., Liu, Y., 2007. Theoretical evidence of d-orbital aromaticity in anionic metal X–3 (X= Sc, Y, La) clusters. Int. J. Quantum Chem. 107, 1886.

Collins, D.J., Zhou, H.C., 2007. Hydrogen storage in metal–organic frameworks. J. Mater. Chem. 17, 3154, and references therein.

Coriani, S., Haaland, A., Helgaker, T., Jørgensen, P., 2006. The equilibrium structure of ferrocene. Chem. Phys. Chem. 7, 245.

Cui, X.-Y., Yang, B.-S., Wub, H.-S., 2010. Ab initio investigation of hydrogenation of $(BN)_{16}$: a comparison with that of $(BN)_{12}$. J. Mol. Struct. (THEOCHEM) 941, 144.

Dag, S., Ozturk, Y., Ciraci, S., Yildirim, T., 2005. Adsorption and dissociation of hydrogen molecules on bare and functionalized carbon nanotubes. Phys. Rev. B 72, 155404.

Das, R., Chattaraj, P.K., 2012. A (T–P) phase diagram of hydrogen storage on $(N_4C_3H)_6Li_6$. J. Phys. Chem. A 16, 3259.

Das, R., Bandaru, S., D'mello, V.C., Chattaraj, P.K., 2013. Effect of microsolvation on hydrogen trapping potential of metal ions. Chem. Phys. 415, 256.

Datta, A., Mallajosyula, S.S., Pati, S.K., 2007. Nonlocal electronic distribution in metallic clusters: a critical examination of aromatic stabilization. Acc. Chem. Res. 40, 213.

Dillon, A.C., Heben, M.J., 2001. Hydrogen storage using carbon adsorbents: past, present and future. Appl. Phys. A 72, 133.

Dillon, A.C., Jones, K.M., Bekkedahl, T.A., Klang, C.H., Bethune, D.S., Heben, M.J., 1997. Storage of hydrogen in single-walled carbon nanotubes. Nature 386, 377.
Dimitrakakis, G.K., Tylianakis, E., Froudakis, G.E., 2008. Pillared graphene: a new 3-D network nanostructure for enhanced hydrogen storage. Nano Lett. 8, 3166.
Dincă, M., Long, J.R., 2008. Hydrogen storage in microporous metal–organic frameworks with exposed metal sites. Angew. Chem. Int. Ed. 47, 6766.
Dincă, M., Dailly, A., Liu, Y., Brown, C.M., Neumann, D.A., Long, J.R., 2006. Hydrogen storage in a microporous metal-organic framework with exposed Mn^{2+} coordination sites. J. Am. Chem. Soc. 128, 16876.
Ding, R.G., Lu, G.Q., Yan, Z.F., Wilson, M.A., 2011. Recent advances in the preparation and utilization of carbon nanotubes for hydrogen storage. J. Nanosci. Nanotechnol. 1, 7.
Duley, S., Giri, S., Chakraborty, A., Chattaraj, P.K., 2010. Bonding, aromaticity and reactivity patterns in some all- metal and non-metal clusters. J. Chem. Sci. 121, 849.
Duley, S., Giri, S., Sathymurthy, N., Islas, R., Merino, G., Chattaraj, P.K., 2011. Aromaticity and hydrogen storage capability of planar N_6^{4-} and N_4^{2-} rings: a conceptual DFT approach. Chem. Phys. Lett. 506, 315.
Durgun, E., Ciraci, S., Zhou, W., Yildirim, T., 2006. Phys. Rev. Lett. 97, 226102.
Eddaoudi, M., Moler, D.B., Li, H., Chen, B., Reineke, T.M., Keeffe, M.O., Yaghi, O.M., 2001. Modular chemistry: secondary building units as a basis for the design of highly porous and robust metal-organic carboxylate frameworks. Acc. Chem. Res. 34, 319.
Eiland, P.F., Pepinsky, R., 1952. X-ray examination of iron biscyclopentadienyl. J. Am. Chem. Soc. 74, 4971.
Ferre-Vilaplana, A., 2005. Ab initio computational investigation of physisorption of molecular hydrogen on achiral single-walled carbon nanotubes. J. Chem. Phys. 122, 214724.
Firlej, L., Roszak, S., Kuchta, B., Pfeifer, P., Wexler, C., 2009. Enhanced hydrogen adsorption in boron substituted carbon nanospaces. J. Chem. Phys. 131, 164702.
Fowler, P.W., Havenith, R.W.A., Steiner, E., 2001. Unconventional ring currents in an 'all-metal aromatic', Al_4^{2-}. Chem. Phys. Lett. 342, 85.
Fowler, P.W., Havenith, R.W.A., Steiner, E., 2002. Ring current and electron delocalisation in an all-metal cluster, Al_4^{2-}. Chem. Phys. Lett. 359, 530.
Frisch, M.J., et al., 2009. Gaussian 09, Revision A.1.
Fuentealba, P., Simon-Manso, Y., Chattaraj, P.K., 2000. J. Phys. Chem. A 104, 3185.
GAUSSIAN, 2003. 03, Revision B.03. Gaussian, Inc., Pittsburgh, PA.
Geerlings, P., De Proft, F., Langenaeker, W., 2003. Conceptual density functional theory. Chem. Rev. 103, 1793.
Giambiagi, M., de Giambiagi, M.S., Grempel, D.R., Heynmann, C.D., 1975. Definition of a bond index with non-orthogonal basis-properties and applications. J. Chim. Phys. 72, 15.
Giambiagi, M., de Giambiagi, M.S., Mundim, K.C., 1990. Definition of a multicenter bond index. Struct. Chem. 1, 423.
Giambiagi, M., de Giambiagi, M.S., dos Santos Silva, C.D., Paiva de Figueiredo, A., 2000. Multicenter bond indices as a measure of aromaticity. Phys. Chem. Chem. Phys. 2, 3381.
Giri, S., Roy, D.R., Duley, S., Chakraborty, A., Parthasarathi, R., Elango, M., Vijayaraj, R., Subramanian, V., Islas, R., Merino, G., Chattaraj, P.K., 2009. Bonding, aromaticity, and structure of trigonal dianion metal clusters. J. Comp. Chem. 31, 1815.
Giri, S., Chakraborty, A., Chattaraj, P.K., 2010. Potential use of some metal clusters as hydrogen storage material-a conceptual DFT approach. J. Mol. Model. 17, 777.
Giri, S., Bandaru, S., Chakraborty, A., Chattaraj, P.K., 2011. Role of aromaticity and charge of a system in its hydrogen trapping potential and vice-versa. Phys. Chem. Chem. Phys. 13, 20602.

Giri, S., Chakraborty, A., Duley, S., Merino, G., Subramanian, V., Chattaraj, P.K., 2012. Analysis of the structure, bonding, aromaticity and existence of possible bond-stretch isomerism in trigonal anionic metal clusters, X_3^- (X = Be, Mg, Ca). In: ICCMSE 2009, AIP Conf. Proc. 1504, p. 160.

Gomes, J.A.N.F., Mallion, R.B., 2001. Aromaticity and ring currents. Chem. Rev. 101, 1349.

Han, S.S., Jang, S.S., 2009. A hydrogen storage nanotank: lithium-organic pillared graphite. Chem. Commun. 28, 5427.

Han, S.S., Deng, W.-Q., Goddard III, W.A., 2007. Improved designs of metal–organic frameworks for hydrogen storage. Angew. Chem. Int. Ed. 46, 6289.

Hashimoto, S., Tsuda, T., Ogata, K., Sugahara, T., Inoue, Y., Ohgaki, K., 2010. Thermodynamic properties of hydrogen + tetra-n-butyl ammonium bromide semi-clathrate hydrate. Paper ID 170819, 5 pages http://dx.doi.org/10.1155/2010/170819.

Hirscher, M., Becher, M., 2003. Hydrogen storage in carbon nanotubes. J. Nanosci. Nanotechnol. 3, 3.

Hoang, T.K.A., Antonelli, D.M., 2009. Exploiting the Kubas interaction in the design of hydrogen storage materials. Adv. Mater. 21, 1787.

Hofmann, W., 1856. On insolinic acid. Proc. Royal Soc. 8, 1–3. http://dx.doi.org/10.1098/rspl.1856.0002.

Hu, Y.H., Zhang, L., 2010. Hydrogen storage in metal–organic frameworks. Adv. Mater. 22, 117.

Hübner, O., Glölss, A., Fichtner, M., Klopper, W., 2004. On the interaction of dihydrogen with aromatic systems. J. Phys. Chem. A 108, 3019.

Hückel, E., 1931. Quantentheoretische beiträge zum benzolproblem. Z. Phys. 70, 204.

Hynek, S., Fuller, W., Bentley, J., 1997. Hydrogen storage by carbon sorption. Int. J. Hydrog. Energy 22, 601.

James, S.L., 2003. Metal-organic frameworks. Chem. Soc. Rev. 32, 276.

Javan, M.B., Ganji, M.D., Sabet, M., Danesh, N., 2011. Incorporation of hydrogen molecules into carbon nitride heterofullerenes: an ab initio study. J. Comp. Theo. Nanosci. 8, 803.

Jeffrey, G.A., 1984. In: Atwood, J.L., Davies, J.E.D., MacNicol, D.D. (Eds.), Inclusion Compounds, 1. Academic Press, London, p. 135.

Jena, P., 2011. Materials for hydrogen storage: past, present, and future. J. Phys. Chem. Lett. 2, 206.

Jiménez, V., Sánchez, P., Díaz, J.A., Valverde, J.L., Romero, A., 2010. Hydrogen storage capacity on different carbon materials. Chem. Phys. Lett. 485, 152.

Juselius, J., Straka, M., Sundholm, D., 2001. Magnetic-shielding calculations on Al_4^{2-} and analogues. A new family of aromatic molecules? J. Phys. Chem. A 105, 9939.

Kandalam, A.K., Kiran, B., Jena, P., 2008. Multidecker organometallic complexes for hydrogen storage. J. Phys. Chem. C 112, 6181.

Kaplan, I.G., Dolgounitcheva, O., Watts, J.D., Ortiz, J.V., 2002. Nondipole bound anions: Be_2^- and Be_3^-. J. Chem. Phys. 117, 3687.

Kealey, T.J., Pauson, P.L., 1951. A new type of organo-iron compound. Nature 168, 1039.

Kekulé, A., 1865. Sur la constitution des substances aromatiques. Bull. Soc. Chim.Fr. (Paris) 3, 98.

Kekulé, A., 1866. Untersuchungen über aromatische Verbindungen. Ann. Chem. Pharm. 137, 129.

Khatua, S., Roy, D.R., Chattaraj, P.K., Bhattacharjee, M., 2007. Synthesis and structure of 1 D Na6 cluster chain with short Na-Na distance: organic like aromaticity in inorganic metal cluster. Chem. Commun. 2, 135.

Khatua, S., Roy, D.R., Bultinck, P., Bhattacharjee, M., Chattaraj, P.K., 2008. Possible aromaticity in alkali cluster chains. Phys. Chem. Chem. Phys. 10, 2461.

Kim, G., Jhi, S.-H., Lim, S., Park, N., 2009. Crossover between multipole Coulomb and Kubas interactions in hydrogen adsorption on metal-graphene complexes. Phys. Rev. B 79, 155437.

Kiran, B., Kandalam, A.K., Jena, P., 2006. Hydrogen storage and the 18-electron rule. J. Chem. Phys. 124, 224703.

Klontzas, E., Mavrandonakis, A., Tylianakis, E., Froudakis, G.E., 2008. Improving hydrogen storage capacity of MOF by functionalization of the organic linker with lithium atoms. Nano Lett. 8, 1572.

Klontzas, E., Tylianakis, E., Froudakis, G.E., 2010. Designing 3D COFs with enhanced hydrogen storage capacity. Nano Lett. 10, 452.

Koopmans, T.A., 1933. Über die Zuordnung von Wellenfunktionen und Eigenwerten zu den einzelnen Elektronen eines Atoms. Physica 1, 104.

Kowalczyk, P., Hołyst, R., Terrones, M., Terrones, H., 2007. Hydrogen storage in nanoporous carbon materials: myth and facts. Phys. Chem. Chem. Phys. 9, 1786.

Kruszewski, J., Krygowski, T.M., 1972. Definition of aromaticity basing on the harmonic oscillator model. Tetrahedron Lett. 36, 3839.

Krygowski, T.M., 1993. Crystallographic studies of inter-and intramolecular interactions reflected in aromatic character of pi-electron systems. J. Chem. Inf. Comput. Sci. 33, 70.

Kubas, G.J. (Ed.), 2001. Metal Dihydrogen and Bond Complexes-Structure, Theory and Reactivity. Kluwer Academic/Plenum Publishing, New York.

Kuc, A., Heine, T., Seifert, G., Duarte, H.A., 2008. H2 adsorption in metal-organic frameworks: dispersion or electrostatic interactions? Chem. Eur. J. 14, 6597.

Kudinov, A.R., Rybinskaya, I., 1999. New triple-decker complexes prepared by the stacking reactions of cationic metallofragments with sandwich compounds. Russ. Chem. Bull. 48, 1615.

Kudinov, A.R., Loginov, D.A., Starikova, Z.A., Petrovskii, P.V., 2002. Dicationic triple-decker complexes with a bridging boratabenzene ligand. J. Organomet. Chem. 649, 136.

Kuznetsov, A.E., Boldyrev, A.I., 2004. A single π-bond captures 3, 4 and 5 atoms. Chem. Phys. Lett. 388, 452.

Kuznetsov, A.E., Boldyrev, A.I., Li, X., Wang, L.S., 2001. On the aromaticity of square planar Ga_4^{2-} and In_4^{2-} in gaseous NaGa4- and NaIn4- clusters. J. Am. Chem. Soc. 123, 8825.

Lang, X., Fan, S., Wang, Y., 2010. Intensification of methane and hydrogen storage in clathrate hydrate and future prospect. J. Nat. Gas Chem. 19, 203.

Li, X., Kuznetsov, A.E., Zhang, H.-F., Boldyrev, A.I., Wang, L.S., 2001a. Observation of all-metal aromatic molecules. Science 291, 859.

Li, X., Zhang, H.F., Wang, L.S., Kuznetsov, A.E., Cannon, A.N., Boldyrev, A.I., 2001b. Experimental and theoretical observations of aromaticity in heterocyclic XAl3−(X= Si, Ge, Sn, Pb) systems. Angew. Chem. Int. Ed. 40, 1867.

Li, Z., Zhao, C., Chen, L., 2008. Structure and aromaticity of, Bi 5 M (M= Li, Na, K) and Bi 5 M +(M= Be, Mg, Ca) clusters. J. Mol. Struct. (THEOCHEM) 854, 46.

Li, C., Li, J., Wu, F., Li, S.-S., Xia, J.-B., Wang, L.-W., 2011. High capacity hydrogen storage in Ca decorated graphyne: a first-principles study. J. Phys. Chem. C 115, 23221.

Liu, C.-S., Zeng, Z., 2010. Boron-tuned bonding mechanism of Li-graphene complex for reversible hydrogen storage. Appl. Phys. Lett. 96, 123101.

Liu, Y., Chi, X.X., Wang, X.B., 2007. N03: PNIP (Poster No 4). In: International Conference on Computational Science, Beijing, May 27-30. http://www.iccs-meeting.org/.

Ma, S., Zhou, H.-C., 2010. Gas storage in porous metal–organic frameworks for clean energy applications. Chem. Commun. 46, 44.

Mallajosyula, S.S., Datta, A., Pati, S.K., 2006. Aromatic superclusters from all-metal aromatic and antiaromatic monomers,$[Al_4]^{2-}$ and $[Al_4]^4$. J. Phys. Chem. B 110, 20098.

Mallion, R.B., 1980. Some comments on the use of the "ring-current" concept in diagnosing and defining "aromaticity". Pure Appl. Chem. 52, 1541.

McClaine, A.W., Breault, R.W., Larsen, C., Konduri, R., Rolfe, J., Becker, F., Miskolczy, G., 2000. Proc. US DOE Hydrogen Prog. Rev. NREL/CP-570-28890.

McNelles, P., Naumkin, F.Y., 2009. A small molecule in metal cluster cages: H 2@ Mg n (n= 8 to 10). Phys. Chem. Chem. Phys. 11, 2858.

Meek, S.T., Greathouse, J.A., Allendorf, M.D., 2011. Metal-organic frameworks: a rapidly growing class of versatile nanoporous materials. Adv. Mater. 23, 249.

Meng, S., Kaxiras, E., Zhang, Z., 2007. Metal-diboride nanotubes as high-capacity hydrogen storage media. Nano Lett. 7, 663.

Mercero, J.M., Ugalde, J.M., 2004. Sandwich-like complexes based on "all-metal" (Al42-) aromatic compounds. J. Am. Chem. Soc. 126, 3380.

Middleton, R., Klein, J., 1999. Production of metastable negative ions in a cesium sputter source: verification of the existence of N2- and CO. Phys. Rev. A 60, 3786.

Minkin, V.I., Glukhovtsev, M.N., Simkin, B.Y., 1994. Aromaticity and Antiaromaticity: Electronic and Structural Aspects. J. Wiley & Sons, New York.

Mondal, S., Ghosh, S., Chattaraj, P.K., 2013a. A molecular dynamics study on SI hydrogen hydrate. J. Mol. Model. 19, 2785.

Mondal, S., Giri, S., Chattaraj, P.K., 2013b. Possibility of having HF doped hydrogen hydrates. J. Phys. Chem. C 117, 11625.

Mondal, S., Srinivasu, K., Ghosh, S., Chattaraj, P.K., 2013c. Isomers of $C_{12}N_{12}$ as potential hydrogen storage materials and the effect of the electric field therein. RSC Adv. 3, 6991.

Mpourmpakis, G., Tylianakis, E., Froudakis, G., 2006a. Hydrogen storage in carbon nanotubes: a multi-scale theoretical study. J. Nanosci. Nanotechnol. 6, 87.

Mpourmpakis, G., Froudakis, G.E., Lithoxoos, G.P., Samios, J., 2006b. Nano Lett. 6, 1581.

Mpourmpakis, G., Froudakis, G.E., Lithoxoos, G.P., Samios, J., 2007. Effect of curvature and chirality for hydrogen storage in single-walled carbon nanotubes: a Combined ab initio and Monte Carlo investigation. J. Chem. Phys. 126, 144704.

Mulliken, R.S., 1955. Electronic population analysis on LCAO–MO molecular wave functions. I. J. Chem. Phys. 23, 1833.

Murray, L.J., Dincă, M., Long, J.R., 2009. Hydrogen storage in metal–organic frameworks. Chem. Soc. Rev. 38, 1294.

Noorizadeh, S., Dardaba, M., 2010. A new NICS-based aromaticity index; NICS-rate. Chem. Phys. Lett. 493, 376.

Oscar, C.J., Jimenez-Halla, J.C., Matito, E., Blancafort, L., Robles, J., Sola, M., 2009. Tuning aromaticity in trigonal alkaline earth metal clusters and their alkali metal salts. J Comput. Chem. 30, 2764.

Padma Malar, E., 2005. Density functional theory analysis of some triple-decker sandwich complexes of iron containing cyclo-P5 and cyclo-As5 ligands. J. Theor. Chem. Acc. 114, 213, and references therein.

Pan, S., Banerjee, S., Chattaraj, P.K., 2012a. Role of lithium decoration on hydrogen storage. J. Mex. Chem. Soc. 56, 229.

Pan, S., Giri, S., Chattaraj, P.K., 2012b. A Computational study on the hydrogen adsorption capacity of various lithium-doped boron hydrides. J. Comput. Chem. 33, 425.

Pan, S., Mondal, S., Chattaraj, P.K., 2013. Cucurbiturils as promising hydrogen storage materials: a case study of Cucurbit[7]uril. New J. Chem. 37, 2492.

Park, N., Hong, S., Kim, G., Jhi, S.H., 2007. Computational study of hydrogen storage characteristics of covalent-bonded graphenes. J. Am. Chem. Soc. 129, 8999.
Parr, R.G., Chattaraj, P.K., 1991. Principle of maximum hardness. J. Am. Chem. Soc. 11, 1854.
Parr, R.G., Pearson, R.G., 1983. Absolute hardness: companion parameter to absolute electronegativity. J. Am. Chem. Soc. 105, 7512.
Parr, R.G., Yang, W., 1984. Density functional approach to the frontier-electron theory of chemical reactivity. J. Am. Chem. Soc. 106, 4049.
Parr, R.G., Yang, W., 1989. Density Functional Theory of Atoms and Molecules. Oxford University Press, New York.
Parr, R.G., Donnelly, R.A., Levy, M., Palke, W.E., 1978. Electronegativity: the density functional viewpoint. J. Chem. Phys. 68, 3801.
Parr, R.G., Szentpaly, L.v, Liu, S., 1999. Electrophilicity index. J. Am. Chem. Soc. 121, 1922.
Parthasarathi, R., Elango, M., Subramanian, V., Chattaraj, P.K., 2005. Variation of electrophilicity during molecular vibrations and internal rotations. Theor. Chem. Acc. 113, 257.
Pearson, R.G., 1987. Recent advances in the concept of hard and soft acids and bases. J. Chem. Educ. 64, 561.
Pearson, R.G., 1997. Chemical Hardness: Applications from Molecules to Solids. Wiley-VCH, Weinheim.
Peng, Q., Chen, G., Mizuseki, H., Kawazoe, Y., 2009. Hydrogen storage capacity of C (60) (OM)(12)(M= Li and Na) clusters. J. Chem. Phys. 131, 214505.
Peterson, V.K., Liu, Y., Brown, C.M., Kepert, C.J., 2006. Neutron powder diffraction study of D2 sorption in Cu3 (1, 3, 5-benzenetricarboxylate) 2. J. Am. Chem. Soc. 128, 15578.
Plimpton, S.J., 1995. Fast parallel algorithms for short-range molecular dynamics. J. Comput. Phys. 117, 1.
Prasad, P.S.R., Sugahara, T., Sum, A.K., Sloan, E.D., Koh, C.A., 2009. Hydrogen storage in double clathrates with tert-butylamine. J. Phys. Chem. A 113, 6540.
Qian-Shu, L., Heng-Tai, Y., Au-chin, T., 1986. Electronic structures of multi-decker transition metal sandwich complexes. Theor. Chim. Acta 70, 379.
Resa, I., Carmona, E., Gutierrez-Puebla, E., Monge, A., 2004. Decamethyldizincocene, a stable compound of Zn (I) with a Zn-Zn bond. Science 305, 1136.
Rosi, N.L., Eckert, J., Eddaoudi, M., Vodak, D.T., Kim, J., Keeffe, M.O., Yaghi, O.M., 2003. Hydrogen storage in microporous metal-organic frameworks. Science 300, 1127.
Rowsell, J.L.C., Yaghi, O.M., 2005. Strategies for hydrogen storage in metal–organic frameworks. Angew. Chem. Int. Ed. 44, 4670.
Rowsell, J.L.C., Millward, A.R., Park, K.S., Yaghi, O.M., 2004. Hydrogen sorption in functionalized metal-organic frameworks. J. Am. Chem. Soc. 126, 5666.
Roy, D.R., Chattaraj, P.K., 2008. Reactivity, selectivity and aromaticity of Be_3^{2-} and its complexes. J. Phys. Chem. A 112, 1612.
Roy, D.R., Duley, S., Chattaraj, P.K., 2008. Bonding, reactivity and aromaticity in some novel all- metal metallocenes. Proc. Ind. Natl. Sci. Acad. Part-A 74, 11.
Saha, D., Deng, S., 2011. Hydrogen adsorption on Pd-and Ru-doped C60 fullerene at an ambient temperature. Langmuir 27, 6780.
Sakintuna, B., L-Darkrimb, F., Hirscherc, M., 2007. Metal hydride materials for solid hydrogen storage: a review. Int. J. Hydrog. Energy 32, 1121.
Schlapbach, L., Züttel, A., 2001. Hydrogen-storage materials for mobile applications. Nature 414, 353.
Schleyer, P.V.R., 2001. Introduction: aromaticity. Chem. Rev. 101, 1115.
Schleyer, P.V.R., Maerker, C., Dransfeld, A., Jiao, H., Hommes, N.J.R.V.E., 1996. Nucleus-independent chemical shifts: a simple and efficient aromaticity probe. J. Am. Chem. Soc. 118, 6317.

Schmidt, W.R., 2001. Proc. US DOE Hydrogen Prog. Prev. NREL/CP-570-30535.
Schulz, R., Boily, S., Zaluski, L., Zaluka, A., Tessier, P., Ström-Olsen, J.O., 1995. Nanocrystalline materials for hydrogen storage. Innov. Metallic Mater. 529–535.
Schüth, F., Bogdanović, B., Felderhoff, M., 2004. Light metal hydrides and complex hydrides for hydrogen storage. Chem. Commun. 20, 2249.
Shevlin, S.A., Guo, Z.X., 2006. Transition-metal-doping-enhanced hydrogen storage in boron nitride systems. Appl. Phys. Lett. 89, 153104.
Shimada, W., Shiro, M., Kondo, H., Takeya, S., Oyama, H., Ebinuma, T., Narita, H., 2005. Tetra-n-butylammonium bromide-water (1/38). Acta Crystallogr. C 61, 65.
Sloan, E.D., Koh, C.A., 2007. Clathrate Hydrates of Natural Gases, third ed. Taylor & Francis – CRC Press, London.
Srinivasu, K., Ghosh, S.K., 2012. Graphyne and graphdiyne: promising materials for nanoelectronics and energy storage applications. J. Phys. Chem. C 116, 5951.
Srinivasu, K., Chandrakumar, K.R.S., Ghosh, S.K., 2008. Quantum chemical studies on hydrogen adsorption in carbon-based model systems: role of charged surface and the electronic induction effect. Phys. Chem. Chem. Phys. 10, 5832.
Srinivasu, K., Chandrakumar, K.R.S., Ghosh, S.K., 2009. Computational investigation of hydrogen adsorption by alkali metal doped organic molecules: role of aromaticity. Chem. Phys. Chem. 10, 427.
Srinivasu, K., Ghosh, S.K., Das, R., Giri, S., Chattaraj, P.K., 2012. Theoretical investigation of hydrogen adsorption in all-metal aromatic cluster. RSC Adv. 2, 2914.
Stanger, A., 2006. Nucleus-independent chemical shifts (NICS): distance dependence and revised criteria for aromaticity and antiaromaticity. J. Org. Chem. 71, 883.
Steriotis, T.A., Charalambopoulou, G.C., Stubos, A.K., 2011. In: Kanellopoulos, N. (Ed.), Nanoporous Materials: Advanced Techniques for Characterization, Modeling and Processing. Taylor and Francis Group, LLC, Boca Raton, FL, USA, p. 513. http://dx.doi.org/10.1201/b10610-20.
Ströbel, R., Garche, J., Moseley, P.T., Jörissen, L., Wolf, G., 2006. Hydrogen storage by carbon materials. J. Power Sources 159, 781.
Strobel, T.A., Koh, C.A., Sloan, E.D., 2007. Hydrogen storage properties of clathrate hydrate materials. Fluid Phase Equilib. 261, 382.
Struzhkin, V.V., Militzer, B., Mao, W.L., Mao, H.-K., Hemley, R.J., 2007. Hydrogen storage in molecular clathrates. Chem. Rev. 107, 4133.
Subrahmanyam, K.S., Kumar, P., Maitra, U., Govindaraj, A., Hembram, K.P.S.S., Waghmare, U.V., Rao, C.N.R., 2011. Directed migration of neural stem cells to sites of CNS injury by the stromal cell-derived factor 1α/CXC chemokine receptor 4 pathway. Proc. Natl. Acad. Sci. U. S. A. 108, 2674–2677. http://dx.doi.org/10.1073/pnas.1019542108.
Sun, Q., Wang, Q., Jena, P., 2005. Storage of molecular hydrogen in BN cage: energetics and thermal stability. Nano Lett. 5, 1273.
Suri, M., Dornfeld, M., Ganz, E., 2009. Calculation of hydrogen storage capacity of metal-organic and covalent-organic frameworks by spillover. J. Chem. Phys. 131, 174703.
Thomas, K.M., 2009. Adsorption and desorption of hydrogen on metal–organic framework materials for storage applications: comparison with other nanoporous materials. Dalton Trans. 9, 1487.
Tsipis, C., 2005. DFT study of "all-metal" aromatic compounds. Coord. Chem. Rev. 249, 2740.
Tylianakis, E., Klontzas, E., Froudakis, G.E., 2011. Multi-scale theoretical investigation of hydrogen storage in covalent organic frameworks. Nanoscale 3, 856.
Wagemans, R.W.P., van Lenthe, J.H., de Jongh, P.E., van Dillen, A.J., de Jong, K.P., 2005. Hydrogen storage in magnesium clusters: quantum chemical study. J. Am. Chem. Soc. 127, 16675.

Wang, J., Ebner, A.D., Ritter, J.A., 2007. Synthesis of metal complex hydrides for hydrogen storage. J. Phys. Chem. C 111, 14917.

Wang, L.K., Sun, Y.-Y., Lucking, M., Chen, Z., Zhao, J.J., Zhang, S.B., 2009. Graphene oxide as an ideal substrate for hydrogen storage. ACS Nano 3, 2995.

Weast, R.C., Astle, M.J., Beyer, W.H., 1983. CRC Handbook of Chemistry and Physics, 64th ed. CRC Press, Boca Raton, FL.

Weck, P.F., Dhilip Kumar, T.J., Kim, E., Balakrishnan, N., 2007. Computational study of hydrogen storage in organometallic compounds. J. Chem. Phys. 126, 094703.

Wen, S.-H., Deng, W.-Q., Han, K.-L., 2008. Endohedral BN metallofullerene M@ B36N36 complex as promising hydrogen storage materials. J. Phys. Chem. C 112, 12195.

Wilkinson, G., Rosenblum, M., Whiting, M.C., Woodward, R.B., 1952. The structure of iron bis-cyclopentadienyl. J. Am. Chem. Soc. 74, 2125.

Wu, X., Gao, Y., Zeng, X.C., 2008. Hydrogen storage in pillared Li-dispersed boron carbide nanotubes. J. Phys. Chem. C 112, 8458.

Wu, G., Wang, J., Zhang, X., Zhu, L., 2009. Hydrogen storage on metal-coated B80 buckyballs with density functional theory. J. Phys. Chem. C 113, 7052.

Wu, H.Y., Fan, X.F., Kuo, J.-L., Deng, W.-Q., 2010a. Carbon doped boron nitride cages as competitive candidates for hydrogen storage materials. Chem. Commun. 46, 883.

Wu, M.M., Wang, Q., Sun, Q., Jena, P., Kawazoe, Y., 2010b. First-principles study of hydrogen adsorption in metal-doped COF-10. J. Chem. Phys. 133, 154706.

Yang, W., Mortier, W.J., 1986. The use of global and local molecular parameters for the analysis of the gas-phase basicity of amines. J. Am. Chem. Soc. 108, 5708.

Yildirim, T., Ciraci, S., 2005. Titanium-decorated carbon nanotubes as a potential high-capacity hydrogen storage medium. Phys. Rev. Lett. 94, 175501.

Yildirim, T., Iniguez, J., Ciraci, S., 2005. Molecular and dissociative adsorption of multiple hydrogen molecules on transition metal decorated C60. Phys. Rev. B 72, 153403.

Yong, L., Chi, X., 2007. Theoretical study on the aromaticity of dianions (X= Zn, Cd, Hg). J. Mol. Struct. (THEOCHEM) 818, 93.

Yoon, M., Yang, S., Wang, E., Zhang, Z., 2007. Charged fullerenes as high-capacity hydrogen storage media. Nano Lett. 7, 2578.

Yue, X., Zhao, J., Qiu, J., 2007. Computational Science—ICCS 2007, Lecture Notes in Computer Science 4488, 280.

Zhan, C.G., Zheng, F., Dixon, D.A., 2002. Electron affinities of Al n clusters and multiple-fold aromaticity of the square Al42-structure. J. Am. Chem. Soc. 124, 14795.

Zhao, X., Xiao, B., Fletcher, A.J., Thomas, K.M., Bradshaw, D., Rosseinsky, M.J., 2004. Hysteretic adsorption and desorption of hydrogen by nanoporous metal-organic frameworks. Science 306, 1012.

Zhao, Y., Kim, Y.-H., Dillon, A.C., Heben, M.J., Zhang, S.B., 2005. Hydrogen storage in novel organometallic buckyballs. Phys. Rev. Lett. 942005, 155504.

Zhao, D., Yuan, D., Zhou, H.-C., 2008. The current status of hydrogen storage in metal–organic frameworks. Energy Environ. Sci. 1, 222.

Zhou, L., 2005. Progress and problems in hydrogen storage methods. Renew. Sust. Energ. Rev. 9, 395.

Zhou, W., Yildirim, T., Durgun, E., Ciraci, S., 2007. Hydrogen absorption properties of metal-ethylene complexes. Phys. Rev. B 76, 085434.

Züttel, A., 2004. Hydrogen storage methods. Naturwissenschaften 91, 157.

Part Three

Hydrogen distribution and infrastructure

Introduction to hydrogen transportation

11

R. Gerboni
Politecnico di Torino, Torino, Italy

Abbreviations

ADR	Accord Européen Relatif au Transport International des Marchandises Dangereuses par Route—European Agreement concerning the International Carriage of Dangerous Goods by Road
CCS	carbon capture and sequestration
CHP	combined heat and power
DOE	US Department of Energy
EIGA	European Industrial Gases Association
ENCOURAGED	Energy Corridor Optimisation for European Markets of Gas, Electricity and Hydrogen - EU Project
EPRI	Electric Power Research Institute
EQHHPP	Euro-Quebec Hydro-Hydrogen Pilot Project
ISO	International Organization for Standardization
JRC	Joint Research Centre
LH_2	liquid hydrogen
LNG	liquefied natural gas
LPG	liquefied petroleum gas
NEEDS	New Energy Externalities Development for Sustainability - EU Project
NFPA	National Fire Protection Agency
REACCESS	Risk of Energy Availability: Common Corridors for Europe Supply Security - EU Project
SOFC	solid oxide fuel cell
WE-NET	World Energy Network

11.1 Introduction

Starting with the obvious consideration that hydrogen is an energy vector, one of the hot topics in the development of a hydrogen economy is the connection between its production site and its final consumption. Hydrogen has the unquestioned virtue of being produced by several primary sources, either fossil or renewables. Several studies have shown that it is usually more convenient from an economical and environmental point of view to transform the primary energy into electricity where the reserves are located, and then electricity is more easily transported to the demand

Compendium of Hydrogen Energy. http://dx.doi.org/10.1016/B978-1-78242-362-1.00011-0

centers. There, the option of producing hydrogen from water using this electricity is viable, although it requires the introduction of an additional efficiency in the cycle. In this configuration the transported vector is electricity and not hydrogen.

However, there are some occasions where the production of electricity may be substituted with other processes, like the gasification of brown coal or biomass, to produce hydrogen and transport it even over long distances. The primary fuel, in this case, is so poor in energy content that its transport overseas or across undemanding countries proves extremely uneconomical and the local transformation into electricity is also questionable.

The renewable sources, like wind, solar, and hydro, are usually exploited to produce electricity that is immediately delivered via the transmission and distribution grid.

However, it is becoming increasingly important to explore the hydrogen option not only as a fuel but also and mainly as an energy storage solution coupled with the renewable energy sources. These are often unpredictably available and their massive adoption in the national energy mixes can create (and already has proven to create) problems in the electrical grid balancing. The transformation of the excess of electricity into hydrogen (instead of storing it in bulk packs of batteries), its storage, and its final distribution to the demand areas is an alternative.

The ENCOURAGED and the REACCESS EU projects addressed the possibility of exploiting high renewable or low energy content potentials (like wind off-shore Morocco, or solar power in the Algerian desert, biomass in Turkey, and lignite in Ukraine) to produce hydrogen and then export it to demanding countries in Europe (Pregger et al., 2011). Figure 11.1 shows the proposed hydrogen corridors.

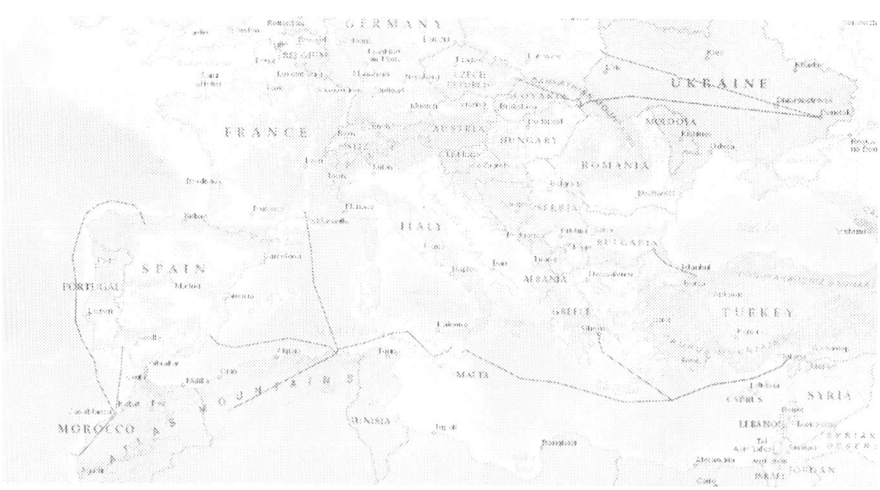

Figure 11.1 Hydrogen corridors to supply Europe as proposed in Encouraged and REACCESS projects.
Adapted from ESRI (2010).

Enlarging the scope from the local/national to the international scale, the famous EQHHPP project, followed by the WE-NET project, has shown that the balancing of hydroelectricity may well pass through the LH_2 solution.

The fact that transporting hydrogen from one place to another is considered difficult also appears evident from the stakeholder opinion assessment, as performed by the European project HyWays in 2004–2007: the project results showed that stakeholders do not see huge problems in transporting hydrogen over short-medium distances, but do worry when it comes to using pipelines (Hugo Seymour et al., 2008). Their respected opinion is that pipelines are expected to be viable no earlier than 2025 and that considerable effort should be paid to improving the existing technology and for granting the penetration in the existing economical pattern. This is surprising since hydrogen pipelines have existed and been operated safely since the 1930s of the last century.

Nonetheless, technology advancements are needed to improve the efficient use of this kind of infrastructure and the optimal integration with the other distribution systems is part of a complex study needed to determine if a hydrogen economy should be implemented, in a radical view, such as that of Bockris in early 1970s, or in a pondered vision, such as that proposed by (Andrews and Shabani, 2012).

11.2 Overview of methods for hydrogen transportation

The following sections deal with the present solutions and expected developments for transporting hydrogen in an affordable and safe way. The solutions are varied according to the distance and quantity to be delivered, with logistics ranging from road/rail, to pipelines to overseas.

11.2.1 Road and rail transportation of gaseous and liquid hydrogen

A very common method of transporting hydrogen gas is to fill it in pressure-proofed seamless vessels. These vessels can be either industrial gas standard size (50–150 l in volume) or larger containers (>150 l) for transportation via tube trailers (Figure 11.2). The small vessels are usually grouped in bundles and are filled at the premises of the centralized production plant and delivered to the final consumer where the empty set of bottles is replaced. Tube trailers can be used either to fill in distributed storages at the demand sites, or as stationary storages (e.g., in refueling stations where a different hydrogen infrastructure such as a pipeline or a local production facility is not available). A typical set-up is nine tubes for 2000 l each (Nexant, 2008). The filling pressure of cylinders is usually around 18–25 MPa, although higher pressure is achievable and indispensable to improve the economical viability of this part of the hydrogen cycle. There are already several examples of manufactured cylinders that can stand pressures up to 70 MPa, but these are not usually intended for hydrogen transportation purposes but rather as onboard storage systems for vehicles. The need for improved

Figure 11.2 Hydrogen tube trailer.
Courtesy of Fiba Canning Inc.

levels of pressure has pushed cylinder's manufacturers to opt for Type II, III, and some Type IV vessels (Figure 11.3). These have a higher investment cost per unit but the cost per MJ of hydrogen transported may result lower if the internal pressure is increased to 40–70 MPa. Tubes (or transport vessels) for tube trailers are typically Type I cylinders. For tube trailers, the target set by the US DOE for the year 2020 to reach economic feasibility is a filling pressure of 52 MPa (US DOE, 2013). This means that composite materials are also under investigation for tubes construction.

At present, the usual material for transport cylinders/tubes is 34CrMo4 steel alloy, and the same material should be used for valves and safety devices.

Although international standards cover the wide plethora of gaseous vessels applications, the European Industrial Gases Association has provided a set of specific

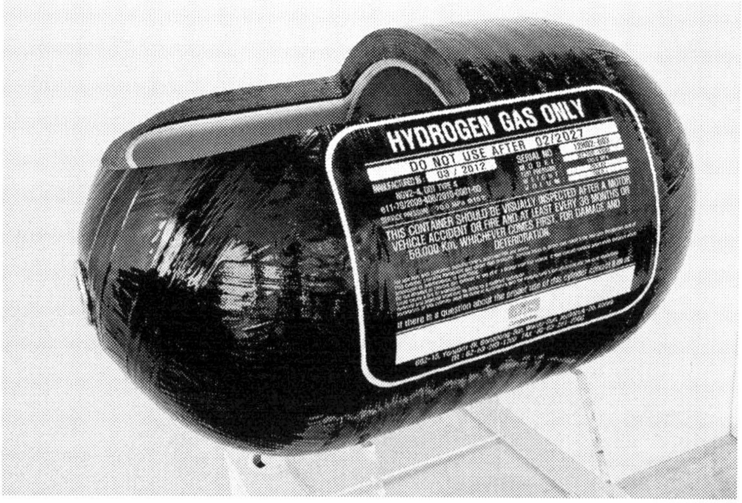

Figure 11.3 Type IV pressure vessel for hydrogen.
Photo courtesy of Stuart Price.

recommendations to be applied to hydrogen cylinders, e.g., the ultimate strength is recommended to be higher than 950 MPa (EIGA, 2011).

As described elsewhere in this book, although energy intensive, hydrogen liquefaction is a good solution for storing (and delivering) large quantities of hydrogen.

As a matter of fact, it represents the main solution for transporting merchant hydrogen over medium/long distances.

Due to the technological challenges imposed and to the large quantity of flammable gas stored, the transport of liquid hydrogen via truck (with a tank trailer) is ruled by local (regional, national) and international regulations concerning dangerous goods transportation—i.e., ADR. The use of trucks for the delivery of liquid hydrogen is monitored with particular care during the trip and in the transfer phase and it is usually recommended to avoid routing through populated areas. As trucks are also used for transporting liquid hydrogen over medium/long distances, a drive stopping for a break is allowed but special attention must be paid to avoid parking the truck nearby possible hazardous storages of other substances such as oxygen or LPG.

The critical point of the chain is the transfer of the liquid fluid from the tank to the final destination as the delta in temperatures between the ambient and the internal environment is extremely high. Flashing may occur, which can cause injuries and damages to the surrounding where workers and equipment operate.

Apart from the flashing hazard, a physiological boil-off occurs in all transportation phases of liquid hydrogen, leading to a net loss in terms of payload. It is said that about 20% of the hydrogen is lost when it is transferred from one small Dewar vessel to another one, and when hydrogen is transferred from a trailer, the loss is said to be around 10%. Therefore, whenever possible, dual-purpose storage/transportation vessels are used.

Tanker trucks have capacities that usually range from 20,000 to 50,000 l (Sherif et al., 1997), although some examples of larger capacity exist, and up to 4000 kg of liquid (Cuni et al., 2008). The pressure inside the tank is usually low (0.6–1 MPa).

Liquid hydrogen can also be transported via rail and the tank has the same size as that used when attached to a trailer. In many areas of the world liquid hydrogen is transported via rail, although it should be remembered that hydrogen is predominantly used in the petrochemical (refining) industry and when this gas is not produced and used onsite, the preferred industrial transport method remains pipelines with some intermediate storage stations.

11.2.2 Ocean transportation of hydrogen

The transport of hydrogen across the sea was studied intensively during the late 1980s to the late 1990s by two research groups with partially similar aims and targets. The leading objective was the ability to transport hydrogen over long distances from cheap production sites to eager areas. The main solution has been individuated in liquid hydrogen, even though a considerable contribution in terms of energy was calculated to be spent to liquefy it, and LH_2 was recognized as having a lower energy density compared to its most proximate competitor, LNG.

The first high-level study concerning sea transport of hydrogen was carried out in the framework of the EQHHPP (Euro-Quebec Hydro-Hydrogen Pilot Project). The project was jointly funded and carried out by the European Commission (JRC Ispra) and the government of Québec, with the support of a multitude of industrial partners, including many actors of the German industrial panorama (Gretz et al., 1994). The basic idea was the exploitation of the excess of cheap and difficult to store hydroelectricity produced in Québec. This electricity was planned to be converted into a "simpler" to manage intermediate energy vector like hydrogen to ship it to one of the most demanding areas of the world, namely the German coast. The project foresaw several phases but was stopped before the end, i.e., before the construction of the vessel, because of the impracticable costs of the operation. Nevertheless, the analyses showed some interesting results: for example, the use of methylcyclohexane as a carrier, alternatively to liquid hydrogen, and the design of a new ship, able to carry 15,000 m^3 of liquid hydrogen, stored in five superinsulated vessels. The choice for the quantity of hydrogen to be transported was not based on the ship capacity but on the power available in Canada (an initial pilot project of 100 MWel was dedicated to the study). The preliminary results showed that liquid hydrogen would have been the cheapest solution among the two (only) when liquid hydrogen was the product requested at the destination (methylcyclohexane frees gaseous hydrogen, therefore, liquefaction should have been performed at the port of arrival) and that theoretically, apart from the obvious environmental advantages in using hydrogen, the delta between the cost of hydrogen per kW and the cost of gasoline would have been limited if all the submerged and invisible costs of the use of the latter were considered. These last costs were represented by the environmental repair and protection costs due to spills and to different accidents and the unpredictable consequences on climate. Later studies, such as ExternE and NEEDS, funded by the European Union, tried to quantify these unpredictable and submerged costs and proved that they are far from being negligible.

The second study was promoted by the Japanese government with a program that was projected to last over a long period of time (1993–2020) and focused on the assessment of the viability of hydrogen as an energy carrier for Japan. The WE-NET (World Energy Network) research program was articulated in many phases and one of them dealt with the design of liquid hydrogen tankers for overseas transport. The capacity of the Japanese tanker was immediately greater than that proposed in the Euro-Canadian project, being closer to a typical LNG tanker (around 200,000 m^3) (Abe et al., 1998). The researchers proposed different architectures for the storage (spherical and prismatic vessels) and also different designs for the tankers (mono or twin hull). While the previous project neglected the boil-off during the simulated trip (50 days), the Japanese study took it into account and proposed reusing the boil-off hydrogen as a fuel for the cargo. As a matter of fact, the boil-off was somehow "required" to have enough fuel for the cargo to reach the destination (an estimated value of 0.2–0.4%/day). The twin-hull design has proven necessary for LH_2 ships because hydrogen, even in its liquid form, is very light and the ship required ballast and a double hull to guarantee stability.

Figure 11.4 Liquefied hydrogen carrier in a rendering by Kawasaki Heavy Industries, Ltd.

In recent years, renewed interest in bulk transportation of hydrogen has occurred, in part, because of the Fukushima nuclear accident and the need to reshape the Japanese energy mix (Figure 11.4 shows a recent rendering of a Japanese LH_2 carrier project).

Stemming from the information gathered during the WE-NET project and promoting the use of a low-energy-intensive and low-price fuel, the new project (Oyama and Kamiya, 2012) predicts the massive gasification of brown coal in Victoria State, Australia, and the subsequent liquefaction of the produced hydrogen coupled with carbon capture and sequestration (CCS). The liquid hydrogen is then transported to Japan via a pair of specially designed tankers (160,000 m^3 each). Apart from the pilot project, the final design would include four spherical tanks (40,000 m^3 each) with the usual boil-off rate of 0.2%/day.

11.2.3 Hydrogen pipelines

Transportation and distribution to single users of compressed hydrogen via pipeline is one of the options that is presently being exploited for the use of hydrogen as an energy vector. As will be seen later, this technology is not always the most convenient and, at the present, the pipeline network for hydrogen transportation is very limited and derived from natural gas technology.

The panorama of hydrogen pipelines is dominated by a few industrial gases companies: Air Products, Air Liquide e Praxair. Other companies share shorter tracks of a few kilometers or operate small networks inside their production plants. Hydrogen is almost always transported to be used in refinery plants or in big chemical plants. The transport of hydrogen to residential consumers is not yet diffused, although some small and specific craft centers, which require hydrogen for their production (e.g., goldsmiths) have started to create local networks to serve their premises with centrally produced hydrogen (usually from electrolysis).

While the most forecasted need for a hydrogen pipeline infrastructure is due to the connection of refueling stations for cars, many have also envisioned the first fuel cell market evolution to head toward residential applications: in this case hydrogen and methane pipelines will be competing, as solid oxide fuel cell (SOFC) is the favorite choice in combined heat and power (CHP) applications. Hydrogen pipelines and storage will also play an important role in future smart grid diffusion in urban areas: as a parallel system to the electrical network and a good energy storage alternative, its adoption will have to be included in the articulated schemes that are proposed for further assessment.

As specified later, codes and standards about hydrogen delivery is still uneven in diffusion: North America, due to the relatively high distribution of hydrogen systems, has provided a number of standards that have local validity. Europe has started a series of initiatives to address safety issues and to develop standards on pipelines, based on the experience gained with natural gas. Single countries have developed local regulations when needed to proceed with demonstration activities and prototypes.

The industrial exploitation of hydrogen in the chemical sector on a large scale started in 1938 in the Ruhr Basin with the construction of a hydrogen pipeline intended to connect several chemical plants.

Western Europe owns the longest network of pipelines, about 1500 km compared to the existing 900 km in the United States. These values may vary according to sources because there is no agreement on the definition of hydrogen pipeline to be accounted for: some analysts do not consider pipelines with a diameter below a certain value and some others only consider the pipelines connecting production plants with external consumers, neglecting the internal pipelines system into the production facility (Gillette and Kolpa, 2007). The European countries that own the largest part of the hydrogen network are France, Germany, and Benelux (Figure 11.5). Hydrogen pipelines with a smaller length exist in other countries, in particular in Great Britain and Sweden, and projects to connect the Netherlands with Sweden are ongoing.

The existing pipelines are constructed with common steels for general construction. There are no known problems connected with the utilization of these pipelines. The operating pressures vary according to the networks and, in general, are between 0.34 and 10 MPa. Their diameter may vary between 10 and 300 mm. More frequently, the operating pressure is about 1–2 MPa and diameters are about 25–30 cm.

If the pressure remains at low values, embrittlement effects are smaller, which is why conventional steels are used today.

To avoid incurring excessive investment costs, there is the possibility of mixing hydrogen with natural gas and to exploit the existing natural gas network. When the hydrogen share in volume is lower than 20% (which might mean 5–7% in energy content), conventional materials may be used without consequence.

To reduce safety and logistical issues, it is possible to bury hydrogen pipelines, as it is usually done with natural gas pipes. In 1983, Air Products realized the first directional perforation (1.6 km long) through a river to install a hydrogen pipeline in Louisiana.

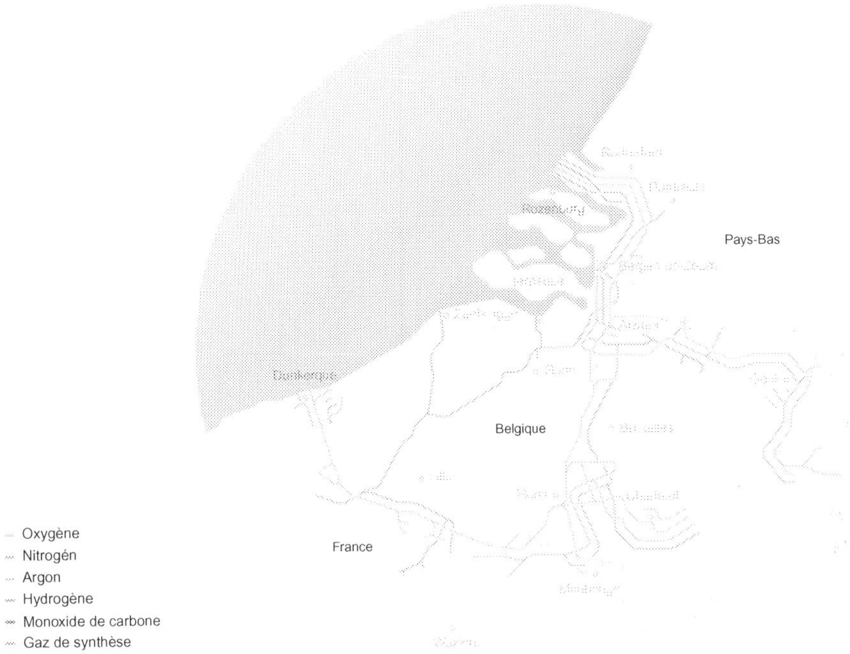

Figure 11.5 European main hydrogen network operated by Air Liquide (map courtesy of Air Liquide).

11.3 Difficulties involved with the transportation of hydrogen

The penetration of hydrogen in the world energy economy is conditioned by several constraints related to the intrinsic peculiarities of this gas more than to technological barriers. The next sections provide an overview of the challenges posed by hydrogen when it comes to its transportation: the energy, economic, and safety issues.

11.3.1 Energy considerations

The capacity of a given pipeline to transport energy is lower when it transports hydrogen than when it transports natural gas. In a pipe of given diameter and internal pressure drop, hydrogen, which has a lower viscosity, can reach a higher speed, and so the flow rate can be higher compared to natural gas. Despite this, the energy content per unit of volume is lower for hydrogen, and the energy consumed for the compression stage is four times higher. This means that the same pipeline can deliver a greater amount of energy in the form of natural gas than in the form of hydrogen (Nexant, 2008).

However, experience has shown that hydrogen pipelines are not very long, compared to the very long energy corridors that deliver natural gas from Russia to Europe, and they seldom need intermediate compression stations. The usual set-up is a high pressure at the production site, both due to high-pressure electrolyzers and to high-pressure steam reforming, and then a recompression of hydrogen at the destination.

As indicative values, the energy requirement for the compression of hydrogen from 20 to 70 bar is about 2.4 MJ, while liquefaction requires 55–65 MJ/kg (Nexant, 2008).

11.3.2 Suitable materials

The choice of the materials employed to transport hydrogen in pipelines depends on its quality: if it is pure and without humidity, there are no corrosion issues and no special treatment is needed for internal pipe surfaces. However, if the hydrogen contains humidity or other components than can be corrosive, the inner surface must be protected with an over layer and/or with suitable paint. Special attention must be paid when hydrogen contains ammonia: in this case, all materials containing copper, copper alloys, zinc, and tin must be avoided since they can be corroded.

It is common practice that distribution lines, joints, and special parts should be manufactured using steel containing a low percentage of carbon to avoid embrittlement (e.g., API 5L X52 and lower strength grades or ASTM A 106 Grade B) (EIGA, 2004) or tempered steels with $R_m < 950$ MPa. No superficial defects must be detected.

In the assembling phase of the network, whenever possible, no part should be welded but only extruded.

In the natural gas distribution network, pressure is low, around 4 bar, and therefore cheaper plastic pipe is usually used. PVC (polyvinyl chloride) and HDPE (high-density polyethylene) are too porous and not usable for transporting hydrogen.

For safety reasons, odorization has been adopted as a good technique to detect natural gas leakages. However, with hydrogen, this solution is hard to apply as the proper odorant should be so penetrating that it is able to alert early when hydrogen reaches 20% of the lower flammability limit (i.e., 0.8% of hydrogen in air).

It should also be noted that many hydrogen applications require hydrogen at an extreme level of purity: the removal of the odor would be an additional complication in the system, apart from being an additional cost. The odor would reduce the reactivity of catalysts and prevent the chemical reaction from completing.

Other detection methods are also proposed and preferred: catalytic combustion sensors, thermal conductivity sensors, electrochemical sensors, semiconducting oxide sensors, mass spectrometers, gas chromatographs, ultrasonic leak detectors, and many others. Today, more sophisticated systems are used such as coating sprays that change in color when in contact with minimal amount of hydrogen (Hoagaland, 2014).

11.3.3 Safety issues

Hydrogen transportation and distribution pose specific issues in terms of safety. The issues are strongly related to the chemical and physical properties of hydrogen: its ability to embrittle materials, its ease in escaping from containment, its wide flammability range, and the limited amount of energy needed to ignite it, all represent barriers to safe use. At the same time, its extremely low density is a guarantee that the gas will likely ascend instead of forming dense dangerous clouds as other hazardous gases do.

Hence, if hydrogen is expected to penetrate economies, in rather industrialized countries but also in emerging ones, the technological development has to be sustained by a parallel development of codes and standards to assure safe use and to allow manufacturers to play in a regulated field that couples the protection of the customer and the competition on the market.

Today, even after the start of the penetration of hydrogen in worldwide energy mixes, some regulations are still missing or not fully applied at national level.

At an international level, ISO Technical Committee 197 is in charge of developing standards related to hydrogen applications. The standards are available for some specific applications (such as those related to onboard systems, fuel tanks and refueling systems for vehicles) and for the production part of the chain (electrolysis and steam methane reformers are already covered), but standards for refueling stations as part of the distribution chain and pipelines are still under development at the time of writing.

As a consequence, local regulations are being implemented, one example being the Italian exercise, which started from the natural gas regulations and was subsequently transformed into a hydrogen specific by adopting ad hoc countermeasures to take into account the peculiarities of this gas. Italy has a long and successful history of natural gas transportation; its network is long and fully interconnected linking more than 100 provinces, in alpine environments as much as other naturally challenging situations (e.g., river crossings, seismic areas, land slides). This experience has encouraged exploiting the existing rules for hydrogen applications.

The adaptation of natural gas regulations on distribution networks was accompanied by field tests developed to understand the safety distances to be imposed between hydrogen installations and the nearest buildings. The need for this accurate evaluation is particularly relevant in the case of hydrogen because the vector is new in application and tough restrictions (such as important barriers or excessive safety distances) may reduce the possibility for the wider public to accept it (Royle and Willoughby, 2011).

To set the most appropriate limits, several groups have performed experimental exercises (Jo and Ahn, 2006; Royle and Willoughby, 2011) or risk analyses (Grasso et al., 2009; Gerboni and Salvador, 2009) using theoretical models.

Both kinds of analyses generally start from the definition of an event tree that allows selecting and concentrating on more representative and risky combinations. The possible initiating events are those that might affect natural gas pipelines (e.g., external events, impacts, mechanical or service failures, etc.). The good buoyancy of hydrogen has been taken into account by analysts in order to be accurate in forecasting the behavior of a gas leakage in the atmosphere and wind direction and speed are here particularly influent. The very wide flammability range does not work in favor of safety but the buoyancy decreases the possibility of cloud formations at low heights (where human receptors are closer). In the event that large clouds are produced, these can be lately ignited and cause explosions. Another possibility is the formation of jet fires due to leakages in pipelines under pressure with an ignition that is not too much delayed. The safety distance for receptors, both humans and buildings, depends on many factors. A good reference for evaluating this distance was presented by Jo and Ahn (2006), and it is proportional to the square root of the steady state pressure in the pipeline and to the diameter of the pipeline.

As in the usual practice of risk prevention, artificial barriers may be inserted to decrease the safety distances from the possible release point to the receptor. In the case of hydrogen, barriers have been studied and proposed for application and vary in size, height, and inclination. For example, NFPA 55 (2005) proposes a 60° inclined barrier to protect from jet fires originating from storages. In Royle and Willoughby (2011) these barriers have been tested against vertical ones and showed mixed response, being more suitable to protect the leakage area against overpressure and heat flux but less efficient to protect the area behind the barrier from the heat flux. A vertical barrier proved more efficient at protecting receptors behind it.

In the case of tube trailers, the quantity of gas stored is limited compared to a pipeline system, but it should be noted that the probability for a tube trailer to be impacted suffers the same frequency as that of road vehicles accidents. Additionally, the probability of the gas cloud or pressurized leakage to be ignited is higher due to the presence of the vehicle parts and to possible sources in the surrounding environment. The presence of houses can also represent involuntary barriers, which may cause overpressure to be increased. When tube trailers are used as storage systems for refueling stations, appropriate safety distances have already been estimated in several regulations. For example, in Italy, where protection boxes around tube trailers are regulated with a maximum volume of stored gas per box equal to 2000 Nm^3. The boxes are to be built in concrete walls, the minimum thickness of which is 15 cm and the height is set to at at least one meter more than the height of the storage system. If more gas is to be stored, more boxes are to be built. The minimum distance between the box and the nearest external building should be 20 m.

Liquid hydrogen spills show an initial behavior that can lead to the formation of a pool. However, the liquid phase is immediately accompanied by the presence of hydrogen vapor, which is quite cold and dense. The temperature of the fluid is so low that it may lead to the solidification of the other air components such as nitrogen and oxygen. The particle of solid oxygen can mix with liquid hydrogen and prepare a potentially explosive mixture that could self-ignite. A jet fire can also originate which again is dependent on the quantity of hydrogen available. Further experimental works are expected on this topic as liquid hydrogen, as seen, is one of the most deployed forms of hydrogen in state-of-the-art distribution systems.

11.3.4 Hydrogen transportation costs

An assessment of the hydrogen transportation and delivery costs is fully dependent on the supply chain chosen. Several studies have focused on the optimization of the pathway, most of them with a specific spatial definition (Yang and Ogden, 2013; Johnson and Ogden, 2012). The need for this definition is due to the fact that the distance to be covered from the production site to the final consumer is crucial for the choice of the transportation method.

It should be remembered that the bulk of hydrogen is used in the petrochemical sector and the distances to be covered are short and within industrial premises. When it comes to the use of merchant hydrogen, delivery costs will become more significant.

If the final use is in fuel-cell-equipped vehicles, fuel stations are the target destination. These may be situated along highways, or in easy to reach strategic crossroads where trucks and trailers can stop and supply hydrogen to local storage. The proper place for a fueling station can also be chosen according to the availability of a nearby natural gas pipeline that can supply a steam reforming unit at the premises of the refueling station. However, if hydrogen is to be conceived as a zero emission fuel for cars, then centralized or distributed electrolyzers are adopted provided electricity is from a renewable source. Hydrogen can be directly delivered to vehicles from the distributed equipment, but it can also be delivered via a pipeline from the centralized electrolyzer. The trade-off is generally solved by the distance to be covered and by the quantity to be delivered.

The DOE has evaluated that stations that supply more than 100 kg/day of hydrogen are expected to be supplied via pipeline or via liquid transport. In this case, the estimated cost of hydrogen at the pump is around 5.70–8.00 $/kg (US DOE, 2013). If the station is small, then a tube trailer is enough to satisfy the demand, but the cost of hydrogen is estimated to be three times higher.

As usual, when a planning exercise is to be performed, assuming the optimization target is the minimum cost of the delivered fuel, several scenarios may be chosen. In the 1990s, a famous study (Amos, 1998) adopted the hypothesis of a (quite high) fixed production rate (which may be equivalent to setting the fixed demand of a refueling station) to be delivered over different distances. Under this hypothesis, although pressures and capital costs are now outdated, the conclusion was that the cost of delivery was proportional to the distance to be covered since the driving time required more trailers per day.

The most recent target set by the DOE is to achieve capacities for tube trailers of 700 kg in 2015 and above 900 in 2020. Accordingly, internal pressure would increase to 400 and 520 bar, respectively. The capital cost of such a vehicle would then rise to $510,000 and $540,000, respectively.

The adoption of liquid hydrogen tank trailers can reduce costs due to the increased quantity of hydrogen delivered, although capital cost is greater than in the case of tube trailers.

The choice of a hydrogen pipeline is affected by several additional constraints, which are similar to those for natural gas pipelines systems: investment costs, cost of installation, maintenance costs, and expenditures related to right of way.

However, the costs related to the investment are higher for hydrogen than for natural gas due to the material choices and costs related to the diameter. The reported average cost for an 8 in. pipeline in 2011 was about 470.000 $/km at a pressure of 150 bar (US DOE, 2013). However, costs due to the layout of the pipeline are not negligible and can rise up to 50% of the capital costs, according to the difficulties related to the terrain or urban environment. Modeling exercises performed to evaluate the best layout pattern in California (Johnson and Ogden, 2012) have considered this important addition. Alternative materials could be used that would lower investment costs: fiber-reinforced pipelines would have capital cost of about 230,000 $/km in 2015 and 150,000 $/km in 2020. Diameters for these pipelines are less than 6 in.

It should also not be neglected that the final destination for the hydrogen could also be represented by stationary applications such as domestic boilers. In this case the last mile of pipeline could be smaller in diameter and in pressure (exactly as in a natural gas distribution network), with diameters no larger than 1 in. and pressures below 30 bars.

Several reports have discussed the use of existing natural gas pipelines to transport hydrogen or hydrogen-natural gas blends. However, it should be noted that embrittlement may be a problem with high-pressure transmission of hydrogen and although some components of a natural gas system could be used for hydrogen transmission, compressors and meters would not be applicable.

11.4 Future trends

In addition to the present transportation systems, much research has been carried out in an effort to improve the performance of the transportation segment of the chain. One of the virtues of hydrogen is its versatility, which allows it to be either considered as a fuel on its own or part of a different chemical compound. The newer transportation schemes tend to make the most of this virtue and to exploit it to reach high economic and energy performance, good stability, and safety. This final section provides some information about futuristic (yet achievable) targets in the hydrogen transportation science and technology.

11.4.1 Alternatives

Alternative methods for transporting hydrogen are not related to the transportation method but rather to the form in which hydrogen is stored.

As seen in other chapters, hydrogen can combine with several materials to form solid compounds. The main purpose of exploring these storage systems is to find a storage solution with an efficient (high) volumetric and, hopefully, gravimetric density. This target is needed mainly for mobile applications, such as automotive, and less for stationary equipment. Although at present the habit of transporting hydrogen in the form of hydrides over long distances is not diffused, the properties that make it a suitable solution for automotive applications are valid also for making it a good option as a transportation method: hydrides would be very easy to transport either by truck or rail.

Hydrides is just the general name of a large family of compounds that are variably suitable for storing hydrogen. Other compounds with different characteristics could be adopted in the future as a batch transportation solution for hydrogen: carbon nanotubes, graphane, and glass balloons are among the solutions investigated, although each of them still has challenges to be solved mainly due to the replicability on a large scale of the results obtained in labs.

While metal hydrides and carbon structures are currently investigated for further improving their performance as onboard storage systems for vehicles, other solutions that make the most of hydrogen capability in chemical binding are iteratively assessed by the scientific community.

Examples are the use of "intermediate" carriers, such as ammonia, methanol, methylcyclohexane, or, as proposed by Teichmann et al. (2012), *N*-ethylcarbazole. According to the authors in Teichmann et al. (2012), the use of these heterocyclic aromatic hydrocarbons has good potential due to their substantial similarities with diesel: the major issues of hydrogen transportation could be then avoided making use of conventional methods, either on roads or via ship. They have also performed a preliminary economic evaluation of these solutions and the use of *N*-ethylcarbazole looks promising when compared to all the pure hydrogen solutions, due also to the relatively high hydrogen concentration in weight into the compound (around 5.3 wt%).

The disadvantage of this kind of solution, as in the case of hydrides, is the need for heat in order to free hydrogen at the point of use. This is an energy expenditure and a complication in the overall system, although in some cases it would be possible to use waste heat produced during the exploitation of hydrogen itself (as in the case of gas turbines, combustion engines, and SOFC).

On the other hand, the carrier production and "enrichment" in hydrogen should not represent an issue as chemical plants and refineries are familiar with those kinds of processes. Once hydrogen is freed, the carrier has to be recovered and returned to the production facility: it is not consumed and can be recycled several times, representing a minor cost for the system.

While these alternatives rely on the chemical characteristics of hydrogen, there is an additional set of solutions for a safe and efficient transportation of hydrogen that exploits its physical properties.

One option is the use of slush hydrogen (a mixture of liquid and solid hydrogen at the triple point of hydrogen: 13.8 K and 7.2 kPa), a technique already explored for space use in the past, and a newer chance is the use of gelled hydrogen (in this case hydrogen is mixed with gellants that are usually other hydrocarbons such as methane): apart from a good energy density of the mixture (improved as the hydrocarbon content increases), gelled hydrogen has the advantage of a greater viscosity which reduces slosh inside the vessels. Gelled hydrogen also reduces the possibility of boil-off, thus increasing the time of containment of the fuel into the vessel. This would be a good property for very long transportation needs, where liquid hydrogen is now normally used but, as has been seen, with a payload that "evaporates" and has to be used as a marine propellant (Palaszewski, 1997).

A third, very promising technique is represented by the cryo-compressing of hydrogen. The DOE is pushing this solution as one of the closest to the cost target achievement. The idea is to cool hydrogen to about 70 K (above liquefaction temperature) and to keep it under pressure (above 35 MPa) (US DOE, 2013). The gas is then delivered at the refueling station via superinsulated tanks and further compressed to fill the vehicle tanks. As hydrogen is consumed, no boil-off is expected as hydrogen is still gaseous and as soon as hydrogen is delivered the internal pressure is decreased as so is the temperature, which helps maintain the quality of the storage.

11.4.2 SuperGrid

The possibility of transporting electricity and hydrogen with the same infrastructure has been an intermittently pursued goal that began in the late 1980s when high-temperature superconductive materials were discovered, but has recently seen a new push. These materials allow for the possibility of coupling their critical temperature with the liquefaction temperature of hydrogen. The research promoted by Electric Power Research Institute (EPRI) and focused on the system architecture predicts the association of hydrogen and electricity production from high-temperature nuclear reactors (EPRI, 2006).

The support to these kinds of studies sometimes depend on their quick need and applicability: the new availability of considerable amounts of shale gas in the United States may have temporarily limited the interest in this frontier.

11.5 Sources of further information and advice

Gupta Ram B., 2008. Hydrogen Fuel: Production, Transport, and Storage. CRC Press.
ExternE EU project www.externe.info.
NEEDS EU project www.needs-project.org.

References

Abe, A., Nakamura, M., Sato, I., Uetani, H., Fujitani, T., 1998. Studies of the large-scale sea transport of liquid hydrogen. Int. J. Hydrog. Energy 23, 115–121.

Amos, W.A., 1998. Cost of Storing and Transporting Hydrogen. National Renewable Energy Laboratory. available at: http://www.nrel.gov/docs/fy99osti/25106.pdf.

Andrews, J., Shabani, B., 2012. Re-envisioning the role of hydrogen in a sustainable energy economy. Int. J. Hydrog. Energy 37, 1184–3020.

Cuni, A., Weber, M., Guerrini, O., Steinberger-Wilckens, R., Trümper, S.C., 2008. Deliverable 2.3—linking distributed European hydrogen production sources. Report for project Roads2HyCon, available at: http://www.ika.rwth-aachen.de/r2h/images/2/2a/Roads2HyCom_R2H2013PU_-_Linking_European_H2_Sources_%28Part_I%29.pdf.

EIGA, 2004. Hydrogen transportation pipelines. IGC Doc 121/04/E.

EIGA, 2011. Hydrogen cylinders and transport vessels. IGC Doc 100/11/E.

EPRI, 2006. The hydrogen—electric energy SuperGrid. White paper available at: www.w2agz.com.

ESRI Light grey canvas (2010), ESRI Inc., Redlands (CA), www.arcgis.com.

Gerboni, R., Salvador, E., 2009. Hydrogen transportation systems: elements of risk analysis. Energy 34, 2223–2229.

Gillette, J.L., Kolpa, R.L., 2007. Overview of Interstate Hydrogen Pipeline System. Argonne National Laboratory, IL, USA and available at US DoE, Tennessee, USA.

Grasso, N., Pilo, F., Ciannelli, N., Carcassi, M.N., Mattei, N., Ceccherini, F., 2009. Fire prevention technical rule for gaseous hydrogen transport in pipelines. Int. J. Hydrog. Energy 34, 4675–4683.

Gretz, J., Drolet, B., Kluyeskens, D., Sandmann, F., Ullman, O., 1994. Status of the hydro-hydrogen pilot project (EQHHPP). Int. J. Hydrog. Energy 19, 169–174.
Hoagaland, B., 2014. Hydrogen leak detection—low cost distributed gas sensors. Available at: http://energy.gov/sites/prod/files/2014/03/f10/webinarslides_element_one_040312.pdf.
Hugo Seymour, E., Murray, L., Fernandes, R., 2008. Key challenges to the introduction of hydrogen—European stakeholder views. Int. J. Hydrog. Energy 33, 3015–3020.
Jo, Y.-D., Ahn, B.J., 2006. Analysis of hazard area associated with hydrogen gas transmission pipelines. Int. J. Hydrog. Energy 31, 2122–2130.
Johnson, N., Ogden, J., 2012. A spatially-explicit optimization model for long-term hydrogen pipeline planning. Int. J. Hydrog. Energy 37, 5421–5433.
Nexant, 2008. Hydrogen delivery infrastructure options analysis. Available at: http://energy.gov/sites/prod/files/2014/03/f11/delivery_infrastructure_analysis.pdf.
NFPA 55, 2005. Standard for the storage, use, and handling of compressed gases and cryogenic fluids in portable and stationary containers, cylinders, and tanks.
Oyama, S., Kamiya, S., 2012. Commercialisation barriers and issues for large scale hydrogen infrastructures. In: Presentation at HySafe Research Priorities Workshop (16 Oct. 2012).
Palaszewski, B., 1997. Gelled Liquid Hydrogen: A White Paper. NASA Lewis Research Center, Cleveland, OH, USA. Available at, http://www.grc.nasa.gov/WWW/Fuels-And-Space-Propellants/GELLED.htm.
Pregger, T., Lavagno, E., Labriet, M., Seljom, P., Biberacher, M., Blesl, M., Trieb, F., O'Sullivan, M., Gerboni, R., Schranz, L., Cabal, H., Lechon, Y., Zocher, D., 2011. Resources, capacities and corridors for energy imports to Europe. Int. J. Energy Sect. Manage. 5 (1), 125–156.
Royle, M., Willoughby, D., 2011. The safety of the future hydrogen economy. Proc. Saf. Environ. 89, 452–462.
Sherif, S.A., Zeytinoglu, N., Veziroglu, T.N., 1997. Liquid hydrogen: potential, problems and a proposed research program. Int. J. Hydrog. Energy 22, 683–688.
Teichmann, D., Arlt, W., Wasserscheid, P., 2012. Liquid organic hydrogen carriers as an efficient vector for the transport and storage of renewable energy. Int. J. Hydrog. Energy 37, 18118–18132.
US DOE, 2013. 3.2 Hydrogen Delivery, 2012. In Fuel Cell Technologies Office Multi-Year Research, Development, and Demonstration (MYRD&D) Plan.
Yang, C., Ogden, J.M., 2013. Renewable and low carbon hydrogen for California—modeling the long-term evolution of fuel infrastructure using a quasi-spatial TIMES model. Int. J. Hydrog. Energy 38, 4250–4265.

Hydrogen transportation by pipelines

12

I.A. Gondal
National University of Sciences & Technology, Rawalpindi, Pakistan

Nomenclature

u	speed of gas flow
ϑ	viscosity of gas
f	friction factor
g	gravity
HHV	higher heating value
LHV	lower heating value
MPa	mega pascal
NG	natural gas
PVC	polyvinyl chloride
Q	gas flow rate
Re	Reynold's number
W_S	wobbe index
Z	compressibility factor
μ	Joule–Thomson coefficient (°C/bar)
ρ	density

12.1 Introduction

The realization of a hydrogen economy is destined to occur with a parallel hydrogen supply chain infrastructure that can meet the demands of users cost-effectively and easily. In the existing energy scenario, electricity is the key energy vector. Hydrogen and electricity are complimentary and compatible; hence in any energy system hydrogen cannot be treated in an isolated manner. The production of hydrogen from renewable energy sources and integrating it with the existing infrastructure is likely to be the pathway to the evolving hydrogen economy (Turner, 2004).

As the fossil fuels exhaust, the options left for filling the gap is either nuclear or renewable energy. The competition between renewable and nuclear energy has been studied in detail by Lee (2012). This study elucidates the competition among renewable and nuclear energy sources for the production of hydrogen. Duu-Hwa Lee has concluded based on certain assumptions that wind energy is the most promising means for producing hydrogen. It is well established in other studies as well that solar thermal energy is also a promising candidate for hydrogen production. The hydrogen economy would invariably result in or precede the development of fuel-cell-based devices,

Compendium of Hydrogen Energy. http://dx.doi.org/10.1016/B978-1-78242-362-1.00012-2

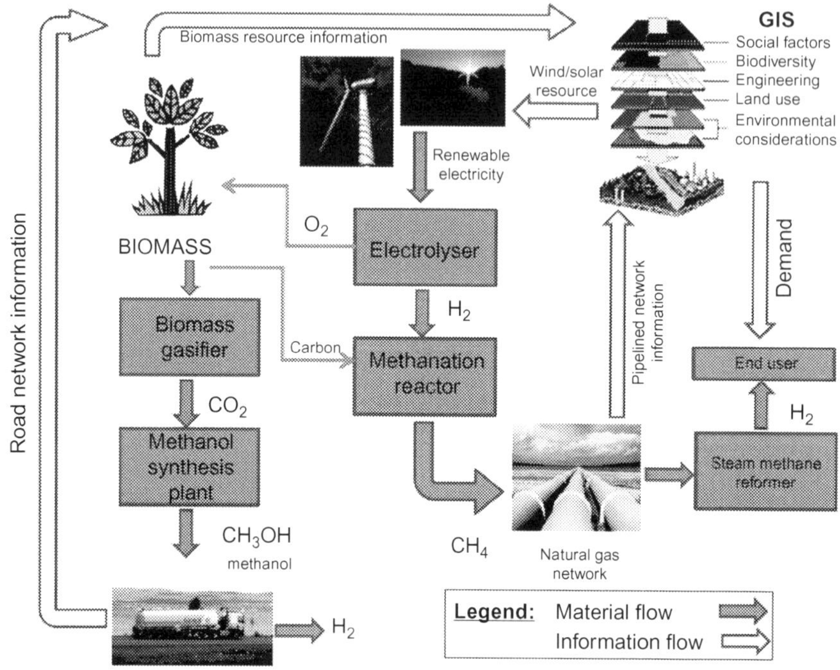

Figure 12.1 GIS-assisted biomass-based renewable hydrogen system (Gondal and Sahir, 2013).

applications, and vehicles for all the energy-consuming sectors. Hence, the delivery of hydrogen in large amounts would be essential to sustaining the economy and society on similar lines as it is today for natural gas or other hydrocarbon fuels. In the absence of hydrocarbons, the hydrogen economy's mainstay would be renewables that can replace the fossil fuels for the production of hydrogen. Thus solar, wind, biomass, geothermal, wave energy resources, etc., would be used for the generation of hydrogen either directly or indirectly. It is also an established fact that wind/solar resources are intermittent in nature and distribution is heterogeneous (DOE, 2011, 2012), hence wind farms and solar power plants are optimally located in distant regions, which may include deserts, coastal areas, offshore, and high-altitude mountainous regions. The renewables-based hydrogen supply chain (HSC) shown in Figure 12.1 depicts the integration of pipeline distribution with the energy system.

12.2 Current hydrogen pipelines

So far the discussion has focused mainly on the necessity of transporting hydrogen as an integral and essential component in the renewable hydrogen economy. Industrial use of hydrogen has many examples where a company produces hydrogen for its own use and then transports it to other locations where hydrogen is consumed. In such cases,

Table 12.1 **World ranking of gas pipelines (The CIA World Factbook 2012, 2011)**

Rank	Country	Gas pipeline length (in km)
1.	The United States	548,665
2.	Russia	158,767
3.	Canada	74,980
4.	China	28,132
5.	Ukraine	33,327
6.	Argentina	28,657
7.	Australia	26,719
8.	Germany	25,094
9.	Mexico	22,705
10.	Iran	19,161

pipelines are used, even though the length is generally very short. However, long pipelines covering large distances exist in quite a few locations throughout the world.

One pipeline measuring 550 km in length and inner diameter of 100 mm, located in France, transports 200×10^6 m^3 of hydrogen annually. Similarly, Belgium has laid a pipeline of 80 km and inner diameter 150 mm that operates at 10 MPa. A 16-km pipeline is laid in England that operates at 5 MPa. Germany has one of the oldest and largest hydrogen pipelines running up to 220 km. It has an inner diameter of 100–300 mm, operates at 2 MPa, and has connected Dusseldorf and Recklinghausen for 60 years and annually transports 1×10^6 m^3 of hydrogen. In Texas, in the United States, a total of 100 km of hydrogen pipeline is laid in the Gulf coast industrial zone. A few pipelines with several kilometers of length are also laid in Iowa and Louisiana in the United States and Alberta, Canada (Tokio Ohta, 2015).

These examples show the extent of hydrogen pipelines in use, but these are minimal compared to the current natural gas infrastructure. Table 12.1 gives the world rankings of gas pipelines by length.

As can be seen, the natural gas pipeline network is very extensively designed all over the world, and the amount of investment required for laying such an infrastructure is huge. Figure 12.2 gives a comparison of the cost of laying out the pipeline infrastructure, ranging from a minimum of US $200,000 per kilometer. The cost of laying a hydrogen pipeline is 10% higher. It's easy to imagine the present worth of the existing pipeline infrastructure, and the likely cost if such a dedicated pipeline network is to be designed for hydrogen. It is thus significant that ways and means be devised for making use of existing network for the transportation of hydrogen.

12.3 Principles of transportation of hydrogen

In this section, the principles for transportation of hydrogen gas will be discussed after a brief overview of the layout of the pipeline infrastructure. Hydrogen gas is considered as a fluid similar to natural gas but certain modifications, precautions, and

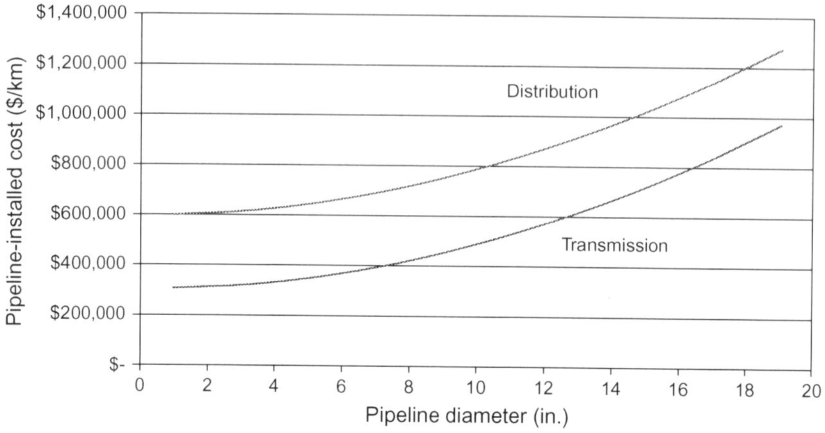

Figure 12.2 Pipeline model installed cost ($/mile) dependence on pipeline diameter (Yang and Ogden, 2007).

infrastructural and material requirements have to be met before its actual transmission through the pipeline network. Hydrogen has its own peculiar set of physical and chemical properties that dictate the employment of various mathematical relations.

12.3.1 Pipeline components

A natural gas transportation system is made up of compressor stations, pipelines, city gate stations, and storage facilities. The compressor station serves as the heart of the transmission system as it supplies the energy required to ensure that the gas continues to flow at a prescribed flow rate and pressure. The onshore and/or offshore pipelines have three types: trunk or gathering, transmission, and transportation and distribution pipelines. Figure 12.3 shows the different types of pipelines according to their function.

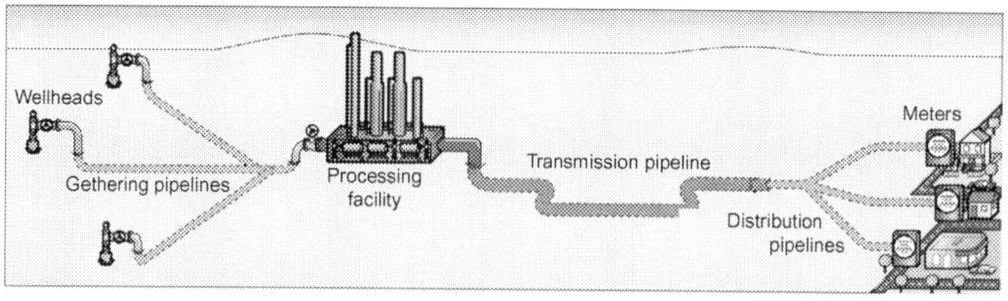

Figure 12.3 How natural gas gets from the well to the consumer (Pipeline and Hazardous Materials Safety Administration, 2013).

Natural gas pipelines primarily serve as a means of moving gas from the field to consumers. Inter- and intrastate pipelines are used for the transportation of natural gas produced from gas fields, either onshore or offshore facilities through gathering systems to commercial, residential, industrial, and utility companies. The pipelines are usually constructed of carbon steel and varying in size from 2 inches (51 mm) to 56 inches (1400 mm) in diameter, depending on the type of pipeline. Pipeline infrastructure originating from the resource field comes from two main networks:

- High-pressure grid-transport network
- Low-pressure grid-distribution network

High-pressure lines are characterized by larger diameter and strong piping with compression stations after regular distance intervals. Low-pressure or distribution networks are identified by pressure reduction stations and a relatively small diameter piping network. The function of the pipeline network is twofold. It serves to supply energy to the consumer in the required quantity, and at the same time, it also provides storage for the fuel gas in a considerable quantity depending upon the dimensional capacity of pipeline and the demand–supply gap. When supply surpasses the consumer demand, the excess gas is "packed" in the pipeline, which is hence called the "linepack."

Transport pipelines are mainly made out of steel because of the high pressures used. Distribution pipelines are made out of cast iron, fibrous cement, PVC-enriched polyethylene (PE), or steel. Currently, mostly PE pipelines are used.

12.3.2 Compression stations

The transport of natural gas through pipelines is based on the principle of pressure-difference induced flow. Due to the viscosity of the gas and the friction with the inner pipeline wall, this flow encounters a resistance, causing a pressure drop. The higher the transport velocity, the higher the pressure drop. Since every pipeline has to maintain a minimum and maximum pressure, recompression is mostly required when long distances have to be covered. This interaction between pressure drop and recompression is shown in Figure 12.4 (Smit and Weeda, 2006). The distance between compression stations is generally 80–100 km. In recent years, piston compressors have been replaced by centrifugal ones, powered by gas engines. Centrifugal compressors are

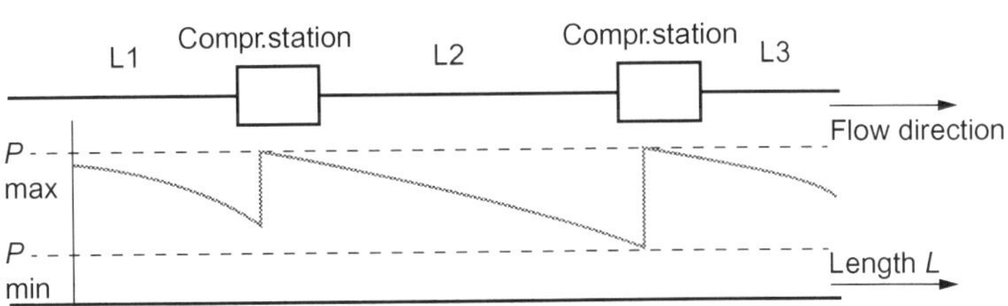

Figure 12.4 Pressure drop in pipelines and the use of compression stations to maintain the pressure within certain boundaries (P_{max} and P_{min}) (Lun and Xie, 2008).

more suitable for high gas volumes although they are less flexible than piston compressors.

12.3.3 Pressure-reduction stations

When two pipelines operate at different pressures, pressure-reduction stations are used to connect them. In these facilities, the pressure is reduced by expanding the gas through a throttle valve. This expansion will change the temperature of the (nonideal) gas. This is called the Joule–Thompson effect, whereby the final temperature of the gas can be calculated as:

$$T_2 = T_1 + \mu(P_2 - P_1)$$

where T_2 is the final gas temperature (°C), T_1 is the initial gas temperature (°C), μ is the Joule–Thomson coefficient (°C/bar), P_2 is the final gas pressure (bar), and P_1 is the initial gas pressure (bar).

All real gases have an inversion temperature at which the Joule–Thompson coefficient changes sign. At room temperature, for most gases, this coefficient is positive, except for hydrogen, neon, and helium. The Joule–Thompson coefficient amounts to 0.5 °C/bar for natural gas and to −0.035 °C/bar for hydrogen. This means that the expansion of natural gas from 80 to 15 bar results in a temperature drop of 32.5 °C. Therefore, to avoid the formation of ice, the gas is often preheated. The effect of various factors on the transportation of hydrogen in pipelines is discussed in the following sections.

12.4 Gas transportation principles

When gas flows through the network, it suffers energy and pressure losses due to the friction between the gas and the inner walls of gas ducts but also due to the heat transfer between gas and the environment. If demanded gas has to be supplied to delivery points with a specified pressure, the undesired pressure drops along the network must be periodically restored. This task is performed by compression stations installed on the network, but these usually consume over 3% or 5% of total gas transported. The pressure of natural gas pipelines was only 2.5 MPa before it rose to 6 MPa in the 1960s. At present, the pressure of pipeline is about 10 MPa (Mokhatab and Poe, 2012).

Understanding the transport of gas through pipelines requires deep knowledge of the loss of head brought about by fluid friction. Other factors include the amount of gas being transported, the length, diameter, and slope of the pipeline. If the transportation is over large distances, it is important that the solution involves the concept of compressible flow. Since the pipelines are laid underground, the assumption of isothermal flow can safely be held. The flow rate in pipelines depends on the physical properties of gas, the length and diameter of the pipelines, the initial gas temperature and pressure as well as the drop in pressure due to friction.

12.4.1 Gas flow analysis

The flow process of gas involves the change of the total energy content of the gas at various points along the pipeline. The energy in gas is comprised of the energy due to position, pressure, and velocity. The application of Bernoulli's equation to the pipeline connects these individual components of the flowing energy resulting in an energy conservation equation. Bernoulli's equation is applied at two points as shown in Figure 12.5.

$$Z_A + \frac{P_A}{\gamma} + \frac{V^2 A}{2g} + H_p = Z_B + \frac{P_B}{\gamma} + \frac{V^2 B}{2g} + H_f \tag{12.1}$$

where H_p is the equivalent head added to the fluid by compressor at point A and H_f is the loss in pressure due to friction between points A and B.

Equation (12.1) is the basic energy equation, which along with the gas laws, is applied to analyze the performance of a pipeline transporting gas. This equation relates the gas properties such as compressibility and gravity with the physical features of pipeline such as length and diameter, and the flow rate and pressures along the pipeline. Hence, if the inlet and outlet pressure of a pipeline segment are known, these can be used to determine the flow rate through the pipeline. While carrying out the analysis certain assumptions are also introduced; these include uniform gas temperature and no heat transfer between gas and the surrounding soil in the buried pipeline. However, in practice, transient situations are encountered when transportation involves long distances and the *isothermal* flow assumption holds good as the gas temperature approaches constant values.

A number of mathematical relationships correlate the properties of gas with its flow rate, pipe length and diameter, and initial and final pressures. A few of these include:

- General flow equation
- AGA equation
- Colebrook–White equation and its modified version
- Weymouth equation

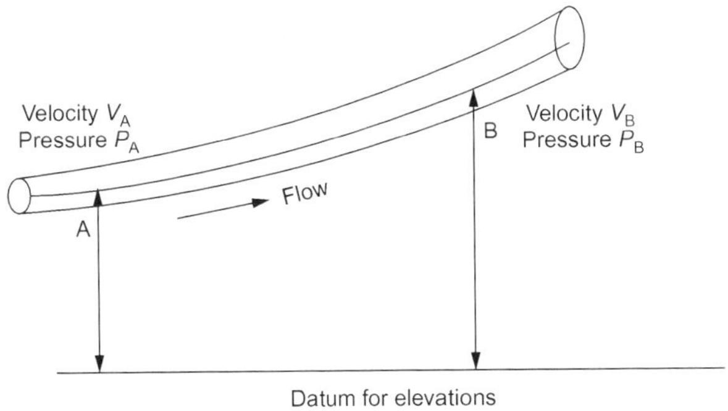

Figure 12.5 Energy of flowing fluid.

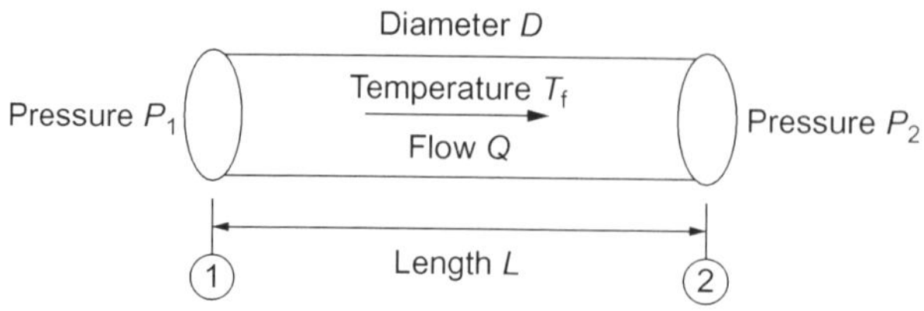

Figure 12.6 Steady flow in gas pipeline.

For a steady and isothermal flow in single phase, as depicted in Figure 12.6, the general flow equation is the primary relationship that relates the pressure drop in the pipeline with the flow rate (Utgikar and Thiesen, 2005):

$$Q = 1.1494 \times 10^{-2}\left(\frac{T_b}{P_b}\right)\left[\frac{P_1^2 - P_2^2}{fGT_aLZ_a}\right]^{0.5} D^{2.5} \tag{12.2}$$

For isothermal flow, gas temperature T_a is assumed to be constant between section 1 and section 2 (Figure 12.6 refers), and Q = the gas flow rate at standard conditions and depends on gas properties, i.e., gravity G and compressibility factor Z. If gravity G is high, the flow rate is reduced. Accordingly, with an increase in the compressibility factor, the flow rate reduces. Similarly, as gas flow temperature T_a increases, the flow rate decreases. Thus, in order to have an increased flow rate, the temperature of the gas must be kept low. The effect of pipeline length and diameter is also evident. For a given pressure P_1 and P_2, the flow rate decreases with increase in pipe length. Accordingly, the flow rate is larger if the diameter is increased. The difference in pressure indicated by $(P_1{}^2 - P_2{}^2)$ defines the rate of flow from upstream to downstream. Obviously when P_1 is equal to P_2, there is no flow. The drop in pressure $(P_1 - P_2)$ occurs because of the friction between the gas and the pipe walls. The extent of pressure drop is dependent on friction factor "f," which represents the internal condition of the pipe, as well as the type of flow, i.e., laminar or turbulent. Equation (12.2) satisfies when the gas pipeline is horizontal, which may not always be the case. If H_C gives the static head of fluid incorporated in Equation (12.2), then for normal slopes, Equation (12.2) can be rewritten as (Schroeder, 2001):

$$Q = 1.1494 \times 10^{-2}\left(\frac{T_b}{P_b}\right)\left[\frac{P_1^2 - P_2^2 - H_C}{fGT_aLZ_a}\right]^{0.5} \tag{12.3}$$

where

$$H_C = \frac{0.0684g(H_2 - H_1)P_a^2}{Z_aT_a} \tag{12.4}$$

H_1 = inlet elevation; H_2 = outlet elevation; (m) and g = gravitational constant m/s^2; and Z_a = compressibility factor, determined from average pressure P_a and average temperature T_a, and P_a is given by (Tabkhi et al., 2008)

$$P_a = \frac{2}{3}\left[(P_1 + P_2) - \left(\frac{P_1P_2}{P_1 + P_2}\right)\right] \tag{12.5}$$

The average temperature is determined from Equation (12.6) (Hydrogen Infrastructure Delivery Reliability R&D Needs, 2015; Tabkhi et al., 2008).

$$T_a = \left[\frac{T_1 - T_2}{\ln\left(\frac{T_1 - T_s}{T_2 - T_s}\right)} \right] + T_s \qquad (12.6)$$

where T_s = soil temperature, T_1=upstream temperature, and T_2=downstream temperature.

12.4.2 Gas velocity in pipeline

The velocity of gas in a pipeline is different than the velocity of an incompressible fluid, which is given by the velocity of molecules moving from one point to another. Gas velocity is steady flow but nonuniform, because it is a function of pressure that varies along the length of the pipeline. Thus, the highest velocity is experienced downstream where the pressure is lowest, while the least velocity occurs upstream where the pressure is highest. Considering a segment of pipeline from point 1 upstream to another point 2, with no gas injection or delivery between these two points, the mass flow rate in case of steady flow is given by:

$$M = Q_1\rho_1 = Q_2\rho_2 = Q_b\rho_b \qquad (12.7)$$

where Q_b = gas flow rate at standard condition and ρ_b = gas density.

Equation (12.7) can be written as

$$Q_1 = Q_b\left(\frac{\rho_b}{\rho_1}\right) \qquad (12.8)$$

Applying the gas law equation

$$\rho_1 = \frac{P_1}{Z_1 R T_1} \qquad (12.9)$$

where T_1 and P_1 are temperature and pressure at point 1 of the section, while at standard conditions

$$\rho_b = \frac{P_b}{Z_b R T_b} \qquad (12.10)$$

Replacing the above values in Equation (12.8):

$$Q_1 = Q_b\left(\frac{P_b}{T_b}\right)\left(\frac{T_1}{T_1}\right)Z_1 \qquad (12.11)$$

For finding the velocity, use the relation:

$$u_1 = \frac{Q_1}{A} \tag{12.12}$$

Thus:

$$u_1 = \frac{Q_b Z_1}{A}\left(\frac{P_b}{T_b}\right)\left(\frac{T_1}{P_1}\right) \tag{12.13}$$

where A = area of cross section.

12.4.3 Elevation effects

Gas flow is affected by elevation but the flow is not the same with incompressible fluids. For incompressible fluids, i.e., liquids, the density or specific weight is far higher than that of a gas, thus the gravitational force on the liquids is quite high when the fluid is ascending. Such elevation effects are quite diminished in the case of gas. The general flow equation can be modified to include the difference in elevation of the gas pipeline as follows:

$$Q = 5.747 \times 10^{-4} F\left(\frac{T_b}{P_b}\right)\left[\left(\frac{P_1^2 - e^s P_2^2}{G T_f L_e Z}\right)\right]^{0.5} D^{2.5} \tag{12.14}$$

where

$$L_e = \frac{L(e^s - 1)}{s} \tag{12.15}$$

The terms "L_e" and "e^s" account for the difference in elevation between upstream and downstream pipe ends. "s" is a parameter depending upon the elevation difference, flow temperature, gas gravity, and compressibility factor and is expressed as:

$$s = 0.0684G\left(\frac{H_2 - H_1}{T_f Z}\right) \tag{12.16}$$

where H_1 = elevation at upstream point 1 and H_2 = elevation at downstream point 2.

12.5 Pipeline transportation of hydrogen gas

Any pipeline network mandated for transportation of any fuel or gas performs two functions. First, it functions to transport an adequate amount of energy to the end user, and second, the pipeline itself serves as storage of gas in the event that the gas supply exceeds the demand. The short-term storage of gas in a pipeline is called the "*line-pack*." The linepack facilitates a continued supply of gas in the network, despite the fluctuations occurring in the demand pattern. In other words, with the help of the linepack, the demand side is able to alter the off-take at its will, independently of the amount of gas injected into the network. A greater linepack means more storage, but it also requires higher pressure. We now consider the two aspects as:

12.5.1 Energetic aspects

While considering the energetic aspects, it should be kept in mind that if hydrogen is to be replaced with natural gas, then the amount of hydrogen supplied to the end user must satisfy the energy content required by it. Hence, considering the equation of flow (12.2), the flow of gas is expressed in normal meter^3/hour, denoted as Nm^3/h. This factor represents mass flow instead of volumetric flow. Flow is dependent on the roughness of the conduit as well as a factor called the Reynold's number given by:

$$Re = \frac{u \cdot D}{\vartheta} \tag{12.17}$$

where ϑ=viscosity of gas in m^2/s and u =speed of gas flow m/s.

To clarify the capacity of hydrogen to carry energy in pipeline equivalent to natural gas, two terms need more explanation:

- Higher Heating Value: The higher heating value (HHV; also known as the gross calorific value or gross energy) of a fuel is defined as the amount of heat released by a specified quantity (initially at 25 °C) once it is combusted and the products have returned to a temperature of 25 °C.
- Lower Heating Value: The lower heating value (also known as net calorific value, net CV, or LHV) of a fuel is defined as the amount of heat released by combusting a specified quantity (initially at 25 °C or another reference state) and returning the temperature of the combustion products to 150 °C.

Some of the relevant physical properties such as HHV, density, and specific gravity are summarized in Table 12.2.

Considering Equation (12.2) and the above value of HHV shows that if the same energy demand is to be satisfied, the volume of H_2 to be transported is thrice that of natural gas. However, the density of natural gas is approximately nine times that of hydrogen (9×0.084), therefore, if the hydrogen flow rate is kept three times, the pressure drop for natural gas and hydrogen would be the same. It may be added here that

Table 12.2 Comparison of physical properties

Gas	HHV (MJ/Nm^3)	Density (kg/m^3)	Specific gravity relative to air
Hydrogen	13	0.084	0.07
Natural gas	40	0.65	0.55

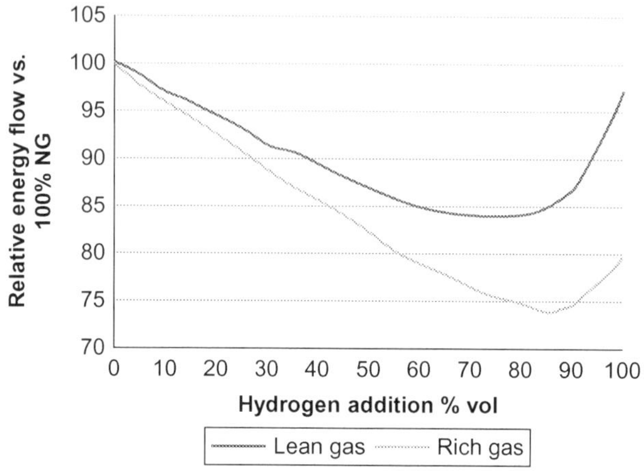

Figure 12.7 Energy flow comparison-H_2 vs. H_2–NG mixture (constant pressure).

pressure drop is one of the most significant parameters in the design of piping infrastructure.

Variables affected by flow rate include "*Z*" and "*f*." Investigations with this aim indicate that for an unchanged pipeline and pressure drop, energy flow by hydrogen is 98% with lean natural gas, while in comparison with rich natural gas, it is 80%. Relative energy flows from 0% to 100% mixture by volume with natural gas are shown in Figure 12.7.

Linepack is one of the significant advantages of any pipeline network, and is affected mainly by rate of flow. The inverse relationship indicates a higher linepack with a reduced flow, while a higher flow rate implies a diminished linepack or storage. Linepack can be analyzed by Equation (12.18) (Menon, 2005).

If $V_{\text{storage},n}$ = storage volume at normal temperature and pressure (273°K and 1 bar) and V_{geom} = volume of pipeline, then

$$V_{\text{storage},n} = V_{\text{geom}} \left[\frac{P_{\text{m}}}{K_{\text{m}}} - \frac{P_{\text{m}'}}{K_{\text{m}'}} \right] \frac{1}{p_n} \frac{T_n}{T} \tag{12.18}$$

where P_{m} = upper mean pressure, $P_{\text{m}'}$=lower mean pressure, and K_{m} = compressibility factor.

The principle of linepack is shown in Figure 12.8. The space between the upper and lower pressure profiles indicates the available linepack.

When H_2 is made to flow in the present natural infrastructure, the storage of H_2 in the form of linepack lies between 65% and 71% of that of natural gas, assuming a rate of flow ranging from 500,000 to 1,500,000 Nm^3/h. These figures represent the line-pack in terms of volume; however, the significant point here is the energy carried by any fuel that has to meet customer demand as a first priority. With this in mind, the energy content of hydrogen is approximately a quarter of that of natural gas. This may hamper the security of supply in the short term. Figure 12.9 shows these values at

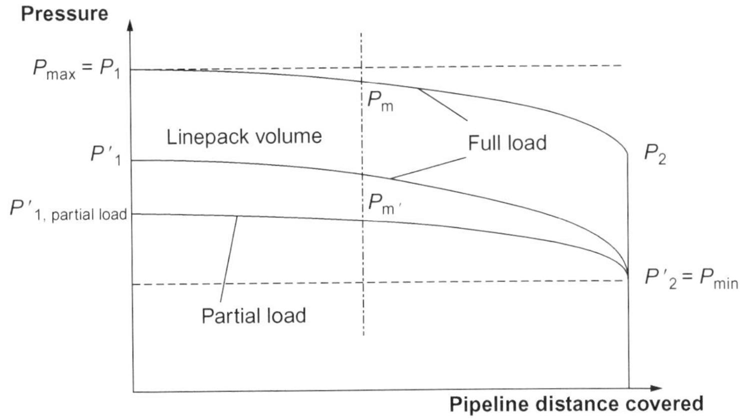

Figure 12.8 Linepack illustration (Yoon et al., 2007).

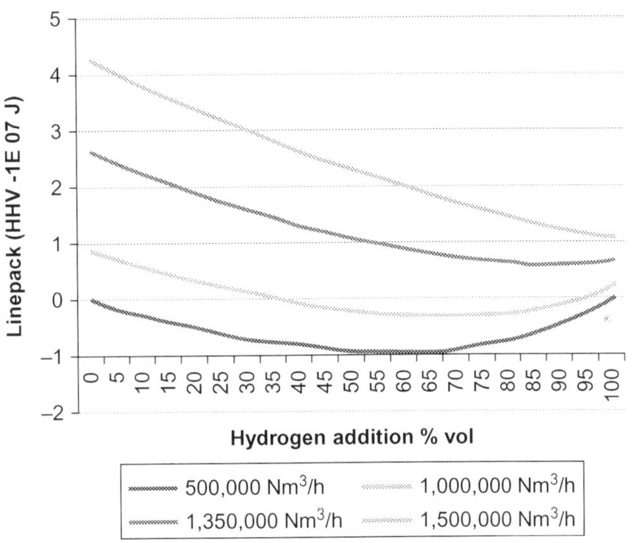

Figure 12.9 Linepack relative to various H_2–NG mixtures.

various mixtures ranging from 0% to 100% hydrogen (for lean natural gas). It is very evident from these graphs that transport of hydrogen in the existing infrastructure is not feasible. The concept of linepack as storage of H_2 does not apply in this context.

12.5.2 Pipeline material aspects in H_2 distribution

Apart from the physical/chemical properties of hydrogen that affect the pipeline network, certain other issues are also need considered for a hydrogen infrastructure; these include leakage and pipe damage.

12.5.2.1 H_2 leakage

Hydrogen is first in the "periodic table" by virtue of being the smallest element, which means it is more likely to escape as compared to natural gas. Basing on their diffusion coefficients hydrogen can diffuse up to four times faster than natural gas. Hence, the components of any pipeline network have to be designed to prevent leakage from valves, seals, and gaskets, thereby raising safety hazards.

Hydrogen leakage though high in volumetric terms is lower in energetic losses. Furthermore, as discussed earlier, the amount of hydrogen leakage depends largely on the pipeline material. Cast iron and fibrous cement pipelines have a greater leakage risk. Currently, polyethylene pipelines are in use in most distribution networks. Leakage/diffusion of hydrogen is five times higher than natural gas; however, it is negligibly small due to its energetic content. Research has shown that the annual loss of hydrogen by leakage is approximately 0.0005–0.001% of the total volume transported (Scholten and Walters, 2002).

Additionally, the compressors installed along the transmission line are another point to consider. Compressors can be either reciprocating or centrifugal. Reciprocating machines can be the piston type or a diaphragm configuration and are being used in hydrogen pipeline networks around the world. Air Liquide, the pioneer in hydrogen technologies, uses volumetric compressors, but their natural gas networks use centrifugal machines. Thus, if hydrogen is to be transported in a high-pressure transmission network, the volume of hydrogen to be transported in high-pressure transmission network, the volume of hydrogen to be transported has to be thrice that of natural gas as already discussed. This results in an increase in compression capability as much as twice that of the current capacity. From the previous discussion it is clear that rotational velocities would have to be increased manifold to match the rate of mass flow. Increased velocities are limited by the material strengths of the compressors, therefore, it may be concluded that the compressors installed in the existing infrastructure are insufficient to handle hydrogen in the same quantum as of natural gas. It may also be mentioned here that the same is not true for distribution infrastructure since compression stations are not employed therein (Polman and Walters, 2002).

Another distinguishing behavior of hydrogen is exhibited when its pressure is reduced. When subjected to pressure reduction natural gas causes a drop in temperature of 0.5 °C with each bar, due to Joule–Thompson effect. However, on the other hand, the temperature of hydrogen rises by 0.35 °C for each increase of a bar pressure. Thus, a rise of 2 °C would occur for a drop in pressure from 80 to 15 bar, which fortunately has no effect on the existing natural gas infrastructure (Utgikar and Thiesen, 2005).

12.5.2.2 Hydrogen embrittlement

Hydrogen embrittlement of pipes is caused by degradation of mechanical properties, which includes surface cracking and propagation of cracks, resulting in pipe failure. However, causes of embrittlement range from the composition of pipe material as well as the operating temperature/pressure conditions. The concentration of hydrogen in a

H_2/NG mixture also determines the extent of degradation/embrittlement (Hydrogen Infrastructure Delivery Reliability R&D Needs, 2015). Pipeline history, especially the intensity and frequency of pressure fluctuations, is also an important determinant of embrittlement predictability. Thus, intensive testing of line, weld, and joints will determine the suitability of existing network for hydrogen fuel gas.

12.5.3 H_2–natural gas mixtures

The profiles of hydrogen mixtures at low pressures are given in Figure 12.10. It can be deduced that mixtures of hydrogen and natural gas up to 40% may not require any major shifts in energy transfer.

The foremost requirement in the initiation of a hydrogen economy is to break the chicken and egg scenario. Therefore, it is not logical to expect an overnight shift of engines, boilers, and burners to fuel cells. Nonetheless, a natural propagation is needed to achieve a full-scale penetration of fuel cells in the market.

It has been shown (Figures 12.7–12.10) that up to 17% mixture of hydrogen by volume in natural gas bulk does not cause a problem (Tabkhi et al., 2008). While a mix of higher percentages requires consideration, the consequent problems are not significant. These include the effects on linepack, pressure drop, and Wobbe index.

The Wobbe index or number is found by dividing the high heating value of the gas by the square root of its specific gravity with respect to air. The greater a gas' Wobbe number, the greater the heating value of the quantity of gas that will flow through a hole of a given size in a given amount of time. Hence, it is a measure of interchangeability of gases and is also applicable to end-use domestic application (Application Data Document, 2015).

$$W_s = \frac{H_s}{\sqrt{d}} \tag{12.19}$$

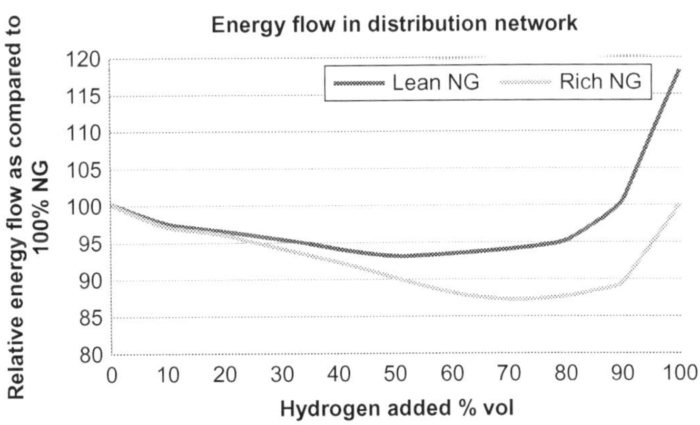

Figure 12.10 Addition of H_2 vs. relative energy flow (NG).

For mixtures of H_2–NG, the Wobbe number is given by:

$$W_{s,mix} = \frac{H_{s,mix}}{\sqrt{d_{mix}}} = \frac{\text{HHV of NG} + \text{HHV of } H_2}{\sqrt{\text{sp.gravity of NG} + \text{sp.gravity of } H_2}} \quad (12.20)$$

For common natural gas burners, the Wobbe number (see Table 12.3) for rich natural gas lies between 48–58 MJ/Nm^3, while that for a lean natural gas is in the range 41–47. It is evident from Figure 12.10 that for lean natural gas value burners, hydrogen injection can be up to 98% by volume, while for rich natural gas it can be within 45% vol. Thus, for user ease, it is preferable to include the complete range, i.e., 41–58% so that the low/high calorific value can be used with equal convenience. To address the issues of flame detection, burner heads, and sealings, multifunctional devices that can run on entire range of H_2/NG mixtures are suggested (Ilbas et al., 2005; Schuan, 1999).

From Figure 12.11 it is clear that for low calorific value, the Wobbe index is lowest at 72% vol mixture, while for high calorific gas, the worst performance range is 75–85%.

12.5.4 Advantages of hydrogen pipeline transportation

There are several advantages of the delivery of hydrogen over extended distances through pipeline. The lifespan of hydrogen pipeline is considerably lengthy, spreading over decades. Hydrogen pipeline network is buried underground, so it is safer and

Table 12.3 Natural gas - Wobbe index

Type	Wobbe index range
Lean NG	41–47
Rich NG	48–58

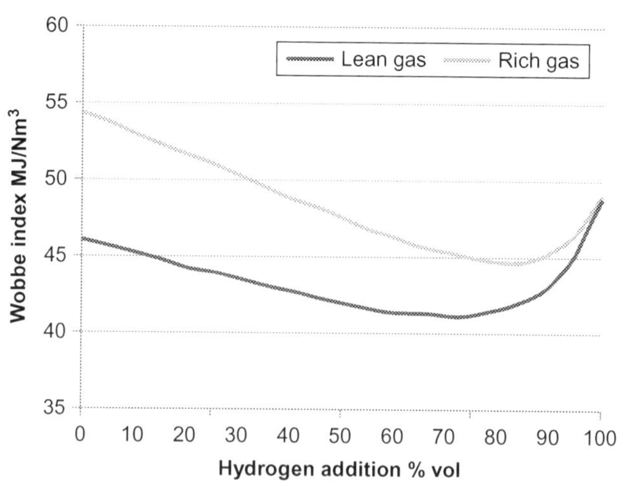

Figure 12.11 Wobbe index behavior with different H_2–NG mixtures.

more dependable than any other mode of transportation. Further, the possibility of occurrence of accidents due to leakage, explosion, or environmental interference could be lower. Moreover, the pipeline does not disrupt and cause congestion on the roads. Though the initial investment of laying the pipeline network is large, the subsequent cost of maintaining and operating the network is low, and the social cost is also not very high. Additionally, as the oil and gas pipelines have been in existence for several decades, initially the mixtures of hydrogen and natural gas can be facilitated through the existing pipeline infrastructure. Pure hydrogen can also be delivered in existing hydrogen pipelines by making modifications such as lowering of carbon content. Low carbon steels are used in the manufacture of new pipelines and hence can be used directly for delivery of pure hydrogen (Gunnar Birgisson Esq and Lavarco Esq, 2004).

12.5.5 Disadvantages of hydrogen pipeline

There are no inherent disadvantages of the hydrogen pipeline network, but the technological challenges and barriers may be considered as disadvantages. Nevertheless as technology evolves, the disadvantages are likely to diminish, thereby allowing hydrogen penetration into the energy market. The challenges foreseen in the development of a hydrogen infrastructure network are discussed in the following sections.

Just as there are power losses in electrical grids, similarly as hydrogen travels through the pipeline network, there is a loss of hydrogen gas. Hence, continuous improvement is required in hydrogen pipelines to improve upon its efficiency.

In the case of natural gas the delivery network operates at a low pressure of 4 bars, therefore, costly steel pipelines can be replaced by cheap polymer materials. However, in the case of hydrogen, the density is quite low, about 1/8th of natural gas, so there is the need to compress it to a high pressure of about 10–20 bars, to enhance its delivery speed. If the pressure is not increased the energy delivery is inefficient. The polymer materials used for gas pipelines have a high porosity and therefore are not appropriate for use in pipelines meant for delivering pressurized hydrogen. Presently, expensive low carbon steels are used for the manufacturing of the majority of hydrogen pipelines. To realize a hydrogen economy, inexpensive materials need to be developed for the manufacture of hydrogen pipelines, to be used for an extensive network. Moreover, since hydrogen causes embrittlement resulting in cracks in pipelines, delivery of hydrogen can seriously hamper the smooth distribution and delivery of hydrogen (Chawla et al., 1986; Thompson, 1977; Zakaria and Davies, 1991). Other related issues are briefly discussed as follows.

12.5.5.1 Safety

The production, storage, distribution, and delivery of hydrogen is associated with significant safety issues. Although hydrogen can be considered as safer than methane or gasoline as a fuel as shown in Table 12.4, safety remains a key concern if hydrogen is to be transported in a pure form or as a mixture with natural gas.

Table 12.4 Safety ranking of fuels

Characteristics	Fuel ranking		
	Gasoline	Methane	Hydrogen
Toxicity of fuel	3	2	1
Toxicity of combustion	3	2	1
Density	3	2	1
Diffusion coefficient	3	2	1
Specific heat	3	2	1
Ignition limit	1	2	3
Ignition energy	2	1	3
Ignition temperature	3	2	1
Flame temperature	3	1	2
Explosion energy	3	2	1
Flame emissivity	3	2	1
Total	30	20	16
Safety factors	0.53	0.8	1

1, safest; 2, less safe; 3, least safe.

Modifications are required in leak-detection systems for pipeline networks meant exclusively for delivery of hydrogen. Leakage caused by cracks, corrosion, embrittlement, and diffusion would have to be detected and remedial measures taken in time to ensure safety in the delivery network. Additionally, mixture-specific devices need to be developed for detecting leaks and cracks if hydrogen is to be transported as a mixture with natural gas.

12.5.5.2 Pig inspections

"Pigs" are inspection devices used to internally inspect the pipelines for any obstruction in pipelines 10 inches in diameter or more. These devices also have to be developed for use with hydrogen pipelines.

12.5.5.3 Sensors

Hydrogen embrittlement, as discussed earlier, can cause rupture of pipelines. To facilitate earlier detection, new ultrasonic sensor technologies need to be developed.

12.5.5.4 Diffusion losses

New seals, gaskets, valves, and fittings need to be developed to prevent diffusion of hydrogen. Current fittings allow diffusion of hydrogen molecule because of its low viscosity and small size. Hydrogen can also permeate and be absorbed into pipe walls.

12.5.5.5 Hydrogen energy density

Hydrogen has a comparatively low volumetric energy density and hence affects the amount of energy transported due to lower volumetric flow rate, particularly in the case of hydrogen–natural gas mixtures. To satisfy the amount of energy needed, high-power compressors are needed to enable higher hydrogen volumetric flow requirements. For natural gas compressors, the energy requirement of typical compressors for every 150 km of pipeline length is approximately 0.3% of the energy content of the volume of delivered natural gas. However, in the case of pure hydrogen, the energy for compression is almost 1.3% of the energy content of hydrogen delivered for an equivalent pipeline length (150 km) (Air Liquide: Hydrogen Delivery Technologies and Systems, 2003). This further requires capacity de-rating, if hydrogen is to be transported in existing pipeline networks as a mixture with natural gas. Accordingly, limitations need to determined, which would be very demand-dependent on the energy system.

12.5.5.6 Leak detection

Leak detection of hydrogen is very difficult because of its colorless, odorless nature and because it burns with an invisible flame. Odorants such as organic sulfides and mercaptans facilitate leak detection in natural gas. However, for hydrogen detection these mercaptans and sulfur-containing compounds are unsuitable, if fuel cells are to be used. This is because sulfur-containing chemicals can affect the fuel cell catalysts and cause premature failure and subsequent replacement of the fuel cell. Thus, a non-sulfur compound as an odorant for hydrogen would require the same dispersion rate as that of hydrogen.

12.5.5.7 Compressors

Most of the compressors in use today with existing natural gas networks are the centrifugal or reciprocating design. New certification procedures need to be established for use with hydrogen pipeline networks, along with a need to develop new technology for hydrogen compression. Moreover, lubricants in use with natural gas infrastructure are also incompatible with hydrogen. Oils designed for use in hydrogen specific equipment need to be developed for use in hydrogen pipelines.

12.5.5.8 Financing

The cost of laying a pipeline transmission and distribution network exclusively for hydrogen is an enormous issue, given the elaborate network required to reach end users. Historically, research has not been funded by gas pipeline operators, therefore, governments may have to fund the development of appropriate technologies for use in hydrogen pipeline networks.

12.6 Conclusion

If a hydrogen economy is to realize, the pipeline infrastructure needs rapid development, not only in the material aspects but also technologies have to be developed to address the associated leakage, embrittlement, and devices to enable high pressure delivery of hydrogen. Table 12.5 shows the significant issues discussed in this chapter as they relate to hydrogen pipeline infrastructure.

The areas of research potentially significant in hydrogen pipeline delivery can be divided into the following categories:

- Inspection technologies
- Materials development
- Remote sensing
- Operational technologies
- Hydrogen metering technologies

A hydrogen pipeline network has considerable implementation and technological barriers that have to be overcome before it can play a part in the energy supply chain. The current natural gas network has evolved over decades to reach a stage of hiccup-free operation. Technology would naturally evolve as prices of fossil fuels increase and governments feel the pressure to plan for an alternate/hydrogen energy system.

12.7 Future trends

Several studies have concluded that an initiation of HSC would be triggered by the rapid and sustained development as well as economic availability of fuel cell applications both for stationary and mobile applications. Infrastructure development in general and hydrogen pipeline in particular would begin when there is a demand for hydrogen and consequent ease of availability at end user. All the components of HSC need to be addressed in detail. It has been emphasized in this chapter that fossil fuels may be the stepping stone for full-fledged development of the hydrogen economy; however, renewable resources are an essential component for any future fuel

Table 12.5 Hydrogen pipeline research and development priorities

Issue	Importance
Safety—odorants, diffusion losses, materials	Very important
Operation and maintenance—leak detection, hydrogen metering, inspection, remote sensing, compression	Important
Transmission and distribution technology—Inspection, hydrogen metering, compression	Very important
Security	Very important

World ranking of gas pipelines; comparison of physical properties; Wobbe index range; safety ranking of fuels; hydrogen pipeline research and development priorities.

supply chain that has to address the issue of greenhouse gases, climate change, and security of supplies. It is anticipated that hydrogen penetration in the energy market may be spread over decades, but the transition phase has already started with the establishment of hydrogen fueling stations. Standalone hydrogen production units would initiate this transition and would later be connected by the developing pipeline infrastructure, finally evolving in a full-fledged hydrogen transmission and distribution network (Zoulias and Lymberopoulos, 2007).

12.8 Further reading

Apart from the references the following publications are recommended for further study:

- "Gas Dynamics"James E. A. John, Theo G. Keith, Pearson Prentice Hall, 2006.
- Renewable Energy: Economics, Emerging Technologies and Global Practices, Andreas Poullikkas, Nova Science Publishers.
- Energy Carriers and Conversion Systems with Emphasis on Hydrogen, Tokio Ohta, Eolss Publishers Company Limited, 2009.

References

"Air Liquide: Hydrogen Delivery Technologies and Systems; Pipeline Transmission of Hydrogen". United States Department of Energy Strategic Initiatives for Hydrogen Delivery Workshop Presentations. Arlington, VA. May 2003.

Application Data Document, 1660AD-5a. EMERSON Process Management.

Chawla, K.K., Rigsbee, J.M., Woodhouse, J.B., 1986. Hydrogen-induced cracking in two line-pipe steels. J. Mater. Sci. 21, 3777–3782.

U.S. DOE, NREL (2011) "Wind Data Details".

U.S. DOE, Energy Information Administration (EIA) (2012) Annual Energy Review 2011.

Gondal, I.A., Sahir, M.H., 2013. Model for biomass-based renewable hydrogen supply chain. Int. J. Energy Res. 37 (10), 1151–1159.

Gunnar Birgisson Esq, W.L.E., Lavarco Esq, W., 2004. An effective regulatory regime for transportation of hydrogen. Int. J. Hydrog. Energy 29, 771–780.

Hydrogen Infrastructure Delivery Reliability R&D Needs, Science applications International Corporation, US Department of Energy.

Ilbas, M., Yilmaz, I., Veziroglu, T., Kaplan, Y., 2005. Hydrogen as burner fuel: modelling of hydrogen-hydrocarbon composite fuel combustion and NOX formation in a small burner. Int. J. Energ. Res. 29, 973–990.

Lee, D.-H., 2012. Toward the clean production of hydrogen: competition among renewable energy sources and nuclear power. Int. J. Hydrog. Energy 37 (20), 15726–15735.

Lun, L., Xie, Y., 2008. Gas turbine cycle recovering pressure energy of natural gas transportation pipelines by vortex tube. In: 6th International Energy Conversion Engineering Conference (IECEC).

Menon, E.S., 2005. Gas Pipeline Hydraulics. CRC Press—Taylor and Francis Group, London.

Mokhatab, S., Poe, W.A., 2012. Handbook of Natural Gas Transmission and Processing. Gulf Professional Publishing, USA.

Pipeline and Hazardous Materials Safety Administration, United States Department of Transportation website (accessed 20.12.13.).

Polman, E., Walters, M., 2002. Pathways to a hydrogen society. In: Proceedings Natural Gas Technologies, Orlando.

Scholten, F., Walters, M., 2002. Hydrogen diffusion through plastic pipes. In: Proceedings of the 17th International Plastic Fuel-gas pipe symposium 2002, San Francisco, pp. 280–284.

Schroeder, D.W., 2001. "A tutorial on pipe flow equations", Unpublished. http://www.psig.org/papers/2000/0112.pdf.

Schuan, T., 1999. Case studies of integrated hydrogen systems, IEA Hydrogen Implementing Agreement, Final report for Subtask A of Task 11.

Smit, R., Weeda, M., 2006. Infrastructure considerations for large hydrogen refueling stations. In: 16th World Hydrogen Energy Conference.

Tabkhi, F., Azzaro-Pantel, C., Pibouleau, L., Domenech, S., 2008. A mathematical framework for modeling and evaluating natural gas pipeline networks under hydrogen injection. Int. J. Hydrog. Energy 33, 6222–6231.

The CIA World Factbook 2012, 2011. Central Intelligence Agency. Skyhorse Publishing Inc., New York.

Thompson, A.W., 1977. Selection of structural materials for hydrogen pipelines and storage vessels. Int. J. Hydrog. Energy 2, 163–173.

Tokio Ohta, Turhan Nejat Veziroglu, "Energy Carriers and Conversion Systems with Emphasis on Hydrogen".

Turner, J.A., 2004. Sustainable hydrogen production. Science 305 (5686), 972–974.

Utgikar, V.P., Thiesen, T., 2005. Safety of compressed hydrogen fuel tanks: leakage from stationary vehicles. Technol. Soc. 27 (3), 315–320.

Yang, C., Ogden, J., 2007. Determining the lowest-cost hydrogen delivery mode. Int. J. Hydrog. Energy 32 (2), 268–286.

Yoon, M., Bruce Warren, C., Adam, S., 2007. Pipeline System Automation and Control. ASME Press, New York.

Zakaria, M.Y.B., Davies, T.J., 1991. Stack cracking by hydrogen embrittlement in welded pipeline steel. J. Mater. Sci. 26, 2189–2194.

Zoulias, E.I., Lymberopoulos, N., 2007. Techno-economic analysis of the integration of hydrogen energy technologies in renewable energy-based stand-alone power systems. Renew. Energy 32 (4), 680–696.

Progress in hydrogen energy infrastructure development—addressing technical and institutional barriers

13

R.K. Dixon[1], *J. Li*[2], *M.Q. Wang*[3]
[1]Global Environment Facility, Washington, DC, USA; [2]Tsinghua University, Beijing, China; [3]Argonne National Laboratory, Chicago, IL, USA

13.1 Introduction

The potential for hydrogen as an energy carrier has been known for more than two centuries. One of the early internal combustion engines (ICEs), e.g., developed by Isaac de Rivaz in 1805, was fueled by hydrogen. About 70 years later, the futurist Jules Verne in 1874 provided his vision of hydrogen energy in the book *L'Ile Mysterieuse*. In 1893, the German inventor Rudolf Diesel developed a diesel engine that was fueled initially by pulverized coal, which involved combustion of carbon and hydrogen (Dixon et al., 2011a).

For about the next 80 years, as economies around the world were industrialized and advances in ICEs and the availability of petroleum-based fuels proliferated, interest in hydrogen as an energy carrier waned. But beginning with the oil price spikes in the 1970s, evaluations of alternative transportation fuels, including hydrogen, began increasing in laboratories around the world (Hoffert et al., 2002). In fact, driven by the need for greater energy security, petroleum-importing countries substantially increased their investment in research, development, and demonstration programs for nonpetroleum transportation fuels (Dixon, 2007).

In the 1980s and 1990s, new concerns emerged about rising levels of urban air pollution and greenhouse gas (GHG) emissions, and because of environmental advantages, interests in hydrogen and fuel cells for transportation applications generated even greater attention, and in some countries hydrogen energy development became a more significant investment priority. In these places, hydrogen infrastructure also became a focus area because the existing suite of traditional energy production, storage, delivery, and fueling technologies were not capable of supporting significant deployments of hydrogen energy systems. Today, there are several developed and developing economies, including Germany, Japan, the United States, and China, that have made national commitments to accelerating the evaluation of hydrogen energy and infrastructure solutions to benefit from their energy and environmental advantages (Sperling and Cannon, 2004; Wang, 2003).

Technology development and market entry have followed a relatively slow and deliberate path for hydrogen energy applications. Hydrogen is currently used

Compendium of Hydrogen Energy. http://dx.doi.org/10.1016/B978-1-78242-362-1.00013-4

primarily as an industrial gas for petrochemical refining and chemical production. While there has been progress in the development and deployment of hydrogen energy systems, these efforts are still in their early stages.

There are several reasons for the relatively slow pace of development. For one, there are still technical and cost barriers to addressing the sustained research, development, and demonstration activities are still needed. For example, hydrogen production is expensive relative to traditional fuels, and the storage and transport of hydrogen face technological and cost challenges because hydrogen is lightweight, active, and a corrosive gas. In addition, fuel cells, which are devices for converting hydrogen to electric power, are still many times more expensive than ICEs and presently do not last as long.

These technical challenges are being addressed by hydrogen energy developers, and at the same time deployment of hydrogen fuel cell vehicles is facing market-driven "headwinds." These include, for example, cost-lowering advances in competing, nonpetroleum transportation technologies such as batteries for electric and hybrid-electric vehicles. Furthermore, recent advances in drilling techniques, coupled with reduced demand for oil due to the global slowdown in economic growth and the adoption of improved fuel economy standards for vehicles, have put downward pressure on oil and natural gas prices, which makes it more difficult for hydrogen and fuel cell vehicles to compete, gain market acceptance, and achieve the economies of scale needed for mass production.

13.1.1 Hydrogen energy development remains an important option

Given the technical barriers, and relatively unfavorable near-term market conditions, why are Shell, BMW, Daimler–Chrysler, Ford, Honda, Hyundai, Nissan, General Motors, and Toyota continuing to invest in hydrogen energy technologies and development (Tucker, 2013)? What is motivating national governments in the US, Japan, Germany, China, and elsewhere to undertake major research, development, and demonstration investments in technologies, tools, and techniques for hydrogen production, delivery, storage, and end-use?

The answers lie in the potential for substantial societal benefits associated with greater use of hydrogen energy systems, and the need to manage risks and uncertainties about the future costs and availability of fossil- and biomass-based transportation fuels (IEA, 2009). Concerns regarding national security, GHG emissions, finite sources of fossil fuels, and environmental quality are partially or wholly resolved if hydrogen energy systems are deployed at scale. Hydrogen production is theoretically limitless, since it is abundant and can be derived from renewable sources (Calderia et al., 2003). Hydrogen can be derived from fossil fuels, biomass, and water, and these alternative sources foster greater energy supply diversity and thus provide an effective risk management solution in a world characterized by high levels of uncertainty about future energy market conditions and possibilities (Turner, 2004). Hydrogen can also be produced from nuclear energy via electrolysis or thermochemical water cracking

(Grant, 2003). Use of hydrogen in fuel cells is much more energy efficient than use of fossil fuels in ICEs as electrochemical conversion in fuel cells is not bound by Carnot efficiency for heat engines.

Technologies that use hydrogen, such as fuel cells and combustion engines, have established safety and performance records that promote protection of the environment and public health. Petroleum imports can be dramatically reduced as hydrogen can be produced from indigenous sources. Economic security is enhanced as energy imports are reduced, infrastructure development creates jobs, and public health risks associated with conventional fuels are diminished.

13.1.2 Technology development activities continue making progress

The three primary barriers to achieving greater adoption of hydrogen energy systems (competitive hydrogen production, efficient storage of hydrogen, and lower cost and more reliable fuel cells) are technically solvable problems (Dixon et al., 2011a). While each of these technological challenges is formidable, public-private research and development partnerships around the world have made progress toward solutions. In addition, these efforts have also focused on addressing the key market barriers and infrastructure development issues. For example, national, state, and local codes and standards for producing, transporting, and storing hydrogen are being developed, and technologies for hydrogen delivery from central and onsite production are making progress. Yet, these standards and systems are still in their early stages of development, and widespread adoption, on scales equivalent to those already in place for petroleum-based fuels, natural gas, and electricity, are still decades away.

A major barrier to address is the need for investment in building infrastructure of hydrogen production, delivery, and storage that would enable consumers to fuel their fuel cell vehicles in a convenient and cost effective manner. Hence, today there is little or no existing hydrogen energy infrastructure and building it will require considerable financial investments over multiple decades (Friedmann and Homer-Dixon, 2004). Commercial, low-cost devices for hydrogen production, storage, and conversion are not yet available, and regulatory issues need to be addressed. Costs for hydrogen production, delivery, and storage technologies need to be reduced. Furthermore, all of the stakeholders including manufacturers, vendors, consumers, and public policy-makers need to gain more experience with hydrogen and fuel cell applications before comparable levels of public acceptance, such as those for fossil fuels, can be achieved.

Fuel cell vehicles, one of the most promising hydrogen energy applications, are being manufactured but in relatively small numbers, and costs and durability remain issues for automakers to address. However, fuel cell technologies (invented in 1879 by William Grover) have improved dramatically in the last several decades, with much reduced precious metal loadings, increased durability, and reduced costs. Because of their central role in converting hydrogen into useful electric power, fuel cells have been called the "microchip of the hydrogen age," as energy efficient and economic applications in the transportation and distributed energy sectors are envisioned.

Cost breakthroughs in hydrogen production, storage, and fuel cells could make hydrogen energy development a higher priority and raise its profile in the global portfolio of energy alternatives.

For all of these reasons it is likely that many nations, energy companies, and automobile manufacturers will maintain their investments in hydrogen energy development. One lesson learned from the global search for energy alternatives over the last several decades is that there is no energy panacea. All energy alternatives offer advantages and disadvantages; therefore investing in a diversified portfolio is essential for proper risk management. Furthermore, because of the many advantages, and despite the remaining barriers, hydrogen energy systems will have an important role to play in global energy research, development, demonstration, and deployment activities.

13.1.3 Purpose of this chapter

This purpose of this chapter is to inform readers about developments in hydrogen energy infrastructure by focusing on two examples that demonstrate recent progress: the United States and China. These examples were chosen because these two nations represent two of the world's leading economies that have been leaders in hydrogen energy research, development, and demonstration for several decades. Other countries are monitoring the progress of the United States and China (as well as of Germany and Japan) and will likely apply the knowledge gained in their own efforts to evaluate, develop, and deploy hydrogen energy technologies and infrastructure.

13.2 Recent progress in hydrogen infrastructure in the United States

The US Department of Energy (US DOE) has been investing in hydrogen energy technologies, modeling tools, and techniques for more than 20 years (US DOE, 2013). Program activities began in the late 1980s with a focus on fuel cell vehicles, and efforts began expanding starting about a decade ago as projects in hydrogen production, delivery, and storage were added to the portfolio. Today, the program consists of a comprehensive set of activities covering research, development, demonstration, modeling and analysis, codes and standards support, and education and outreach (US DOE, 2013). At its peak, the US DOE annual research and development (R&D) budget reached more than $300 million.

Driven by national policy objectives for greater energy security, environmental quality, and economic growth, the program conducts hydrogen projects with industry, universities, national laboratories, and other collaborators. The American Recovery and Reinvestment Act of 2009 provided additional resources for major demonstration projects. The Recovery Act provided approximately $41.9 million for hydrogen projects, which was matched by approximately $54 million in cost-share funding from industry. These funds have been used to deploy more than 1100 hydrogen fuel cells, primarily in backup power and forklift applications. As of October 2012, more than 1 million hours of operation had been achieved (US DOE, 2012).

US DOE investments in hydrogen energy development have been producing results and progress toward research and development targets, demonstrating the viability of various technologies and systems, building public awareness toward customer acceptance, improving modeling and analysis of technologies and market uptake pathways, and developing appropriate codes and standards to guide the safe and cost-effective adoption of hydrogen energy systems. Hydrogen infrastructure is an important program priority for US DOE and key areas of progress and program have achieved accomplishments.

13.2.1 Recent US DOE progress in hydrogen production, delivery, and storage systems

The focus of DOE activities in hydrogen production involves developing technologies to enable the long-term viability of hydrogen as an energy carrier for a range of applications, including stationary power, distributed power generation, back-up power, specialty vehicles, transportation, and portable power. Key technology development areas include biomass gasification, reforming of bio-derived liquids, electrolysis, solar-thermochemical hydrogen production, photo-electrochemical (PEC) hydrogen production, and biological hydrogen production (US DOE, 2012). Examples of recent progress include:

- Development of electrolyzer systems that incorporate low-cost stack components into high-efficiency hydrogen production systems. A system recently completed over 100 h of field testing at the National Renewable Energy Laboratory test facility for renewable energy integration.
- Demonstration of extended durability for photo-electrochemical hydrogen production technologies from about 20 h up to more than 100 h. Enhanced stability was also recently demonstrated under operating conditions consistent with solar-to-hydrogen conversion efficiencies exceeding 10%.
- Achievement of improved hydrogen fermentation rates in bioderived liquids by optimizing reactor design and operating conditions, resulting in a two-fold increase in hydrogen production through higher cellulose feedstock loading.

Table 13.1 shows the results of recent US DOE analysis that assesses the potential for future progress toward hydrogen production cost targets (US DOE, 2012). The figure-of-metric evaluated in the table is gallons of gasoline equivalent (gge). As shown, substantial improvements in hydrogen production cost targets are still needed for 2020 targets to be achieved.

The focus of US DOE activities in hydrogen delivery involves reducing the cost and increasing the energy efficiency of hydrogen delivery for enabling greater deployment of hydrogen energy systems (US DOE, 2012). Examples of recent progress include:

- Design and construction of a custom-built trailer (shown in Figure 13.1) capable of holding four 40-foot pressure vessels and an additional 30-foot pressure vessel. This new design has the potential to increase overall capacity by roughly 18%, from about 615 kg in the current US Department of Transportation (US DOT) approved design, to more than 725 kg.

Table 13.1 US DOE cost status and targets for hydrogen production (US DOE, 2012)[1]

	$/gge (production costs only)	2011 status	2015 target	2020 target	Ultimate target
Distributed	Electrolysis	$2.50–5.30	$3.90	$2.30	Production: $1–2 Dispensing: $2–4
	Natural gas steam methane reforming	$1.33–2.30	$2.30	$2.30	
	Bio-derived liquids (based on ethanol reforming case)	$2.70–6.40	$5.90	$2.30	
Central	Electrolysis	$3.10–6.10	$3.00	$2.00	
	Nuclear	$2.30–4.20	$2.00	$2.00	
	Biomass gasification	$1.80–3.50	$2.10	$2.00	
	Solar thermochemical	NA	$14.80	$3.70	
	Photo-electrochemical	NA	$17.30	$5.70	
	Biological	NA	NA	$9.20	

[1]Based on US DOE analysis published at http://www.hydrogen.energy.gov/pdfs/12002_h2_prod_status_cost_plots.pdf.

Figure 13.1 US DOE's Lincoln carbon fiber composite tube trailer pressure (US DOE, 2012).

- Analysis of the cost and power requirements of refueling station compression and pumping technologies, and of the various configurations of high-pressure tube-trailers within DOT-specified weight and size constraints. Two compression options to reduce station capital costs by at least 15% were identified: (1) a high-pressure (900-bar) liquid pump combined with an evaporator to gasify hydrogen before dispensing; and (2) a high-pressure tube trailer that can reduce compression needs at the station, especially in early market applications where the utilization of station compressors would be low.

Table 13.2 shows the results of recent US DOE analysis that assesses the potential for future progress toward hydrogen delivery cost targets (US DOE, 2012). As shown,

Table 13.2 US DOE status of progress toward hydrogen delivery cost targets (US DOE, 2012)

Delivery element	Targets (2015/2020)	Status
Tube trailers	Reduce capital cost to < \$728/kg by 2015 and < \$574/kg by 2020; Increase capacity to 700 kg by 2015 and 940 kg by 2020	Capital cost: \$510/kg Capacity: 726 kg
Pipelines	Reduce cost/mile (8 in. diameter pipe installed) to < \$735,000 by 2015 and < \$710,000 by 2020	Installed steel pipeline cost: \$3 million/mile; Pipeline cost contribution: \$1.7/kg H_2 Compressor cost contribution: \$0.1/kg H_2
Forecourt Compression (1000 kg/day station)	Reduce uninstalled capital cost to \$400,000 by 2015 and \$240,000 by 2020 for 700 bar dispensing	Capital cost: \$530,000 for 700 bar compression
Pipeline Compression	Reduce the uninstalled capital cost for 3000 kW compressor to < \$2.3 million by 2015 and < \$1.9 million by 2020	Capital cost: \$2.7 million
Forecourt Storage (1000 kg/day station)	Reduce high pressure tank cost to \$1200/kg of stored H_2 by 2015 and \$1000 by 2020	Storage tank cost: \$1450 per kg of stored H_2

there is still progress that needs to be made to achieve 2015/2020 program targets for hydrogen energy delivery.

The focus of DOE activities in hydrogen storage involves further advancements in new and novel hydrogen storage materials, including metal hydrides, chemical hydrogen storage, and sorbents, in addition to advanced tank development and total systems engineering to meet DOE onboard hydrogen-storage targets (US DOE, 2012). Examples of recent progress include:

- Validation of hydrogen excess uptake in a metal organic framework material synthesized by Northwestern University (NU-100), Illinois, US. The validated excess capacity of ~8 weight percentage at 50 bar and \$77,000 for the NU-100 metal organic framework is among the highest confirmed to date.
- Achievement of a 30% improvement in hydrogen weight percentage uptake when normalized to surface area through boron incorporation into porous carbon.
- Conducting of failure modes and effects analysis for both adsorbent and chemical hydrogen material systems, identifying potential failure modes requiring further consideration.
- Development of advanced composite pressure vessels for cryo-sorbents with 11% lower weight, 4% greater internal volume, and 10% lower cost (compared with the baseline established during phase I of the Hydrogen Storage Center of Excellence in 2011).

- Demonstration of carbonized fiber from low-cost, textile-grade, polyacrylonitrile blended with methyl acrylate comonomer, which meets the 2012 milestone of at least 300 kilopound-force per square inch (KSI) for strength and 30 million pounds per square inch (MSI) for modulus.
- Development of a pressure vessel design to achieve a 20% reduction in carbon fiber requirements.

13.2.2 Recent US DOE progress in hydrogen manufacturing research and development, technology validation, codes and standards, education, and market transformation

The focus of US DOE activities in manufacturing involves novel electrode deposition processes for membrane electrode assembly (MEA) fabrication, reduction in the number of assembly steps to produce lower-cost MEAs, flow field plate manufacturing variability and its impact on performance, and fabrication technologies for high-pressure composite hydrogen-storage tanks (US DOE, 2012). Examples of recent progress include:

- Scale-up of microporous layer ink in MEA manufacturing for full-length and full-width roll coating was found to cause severe bubble formation, leading to variable viscosities in the ink. By modifying additives and processes, the problem was solved and the cost of the microporous layer was reduced by 37% compared with the benchmark. This also resulted in a three times increase in capacity.
- Improved imaging of polymer electrolyte membrane thickness and discrete defects (e.g., bubbles, scratches, divots) using optical diagnostics on a full-scale webline, detecting defects on the order of $\sim$10–100 μ in membranes at standard web speeds of 30 feet per minute.

Technology validation demonstrates, tests, and validates hydrogen, fuel cell, and infrastructure technologies and uses the results to provide feedback to US DOE's hydrogen program activities. The focus has involved demonstrations that emphasize codevelopment and integration of hydrogen infrastructure with fuel cell vehicles (FCVs) to permit industry to assess progress toward technology readiness (US DOE, 2012).

For example, the Fountain Valley Renewable Energy Tri-Generation Station—the world's first facility capable of coproducing hydrogen, heat, and power—has operated for more than 1000 h in power and power-and-hydrogen modes, and over 5 million standard cubic feet of digester gas has been processed to produce more than 5000 kg of hydrogen and over 1 million kWh of electricity. The system has achieved an overall efficiency of 54% when coproducing hydrogen and power (US DOE, 2012).

The focus of US DOE codes and standards activities involves providing the technical basis for codes and standards that address fuel specifications, separation distances, materials and components compatibilities, and hydrogen sensor technologies. Activities include collaborative efforts with standards setting organizations and other stakeholders to achieve greater harmonization of regulations, codes, and standards (US DOE, 2012). Recent examples of progress include:

- Publication of the compressed hydrogen materials compatibility (CHMC) testing and data application standard by the Canadian Standards Association.
- Development of accelerated test methods for measurement of hydrogen-assisted fatigue crack growth; this accelerated test greatly reduces the cost barriers that prevent qualification of new materials for hydrogen energy applications.
- Conducting of two fire training classes at the Los Angeles City, California, US and County Fire Department, with approximately 300 first responders in attendance. More than 23,000 code officials and first responders have participated to date.

The focus of US DOE hydrogen education activities include providing technically accurate and objective information to target audiences who are involved either directly or indirectly in hydrogen fuel cell and infrastructure projects (US DOE, 2012). Recent examples of progress include:

- Initiation of a cluster group in the Northeast region of the US for collaboration in developing technology development and deployment roadmaps in seven states.
- Continuation of training for middle and high school teachers through the "H2 Educate!" program. This program has reached about 9700 teachers in 35 states. About 90% of participants believe the training resources increased the effectiveness of their lesson plans.

The focus of US DOE hydrogen market transformation activities include efforts to facilitate the growth of early markets for fuel cells used in portable, stationary, and specialty-vehicle applications. They are aimed at addressing several of the key market barriers particularly relatively high fuel cell costs by encouraging economies-of-scale through early market deployments (US DOE, 2012). Recent examples of progress include:

- Validated over 500 fuel cell forklifts with over 1.4 million hours of operation, 250,000 hydrogen fills, and 185,000 kg hydrogen dispensed.
- Demonstration and validation of a fuel cell, mobile, lighting system that combines high-pressure (5000-psi, equals to 344.74 bar) hydrogen storage, efficient lighting, and a 5 kW fuel cell. The mobile lighting system was field tested at industry and government installations and was demonstrated at major media events, including the Academy Awards, the Golden Globes, and the Screen Actors Guild Awards.
- Publication of guidelines for federal facility managers for procuring energy supplies from stationary fuel cell power systems, including the use of innovative financing mechanisms that require little or no capital investment.

13.2.3 Summary of progress in the United States

US DOE hydrogen energy and infrastructure development activities are making progress toward technical cost and performance targets. Further progress is needed for the targets to be achieved and in most cases the technical challenges and targets appear to be within reach. Reaching the targets will help to make hydrogen energy a more commercially viable alternative in the US and global energy markets.

13.3 Recent progress in hydrogen infrastructure and fuel cell vehicle and fuel cell bus demonstrations in China

Hydrogen energy development is part of China's overall approach to meeting its growing energy and environmental needs and to leap forward the Chinese auto industry from ICE technologies to advanced vehicle technologies (GEF, 2009). Chinese demand for cars, motorcycles, buses, and trains is growing substantially and urban transportation is becoming a leading contributor to air pollution in the major cities and is raising significant concerns for public health and safety. For example, China's transportation fleet is projected to expand from 16 to 94 million vehicles between 2000 and 2020, with liquid and electricity transport fuel demand growing from about 5 Quadrillion British thermal units (Quads) to over 20 Quads in 2035 (Hao et al., 2011; Wang et al., 2006). As a result, Chinese government agencies, universities, and private companies are involved in research, development, and demonstration programs for cleaner and more energy efficient transportation systems (Dixon et al., 2011a).

Chinese investments include research, development, and demonstration projects in FCVs, fuel cell buses (FCBs), and associated infrastructure in several Chinese cities. For example, over the past decade, three generations of FCVs have been developed, demonstrated, and deployed in China. In addition, several generations of FCBs have been developed, demonstrated, and deployed. Collectively, these efforts have involved more than 200 FCBs and FCVs. In fact, large-scale FCV and FCB demonstrations, including those at the 2008 Beijing Olympic Games and the World Exposition 2010 Shanghai China (EXPO), have been successfully implemented and the results have been documented (Dixon et al., 2011a,b).

These demonstrations have included development of hydrogen infrastructure, including hydrogen production facilities, fueling stations, and maintenance stations that were operated to support the FCV and FCB fleets. The experiences gained from these demonstration activities serve as the foundation for scaling-up infrastructure and fleet demonstration projects in other Chinese cities.

13.3.1 *Hydrogen infrastructure demonstrations in Beijing and Shanghai, China*

Three different hydrogen sources were used for fueling FCVs and FCBs in Beijing and Shanghai: (1) hydrogen from coke oven gases, (2) hydrogen from steam methane reforming, and (3) hydrogen from water using electrolysis. Operation data for hydrogen refueling stations and FCVs and FCBs were collected, analyzed, and archived at a central data processing facility so that technical performance could be continuously monitored and evaluated.

Infrastructure development for operating and maintaining FCVs and FCBs in Beijing started in 2005. The first hydrogen refueling station was completed for operation in the Haidian District, Beijing on November 18, 2006, and was later named as Beijing Hydrogen Park (Figure 13.2). Hydrogen was delivered to the station by trailers. On July 1, 2008, an onsite natural gas reformer was put into service.

Figure 13.2 Hydrogen trailers, natural gas reformer, and refueling pump (from left to right) at the Beijing hydrogen park

The refueling station supported the first phase of FCB demonstration sponsored by the Global Environmental Facility (GEF). There were three DaimlerChrysler first-generation Citaro FCBs that were delivered to Beijing with the financial support of the GEF. These three FCBs were delivered to a bus company on November 23, 2005, and were put into service on June 20, 2006 (Figure 13.3). By October 24, 2007, the bus fleet was in service on Bus Line 384 and operated for a total of 5,699 h with total distance traveled of 92,116 km and total passengers of 56,793. A total amount of 15,813 kg of hydrogen was used from June 2006 to October 2007.

Infrastructure development for operating and maintaining FCVs and FCBs was a critical component of the demonstration project during the 2008 Beijing Olympics (Dixon et al., 2011a,b). For this project, vehicle maintenance, fueling, and onsite hydrogen production facilities were built in the Beijing Hydrogen Park by BP (Figures 13.4 and 13.5).

In total, there were three FCBs and 20 FCVs for transporting athletes, event goers, and onsite personnel at the 2008 Beijing Olympic Games (Figure 13.6). The three

Figure 13.3 FCB in service in 2006–2007 in Beijing.

Figure 13.4 Beijing hydrogen park during the 2008 Beijing Olympic games.

Figure 13.5 Refueling station and fuel cell bus during the 2008 Beijing Olympic games.

FCBs, manufactured by the Beijing Foton Bus Company, served on the same Bus Line 384 as the DaimlerChrysler FCBs did. From July 1, 2008 to July 1, 2009, with support by the second phase of the GEF FCV program, the FCBs demonstrated for a distance of about 60,000 km, total working hours of 3646, total hydrogen consumption of 5753 kg, and total passenger count of 39,995.

The complete technical specifications for the hydrogen refueling station during the Beijing Olympic Games are given in Table 13.3.

Following the successful demonstration of FCBs and FCVs at the 2008 Beijing Olympic Games, an expanded program was implemented at the World EXPO 2010 in Shanghai China, which was held in May to October, 2010. More than 1000 alternative fueled vehicles were deployed and demonstrated at the EXPO. They included 140 battery-powered electric cars, 120 battery-powered electric buses, 90 fuel cell

Figure 13.6 Fuel cell buses demonstrated in service during the 2008 Beijing Olympic games.

Table 13.3 Technical specifications of hydrogen refueling station during the 2008 Beijing Olympic games

2008 Beijing Olympic games	
Refilling times a week	3–4 per bus
Number of buses refilled each time	8–10
Maximum refilling capacity per bus	40 kg of H_2
Refilling capacity per week	50 times
Daily average/maximum supply capacity	240/300 kg of hydrogen (based on 10 buses)
Refilling time	<15 min
Maximum refilling interval	10 min
Maximum gas temperature allowed in hydrogen tank after refilling	85 °C (Celcius degree)
Maximum pressure allowed after refilling	438 bar at 85 °C (Celcius degree)
Maximum pressure in bus $_{\text{hydrogen}}$ tank	350 bar

(FC) cars, 6 FCBs, 100 FC tourist vehicles, 61 Super Capacitor buses, 350 hybrid electric vehicles, and 150 hybrid electric buses (Dixon et al., 2011a).

This fleet of vehicles at the EXPO was built, operated, and maintained by domestic and international manufacturers, universities, and transit agencies, including Beijing Public Transit Company, Tsinghua University in Beijing, Tongji University in Shanghai, Shanghai Automotive Industry Corporation, the Shanghai Fuel Cell Vehicle Powertrain Company, Nanjing Special Purpose Vehicle Company, DaimlerChrysler, China National Fuel Cell Vehicle and Powertrain Research and Engineering Center, and Beiqi Foton Motor Company. The private sector and public-private partnerships were important contributors to this team effort.

Two permanent and two mobile hydrogen refueling stations were constructed and operated at the EXPO to serve the FCV and FCB fleets (Figures 13.7, 13.8, and 13.9). The complete technical specifications for the hydrogen refueling stations at the EXPO are given in Table 13.4. Tongji University in Shanghai and Shanghai Sunrise Energy

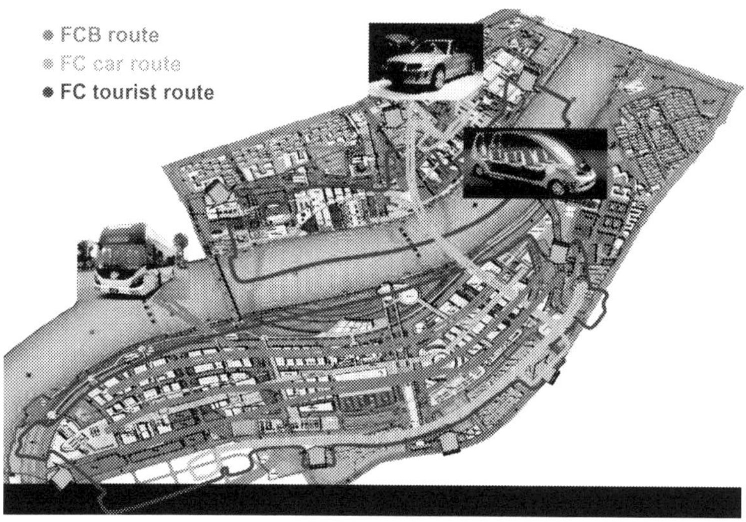

Figure 13.7 Demonstration routes of fuel cell vehicles and fuel cell buses in the world exposition 2010 Shanghai China.

Figure 13.8 Fuel cell buses in the world exposition 2010 Shanghai China.

Figure 13.9 Hydrogen refueling station and FCBs demonstrated in Singapore during the 2010 youth Olympic games.

Table 13.4 **Technical specifications of hydrogen refueling stations for the world exposition 2010 shanghai China**

World exposition 2010 Shanghai China	
Refilling times a week	60
Number of buses refilled each time	1–3
Maximum refilling capacity per bus	30 kg of hydrogen
Refilling capacity per week	85 kg of hydrogen
Daily average/maximum supply capacity	12/300 kg of hydrogen
Refilling time	5 min for FCVs and 20 min for FCBs
Maximum temperature allowed in hydrogen tank after refilling	85 °C (Celcius degree)
Maximum pressure allowed after refilling	380 bar
Maximum pressure in $_{\text{FCB hydrogen}}$ tank	350 bar

Systems Company developed the mobile hydrogen refueling systems. A maintenance facility was also successfully operated for FCVs and FCBs at the EXPO site. The EXPO's goals of zero emissions in the EXPO site and low emissions on the perimeter of the site were achieved, due, in part, to the integrated design and demonstration of FCVs, FCBs, and other electric drive vehicles. For example, based on monitoring data, six FCBs at the EXPO avoided over 700 metric tons of CO_2 emissions (Dixon et al., 2011a).

Under the collaboration between Tsinghua University in Beijing and Nanyang Technology University of Singapore, one FCB built by the King Long Company in China was shipped to Singapore and demonstrated during the first Youth Olympic Games (2010) in Singapore. A mobile refueling station was constructed in the campus of Nanyang Technology University.

Looking forward, scaling up the FCV and FCB programs to additional cities in China is part of the nation's energy plans (Figure 13.10 and Table 13.5). As a

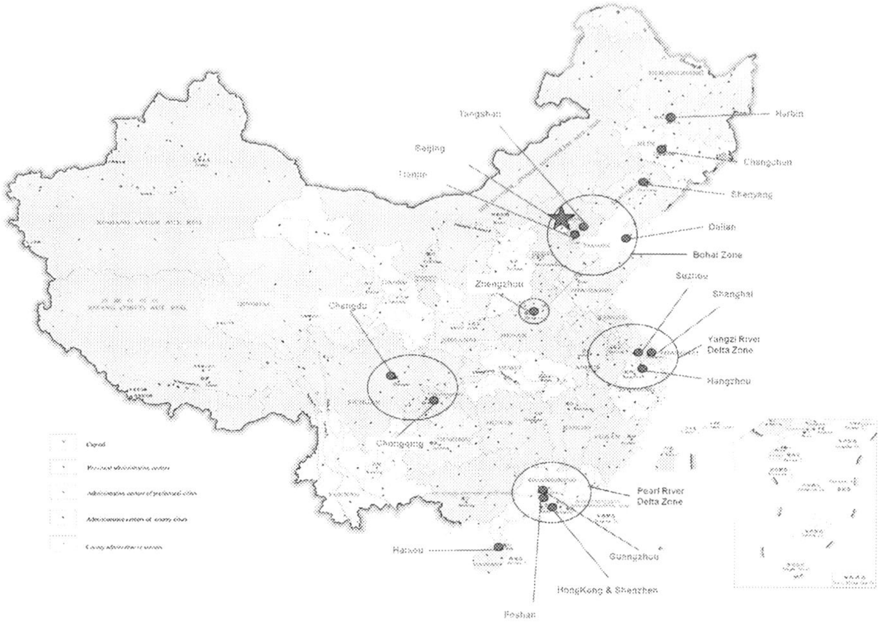

Figure 13.10 Network of future fuel cell vehicle and fuel cell bus demonstrations in China.

Table 13.5 **Hydrogen station construction plans in China**

Year	Cities	Number of stations	H_2 refueling capacity
2015	Beijing	2	450 kg
2015	Shanghai	2	450 kg
2015	Guangzhou	1	To be determined
2015	Foshan	1	
2015	Zhengzhou	1	
2015	Dalian	1	
2020	10 Cities	60	45,000 kg
2025	Whole Country	300	450,000 kg

follow-up to the EXPO, plans call for about 200 FCVs to be deployed in Chongming Island (in Shanghai's Delta Region) for long-term use as part of its "green island" strategy. The city of Dalian proposes to host a third field demonstration of FCVs that includes vehicle-to-grid technologies in support of an overall smart grid plan. To strengthen hydrogen infrastructure across the country, fueling stations will be built in the Beijing–Tianjin–Tangshan–Dalian region (the Bohai Delta Zone), the Shanghai–Suzhou–Hangzhou region (the Yangzi River Delta Zone), and the Guangzhou–Foshan–Shenzhen region (the Pearl River Delta Zone). These three regions are leading economic developing regions in China, but are heavy pollution areas that need clean vehicles. The other regions such as Hainan, Zhengzhou, Chengdu, and Chongqing are also considering development of FCVs and hydrogen infrastructure. These areas are collaborating to form a cooperative network for FCV and FCB development and demonstrations. Table 13.5 presents the plan of refueling stations of hydrogen in China from 2015 to 2025.

Raising public awareness is a significant aspect of infrastructure development and was an important part of the Chinese FCB and FCV demonstration programs. Public acceptance is essential for program expansion and scale-up. Before the demonstration programs were launched there was little awareness or understanding of FCVs, FCBs, and infrastructure for production, delivery, and fueling of hydrogen in China. Post-event surveys showed that during the 2008 Beijing Olympic Games, more than 70% of Beijing residents had some knowledge of the local FCB demonstration and less than 20% of the Beijing FCB passengers expressed safety or other concerns with the FCBs. In Shanghai, more than 70 million people visited the EXPO and witnessed the routine operation of FCBs and FCVs that were key parts of the event's transportation network (GEF, 2010).

The two Chinese FCV and FCB demonstration projects accumulated valuable experience and knowledge that have contributed to further advancements of hydrogen energy infrastructures in China and elsewhere (Campbell, 2010; Chen et al., 2008). The FCV and FCB trials also uncovered technical, financial, and policy impediments (Romm, 2004). With regard to technical barriers, future research and development in China (and elsewhere) is focusing on hydrogen storage, fuel cell component development and manufacturing, power system integration, overall vehicle development, and systematic measurement and evaluation of results (Turner, 2004; Ziittle et al., 2002).

As a result, a number of new technologies to enhance the viability and reduce costs of FCBs have been developed, but as in the United States, further improvements are needed before their market-readiness reaches viable levels. Future research and development will emphasize bipolar plates and membrane electrode assemblies for fuel cell stacks, a new generation of vehicle controllers, improved transmission systems, a newly designed fuel cell vehicle power train platform, fuel cell battery management systems, integrated water cooling systems, hydrogen and electric system crash safety tests, and energy recovery systems (Dixon et al., 2010).

Public–private partnerships have been important components of China's FCV and FCB investments (Yuan and Lin, 2010). Both domestic and international partnerships have contributed to the success of these investments. Innovation and development of new and improved technologies have had success as a result of partnerships activities. Manufactured under a number of product names (e.g., Foton, Sunwin, and Kinglong), the costs of this new generation of FCBs have been reduced by more than 50% (Ministry of Finance, 2010). The Ministry of Science and Technology's (2010) subsidies to local cities for FCB purchases were about US$100,000 per bus over the past decade, but as manufacturing costs decline the subsidies are shrinking.

China's growing need for sustainable urban transportation systems presents challenges and opportunities in technology innovation and transfer that transcends the energy sector (Ming et al., 2008). Rapidly growing cities in China are near-term, high-growth markets for vehicle fleets (including buses, taxis, public service vehicles, delivery vans, as well as personal vehicles) that are quiet, reliable, and pollution-free. These cities provide a focal point for early-market hydrogen infrastructure development, as well as the development and deployment of complementary next-generation electric sector smart-grid systems (IEA, 2008, 2010a). Construction and operation of innovative and complementary hydrogen and electric power systems has the potential to concomitantly provide clean, cost-efficient energy for China's transport, buildings, and industrial sectors. Hydrogen production and storage can be designed to complement and provide back-up power for existing and future electric systems. For example, hydrogen can power FCVs, and the FCVs integrated in buildings or industrial facilities can provide distributed power generation, back-up power, and/or remote power generation (Dixon et al., 2011a; IEA, 2005).

Current and future generations of hydrogen powered FCVs and FCBs provide options for China's efforts to develop a portfolio of sustainable urban transportation systems (Sperling and Cannon, 2004). Hydrogen and battery-powered electric vehicles seem to be on different but complementary development pathways in China. Infrastructure development will be an important element of China's future energy sector investments if hydrogen, electric, and hybrid-powered vehicles are fully utilized (Demirdöven and Deutch, 2004). For example, electric and hydrogen-powered vehicles will depend on charging or refueling stations built for passenger cars and buses. A network of fast-charging and/or battery replacement and hydrogen production stations will also be needed. Hydrogen and fuel cell vehicles will require hydrogen production, delivery, and storage facilities, including large numbers of hydrogen refueling stations conveniently located for bus and vehicle fleet operators (Harris et al., 2004). Progressive Chinese energy policies, such as dynamic energy pricing, advanced metering, and automated energy distribution systems that encourage vehicle fueling during off-peak periods, are now complementing and encouraging infrastructure development.

13.3.2 Summary of progress in China

Energy security, environmental protection, and economic development are significant challenges for China (IEA, 2010b; Leung, 2010). Aggressive laws, regulations, and plans to move toward a sustainable transportation sector are being developed, but today 98% of vehicles are powered by fossil fuels (Hu et al., 2010). As a result, China has become increasingly dependent on petroleum imports, and urban air pollution from transportation is a growing threat to human health and the environment in most Chinese cities. While China's efforts to develop FCVs, FCBs, and hydrogen production, delivery, and storage technologies have been informative, scale-up and expansion of hydrogen energy systems must continue to grow in the coming years if they are to make a substantial impact in the years ahead (Dixon et al., 2011a; IEA, 2005, 2008, 2010a).

13.4 Conclusions

Hydrogen energy development continues to be an important research, development, and demonstration pathway for major economies around the world. With respect to hydrogen energy products and services, the focus is on FCVs, FCBs, and industrial equipment like forklifts. However, addressing the needs for lower cost and better performing hydrogen energy infrastructure for these products is also an important priority. While progress has been made in areas such as hydrogen production, storage, delivery, codes and standards, and customer acceptance, cost and other targets necessary for commercialization and market scale-up have not been reached.

Recently, following the global economic downturn and increases in the production of petroleum and natural gas due to advances in drilling techniques, many questions have been raised about the prospects of long-term market success for hydrogen energy systems (Pacala and Socolow, 2004). Infrastructure development costs remain an important concern. It is important to ask questions and focus ongoing research, development, and demonstration activities on the key barriers so that the most important technical targets are addressed and hopefully reached in a timely and cost-effective manner.

Nevertheless, one of the lessons learned in energy markets over the last several decades is the variability and uncertainty associated with the price and availability of various types of fossil and bio-based fuels. Not only does hydrogen energy offer the potential for near limitless supplies and clean conversion to electricity using fuel cells, it also provides risk management advantages as it can be derived from many locally-supplied sources, including water. These advantages continue to make hydrogen energy a valuable investment opportunity in the energy portfolio of many countries around the world.

Acknowledgments

Richard Scheer, Sun Young Lim, and Hang Yin assisted with manuscript preparation.

References

Calderia, K., Jain, A.K., Hoffert, M.I., 2003. Climate sensitivity uncertainty and the need for energy without CO_2 emission. Science 299, 2052–2054.

Campbell, R.J., 2010. China and the United States – a comparison of green energy programs and policies. Congressional Research Service, Washington, DC.

Chen, J., Wang, J., Lun, J., 2008. The current situation and trend of research and development in fuel cell vehicle technology in the world. Automot. Eng. 5, 11–17 (in Mandarin).

Demirdöven, N., Deutch, J., 2004. Hybrid cars now, fuel cell cars later. Science 305, 974–975.

Dixon, R.K., 2007. Advancing towards a hydrogen energy economy: status, opportunities and barriers. Mitig. Adapt. Strat. Glob. Chang. 12, 325–341.

Dixon, R.K., McGowanb, E., Onysko, G., Scheer, R.M., 2010. US energy conservation and efficiency policies: Challenges and opportunities, 38 Elsevier, 6398–6408.
Dixon, R.K., Wang, X., Wang, M.Q., Wang, J., Zhang, Z.H., 2011a. Development and demonstration of fuel cell vehicles and supporting infrastructure in China. Mitig. Adapt. Strateg. Glob. Change 16, 775–789. http://dx.doi.org/10.1007/s11027-011-9293-y.
Dixon, R.K., Scheer, R.M., Williams, G.T., 2011b. Sustainable energy investments: contributions of the Global Environment Facility. Mitig. Adapt. Strateg. Glob. Change 16, 83–93.
Friedmann, S.J., Homer-Dixon, T., 2004. Out of the energy box. Foreign Aff. 83, 72–83.
Global Environment Facility (GEF), 2009. Investing in Sustainable Transport and Urban Systems: the GEF Experience. Global Environment Facility, Washington, DC.
Global Environment Facility (GEF), 2010. Greening Opportunities at World Events: GEF Investment Experiences. Global Environment Facility, Washington, DC.
Grant, P., 2003. The hydrogen economy: nuclear renaissance? Nature 424, 129–130.
Hao, H., Wang, H., Yi, R., 2011. Hybrid modeling of China's vehicle ownership and projection through 2050. Energy 36, 0360–5442.
Harris, R., Book, D., Andersen, P., Edwards, P., 2004. Hydrogen storage, the great challenge. Fuel Cell Rev. 1, 17–23.
Hoffert, M.I., et al., 2002. Advanced technology paths to global climate stability: energy for a greenhouse planet. Science 298, 981–987.
Hu, X., Chang, S., Li, J., Qin, Y., 2010. Energy for sustainable road transportation in China: challenges, initiatives and policy implications. Energy 35, 4289–4301.
International Energy Agency (IEA), 2005. Prospects for Hydrogen and Fuel Cells. International Energy Agency, Paris.
International Energy Agency (IEA), 2008. Energy Technology Perspectives 2008. International Energy Agency, Paris.
International Energy Agency (IEA), 2009. IEA Statistics: CO_2 Emissions from Fuel Combustion, 2009th ed. International Energy Agency, Paris.
International Energy Agency (IEA), 2010a. Energy Technology Perspectives 2010. International Energy Agency, Paris.
International Energy Agency (IEA), 2010b. World Energy Outlook. International Energy Agency, Paris.
Leung, G.C.K., 2010. China's oil use, 1990—2008. Energy Policy 38, 932–944.
Ming, P., Lun, J., Mytelka, L.K., 2008. Hydrogen and fuel cell activities in China. In: Mytelka, L.K., Boyle, G. (Eds.), Making Choices About Hydrogen: Transport Issues for Developing Countries. UNU Press, Tokyo.
Ministry of Finance, 2010. Global Environment Facility in China. Ministry of Finance of People's Republic of China, Beijing.
Ministry of Science and Technology, 2010. Fuel Cell Vehicles in China. Ministry of Science and Technology of People's Republic of China, Beijing.
Pacala, S., Socolow, R., 2004. Stabilization wedges: solving the climate problem for the next 50 years with current technologies. Science 305, 968–972.
Romm, J.J., 2004. The Hype About Hydrogen. Island Press, Washington, DC.
Sperling, D., Cannon, J.S., 2004. The Hydrogen Energy Transition: Moving Toward the Post Petroleum Age in Transportation. Elsevier Academic Press, London.
Tucker, W., 2013. Are Hydrogen Cars the Future—Again? Fuel Freedom Website http://www.fuelfreedom.org/blogs/over-a-barrel/.
Turner, I., 2004. Sustainable hydrogen production. Science 305, 972–976.
US DOE, 2012. 2012 Annual Progress Report—Hydrogen and Fuel Cells Program. US Department of Energy, Washington, DC.

US DOE, 2013. Fuel Cell Technologies Office Multi-Year Research, Development and Demonstration Plan and updates. US Fuel Cell Technologies Office, Washington, DC.

Wang, M.Q., 2003. Fuel choices for fuel-cell vehicles: well-to-wheels energy and emission impacts. J. Power Sources 112, 307–312.

Wang, M., Huo, H., Johnson, L., He, D., 2006. Projection of Chinese motor vehicle growth, oil Demand, and CO_2 emissions through 2050. ANL/ESD/06-6, Center for Transportation Research. Argonne National Laboratory, IL.

Yuan, K., Lin, W., 2010. Hydrogen in China: policy, programs and progress. Int. J. Hydrog. Energy 35, 3110–3113.

Züttle, A., Sudan, P., Mauron, P.H., Kiyobayashi, T., Emmenegger, C.H., Schlapbach, L., 2002. Hydrogen storage in carbon nanostructures. Int. J. Hydrogen Energy 27, 203–212.

Designing optimal infrastructures for delivering hydrogen to consumers

14

F.G.N. Li[1], W. McDowall[1], P. Agnolucci[1], O. Akgul[2], L.G. Papageorgiou[2]
[1]UCL Energy Institute, University College London, London, United Kingdom; [2]Centre for Process Systems Engineering, University College London, London, United Kingdom

Nomenclature and abbreviations

BG	biomass gasification
CCS	carbon capture and storage
CG	coal gasification
CO_2	carbon dioxide
DfT	United Kingdom Department for Transport
DRD	Northern Ireland Department for Regional Development
EFOM	energy flow optimization model
FCV	fuel cell vehicle
GET	global energy transition model
GH_2	gaseous hydrogen
GIS	geographic information system
GROS	general register office for Scotland
HSC	hydrogen supply chain
IEA	International Energy Agency
LH_2	liquid hydrogen
MARKAL	MARKet ALlocation model
MESSAGE	model for energy supply systems and their general environmental impact
MILP	mixed-integer linear programming
NISRA	Northern Ireland Statistics and Research Agency
NUTS	Nomenclature des Unités Territoriales Statistiques (Nomenclature of Territorial Units for Statistics)
ONS	United Kingdom Office of National Statistics
SMR	steam methane reforming
TIMES	The Integrated MARKAL-EFOM System model

14.1 Introduction

The optimum design of infrastructure for the delivery of hydrogen to consumers in a market without an existing delivery system represents a complex area of analysis. Hydrogen system planners face important questions when determining how to phase the deployment of new supply facilities in terms of their geographic proximity to

Compendium of Hydrogen Energy. http://dx.doi.org/10.1016/B978-1-78242-362-1.00014-6

demand and their scale of production. In the early years of system rollout, will it be more cost-effective to produce hydrogen from small-scale facilities located close to demand centers? Or will economies of scale make the cost optimal approach one that involves a smaller number of larger facilities, with hydrogen being transported over longer distances? As the level of demand on the system grows over time, will there be a tipping point when one approach becomes preferred over the other? These questions are complicated to analyze and the answers are likely to vary with different mixtures of demand conditions, supply chain costs, and technology performance levels.

As with the development of any new energy vector, the physical infrastructure for hydrogen supply must be built out based on future projections of demand. The rate of demand growth for hydrogen and its spatial distribution are therefore key factors that must be taken into account. Future hydrogen demand in many regions may be split between the transport, industry, and buildings sectors. Therefore, an understanding of demand factors that could affect uptake in these markets should ideally be considered. In addition to uncertainty over where demand will grow geographically and how quickly, there are several different technical options to consider for storing hydrogen and delivering it to end users. Finally, there are cost and performance considerations surrounding the manufacture of the hydrogen itself from production facilities using different technologies and at different scales. All of these elements combine to form a complex techno-economic optimization problem, where the trade-offs between different supply strategies and how they could change through time in response to demand shifts is not immediately obvious.

The optimization of the technical system design to meet future demand growth has been tackled by a number of different studies using computational models at different scales. However, all the approaches used to date have made significant compromises in their representation of the problem in order to facilitate computation. While each study individually achieves new insights into the specific problem domain considered, it is difficult to draw robust and general conclusions about different infrastructure strategies from work produced to date.

This chapter begins by giving a brief overview of the different technical options for production, storage, and delivery of hydrogen to end users. This is then followed by a literature review of the state of the art concerning the planning of hydrogen infrastructure using spatially explicit models. A series of case study scenarios are then explored for the UK transport sector using an advanced spatially explicit model. The results are then used to inform discussion of some of the key trade-offs between centralized and decentralized supply strategies, matching production capacity to demand over time, and the role of key technologies such as bioenergy with carbon capture and sequestration (CCS), and the key methodological challenges. The discussion will be of interest to policymakers, researchers, and industry stakeholders with an interest in the future development of hydrogen supply systems.

14.2 Building blocks of hydrogen infrastructure

The core features of hydrogen infrastructure include production technologies, storage and distribution mechanisms, and dispensing systems at refueling stations or other end-use delivery points. The modeling frameworks discussed in this chapter aim to provide insights into how these infrastructure building blocks can best be configured. This section provides a brief overview of the main technologies used in infrastructure optimization planning models.

14.2.1 Freight deliveries

Bulk freight delivery by vehicle is likely to be a key method of hydrogen distribution to end users in the early stages of any infrastructure development, when demand may be relatively low. Hydrogen can either be stored as a compressed gas (GH_2) and transported by tube trailer or stored as a cryogenically cooled liquid (LH_2) and transported by tanker. Storing hydrogen as LH_2 enables much higher energy densities, which lowers the per-unit cost of transport, but at the same time is more expensive in terms of production costs. This is because creating LH_2 involves the additional purchase of liquefaction units with high capital costs as well as the use of significant process energy in supercooling the gas to a liquid state. An emerging method of transporting hydrogen is to absorb the gas into powdered solids, known as metal hydrides. While this is not yet a technologically mature approach, it could become significant in the future.

14.2.2 Pipeline distribution

At some level of demand volume, it becomes less expensive to construct dedicated pipelines to carry compressed H_2 gas directly from the production point to the end user (Yang and Ogden, 2007). New hydrogen pipelines are generally constructed from epoxy lined steel, with costs similar to those for natural gas pipelines (Decourt et al., 2014; van der Zwaan et al., 2011). Much of the costs associated with pipeline deployment are independent of the pipe composition itself, and instead are a function of the complexity of the installation and the securing of land access rights (Mintz et al., 2002; Parker, 2004). In areas with existing natural gas infrastructure, it may be possible to repurpose some or all of the existing pipeline network to carry hydrogen, which could avoid a significant amount of the otherwise incurred deployment costs (Dodds and Demoullin, 2013; Dodds and McDowall, 2013; Dodds et al., 2014; Dougherty et al., 2009; Haeseldonckx and D'haeseleer, 2007, 2011; Smit et al., 2007).

14.2.3 Production facilities

Hydrogen production facilities are available at a variety of scales, from packaged units no larger than a single shipping container to industrial plants covering the equivalent of tens of city blocks. Adding to the complexity is the fact that hydrogen can be produced from a variety of chemical processes using different technologies. Much has been written elsewhere on different production pathways for hydrogen and the associated environmental and economic trade-offs (Andrews and Shabani, 2012; Bartels et al., 2010; Chaubey et al., 2013; Dougherty et al., 2009; Hetland and Mulder, 2007; Lemus and Martínez Duart, 2010; Mueller-Langer et al., 2007). For the purpose of considering the spatial planning of new supply infrastructure, issues of scale and cost are perhaps of the most immediate relevance. Some technologies are optimized for large-scale central production, with significant spatial footprints, while others can scale down to smaller facilities that could be easily distributed over a wide area. In practice there are likely to be real-world constraints on the siting of production facilities relating to planning legislation (i.e., zoning laws), land availability, and cost. System operators therefore are likely to face a choice between:

- Constructing centralized production facilities that produce the hydrogen more efficiently and at lower cost, but require more extensive delivery infrastructure (i.e., more vehicle deliveries or a more extensive pipeline network);
- Decentralized production, where production costs per unit of hydrogen are higher, but delivery infrastructure costs are lower because the production facilities are closer to demand outlets, or are even colocated together.

A number of mature production technologies are forerunners for consideration in the near term for supplying hydrogen infrastructure: steam methane reformation (SMR), gasification of coal or biomass, and electrolysis. SMR is the dominant form of hydrogen production worldwide (Chaubey et al., 2013), while other hydrocarbon feedstocks such as coal or biomass can also be used via gasification processes. The use of CCS techniques in conjunction with hydrogen production is an interesting future avenue for technology development as it offers a potential low emission pathway for producing hydrogen. Finally, electrolysis of water is also important to consider as it is the main low emission alternative to fossil-based hydrogen production with CCS when used in conjunction with renewable or nuclear electricity.

In very broad terms, the literature on hydrogen production generally presents SMR and biomass gasification as being viable at a range of scales with varying efficiencies and costs, coal gasification and direct nuclear electrolysis are generally discussed as large-scale processes only, and electrolysis with grid electricity (renewable or otherwise) is considered at medium or small scales (Dodds and McDowall, 2012a). CCS is understood to generally only be viable in medium- or large-scale facilities due to the additional expense of constructing a CO_2 pipeline network to many small stations (Mueller-Langer et al., 2007; Ogden, 2001).

14.2.4 Refueling stations

Final distribution of hydrogen to motorists occurs at refueling stations. Hydrogen fuel cell vehicles (FCVs) make use of hydrogen in gaseous form, typically delivered at pressures of 350 or 700 bar. Consumers resupply their vehicles from fueling nozzles that resemble conventional station pumps that dispense automotive fuels such as diesel or gasoline. Hydrogen is stored either in gaseous or liquid form, with deliveries of GH_2 made via pipelines or truck trailers, and LH_2 via road tankers. Larger stations may also be equipped with on-site production facilities. Accepting LH_2 deliveries or storing liquid hydrogen at refueling stations incurs additional capital and operational expenses, since hydrogen for FCVs has to be converted back into gaseous form for sale to customers.

14.3 Review of hydrogen infrastructure models

The existing approaches to modeling hydrogen infrastructure can be separated into the following broad categories:

- National scale energy system models
- Regional scale hydrogen supply chain (HSC) models
- Local scale models
- Hybrid approaches

A brief overview of each model category follows.

14.3.1 National scale energy system models

Many studies have analyzed the prospects for hydrogen technologies using national or global energy system optimization models. Most of these models use linear or mixed-integer linear programming (MILP) techniques to find the energy system that meets energy service demands at the least cost. These studies typically include detailed HSCs, with representation of various hydrogen production technologies, feedstocks, and modes of distribution. Hydrogen end-use technologies compete with other technologies (such as battery electric vehicles) to meet energy service demands (e.g., demands for car transport), and thus hydrogen demand and supply are both endogenously optimized. This feature, the simultaneous optimization of both hydrogen demand and supply within an energy system context, is the key strength of these models. Other approaches, such as regional scale and local scale models, generally represent hydrogen demand as an exogenous input in their analysis.

One of the most commonly used analysis frameworks is the MARKAL (MARKet ALlocation)/TIMES family of energy system models, developed in the 1970s by the International Energy Agency (IEA), and made available through the IEA Energy Technology Systems Analysis Program (Loulou et al., 2004). This framework has been used to examine the prospects for hydrogen in China (Rits et al., 2004), the United States (Tseng et al., 2005; Yeh et al., 2008), the United Kingdom (Dodds and McDowall, 2014; Strachan et al., 2009), Italy (Contaldi et al., 2008), Spain (Contreras et al., 2009), Norway (Rosenberg et al., 2010), Japan (Endo, 2007), and at European and global scales (Gül et al., 2009; Martinus et al., 2005; Rösler, 2012). Other, similar frameworks that have been applied to hydrogen include MESSAGE (Barreto et al., 2003), GET (Grahn and Williander, 2009), GRAPE (Ishimoto et al., 2012), and REDGEM70 (Takeshita, 2012). Despite their strengths in matching supply and demand, these models suffer a series of widely recognized weaknesses when it comes to representing and examining hydrogen infrastructure:

- *Weak representation of transport market structure and behavioral dynamics*: Most studies using the models above rely on approaches that assume that consumers are well informed and homogeneous in their preferences, and their vehicle adoption decisions are guided by a central planner who seeks to optimize the energy system with perfect foresight. The result is that the costs of transition are understated, and the rate of transition is likely to be overstated.

- *Weak representation of consumer and investor risk-aversion to new technologies*: In the real world, investors and consumers are resistant to purchasing new, untried technologies, because of perceived technology risk. This results, effectively, in a high cost of capital for the purchase of such technologies. This is often represented in energy system models through the application of a "hurdle rate" to new, risky technologies such as hydrogen vehicles. However, such rates can only be specified exogenously, whereas they should ideally be dependent on the deployment of the new technology.
- *Absence of spatial structure below the nation-state scale*: Few models of this kind represent spatial heterogeneity within the nation-state. Instead, most make explicit or implicit decisions about the average distances of distribution in order to calculate the costs associated with hydrogen delivery. This is problematic, since the costs of hydrogen delivery are highly dependent on the distribution distances and flow rates at which it is delivered. Some modelers apply exogenous (and typically static) constraints to represent at least some level of spatial heterogeneity, for example, requiring that a proportion of hydrogen supply be derived from distributed sources that meet demands in remote locations with relatively low demand densities (e.g., Shay et al., 2006).
- *Weak representation of returns-to-scale in hydrogen infrastructure provision*: There are many processes that lead to returns-to-scale in the course of developing a hydrogen infrastructure. Some of these are related to consumer adoption behavior (greater familiarity with a technology leads to heightened willingness to adopt, as discussed in Rogers, 2003), but also to volume manufacturing, to scale economies in infrastructure (a larger steam-methane reformer is cheaper, per unit of hydrogen produced, than a smaller one), and higher capacity factors (early-stage infrastructure will be under-utilized). Many of these models can be used with "endogenous technology learning" applied to certain infrastructure costs, by which costs reduce as a function of cumulative deployment, but this has been rarely done for hydrogen (for exceptions, see Krzyzanowski et al., 2007; Muncaster, 2008). Instead, the modeler must make assumptions about the appropriate scale at which to assume hydrogen is deployed, or must model the cost reductions associated with a transition as a function of time, thus implicitly assuming that a transition will take place.

14.3.2 Regional scale HSC models

Regional scale models depicting spatially optimized HSCs are a growing branch of the hydrogen literature. Typically, the level of demand for hydrogen and its geographical position within the region being considered are specified as exogenous inputs, with the model problem formulation focusing on the optimal configuration of supply side components. Modeling of HSCs can consist of a relatively simple quantification implemented through a spreadsheet model or by adopting a formal optimization methodology, notably dynamic programming (DP) or MILP.

Compared to the more common MILP approach, models using DP are relatively simple and a number of simplifying constraints are typically added to reduce the feasible solution set to a manageable size (Lin et al., 2008a). Among the work adopting MILP, a number of authors present limited information on the model used in the analysis. This group includes Ball et al. (2007), Hugo et al. (2005), Murthy Konda et al. (2011), Kamarudin et al. (2009), and Stiller et al. (2010). Other authors present models, that although not fully optimized in the MILP framework, include some iterative "optimization" (Johnson et al., 2008; Stiller and Bünger, 2009). There are three main models used to represent regional HSCs that are fully described in the literature:

Parker et al. (2010), Johnson and Ogden (2012), and Almansoori and Shah (2009). The latter, which presents a refinement of the model originally discussed in Almansoori and Shah (2006), is the most important model in this branch of the literature, as the equations presented have been widely used by other researchers.

Almansoori and Shah (2009) model primary energy sources, production plants (SMR, coal gasification, biomass gasification, and water electrolysis), storage, and transportation through tanker truck or railway tank car. The framework does not allow the model to determine the mix between liquid and compressed gas hydrogen endogenously so it is assumed exogenously. The model takes into account geographical granularity by representing all relevant variables for each geographical area. However, penetration rates of hydrogen are assumed to be constant across regions. The objective function in the model comprises production, storage, transportation, and primary energy costs. The authors sum costs occurring in different years rather than discounting them to a common year before summing them up, which would be more appropriate for assessing investments. The MILP model allows the user to specify constraints relating to the availability of primary energy sources, demand and production of hydrogen, transportation of hydrogen between spatial grids, and time constraints that describe how production and storage plants transfer across time periods. The model in Almansoori and Shah (2009) has been improved by a number of authors by introducing modifications such as additional delivery modes, including pipelines and ships (Han et al., 2012; Kamarudin et al., 2009; Kim et al., 2008; Sabio et al., 2012), multiobjective optimization (Brey et al., 2006; Hugo et al., 2005; Kim and Moon, 2008; Li et al., 2008; Sabio et al., 2012), discounting of costs (Sabio et al., 2012), and introducing stochasticity (Kim et al., 2008).

The SHIPmod model (Agnolucci et al., 2013) represents a recent refinement of Almansoori and Shah's base HSC framework. SHIPmod employs a discount factor formulation like the one discussed in Sabio et al. (2012), with this accounting for multiyear time steps, as in Akgul et al. (2012). Unlike most regional scale HSC optimization models, SHIPmod simultaneously represents both liquid (LH_2) and compressed gaseous hydrogen (GH_2), with the split between the two being determined endogenously rather than assumed exogenously. SHIPmod is one of few models that represents both CCS technology at hydrogen production sites and also represents the CO_2 delivered by pipes to a number of existing storage reservoirs. Demand growth is based on a logistic function at a rate informed by analysis of historical precedents (McDowall, forthcoming). In addition, the model allows for different demand penetration rates across geographical areas, with the demand for each area determined by factors such as income, cars per household, and education. The SHIPmod study is one of the only examples of work that combines analysis of the factors determining the possible spatial and temporal pattern demand for hydrogen with the optimization of the infrastructure needed to deliver it (Agnolucci et al., 2013).

14.3.3 Local scale models

To date, local scale spatial models of hydrogen infrastructure have focused on determining the optimal location of hydrogen refueling stations in a relatively small geographical space, normally a big city or a county region, but sometimes implemented at the state level (Upchurch and Kuby, 2010). Other studies adopt heuristic methods, such as the Idealized City Model developed by Johnson et al. (2005), or other *ad hoc* rules (Melaina, 2003; Melendez and Milbrandt, 2006). Work employing optimization techniques is usually based on a MILP framework. In terms of problem formulation, there are two main approaches, one based on the assumption that customers want to refuel their car near their home (*p*-median technique), and the other based on the idea that they want to refuel along their usual driving routes (flow model technique). The *p*-median approach consists of locating *P* facilities, in this case hydrogen filling stations, to minimize the average driving time or distance to the stations weighted by the population in the geographical area (Lin et al., 2008b; Nicholas, 2004). Flow models, on the other hand, attempt to optimally locate refueling stations to maximize their proximity to passing traffic (Kim and Kuby, 2012; Kuby and Lim, 2005; Upchurch et al., 2009).

14.3.4 Hybrid approaches

This final section describes the small number of studies that have attempted to combine models at different scales, linking the strengths of a national energy system optimization approach with a spatially detailed infrastructure optimization model. A fairly straightforward approach was used by Strachan et al. (2009) that couples a GIS (geographic information system) model to the UK MARKAL model. First, a GIS approach is used to assess possible transmission infrastructure options between a set of demand and supply centers. These configurations are input as technologies in the energy system model. By minimizing the cost of meeting a certain level of transport demand the energy system model can choose any of the configurations discussed above and by doing so it "builds" the related spatial infrastructure. The approach does not assess distribution infrastructure, focusing only on transmission.

Rosenberg et al. (2010) is another example of research that endogenously forecasts hydrogen demand while optimizing the related spatial infrastructure. The authors link a Norwegian MARKAL energy systems model to a regional HSC model, H2Invest, which is the framework described in Stiller and Bünger (2009). The resulting model is not fully optimized in the MILP framework, but rather it works on the basis of an iterative procedure that is hoped to provide a solution close to the mathematically optimal one. Coupling the two models required passing the outputs from one as inputs into the other and iterating toward a convergence solution in the two models.

Finally, the SHIPmod HSC model (Agnolucci et al., 2013), described in Section 14.3.2, has been soft-linked to the outputs of the UK MARKAL energy system model (Kannan et al., 2007). This approach places the optimization of HSCs within the wider context of optimal energy system evolution.

14.4 Case study: Decarbonizing UK transport demand with hydrogen vehicles

Having reviewed key literature in Section 14.3, the remainder of this chapter focuses on a case study scenario to illustrate key issues and highlight emerging research questions relevant to the optimal design of hydrogen infrastructure. The results of the UK scenario modeling carried out in Agnolucci et al. (2013) for the transport sector are presented. Agnolucci et al. (2013) uses the SHIPmod model, which remains one of the most advanced studies of hydrogen infrastructure deployment, due to its linkage of a detailed HSC model to the outputs from a full national energy system model, and its explicit consideration of factors that affect the spatial distribution of demand. The SHIPmod model focuses on future transport sector demand scenarios, and therefore, hydrogen use in buildings and industry for heat or electricity generation are not explicitly considered. SHIPmod also does not include a representation of pipelines for hydrogen distribution, with all deliveries being made by vehicle freight.

The case study presented here describes the model formulation, the key elements of the model, the process of soft-linking the infrastructure model to a national energy system model, demand assumptions, and finally the results and insights generated.

14.4.1 Model formulation

SHIPmod's full mathematical specification is detailed in Agnolucci et al. (2013). The model aims to minimize the total supply chain system cost by determining the optimal:

- Locations, scales, and types of hydrogen production plants, storage facilities, and filling stations;
- Hydrogen production rates at the production plants, stored amounts in the storage facilities;
- Flows of hydrogen between regions;
- Modes of transport for delivery of hydrogen;
- Flows of captured CO_2 between regions as well as between collection points and reservoirs;
- CO_2 inventory levels of the reservoirs;
- Locations and sizes of the onshore and offshore CO_2 pipes;

Subject to:

- Hydrogen demand in each region and time period;
- Types of hydrogen production technologies, storage, and filling stations;
- Transport logistics characteristics (costs, modes, distances, and availabilities);
- Capital investment and operating costs for each hydrogen production technology, storage facility, and filling station type;
- Remaining values for each hydrogen production technology, storage facility, filling station type as well as for onshore and offshore CO_2 pipes in each time period;
- Carbon tax per unit of CO_2 emitted in each time period, carbon emission and capture factors for each hydrogen production technology type;
- Locations of the CO_2 collection points and reservoirs, reservoir capacities and their connectivities to the collection points; and
- Linear capital cost and CO_2 flow correlations for onshore and offshore CO_2 pipes.

14.4.2 Overview of model components

The components of SHIPmod are:

- 35 UK regions, based on NUTS2[1] geocode classifications;
- Two physical forms of hydrogen, both liquid (LH_2) and compressed gas (GH_2);

[1]European Union "Nomenclature des Unités Territoriales Statistiques" (NUTS).

- 30 different types of hydrogen production facilities, with 4 core technologies including SMR, coal gasification (CG), biomass gasification (BG), and electrolysis. In the case of SMR, CG, and BG, the model incorporates plants with and without carbon capture and sequestration (CCS). Plant configurations producing GH_2 and LH_2 are available for all technology types, and facilities can be represented at four scales: distributed, small, medium, and large (although not all technologies are available at all scales). Each technology has different levels of CO_2 emissions associated with it, based on the carbon content of the feedstocks used (i.e., gas, coal, biomass, electricity) and the efficiency of the plant.
- Six types of storage facility in total, using three sizes (small, medium, and large), with variants for each size storing GH_2 or LH_2;
- Two types of transportation mode, either trailers transporting GH_2 and tankers transporting LH_2. Capital costs for tankers are almost twice that of trailers although because of their significantly increased carrying capacity, they are much less expensive per unit of hydrogen transported;
- Seven types of hydrogen vehicle filling stations. Filling stations are available in three sizes, small medium, and large, with variants for each size accepting deliveries by GH_2 trailers or LH_2 tankers, and a final variant with an onsite production plant (large-sized stations only);
- Three CO_2 capture aggregation station locations and three associated storage reservoirs, with associated storage capacities and infrastructure costs (i.e., pipelines).

A more detailed description of the input parameters used to define the above components is included in the Appendix.

14.4.3 Soft-linked inputs from UK MARKAL

In this case study, three main inputs to SHIPmod were taken from the UK MARKAL energy system model, based on the results of the runs presented in Dodds and McDowall (2012b):

- The CO_2 intensity of grid electricity varies over time as the model progressively decarbonizes the UK power sector to hit legislated national emissions targets for the year 2050. This has the effect of lowering the CO_2 emissions associated with electricity-using processes in SHIPmod over time.
- A tax on CO_2 emissions is introduced, with the level of the tax corresponding to the marginal abatement cost within a least-cost energy system transition that meets the UK's carbon reduction targets. This has the effect of increasing the production costs from carbon intensive hydrogen production pathways over time. The level of the tax across years is shown in the Appendix.
- The timing of the market entry of FCVs is based on the outputs from the energy system model. This way, the transition is consistent with a broader analysis of cost-optimal decarbonization trajectories. In this case, UK MARKAL determines that hydrogen FCVs only become cost-effective compared to alternatives from around the year 2040 onward. It should be noted that as some consumers are likely to be less price-sensitive and eager to adopt new, innovative technologies beforehand. Transitions from energy system models like MARKAL/TIMES are likely to be conservative (McDowall, forthcoming), and therefore, actual deployment in SHIPmod for this case study starts from around 2035 (see below).

14.4.4 Core demand assumptions

Many studies involving regional scale models, as described in Section 14.1, have assumed constant penetration rates of demand across different spatial regions. This is a limitation of past approaches that is unlikely to represent the real-world diffusion of any product, let alone one that requires an expensive supporting supply infrastructure. In order to generate a plausible scenario of the diffusion of hydrogen into the transport sector, a logistic diffusion model (Rogers, 2003) was employed for the case study scenarios. Following the main view from the literature (e.g., Almansoori and Shah, 2009; or Kim and Moon, 2008) it is assumed that hydrogen vehicles can ultimately reach 100% of the vehicle stock. Following the insights of Agnolucci and McDowall (2013), the optimism in the literature surrounding uptake rates is tempered by selecting a hydrogen demand scenario (namely, the "high policy support, modest learning scenario" scenario from the HyWays European Hydrogen Roadmap (2008)) that does not postulate introduction of hydrogen unfolding at a quicker pace than those observed in historical analogies (for a discussion of rates of transition for alternative fuelled vehicles, see McDowall (forthcoming)). Studies on the diffusion of innovations (Rogers, 2003) have suggested that around 2.5% of consumers are likely to act as "innovators." Therefore, this case study assumes that a 2.5% market share of such "innovators" can be reached in 2035 and a logistic curve is proposed for uptake, with the parameters estimated from the HyWays scenario.

A number of factors related (see Table 14.1) to the technological specification of the vehicles and the socioeconomic characteristics of the adopters are expected to be relevant in the adoption of FCVs (Ewing and Sarigöllü, 2000). Among the factors discussed in Melendez and Milbrandt (2006), the case study considers access to cars, education, commuting distance, and household income. All of these attributes are expected to have a positive impact on the diffusion of FCVs. The authors also believe that the diffusion of FCVs will be facilitated by high population density, as well as the total size of the population, which can be considered as a proxy for market size (Dunning, 1980).

The data used to input the relevant socioeconomic attributes were collected from the latest available UK census.[2] Following Melendez and Milbrandt (2006), scores from 1 (least favorable to hydrogen) to 5 (most favorable to hydrogen) for each attribute used in the study were constructed (by using the ClassInt package in R) for each geographical area. To combine these scores into one single mark for each area, a simple average was computed on the grounds that there is not enough research in the literature to warrant different weights across attributes on the basis of their importance.[3] The results from the scoring exercise are shown graphically in Figure 14.1.

[2]Data can be found in (ONS, 2012) for England and Wales, (NISRA, 2012) for Northern Ireland, and (GROS, 2012) for Scotland). As each attribute implies the use of three different variables defined in the census, one for each group of countries comprised in the United Kingdom, the detailed sources of the variable used in the study are not given here, although they are available upon request.

[3]In addition, an approach giving more weight to income and number of cars per households as done in (Melendez and Milbrandt, 2006) would have given too much importance to a single factor, as in the sample the correlation between income and access to cars is considerably high, with a value of about 0.70.

Table 14.1 Socioeconomic attributes thought to influence the adoption of hydrogen vehicles and related variables

Attribute	Variable
Access to cars	Percentage of households with two or more vehicles
Education	Percentage of population with higher-level qualifications
Commuting distance	Average commuting distance per person in miles
Household income	Gross disposable household income per capita at 2001 base prices
Population	Number of persons
Population density	Number of persons per hectare

Under this case study scenario, hydrogen is therefore expected to penetrate the passenger transport sector first in the southeast of England and then to develop along a corridor going from Manchester to London, including all the areas in between, with the exception of West Midlands. The third group of areas to adopt hydrogen vehicles includes Wales, some parts of northern England, and the West Midlands. The fourth uptake wave comprises large parts of Scotland, South Yorkshire in the north, Devon in the southwest, and Northern Ireland. Finally, the last group of areas comprises the far southwest and the far north of England as well as northernmost Scotland. Interestingly, this ranking, based on socioeconomic factors, generates an uptake scenario where hydrogen demand diffuses outward from the southeast of England in a broadly spatially contiguous fashion, although this was by no means guaranteed by the adopted approach.

Information from the ranking above is used to assign a set of five logistics to the geographical areas described. Hydrogen is introduced in the most promising areas in 2020 and in the least promising ones 10 years later. Based on evidence of the typically faster rate of diffusion in late adopting regions (Grübler et al., 1999), catching up occurs through a higher growth rate in the logistics for the area where hydrogen is introduced at a later stage. To compute hydrogen demand, passenger kilometers are estimated for each area by allocating traffic figures from DfT (2012) and DRD (2009) for Great Britain and Northern Ireland, respectively, on the basis of commuting distance data. Given the traffic figures for each area, the logistics have been applied to identify the passenger kilometers traveled by using hydrogen, from which one can easily compute hydrogen demand by using efficiency data for FCVs from Dodds and McDowall (2012a). The result of this procedure is shown in Figure 14.2.

14.4.5 Overview of case study scenarios

Scenarios are a widely applied approach for dealing with deep uncertainty in system modeling. This case study employed a number of scenario variants. The baseline scenario uses the hydrogen fuel demand projections, resource costs, and technology characteristics outlined above. In addition to the base case, four

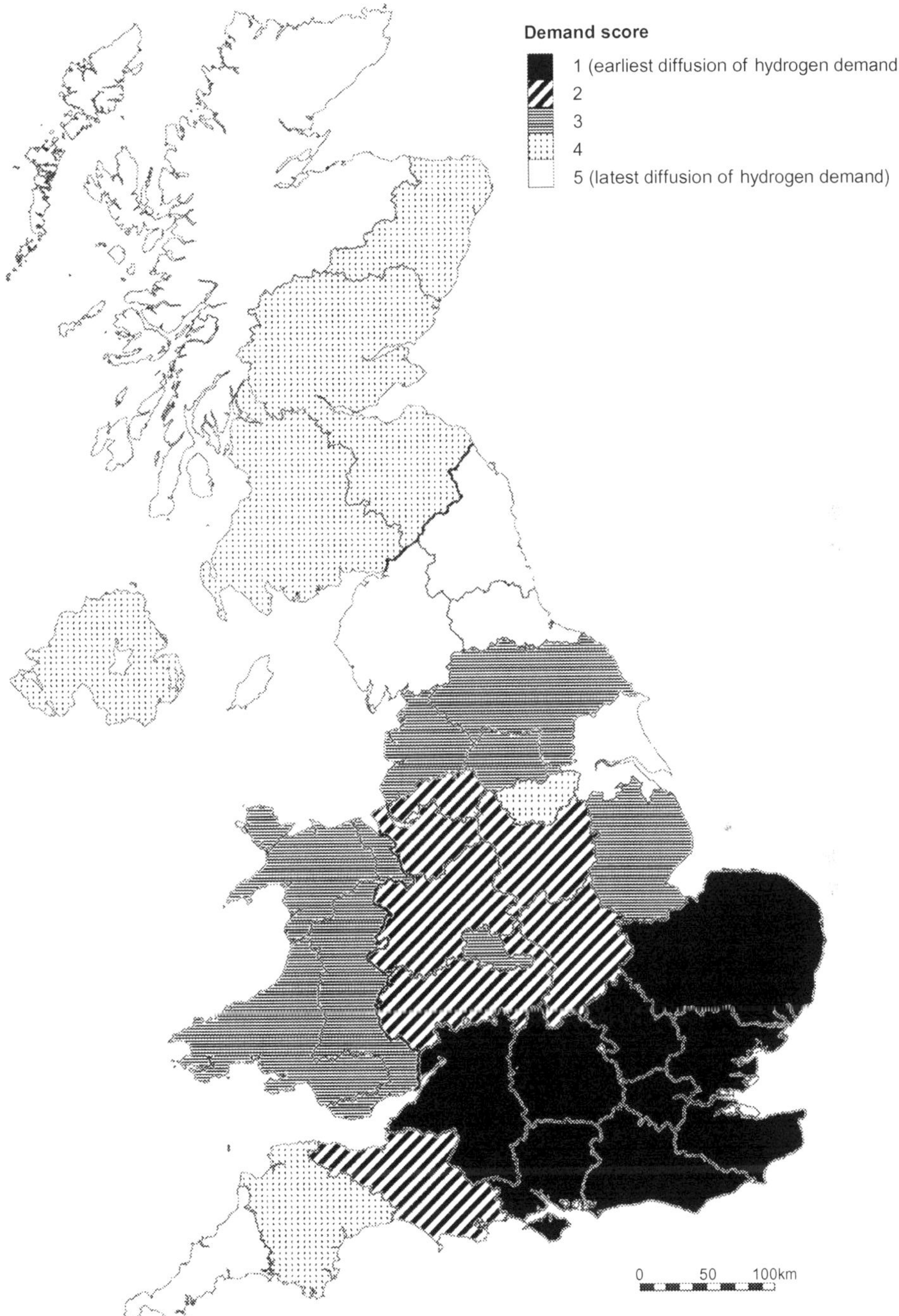

Figure 14.1 Rates of hydrogen diffusion for the study scenarios in different UK regions.

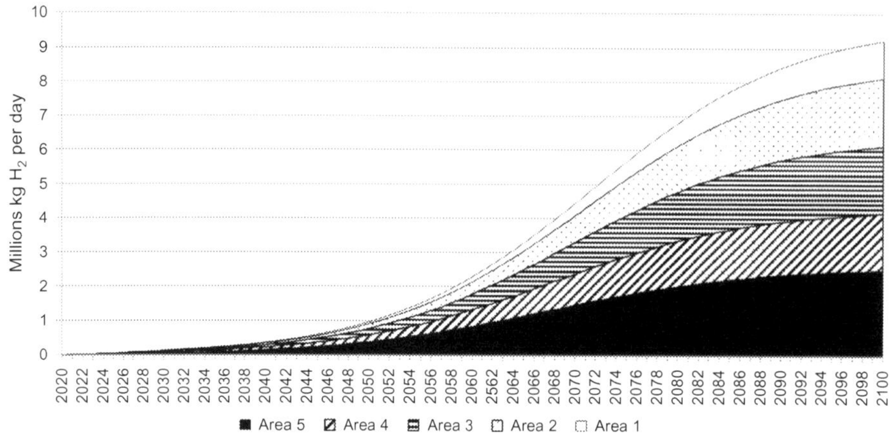

Figure 14.2 Daily demand for hydrogen split according to order of areas penetrated by hydrogen.

alternative scenarios as described in Table 14.2 were generated to examine two basic sources of uncertainty:

- Hydrogen demand characteristics and evolution
- Technology and resource availability

14.5 Results

The results from each model scenario are discussed in detail in Agnolucci et al. (2013). This case study illuminates three main areas of interest directly relevant to hydrogen system planners: the trade-offs between production scale and transport costs, the tension between production capacity and capital utilization, and the role of technological uncertainties in hydrogen production.

14.5.1 Trade-offs between production scale and transport costs

The spatial pattern of hydrogen demand results in trade-offs between production and transportation costs. Larger production plants produce hydrogen at a lower cost (incorporating capital, operating and fuel costs) because of scale economies, but incur higher transportation costs. The opposite occurs for distributed and small hydrogen production facilities. In this case study, the model used showed a preference for deploying large production facilities in central regions, or close to regions with high demand, where supply can occur over relatively short transport distances. Small and distributed production facilities were generally only established in peripheral regions where transport costs from the main production centers were determined to be prohibitive.

This is clearly illustrated in the base case scenario (see Figure 14.3). Small, relatively costly, production facilities (such as distributed and small-scale SMR) are built

Table 14.2 Case study scenarios, their characteristics, and their rationale for inclusion

Scenario name	Scenario description	Rationale for inclusion
Base case	The base case scenario uses the demand assumptions and technologies described in Sections 14.3.2–14.3.4	This is the baseline case against which other scenarios can be compared
Diffuse demand	Total demand for hydrogen is the same as in the base case, but in this scenario demand is equally apportioned to each of the 35 regions based on population, which results in less concentrated demand	These two scenarios tests the sensitivity of geographical dispersion/clustering of demand on the optimum system configuration
Clustered demand	Total demand for hydrogen is the same as the base case, but most of this demand is spatially clustered in the "leading" regions, i.e., the four major urban centers of London, the West Midlands, Southwest Scotland, and Manchester–Merseyside. Demand outside of these regions is built up later and more slowly	
High demand	Demand for hydrogen is increased fivefold, with the same spatial distribution as the base case. This results in a demand uptake trajectory which is within the range of those typically discussed in much of the literature around hydrogen FCV uptake, with a much faster rate of deployment than considered in the base case	This scenario is used to assess the impact of varying total demand on the optimal configuration of the system
No biomass	This scenario is identical in terms of demand uptake as the base case, with the exception that biomass technologies are not available for hydrogen production	This scenario is included as a sensitivity check, as low carbon hydrogen production from biomass was observed to be important in early model runs

in more isolated regions, such as Northern Ireland, the southwest (Cornwall, Devon, etc.), and Wales. In central England, close to the major demand centers of London and the West Midlands, larger production plants are built using SMR with CCS and medium-sized biomass gasification plants, with hydrogen distributed to surrounding regions. No hydrogen is produced via electrolysis or from coal with CCS in this scenario.

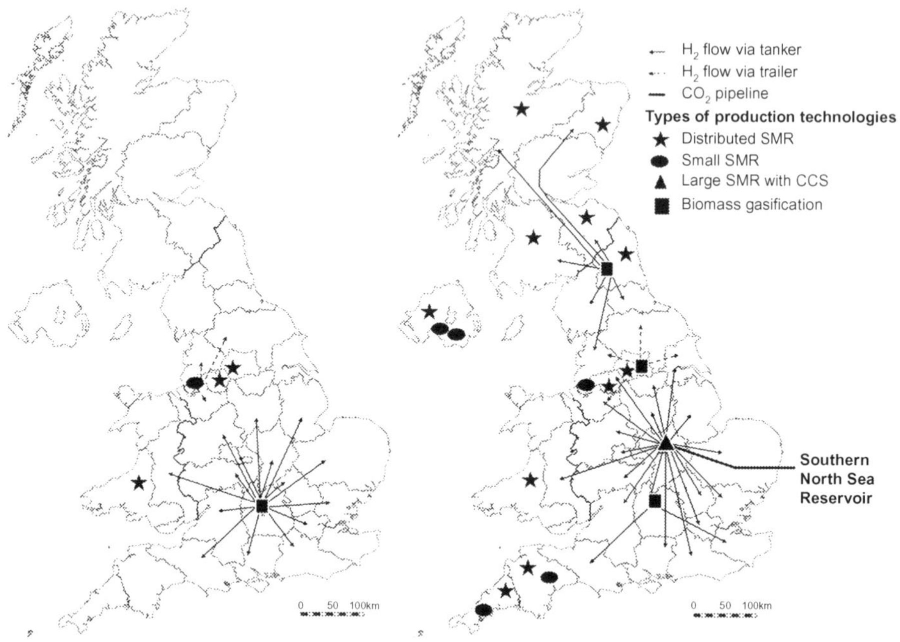

Figure 14.3 Evolution of supply in the base case scenario (first and last model period).

Most of the other scenarios considered, including the "diffuse," "clustered," and "no biomass" cases, show a similar overall pattern for hydrogen production to the base case, i.e., with large plants being built in central England, with distributed SMR serving peripheral regions. The exception is the "high demand" case. In this scenario, the majority of the hydrogen is produced from medium-sized biomass plants, with one large biomass plant constructed in central England. Northern Ireland and Cornwall are the only regions that see deployment of distributed SMR, which produces only very small quantities of hydrogen in comparison to the larger biomass installations. The higher levels of demand in this scenario enable the model to cost-effectively build an increased number of larger-scale plants relative to the base case.

The transport of hydrogen between regions plays a significant role in all scenarios. Examining hydrogen flows between regions as a proportion of total hydrogen production (Figure 14.4) shows that most hydrogen in the model is not produced within each NUTS2 region but instead delivered from outside regions by tanker or trailer. The exception is the "clustered demand" scenario, which sees no trucked hydrogen in the first period, because all demand is located in the regions where hydrogen is first deployed, i.e., regions containing the UK's largest urban centers. The importance of distribution grows over time in this scenario, as demand emerges in peripheral regions and is supplied by hydrogen produced in the major demand centers. In the "high demand" scenario, the presence of scale economies enables the establishment of cost-effective, larger scale local plants, rather than either expensive small plants (such

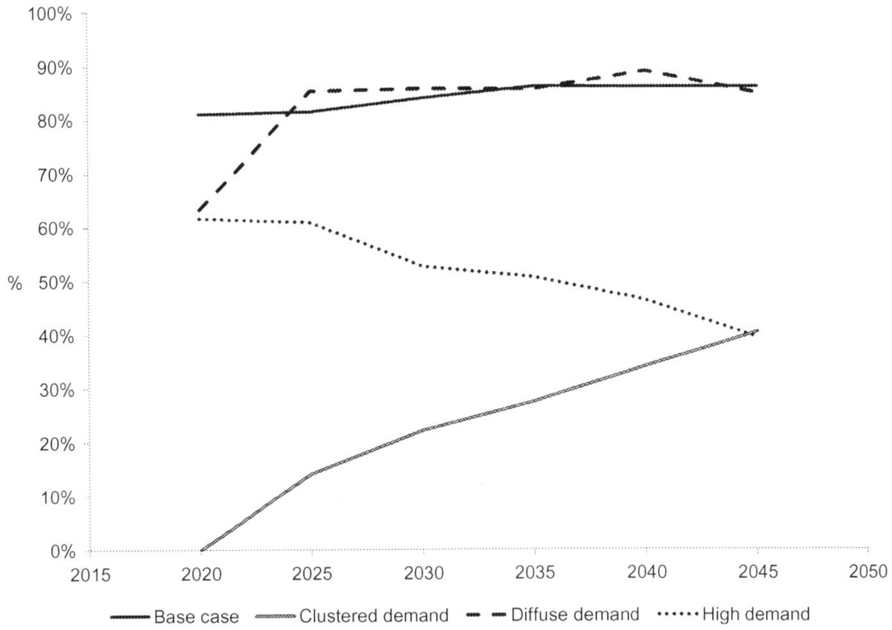

Figure 14.4 Proportion of production that is transported between regions (%) rather than produced locally, in different scenarios.

as distributed SMR) or imports, leading to a declining share of trucked hydrogen as time goes by.

The importance of transportation, and in particular transportation costs, is also highlighted by the hydrogen form chosen by the model in this case study. As discussed in Section 14.2, LH_2 production plants are more expensive than equivalent plants using the same technology producing GH_2. However, the costs of distributing LH_2 are much lower on a per unit basis than distributed GH_2, because the energy density of stored LH_2 is much higher. As most scenarios in the case study considered here are dominated by LH_2 produced in large centralized plants, the additional transportation costs of GH_2 are clearly more important than the additional liquefaction costs, with the exception of peripheral regions such as Northern Ireland and Cornwall, where small quantities of GH_2 are produced in distributed plants. The two scenarios show revealing exceptions to this overall trend:

- In the "high demand" scenario there is sufficient demand in a number of regions to support a medium-sized biomass gasification plant within the regions themselves rather than putting larger-scale plant in adjacent areas. As more of the total demand is met locally, there is a lower requirement to import hydrogen from other regions, and the model prefers to build these production plants in configurations that produce the cheaper GH_2 rather than LH_2;
- In the "clustered demand" scenario, a greater proportion of demand is located in the major early-adopter regions, and the model builds the relatively cheaper GH_2 production plants in preference to LH_2 plants, as less hydrogen is transported long distances. Demand in

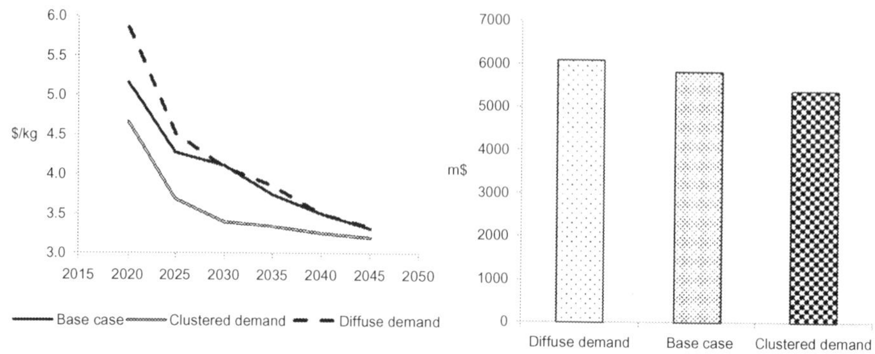

Figure 14.5 Undiscounted costs of delivered hydrogen over time in different scenarios (left) total discounted costs across the model time horizon (right).

late-comer regions is met either by local GH_2 production from small distributed SMR plants in each region, or from two larger LH_2 plants, one built in the north of England and another located in south-central England.

Given the importance of the trade-off between scale-economies and transport costs, the spatial pattern of demand across regions also has a strong effect on total costs nationally, as illustrated in Figure 14.5. The total discounted costs of hydrogen supply are 10% higher in the "diffuse" scenario compared with the "clustered" one. This cost differential is particularly large in the early periods, with the costs per kilogram of hydrogen in the diffuse scenario 25% greater than in the clustered scenario.

14.5.2 Trade-offs between production capacity and capital utilization

For the same types of hydrogen production facility, economies of scale result in larger-scale plants with lower production costs than smaller-scale plants. However, large-scale plants require significant additional capital investment to deploy and may not run at full capacity if demand for hydrogen is low, which happens in all scenarios during the early years. An interesting question for hydrogen system planners therefore, from a capital utilization perspective, is whether it is more resource efficient to build a larger number of smaller-scale, cheaper plants that are more fully utilized in all time periods, or a smaller number of larger, more expensive plants, that run below full capacity?

In this case study, the trade-off between production costs and transport costs, the low level of demand, and the broad spatial dispersion of demand results in the model leaves significant production capacity unused in all scenarios. Due to the large difference between minimum and maximum production capacities, the strategy of building larger-scale plants ultimately becomes more cost-effective over time compared to smaller plants despite this approach leaving a considerable amount of production capacity unused. In addition, transport costs prevent the model from simply building a single large plant and using it to maximum capacity by exporting hydrogen to all other regions.

In the case study, demand grows over time and the model crosses different thresholds at which larger hydrogen production plants become cost-effective, triggering their periodic deployment. This results in a pattern by which spare capacity falls as demand grows until a new threshold is crossed for an additional investment in a large new plant, which again increases the space capacity (see Figure 14.6). It is possible that the pattern of these transitions and the preference for larger, more capital intensive plant is driven by the particular mixture of performance and cost characteristics used for the production technologies considered in this case study, but it is nevertheless an interesting observation.

As noted previously, the version of the SHIPmod model used in the case study did not include hydrogen pipelines as a transport option along with vehicle freight. Pipelines could potentially enable more cost-effective use of large, centralized hydrogen production technologies by reducing long-distance, high-volume transport costs. While the precise impact of including pipelines would need to be tested, the authors speculate that their omission may have caused an overstatement of the costs of hydrogen presented in the case study, as well as the levels of unused production capacity.

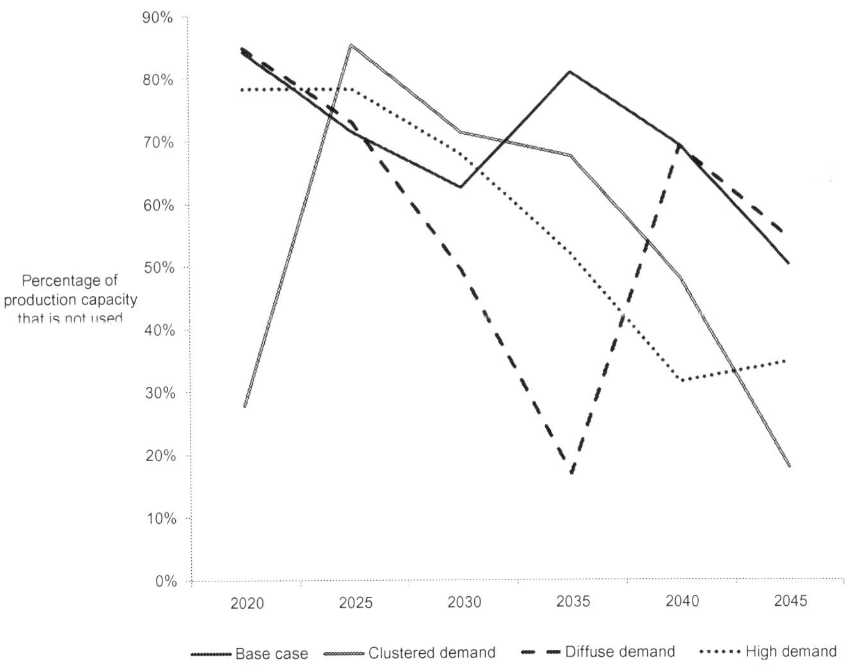

Figure 14.6 Spare capacity as a proportion of total capacity under different modeled scenarios.

14.5.3 Technological uncertainties: roles of bioenergy and CCS

The "no biomass" scenario results in a complete reliance on natural gas for hydrogen production, with SMR plants of various sizes built across the country. In this scenario, because biomass energy is not available, the model is dependent on fossil fuels with CCS to produce low carbon hydrogen for consumers. As a result, production facilities with CCS are introduced much earlier than in other scenarios, and at a smaller scale, building two medium-sized SMR–CCS plants by 2025, as well as a single large SMR–CCS plant later on. This is perhaps unsurprising: unabated small and medium SMR plants would incur excessive carbon costs, and electrolysis remains expensive and still incurs relatively high carbon costs until the power grid becomes significantly decarbonized from around 2030. The evolution of CCS plant and pipeline capacity is illustrated in Figure 14.7, where black lines represent pipelines, light shaded regions contain a medium-sized SMR–CCS plant, and the dark-shaded region contains a large SMR–CCS plant.

The evolution of the CCS network is clear: an initial medium-scale SMR–CCS plant is built between the major urban centers of Birmingham and London in 2020, with a pipeline taking CO_2 to a storage reservoir in the southern North Sea. In 2025, an additional medium SMR–CCS plant is constructed in Lancashire in the northwest of England. By 2035, sufficient additional demand has developed to justify a third plant, and a large-scale SMR–CCS plant is deployed in central England. This additional plant makes use of the existing CO_2 pipeline capacity and is constructed on the route of the pipeline to the southern North Sea reservoir.

14.6 Conclusions

14.6.1 Limitations of hydrogen infrastructure modeling

The literature review described in Section 14.3 highlights a number of limitations within the field of hydrogen infrastructure modeling. First, there is the general caution that, as with all studies exploring possible future scenarios, the input data is all highly uncertain, and small changes to performance or costs variables could conceivably

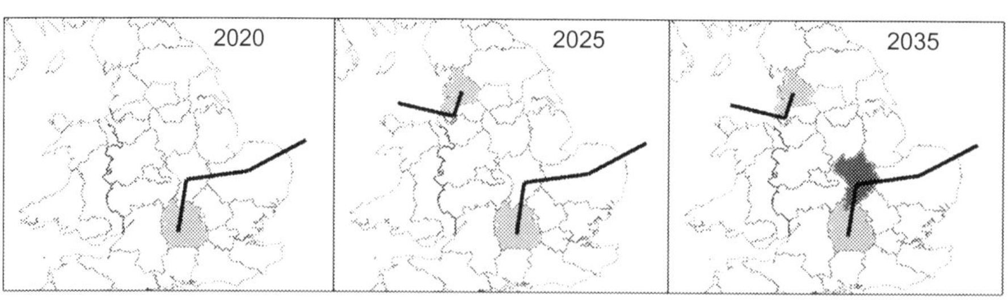

Figure 14.7 Evolution of the CCS network in the "no biomass" scenario over the 2020–2035 time period.

produce large shifts in terms of the optimum system configuration. Second, relatively few models include representations of hydrogen pipelines, a major option for hydrogen infrastructure, CCS infrastructure, a major option for low carbon hydrogen production, or future hydrogen demand from buildings and/or industry, which are potentially significant markets for hydrogen technologies (Li and Agnolucci, 2014).

More fundamentally, the reliance in many models on an exogenous hydrogen demand curve is a clear limitation. Fuel demand is strongly influenced by its costs, relative to the other options, but the use of an exogenous demand forecast prevents any feedback between supply and demand. The model used in the case study presented above attempted to improve on previous modeling practice of HSCs by deriving demand assumptions from a coherent analysis of energy system possibilities, and by ensuring that demand assumptions were consistent with other key parameters such as carbon prices and the carbon intensity of electricity through the use of a national scale energy system model. A more advanced approach would be to integrate regional supply chain modeling with national energy system modeling so that H_2 demand can respond to the infrastructure costs generated by the combined approach. This would require iteration between the two models, building on the approach previously adopted by Rosenberg et al. (2010).

A final point concerns the perfect-foresight formulation of most of the optimization models found in the literature. Typically models of this kind optimize over the full-time horizon, with the least-cost pathway solution produced on the basis of full knowledge of future system cost and performance conditions. However, in the real world, decision-makers do not act with perfect foresight. Rather, they hedge in the face of uncertainty, and those making risky investments require compensation for the risk they take on, resulting in higher costs of capital. In effect, these risk effects may result in returns to scale: as a transition unfolds, investor confidence in the future success of hydrogen builds, and the costs of capital fall. Alternatives include myopic formulations, in which no knowledge of the future is assumed, or stochastic optimization formulations.

14.6.2 Key insights for hydrogen system planners

This chapter presented a case study using an optimization-based framework for the design and evolution of HSC and CCS pipeline networks over a long planning horizon. The overall problem has been formulated as a multiperiod, MILP model. The case study results highlight a number of important questions that need to be resolved by hydrogen system planners as well as providing insights into the role of key unknowns, such as the future availability of bioenergy resources.

First, despite the enthusiasm in some of the literature on the potential for hydrogen to facilitate a highly decentralized energy system, the case study discussed above found that the most cost-effective approach was to deploy a smaller number of large-scale production facilities, even when this strategy left large amounts of supply capacity underutilized. Small and distributed production facilities were established only in peripheral regions where transport costs became prohibitive. The trade-off between production costs and transportation costs is an important factor in determining the preference for large plants. The modeling work also demonstrated that the per unit lower transportation costs of liquid hydrogen more than compensate for the increased production costs.

Second, the case study revealed that varying the spatial pattern of demand has significant impacts on both the structure of the optimal supply system and on the overall costs of delivered hydrogen. Highly clustered demand is, unsurprisingly, rather cheaper to service than highly diffuse demand. In addition, clustered demand shifts the preference of the model to gaseous hydrogen rather than liquid hydrogen, due to the lower importance of transport costs caused by the shorter length of the average delivery haul. Depending on the number of clusters and their relative size, medium-sized production plants can sometimes become more cost-effective than large plants because of the reduced requirement for transportation.

Third, the model shows that the overall level of demand can play a similar role to the degree of spatial clustering of demand. With high demand, as with highly spatially clustered demand, medium-sized production becomes cost-effective and hydrogen tends to be produced in gaseous form because of the relatively small catchment areas for each plant. These can be considered important results because demand assumptions, particularly the spatial pattern of demand, tend to be downplayed in the literature, despite having clear implications for marketing and transition strategies of hydrogen in the passenger vehicle sector.

Of course, all of these conclusions are conditionally applicable to the case study scenarios described, with the particular combination of cost and performance levels used, the demand uptake pattern assumed, and the spatial layout of population centers that are specific to the UK. Hydrogen system planners and researchers will need to draw their own inferences about how these findings map to the markets and geographies where they focus their own efforts. However, the authors of this chapter hope that the work presented highlights some of the key issues, illuminates some of the key unknowns, and demonstrates relevant and useful techniques that can be applied by others to their own analyses when designing hydrogen infrastructure.

Appendix

Production technologies

Values related to minimum and maximum production capacities of the plants are presented in Table 14.4. GH_2 capital costs (Table 14.3) are taken from NAE and NRC (2004) and NRC (2008) with the exception of the values for Medium SMR and Small BG which are taken from Iaquaniello et al. (2008) and Krewitt and Schmid (2005), respectively. LH_2 capital costs (Table 14.3) comprise the capital of the production and of the liquefaction plant. Costs for liquefaction units are taken from Krewitt and Schmid (2005). All values have been scaled to the maximum capacity of each plant in Table 14.4 based on the size factors from NAE and NRC (2004). In terms of Unit Production Cost (i.e., the sum of fuel and operating costs per unit production), the techno-economic analysis described in Appendix C of Almansoori (2006) is used.

Table 14.3 Capital costs of hydrogen production plants in $m

Product type (*i*)	Plant size (*p*)	Production technology (*j*)						
		SMR	SMR–CCS	CG	CG CCS	BG	BG CCS	Electrolysis
LH_2	Distributed	–	–	–	–	–	–	–
	Small	50	–	–	–	93	–	64
	Medium	280	330	–	–	329	379	553
	Large	860	910	1587	1637	1572	1622	–
GH_2	Distributed	4.9	–	–	–	–	–	7.1
	Small	14.5	–	–	–	48.2	–	29
	Medium	127	177	–	–	175	225	399
	Large	453	503	1152	1202	1165	1215	–

Table 14.4 Minimum/maximum production capacities for hydrogen production plants in thousand kg d^{-1}

Product type (*i*)	Plant size (*p*)	Production technology (*j*)						
		SMR	SMR–CCS	CG	CG CCS	BG	BG CCS	Electro lysis
LH_2	Distributed	–	–	–	–	–	–	–
	Small	1.6/10	–	–	–	1.6/14	–	1.6/10
	Medium	10/ 150	–	–	–	15/ 150	–	10/150
	Large	200/ 1100	200/ 1100	200/ 1200	200/ 1200	200/ 1100	200/ 1100	–
GH_2	Distributed	0/1.5	–	–	–	–	–	0/1.5
	Small	1.6/10	–	–	–	1.6/14	–	1.6/10
	Medium	10/ 150	–	–	–	15/ 150	–	10/150
	Large	200/ 1100	200/ 1100	200/ 1200	200/ 1200	200/ 1100	200/ 1100	–

The values in Almansoori (2006) were updated to include the capital costs described above as well as prices from primary sources that are more reflective of the current view of expected future market conditions. The natural gas price used in the analysis is 1.9 p/kWh, i.e., the average price paid by UK interruptible consumers over the period 2008–2011 according to UK Department of Energy and Climate Change (DECC) quarterly price statistics. Note that this implies a price of 8.2 dollars/million BTU against the 2.5 used in Almansoori (2006). The electricity price used in the computation is 5.4 p/kWh from the relevant DECC quarterly price statistics, which implies about 0.08 USD per kWh against the 0.05 assumed in Almansoori (2006) (Table 14.5).

Table 14.5 Unit production costs for hydrogen production plants in $ kg^{-1}

		Production technology (*j*)						
Product type (*i*)	**Plant size (*p*)**	**SMR**	**SMR–CCS**	**CG**	**CG CCS**	**BG**	**BG CCS**	**Electrolysis**
LH_2	Distributed	–	–	–	–	–	–	–
	Small	6.09	–	–	–	–		8.94
	Medium	2.94	3.17	–	–	2.98	3.18	6.37
	Large	2.36	2.45	2.11	2.18	2.3	2.38	–
GH_2	Distributed	3.87	–	–	–	–	–	5.86
	Small	3.28	–	–	–	–	–	5.62
	Medium	1.94	2.17	–	–	1.89	2.09	5.09
	Large	1.79	1.88	1.42	1.5	1.64	1.71	–

Table 14.6 CO_2 emissions from electricity

Year	Emission factors (gCO_2/kWh)
2020	391
2025	235
2030	168
2035	102
2040	69
2045	45
2050	26
2055	26
2060	26

CO_2 emissions from hydrogen production depend on the carbon content per MJ of the energy sources used in the production process, the efficiency of the plants—mainly sourced from NAE and NRC (2004), the electricity consumption of the plant, whether the hydrogen is produced in liquid or compressed gas form, and finally, whether CO_2 is being sequestered or not. Table 14.6 shows the emission factors of electricity which were taken from the UK MARKAL scenario presented in Dodds and McDowall (2012b) and the emission factors of the energy sources used to produce hydrogen.

Transportation modes

Most of the parameters from Table 14.7 are taken from Almansoori and Shah (2009) with the exception of the fuel price, which is set at the dollar equivalent of 1.50 British pounds per liter (reflecting contemporary UK market pricing for road transport fuel), the minimum flow rate, which is set equal to the size of a single unit as described in

Table 14.7 **Parameters for transportation modes**

Parameter	Measurement unit	LH_2 tanker	GH_2 trailer
Driver wage	$\$ h^{-1}$	23	23
Fuel economy	$km\ l^{-1}$	2.3	2.3
Fuel price	$\$ l^{-1}$	2.25	2.3
General expenses	$\$ d^{-1}$	8.22	8.2
Load/unload time	h	2	0.25
Maintenance expenses	$\$ km^{-1}$	0.1	0.1
Minimum/maximum flow rate	$kg\ d^{-1}$	3370/1,100,000	250/1,100,000
Average speed	$km\ h^{-1}$	55	55
Capacity	$kg\ mode^{-1}$	3370	250
Local availability	$h\ d^{-1}$	15	15
Regional availability	$h\ d^{-1}$	18	18
Capital cost	$\$ mode^{-1}$	775,000	460,000

Krewitt and Schmid (2005), and the capital costs which were also sourced from Krewitt and Schmid (2005).

Storage plants

Storage parameters have been sources from the US H_2A database (Steward et al., 2008) which presents values for both LH_2 and GH_2. As one can see in Table 14.8, storing GH_2 is considerably more expensive than storing LH_2, a factor which helps offset the cost of liquefaction needed to produce LH_2.

Table 14.8 **Parameters for storage facilities**

Parameter	Size	Measurement unit	LH_2	GH_2
Minimum capacity	Small	$kg\ d^{-1}$	0	0
	Medium		10,000	380
	Large		200,000	5010
Maximum capacity	Small	$kg\ d^{-1}$	9500	370
	Medium		150,000	5000
	Large		540,000	25,000
Capital costs	Small	\$	2,069,829	639,000
	Medium		7,862,044	7,851,000
	Large		25,526,292	38,868,000
Unit cost	Small	$\$ d^{-1}$	0.02698	0.27926
	Medium		0.00635	0.18972
	Large		0.00569	0.18712

Table 14.9 **Parameters for filling stations**

Parameter	Measurement unit	Size	LH_2 (tanker)	GH_2 (trailer)	GH_2 (distributed)
Maximum capacity	$kg\ d^{-1}$	Small	325	325	–
		Medium	750	750	–
		Large	1500	1500	1500
Capital costs	$	Small	318,000	234,000	–
		Medium	637,000	499,000	–
		Large	1,274,000	998,000	2,607,000

Filling stations

Three types of filling stations retailing hydrogen to owners of hydrogen FCVs are considered in SHIPmod, namely stations receiving LH_2 by tanker, stations receiving GH_2 by trailer, and finally stations with an on-site production plant. In all cases, hydrogen is retailed in GH_2 form for use in passenger vehicles. Filling stations are available in three sizes, servicing 72, 167, or 333 cars per day. Only the largest filling stations have the option to utilize on-site production plant. As one can see in Table 14.9, stations receiving LH_2 are considerably pricier than stations receiving GH_2, due to the former requiring high pressure storage, LH_2 storage, evaporators, and cryogenic compressors which are considerably more expensive than the compressors required by stations receiving GH_2 through trailers. Stations with on-site production are even more expensive due to the required low pressure storage. Note that the cost of the hydrogen production technologies that must be installed adjacent to stations with on-site production is not included in the capital cost of the station, but rather in the cost of the production technologies shown above in Table 14.3. The technical specification of the filling stations can be seen in Table 14.10.

CO_2 capture and storage infrastructure

In order to sequester CO_2, SHIPmod assumes that one has to build onshore pipes from the plant up to the collection points, and offshore pipes from the collection points to the reservoirs. The capital cost of onshore and offshore CO_2 pipes was modeled through a linear relationship between the cost per km and the diameter of the pipelines which was obtained from IPCC (2005). This source describes two curves (high and low) for offshore and onshore pipes. After obtaining the slope and intercept of the curve an average of the parameters was computed and used in the modeling. Collection points are onshore locations near the reservoirs from where offshore pipelines reaching the reservoirs begin. Following DTI (2006), this chapter takes into account three CO_2 reservoirs around the UK. The maximum capacity for each reservoir was sourced from DECC (2010). Table 14.11 shows the CO_2 reservoirs modeled in this study and the regions where collection points for each reservoir are located.

Table 14.10 Technological specifications of filling stations

	LH_2 (tanker)			GH_2 (distributed)	GH_2 (trailer)		
	Small	Medium	Large	Large	Small	Medium	Large
Maximum throughput (kg d^{-1})	325	750	1500	1500	325	750	1500
Served cars per day	72	167	333	333	72	167	333
Dispensers	2	3	6	6	2	3	6
Gas compressors required	0	0	0	2	1	1	2
Gas compressor throughput (kg/h/comp-essor)	0	0	0	63	27	63	63
High-pressure storage (kg)	38	75	150	150	0	0	0
Low-pressure storage (kg)	0	0	0	1500	0	0	0
Liquid H_2 storage (kg)	2250	4500	9000	0	0	0	0
Evaporator unit (kg d^{-1})	375	750	1500	0	0	0	0
Cryogenic compressors	1	1	1	0	0	0	0
Cryogenic compressor power (kW)	17.5	35	70	0	0	0	0

Table 14.11 CO_2 **reservoirs modeled in this study and related collection points**

Reservoir	Collection points
UK Northern and Central North Sea	North Eastern Scotland
UK Southern North Sea	East Anglia
East Irish Sea	Merseyside

Acknowledgments

This chapter draws heavily on work previously published by Agnolucci and McDowall (2013) and Agnolucci et al. (2013).

References

Agnolucci, P., McDowall, W., 2013. Designing future hydrogen infrastructure: insights from analysis at different spatial scales. Int. J. Hydrog. Energy 38, 5181–5191. http://dx.doi.org/10.1016/j.ijhydene.2013.02.042.

Agnolucci, P., Akgul, O., McDowall, W., Papageorgiou, L.G., 2013. The importance of economies of scale, transport costs and demand patterns in optimising hydrogen fuelling infrastructure: an exploration with SHIPMod (spatial hydrogen infrastructure planning model). Int. J. Hydrog. Energy 38, 11189–11201. http://dx.doi.org/10.1016/j.ijhydene.2013.06.071.

Akgul, O., Shah, N., Papageorgiou, L.G., 2012. An optimization framework for a hybrid first/second generation bioethanol supply chain. Comput. Chem. Eng. 42, 101–114. http://dx.doi.org/10.1016/j.compchemeng.2012.01.012.

Almansoori, A.S.A.I., 2006. Design and Operation of a Future Hydrogen Supply Chain. Imperial College London, London, UK.

Almansoori, A., Shah, N., 2006. Design and operation of a future hydrogen supply chain. Chem. Eng. Res. Des. 84, 423–438. http://dx.doi.org/10.1205/cherd.05193.

Almansoori, A., Shah, N., 2009. Design and operation of a future hydrogen supply chain: multi-period model. Int. J. Hydrog. Energy 34, 7883–7897. http://dx.doi.org/10.1016/j.ijhydene.2009.07.109.

Andrews, J., Shabani, B., 2012. Where does hydrogen fit in a sustainable energy economy? Proc. Eng. 49, 15–25. http://dx.doi.org/10.1016/j.proeng.2012.10.107.

Ball, M., Wietschel, M., Rentz, O., 2007. Integration of a hydrogen economy into the German energy system: an optimising modelling approach. Int. J. Hydrog. Energy 32, 1355–1368. http://dx.doi.org/10.1016/j.ijhydene.2006.10.016.

Barreto, L., Makihira, A., Riahi, K., 2003. The hydrogen economy in the 21st century: a sustainable development scenario. Int. J. Hydrog. Energy 28, 267–284. http://dx.doi.org/10.1016/S0360-3199(02)00074-5.

Bartels, J.R., Pate, M.B., Olson, N.K., 2010. An economic survey of hydrogen production from conventional and alternative energy sources. Int. J. Hydrog. Energy 35, 8371–8384. http://dx.doi.org/10.1016/j.ijhydene.2010.04.035.

Brey, J.J., Brey, R., Carazo, A.F., Contreras, I., Hernández-Díaz, A.G., Gallardo, V., 2006. Designing a gradual transition to a hydrogen economy in Spain. J. Power Sources 159, 1231–1240. http://dx.doi.org/10.1016/j.jpowsour.2005.12.089.

Chaubey, R., Sahu, S., James, O.O., Maity, S., 2013. A review on development of industrial processes and emerging techniques for production of hydrogen from renewable and sustainable sources. Renew. Sustain. Energy Rev. 23, 443–462. http://dx.doi.org/10.1016/j.rser.2013.02.019.

Contaldi, M., Gracceva, F., Mattucci, A., 2008. Hydrogen perspectives in Italy: analysis of possible deployment scenarios. Int. J. Hydrog. Energy 33, 1630–1642. http://dx.doi.org/10.1016/j.ijhydene.2007.12.035.

Contreras, A., Guervós, E., Posso, F., 2009. Market penetration analysis of the use of hydrogen in the road transport sector of the Madrid region, using MARKAL. Int. J. Hydrog. Energy 34, 13–20. http://dx.doi.org/10.1016/j.ijhydene.2008.10.031.

DECC, 2010. CO2 Storage in the UK—Industry Potential. UK Department of Energy and Climate Change (DECC), London, UK.

Decourt, B., Lajoie, B., Debarre, R., Soupa, O., 2014. Leading the Energy Transition: Hydrogen-Based Energy Conversion. Schlumberger Business Consulting (SBC) Energy Institute, Paris, France.

DfT, 2012. Road Traffic Estimates 2011. UK Department for Transport (DfT), London, UK.

Dodds, P.E., Demoullin, S., 2013. Conversion of the UK gas system to transport hydrogen. Int. J. Hydrog. Energy 38 (18), 7189–7200. http://dx.doi.org/10.1016/j.ijhydene.2013.03.070.

Dodds, P.E., McDowall, W., 2012a. Hydrogen transitions in the UK: an economic appraisal from an energy systems perspective. In: World Hydrogen Energy Conference (WHEC) 2012. Toronto, Canada.

Dodds, P.E., McDowall, W., 2012a. UKSHEC Working Paper No. 6: A review of hydrogen production technologies for energy system models (No. 6), UKSHEC Working Paper. London, UK.

Dodds, P.E., McDowall, W., 2013. The future of the UK gas network. Energy Policy 60, 305–316. http://dx.doi.org/10.1016/j.enpol.2013.05.030.

Dodds, P.E., McDowall, W., 2014. Methodologies for representing the road transport sector in energy system models. Int. J. Hydrog. Energy 39 (5), 2345–2358. http://dx.doi.org/10.1016/j.ijhydene.2013.11.021.

Dodds, P.E., Ekins, P., Hawkes, A.D., Li, F., McDowall, W., Grünewald, P., Kansara, T., Agnolucci, P., 2014. The role of hydrogen and fuel cells in providing affordable, secure low-carbon heat. In: 20th World Hydrogen Energy Conference (WHEC 2014). Gwangju, South Korea.

Dougherty, W., Kartha, S., Rajan, C., Lazarus, M., Bailie, A., Runkle, B., Fencl, A., 2009. Greenhouse gas reduction benefits and costs of a large-scale transition to hydrogen in the USA. Energy Policy 37, 56–67. http://dx.doi.org/10.1016/j.enpol.2008.06.039.

DRD, 2009. Traffic and Travel Information 2009. Northern Ireleand Department for Regional Development (DRD), Belfast, Northern Ireland, UK.

DTI, 2006. Industrial carbon dioxide emissions and carbon dioxide storage potential in the UK (Report No. COAL R308 DTI/Pub URN 06/2027). UK Department of Trade and Industry (DTI), London, UK.

Dunning, J.H., 1980. Toward an eclectic theory of international production: some empirical tests. J. Int. Bus. Stud. 11, 9–31. http://dx.doi.org/10.1057/palgrave.jibs.8490593.

Endo, E., 2007. Market penetration analysis of fuel cell vehicles in Japan by using the energy system model MARKAL. Int. J. Hydrog. Energy 32, 1347–1354. http://dx.doi.org/10.1016/j.ijhydene.2006.10.015.

Ewing, G., Sarigöllü, E., 2000. Assessing consumer preferences for clean-fuel vehicles: a discrete choice experiment. J. Public Policy Mark. 19, 106–118. http://dx.doi.org/10.1509/jppm.19.1.106.16946.

Grahn, M., Williander, M., 2009. The role of ICEVs, HEVs, PHEVs, BEVs and FCVs in achieving stringent CO2 targets: results from global energy systems modeling. In: 24th Electric Vehicle Symposium (EVS24 2009). Stavanger, Norway.

GROS, 2012. Scotland's Census Results OnLine (SCROL) [WWW Document]. URL http://www.scrol.gov.uk/scrol/common/home.jsp (accessed 6.28.12.).

Grübler, A., Nakićenović, N., Victor, D.G., 1999. Dynamics of energy technologies and global change. Energy Policy 27, 247–280. http://dx.doi.org/10.1016/S0301-4215(98)00067-6.

Gül, T., Kypreos, S., Turton, H., Barreto, L., 2009. An energy-economic scenario analysis of alternative fuels for personal transport using the global multi-regional MARKAL model (GMM). Energy 34, 1423–1437. http://dx.doi.org/10.1016/j.energy.2009.04.010.

Haeseldonckx, D., D'haeseleer, W., 2007. The use of the natural-gas pipeline infrastructure for hydrogen transport in a changing market structure. Int. J. Hydrog. Energy 32, 1381–1386. http://dx.doi.org/10.1016/j.ijhydene.2006.10.018.

Haeseldonckx, D., D'haeseleer, W., 2011. Concrete transition issues towards a fully-fledged use of hydrogen as an energy carrier: methodology and modelling. Int. J. Hydrog. Energy 36, 4636–4652. http://dx.doi.org/10.1016/j.ijhydene.2011.01.113.

Han, J.-H., Ryu, J.-H., Lee, I.-B., 2012. Modeling the operation of hydrogen supply networks considering facility location. Int. J. Hydrog. Energy 37, 5328–5346. http://dx.doi.org/10.1016/j.ijhydene.2011.04.001.

Hetland, J., Mulder, G., 2007. In search of a sustainable hydrogen economy: how a large-scale transition to hydrogen may affect the primary energy demand and greenhouse gas emissions. Int. J. Hydrog. Energy 32, 736–747. http://dx.doi.org/10.1016/j.ijhydene.2006.08.011.

Hugo, A., Rutter, P., Pistikopoulos, S., Amorelli, A., Zoia, G., 2005. Hydrogen infrastructure strategic planning using multi-objective optimization. Int. J. Hydrog. Energy 30, 1523–1534. http://dx.doi.org/10.1016/j.ijhydene.2005.04.017.

HyWays, 2008. HyWays: The European Hydrogen Roadmap. The HyWays Consortium, co-funded by the European Commission under the Framework Programme 6.

Iaquaniello, G., Giacobbe, F., Morico, B., Cosenza, S., Farace, A., 2008. Membrane reforming in converting natural gas to hydrogen: production costs, part II. Int. J. Hydrog. Energy 33, 6595–6601. http://dx.doi.org/10.1016/j.ijhydene.2008.06.026.

IPCC, 2005. Carbon Dioxide Capture and Storage: Special Report of the Intergovernmental Panel on Climate Change. Cambridge University Press, Cambridge, UK.

Ishimoto, Y., Kurosawa, A., Sasakura, M., Sakata, K., 2012. Study of global demand for low carbon hydrogen toward 2050 (Poster). In: 19th World Hydrogen Energy Conference (WHEC 2012). Toronto, Canada.

Johnson, N., Ogden, J., 2012. A spatially-explicit optimization model for long-term hydrogen pipeline planning. Int. J. Hydrog. Energy 37, 5421–5433. http://dx.doi.org/10.1016/j.ijhydene.2011.08.109.

Johnson, N., Yang, C., Ni, J., Johnson, J., Lin, Z., Ogden, J., 2005. Optimal design of a fossil fuel-based hydrogen infrastructure with carbon capture and sequestration: case study in Ohio. In: National Hydrogen Association (NHA) Annual Hydrogen Conference: Partnering for the Global Hydrogen Future. University of California, Davis, Washington, DC, USA.

Johnson, N., Yang, C., Ogden, J., 2008. A GIS-based assessment of coal-based hydrogen infrastructure deployment in the state of Ohio. Int. J. Hydrog. Energy 33, 5287–5303. http://dx.doi.org/10.1016/j.ijhydene.2008.06.069.

Kamarudin, S.K., Daud, W.R.W., Yaakub, Z., Misron, Z., Anuar, W., Yusuf, N.N.A.N., 2009. Synthesis and optimization of future hydrogen energy infrastructure planning in Peninsular

Malaysia. Int. J. Hydrog. Energy 34, 2077–2088. http://dx.doi.org/10.1016/j.ijhydene.2008.12.086.

Kannan, R., Strachan, N., Pye, S., Anandarajah, G., Balta-Ozkan, N., 2007. UK MARKAL Model Documentation. London, UK.

Kim, J.-G., Kuby, M., 2012. The deviationflow refueling location model for optimizing a network of refueling stations. Int. J. Hydrog. Energy 37, 5406–5420. http://dx.doi.org/10.1016/j.ijhydene.2011.08.108.

Kim, J., Moon, I., 2008. Strategic design of hydrogen infrastructure considering cost and safety using multiobjective optimization. Int. J. Hydrog. Energy 33, 5887–5896. http://dx.doi.org/10.1016/j.ijhydene.2008.07.028.

Kim, J., Lee, Y., Moon, I., 2008. Optimization of a hydrogen supply chain under demand uncertainty. Int. J. Hydrog. Energy 33, 4715–4729. http://dx.doi.org/10.1016/j.ijhydene.2008.06.007.

Krewitt, W., Schmid, S.A., 2005. Fuel cell technologies and hydrogen production/distribution options, CASCADE MINTS. Cologne, Germany.

Krzyzanowski, D.A., Kypreos, S., Barreto, L., 2007. Supporting hydrogen based transportation: case studies with Global MARKAL Model. Comput. Manag. Sci. 5, 207–231. http://dx.doi.org/10.1007/s10287-007-0040-5.

Kuby, M., Lim, S., 2005. The flow-refueling location problem for alternative-fuel vehicles. Socioecon. Plann. Sci. 39, 125–145. http://dx.doi.org/10.1016/j.seps.2004.03.001.

Lemus, R.G., Martínez Duart, J.M., 2010. Updated hydrogen production costs and parities for conventional and renewable technologies. Int. J. Hydrog. Energy 35, 3929–3936. http://dx.doi.org/10.1016/j.ijhydene.2010.02.034.

Li, F., Agnolucci, P., 2014. Heat markets. In: Dodds, P.E., Hawkes, A. (Eds.), The Role of Hydrogen and Fuel Cells in Providing Affordable, Secure Low-Carbon Heat. H2FC Supergen, London, UK, pp. 47–59.

Li, Z., Gao, D., Chang, L., Liu, P., Pistikopoulos, E.N., 2008. Hydrogen infrastructure design and optimization: a case study of China. Int. J. Hydrog. Energy 33, 5275–5286. http://dx.doi.org/10.1016/j.ijhydene.2008.06.076.

Lin, Z., Chen, C.-W., Ogden, J., Fan, Y., 2008a. The least-cost hydrogen for Southern California. Int. J. Hydrog. Energy 33, 3009–3014. http://dx.doi.org/10.1016/j.ijhydene.2008.01.039.

Lin, Z., Ogden, J., Fan, Y., Chen, C.-W., 2008b. The fuel-travel-back approach to hydrogen station siting. Int. J. Hydrog. Energy 33, 3096–3101. http://dx.doi.org/10.1016/j.ijhydene.2008.01.040.

Loulou, R., Goldstein, G., Noble, K., 2004. Documentation for the MARKAL Family of Models. Paris, France

Martinus, G., Smekens, K., Rösler, H., 2005. Modelling the transition to a hydrogen economy. In: 5th BIEE (British Institute of Energy Economics) Academic Conference—European Energy: Synergies and Conflicts. Oxford, UK.

McDowall, W., forthcoming. Are scenarios of hydrogen vehicle adoption optimistic? A comparison of hydrogen scenarios with historical analogies. Environ. Innov. Soc. Transitions.

Melaina, M., 2003. Initiating hydrogen infrastructures: preliminary analysis of a sufficient number of initial hydrogen stations in the US. Int. J. Hydrog. Energy 28, 743–755. http://dx.doi.org/10.1016/S0360-3199(02)00240-9.

Melendez, M., Milbrandt, A., 2006. Geographically based hydrogen consumer demand and infrastructure analysis (NREL/TP-540-40373). Golden, CO, USA.

Mintz, M., Molburg, J., Folga, S., Gillette, J., 2002. Hydrogen Distribution Infrastructure. Article prepared for Argonne National Laboratory, Center for Transportation Research and Decision and Information Sciences Division, Argonne, IL, USA.

Mueller-Langer, F., Tzimas, E., Kaltschmitt, M., Peteves, S., 2007. Techno-economic assessment of hydrogen production processes for the hydrogen economy for the short and

medium term. Int. J. Hydrog. Energy 32, 3797–3810. http://dx.doi.org/10.1016/j.ijhydene.2007.05.027.
Muncaster, K.A., 2008. Can Hydrogen Win?: Exploring Scenarios for Hydrogen Fuelled Vehicles. Simon Fraser University, Canada.
Murthy Konda, N.V.S.N., Shah, N., Brandon, N.P., 2011. Optimal transition towards a large-scale hydrogen infrastructure for the transport sector: the case for the Netherlands. Int. J. Hydrog. Energy 36, 4619–4635. http://dx.doi.org/10.1016/j.ijhydene.2011.01.104.
NAE, NRC, 2004. The Hydrogen Economy: Opportunities, Costs, Barriers, and R&D Needs. The National Academies Press, Washington, DC, USA.
Nicholas, M.A., 2004. Hydrogen Station Siting and Refueling Analysis Using Geographic Information Systems: A Case Study of Sacramento County. University of California, Davis.
NISRA, 2012. Census [WWW Document]. URL http://www.nisra.gov.uk/Census.html (accessed 6.28.12.).
NRC, 2008. Transitions to Alternative Transportation Technologies: A Focus on Hydrogen. The National Academies Press, Washington, DC, USA.
Ogden, J.M., 2001. Review of Small Stationary Reformers for Hydrogen Production. Center for Energy and Environmental Studies, Princeton University, Princeton, NJ, USA, for the International Energy Agency (IEA), Paris, France.
ONS, 2012. Neighbourhood Statistics [WWW Document]. URL http://neighbourhood.statistics.gov.uk/dissemination/LeadHome.do?m=0&s=1404121404148&enc=1&nsjs=true&nsck=false&nssvg=false&nswid=1296 (accessed 6.28.12.).
Parker, N., 2004. Using Natural Gas Transmission Pipeline Costs to Estimate Hydrogen Pipeline Costs. University of California, Davis.
Parker, N., Fan, Y., Ogden, J., 2010. From waste to hydrogen: an optimal design of energy production and distribution network. Transp. Res. Part E Logist. Transp. Rev. 46, 534–545. http://dx.doi.org/10.1016/j.tre.2009.04.002.
Rits, V., Kypreos, S., Wokaun, A., 2004. Evaluating the diffusion of fuel cell cars in the China markets. Tokyo, Japan.
Rogers, E.M., 2003. Diffusion of Innovations, 5th ed. Free Press, New York, USA.
Rosenberg, E., Fidje, A., Espegren, K.A., Stiller, C., Svensson, A.M., Møller-Holst, S., 2010. Market penetration analysis of hydrogen vehicles in Norwegian passenger transport towards 2050. Int. J. Hydrog. Energy 35, 7267–7279. http://dx.doi.org/10.1016/j.ijhydene.2010.04.153.
Rösler, H., 2012. When and how to decarbonise the transport sector? In: 31st International Energy Workshop (IEW 2012). Cape Town, South Africa.
Sabio, N., Kostin, A., Guillén-Gosálbez, G., Jiménez, L., 2012. Holistic minimization of the life cycle environmental impact of hydrogen infrastructures using multi-objective optimization and principal component analysis. Int. J. Hydrog. Energy 37, 5385–5405. http://dx.doi.org/10.1016/j.ijhydene.2011.09.039.
Shay, C.L., Yeh, S., Decarolis, J., Loughlin, D.H., Gage, C.L., Wright, E., 2006. EPA U.S. National MARKAL Database: Database Documentation (EPA/600/R-06/057). Washington, DC, USA.
Smit, R., Weeda, M., de Groot, A., 2007. Hydrogen infrastructure development in the Netherlands. Int. J. Hydrog. Energy 32, 1387–1395. http://dx.doi.org/10.1016/j.ijhydene.2006.10.044.
Steward, D., Ramsden, T., Zuboy, J., 2008. H2A Production Model, Version 2 User Guide (NREL/TP-560-43983). Golden, CO, USA.

Stiller, C., Bünger, U., 2009. NorWays project deliverable no. 8: description and results of the infrastructure model H2INVEST. Trondheim, Norway.
Stiller, C., Bünger, U., Møller-Holst, S., Svensson, A.M., Espegren, K.A., Nowak, M., 2010. Pathways to a hydrogen fuel infrastructure in Norway. Int. J. Hydrog. Energy 35, 2597–2601. http://dx.doi.org/10.1016/j.ijhydene.2009.04.003.
Strachan, N., Balta-Ozkan, N., Joffe, D., McGeevor, K., Hughes, N., 2009. Soft-linking energy systems and GIS models to investigate spatial hydrogen infrastructure development in a low-carbon UK energy system. Int. J. Hydrog. Energy 34, 642–657. http://dx.doi.org/10.1016/j.ijhydene.2008.10.083.
Takeshita, T., 2012. Assessing the effects of internalizing externalities on the road transport sector. In: 31st International Energy Workshop (IEW 2012). Cape Town, South Africa.
Tseng, P., Lee, J., Friley, P., 2005. A hydrogen economy: opportunities and challenges. Energy 30, 2703–2720. http://dx.doi.org/10.1016/j.energy.2004.07.015.
Upchurch, C., Kuby, M., 2010. Comparing the p-median and flow-refueling models for locating alternative-fuel stations. J. Transp. Geogr. 18, 750–758. http://dx.doi.org/10.1016/j.jtrangeo.2010.06.015.
Upchurch, C., Kuby, M., Lim, S., 2009. A model for location of capacitated alternative-fuel stations. Geogr. Anal. 41, 85–106. http://dx.doi.org/10.1111/j.1538-4632.2009.00744.x.
Van der Zwaan, B.C.C., Schoots, K., Rivera-Tinoco, R., Verbong, G.P.J., 2011. The cost of pipelining climate change mitigation: an overview of the economics of CH4, CO2 and H2 transportation. Appl. Energy 88, 3821–3831. http://dx.doi.org/10.1016/j.apenergy.2011.05.019.
Yang, C., Ogden, J.M., 2007. Determining the lowest-cost hydrogen delivery mode. Int. J. Hydrog. Energy 32, 268–286. http://dx.doi.org/10.1016/j.ijhydene.2006.05.009.
Yeh, S., Farrell, A., Plevin, R., Sanstad, A., Weyant, J., 2008. Optimizing U.S. mitigation strategies for the light-duty transportation sector: what we learn from a bottom-up model. Environ. Sci. Technol. 42, 8202–8210. http://dx.doi.org/10.1021/es8005805.

Investment in the infrastructure for hydrogen passenger cars—New hype or reality?

15

N. Bento
DINÂMIA'CET, ISCTE – University Institute of Lisbon, Lisboa, Portugal

Abbreviations

AFCC	Automotive Fuel Cell Cooperation
BEV	battery electric vehicle
CaFCP	California Fuel Cell Partnership
CARB	California Air Resources Board
CEP	Clean Energy Partnership
CHP	combined heat and power
DOE	United States of America, Department of Energy
EU	European Union
FC	fuel cell
FCH JU	European Fuel Cells and Hydrogen Joint Undertaking
FCV	fuel cell vehicle
H2 or H_2	hydrogen
HEV	hybrid electric vehicles
HRS	hydrogen refueling stations
HySUT	Japanese Research Association of Hydrogen Supply/Utilization Technology
LH2	cryogenic (or liquid) hydrogen
METI	Japanese Ministry of Economy, Trade and Industry
MoU	memorandum of understanding
NGO	nongovernmental organization
NIP	German National Innovation Programme for Hydrogen and Fuel Cell Technology
NOW	German National Association for the Advancement of Hydrogen and Fuel Cells
OEM	original equipment manufacturer
PEM	proton exchange membrane
PZEV	partial zero-emission vehicle
R&D	research and development
RD&D	research, development and demonstration
SHHP	Scandinavia Hydrogen Highway Partnership
SMR	steam methane reformer
US	United States of America
ZEV	California "Zero-Emission Vehicle" mandate

Compendium of Hydrogen Energy. http://dx.doi.org/10.1016/B978-1-78242-362-1.00015-8

15.1 Introduction

Hydrogen can be produced from an unlimited flow of renewable energies and combined with oxygen in a fuel cell to power cars, with the release of only water vapor. Hence, hydrogen fuel cell vehicles (FCVs) have been promoted as a long-term fuel option to reduce greenhouse gas emissions and oil dependence in transportation (NRC, 2013, 2008; Hoffmann, 2012). Tens of thousands of those vehicles have been on the roads in demonstration in the last decade (Bakker et al., 2012; Bakker, 2010a), but commercialization still requires scaling up production to tens of thousands units per year. Although some progress has been made in fuel cells that could indicate that technology is approaching readiness (Reuters, 2014; DOE, 2013), the absence of a network of hydrogen distribution impedes the start of the transition (Ogden et al., 2011). The construction of the early infrastructure is a major challenge because consumers will only begin using the first hydrogen cars when there are enough refueling stations to refuel them conveniently, whereas fuel suppliers will deploy a network of refueling stations provided that there are enough vehicles to justify the high investments.

The implementation of hydrogen FCVs implies radical changes both at the level of the car and at the level of the infrastructure. The key challenge for transitioning to such a radical innovation is the fact that a number of different actors have to come together (at the same time) and coordinate their actions to provide the technology (e.g., fuel cells, hydrogen storage devices), market the car, establish a new business model for the distribution of the fuel, and set up the institutional conditions (e.g., codes and standards) (Konrad et al., 2012). This adds to the high uncertainties (e.g., technology costs, performances, choices, market uptake, regulatory conditions) organizations already face in the early years, when no cars are commercialized and no infrastructure distributes hydrogen fuel to consumers (Bento, 2010a). Within this context, actors have to take their decisions based on prospects—or what the theories of "sociology of expectations" in science and technology call *collective expectations*, i.e., expectations that are shared by a broad range of stakeholders (van Lente, 1993; van Lente and Rip, 1998; Borup et al., 2006; Bakker et al., 2011, 2012)—rather than solid information. This increases the risk of technological promises and disappointments.

Hydrogen and fuel cell have actually experienced several hype and disappointment cycles during the past 40 years. More recently, the expectations on the potential of the technology started to increase at the end of the 1990s and peaked in early 2000s with a focus on mobile applications (Bakker, 2010a,b; Ruef and Markard, 2010; Bakker and Budde, 2012; Konrad et al., 2012; Romm, 2004). The hype was largely triggered by strong statements from Daimler—followed by other companies such as Honda, Toyota, and GM—about the commercialization of hydrogen cars in 2004.[1] However, soon after, automakers decided to postpone the market

[1]Bakker (2010a,b) ironically notes that "Contrary to popular belief, hydrogen is not always ten years away." because at this point it would be "only two years into the future."

launch, under the argument of technical problems and high costs of fuel cells as well as the lack of hydrogen filling stations, leading to a generalized disappointment concerning hydrogen-powered cars. Indeed, the succession of overoptimistic and modest announcements in the same year could have served the purpose of automakers to cool off the intention of governments in California and Europe to adopt stricter emissions' regulations in the belief that the technology was "ready," as noted by Bakker (2010a,b, p. 6543).

A concept often employed to analyze hype-disappointment dynamics is the empirically observed "Gartner hype cycle," which is used by Gartner consultants to define the timing of strategic investment in emerging innovations (Fenn and Time, 2007). The basic regularity starts with a technology trigger that rapidly increases public attention in what culminates in the "hype", i.e., a peak of inflated expectations. This is followed by a strong disappointment and decline of expectations, which leads to the trough of disillusionment. Then, the technology gradually enters into a slope of enlightenment (i.e., a less visible period of slower, but surer, progress), preparing it for the plateau of productivity that enables commercialization. The underlying rationale of the Gartner's cycle is that the damage of disillusionment should be relativized because most emerging technologies follow their maturity curve to reach the market at the end (Fenn and Time, 2007; Ruef and Markard, 2010).

Hydrogen FCVs came back on the energy agenda during the past two years. Several factors contributed to this "revival," such as the (announced) falling costs of FCVs and the involvement of automakers in the installation of the infrastructure (Bullis, 2013). In addition, the slow penetration of battery electric vehicles (BEVs) in the market and their technical problems helped to increase enthusiasm around FCVs. The latter surpassed the former in terms of perceived importance for the future of the automobile industry according to the KPMG global automotive survey 2013, and even enlarged its advantage in the 2014 survey (69% compared to 59%) (KPMG, 2014). This resurgence raises the question on whether the technology is finally arriving at the plateau of productivity or beginning another hype-disappointment cycle.

This study unveils not only the characteristics of the "revival" of an innovation after the disappointment, but also the strategies and motivations that permitted fuel cells to gain again visibility and come back to create the car of the future.

The chapter reviews the challenges posed by the investment in hydrogen infrastructure after the hype. This is a central question that has to be answered to enable the transition toward a hydrogen economy. This chapter also examines the strategies that have been recently followed by pioneer countries to address these barriers.

The remainder of this chapter is structured as follows. Section 15.2 presents the principal barriers to the investment in hydrogen fuel infrastructure. Section 15.3 provides evidence from the study of three cases (California, Japan, and Germany), which have ongoing initiatives to build up the network of refueling stations, by providing a brief contextualization of each of them and a review of the mechanisms used to meet the challenges. Section 15.4 discusses the current and future trends concerning hydrogen FCVs, before concluding the chapter.

15.2 Uncertainties surrounding the investment in hydrogen infrastructure

Hydrogen and FCV is a complex technology system composed of many elements that are still in the research, development, and demonstration (RD&D) stage. Progress toward the long-awaited goal of widespread commercialization goes hand-in-hand with advancements in hydrogen production, storage, delivery, and fuel cell technologies. The latter is particularly important as it provides the energy service directly to end users. This section analyzes the current obstacles facing the development of a hydrogen infrastructure, including transition issues in early commercialization.

15.2.1 Challenges of building a new infrastructure for hydrogen

The availability of a hydrogen infrastructure is a precondition for the introduction of FCVs in the market. An early network of refueling stations, with acceptable coverage, is needed to allow for the development of demand for hydrogen-powered cars, but fuel providers will only provide a network if there are enough vehicles to use the refueling stations (Ogden et al., 2011). Indeed, the large irreversible ("sunk") costs and the uncertainties on utilization (especially in early years) reduce the incentives of firms to invest in such infrastructures (Bento, 2010a).

Hydrogen is the most plentiful gas in the world but it only exists in nature along with other elements from which it has to be extracted. The annual production of hydrogen is approximately 80 tons per year, of which about 40% is manufactured and used in refineries to make gasoline and diesel with 96% of the gas (used in refining) produced by steam reforming of natural gas.[2] Therefore, like electricity, hydrogen is an energy carrier, and it must be produced from a primary source (hydrocarbons, renewable electricity, nuclear) and transmitted to the place of consumption to deliver an energy service (e.g., stationary, mobile, portable) using fuel cell technology for higher efficiency. Hence, an infrastructure for hydrogen production and delivery is required. In the case of mobile applications, hydrogen refueling stations (HRSs) are also needed in sufficient number and strategically located to reduce capital needs.[3]

The deployment of the infrastructure should be gradual and coevolve with the development of demand over time in order to minimize transition costs (Ogden et al., 2011). Initially, onsite production of hydrogen is the most likely way to supply the low levels of demand—which are insufficient to justify the investments in central production and delivery—by making use of the energy infrastructure already in place. Therefore, small steam methane reformers are a strong option whenever the cost of

[2]https://www.eni.com/en_IT/innovationtechnology/technological-focus/produzione-idrogeno/produzione-idrogeno.shtml (last accessed 30.05.14).

[3]The cost per station is estimated to be around $1 million in Europe (cf. German "H2Mobility" initiative), $1–2 million in the United States (California), and $5–6 million in Japan (cf. Hara, 2013).

hydrogen production is the criteria. Alternatively, onsite electrolysis of renewable electricity is a possibility where the environmental benefits are part of the investment decision. As the number of FCVs grows, and therefore the demand for hydrogen, it may become economical at some point to invest in a large central plant to reap the economies of scale in production that compensate for the additional costs with the delivery system (by trucks or pipelines). Geographical specificities in terms of primary energy endowments (and feedstock prices) and demand density will determine the choice of supply strategies. Thus, the technologies needed for the production, delivery, and distribution, i.e., the hydrogen "pathway," may look different from region to region. In the past decade, several studies have estimated the capital costs needed to build the infrastructure for hydrogen fuel. Ogden et al. (2011) provides a review of the results for the United States and Bento (2008, 2010d) synthesizes the findings for Europe. The project HyWays (2008) estimated €60 billion for the total investment cost needed for the infrastructure (production, delivery, and distribution) required to assist 16 million cars (8% of the fleet) in Europe by 2030. In the United States, the infrastructure is expected to cost around $70 billion ($2005) (Ogden and Yang, 2009). However, these projections should be updated with more recent data, namely from the first demonstration projects, to better inform about actual costs of technologies as well as likely opportunities for cost reductions through economies of scale and learning.

The key issue in this transition period is to reconcile the need for a reliable and convenient refueling network—especially in the early years when the number of vehicles is small—with the costs of building (small and under-utilized) stations that are unprofitable in the short term. In other words, hydrogen supply and demand should co-evolve as much as possible to reduce investment risks and costs. The complexity of this "chicken-or-egg" problem stems from the fact that several stakeholders—in fuel distribution, hydrogen production, and vehicle manufacturing—have to rely on each other's investments to start the system successfully (Bento, 2010b; Ogden et al., 2011). Strategies exist to help the establishment of the first hydrogen fuel infrastructures—independently of the strategy, early investments often receive the support of the government given their spillovers for the development of the network (Bento, 2008). The traditional approach consists on the gradual construction of a network of refueling stations in major agglomerations and highways. This strategy is currently pursued by the German initiative "H2 Mobility" or the Japanese HySUT project. Another strategy is the creation of early dispersed mini-networks, or "lighthouse" cities, where the first fleets are demonstrated and cars deployed together with the refueling infrastructure (Ogden et al., 2011). A small number of refueling stations would be enough to assist the initial demand in these "clusters," which could serve as starting points to scale up the network at national level. California is actually funding the development of five urban clusters to prepare the commercial launch of FCVs which is expected in 2015 (CaFCP, 2012). This strategy replaced the precedent vision that wanted to start the transition around a "California Hydrogen Highway Network."

In addition to the infrastructure challenge, technology costs must be substantially reduced in the coming years to make hydrogen FCVs a credible alternative to the incumbent technology.

15.2.2 Need for reliable and affordable technologies

Fuel cells can use hydrogen produced from multiple domestic renewable and low-carbon sources to power diverse applications: stationary, portable, and transport. They started to be used in space programs by NASA in the 1960s. Today, fuel cells are gradually adopted by public and private sectors as the new generations of the technology are becoming more affordable, reliable, and durable (Sharaf and Orhan, 2014). Even though they are still not mature and improvements are needed to make them competitive with conventional technologies, fuel cells are closer to commercialization in some niche markets where the benefits derived from their use are higher enough to compensate for the disadvantages in terms of cost and durability (IPHE, 2013). This is the case with forklift trucks where the use of fuel cells provides silent operation and zero emissions, which are significant advantages over conventional technologies. Furthermore, they can be operated almost continuously with lower operational costs and require less equipment (less space) in comparison with battery electrical trolleys. Other early markets for fuel cells comprise stationary applications in residential combined heat and power and backup and remote power generation (IPHE, 2013; Carter and Wing, 2014).

Figure 15.1 (leftmost graph) shows the gradual take-off of the number of fuel cells shipped per year since the early 2000s. The annual number passed from a few hundred

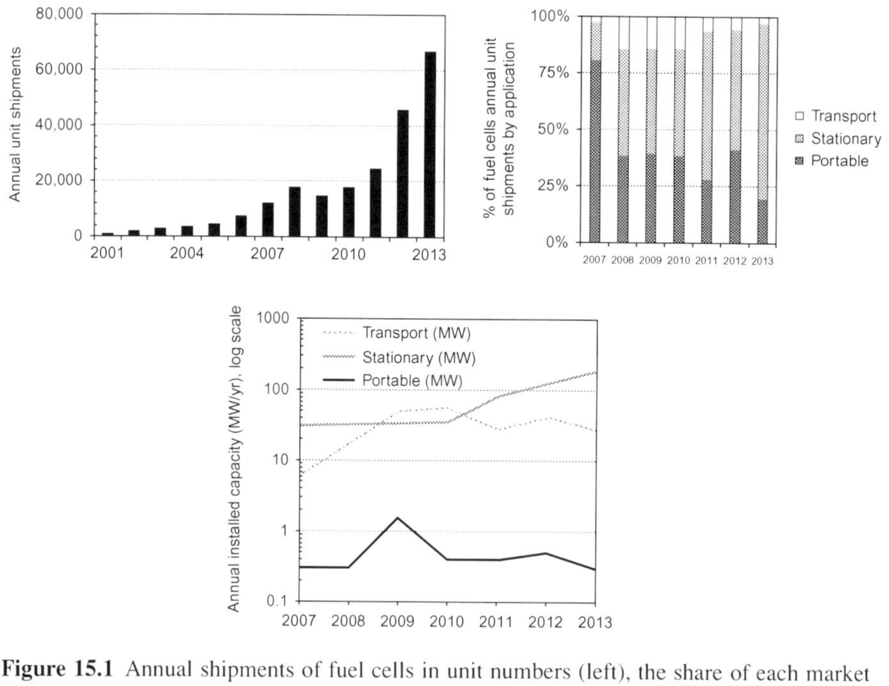

Figure 15.1 Annual shipments of fuel cells in unit numbers (left), the share of each market segment in the units shipped (right) and the growth of annual installed capacity by application (centre, below).

Source: Carter and Wing (2014), Adamson and Crawley (2006), and Fuel Cell Today (various): http://fuelcelltoday.com/analysis/industry-review.

fuel cells in 2001 to more than 66,000 in 2013 (Carter and Wing, 2014). The data also reveals a break in the upward trend during the years 2009 and 2010, resuming the previous trend in 2011. Sales were pushed by the growth in stationary applications, both in terms of unit numbers (rightmost graph) and capacity (graph below). The market is mostly driven by the Japanese residential fuel cell microcombined heat and power (micro-CHP), "Ene-Farm" program, which had subsidized the installation of more than 40,000 units (IPHE, 2013).[4] However, in terms of capacity, the major part of the capacity was installed in the power generation segment which is fundamentally shared between three companies (FuelCell Energy, Bloom Energy and ClearEdge Power—ex-UTC Power Technology[5]) and concentrated in two countries: the United States and Korea (Carter and Wing, 2014). Other stationary applications, such as back-up and remote power generation (e.g., for communication towers), are experiencing increased interest in the telecom sector, especially after the 2009 Recovery Act funding in the United States. Conversely, the prospects of growth in portable applications have not been confirmed and, as a consequence, the share of this market has been falling over the years. As for transport applications, the sales of fuel cell systems for material handling equipment have dominated the market—apart from the limited number of fuel cells installed in small fleets of cars and buses for demonstration purposes. This market has shown interesting developments in the United States since the Recovery Act and funding, and more units are sold today even without the subsidies (IPHE, 2013).[6]

Within this context, the commercialization of FCVs is key to dramatically increasing the number of systems shipped per year. The large-scale production of fuel cells would create spillovers in terms of cost reductions and technology improvement that could boost the development of the technology in the other applications as well.

Table 15.1 shows the progress of hydrogen and fuel cell technologies toward the goals set for commercialization in mobile applications, comparing the current state with that of five years ago (when disappointment dominated) as well as with official targets. The price of the hydrogen at the station, on the one hand, is still above the

[4]This scheme has recently inspired the European project "Ene.field" that aims to install 1,000 micro-CHP units across 12 Member States by 2017. In Germany, the Callux project has deployed 350 systems between 2008 and 2012 and there are plans for 200 more by 2015 (IPHE, 2013).

[5]In a recent development, ClearEdge abruptly closed operations, laying off 268 employees and announcing the intention to file for bankruptcy. The company's manager pointed to the delay of a significant contract and problems in collecting money that was owed by the customers as the main reasons for the decision. This case again shows the financial fragility of fuel cells companies that face strong expenditures in the short term and extremely volatile and narrow markets. See: "ClearEdge confirms closure, bankruptcy intent," *Hartford Business Journal*, 4/29/2014. Available at: http://www.hartfordbusiness.com/apps/pbcs.dll/article?AID=/20140429/NEWS01/140429939.

[6]The announcement of an important Wal-Mart order to Plug Power's fuel cell forklifts and the perspectives of profits for FuelCell Energy have recently spurred a surge in the stocks of these companies. Several financial analysts remembered the hype around fuel cells in the past, and pointed to the automobile market and "whether automakers ultimately start rolling out fuel cell vehicles en masse" as a possible game changer. See: Wile R., "Wall Street Is Going Crazy For A Revolutionary Technology That Could Change The Energy Market As We Know It," March 11, 2014. Available at: http://www.businessinsider.com/fuel-cell-rally-2014-3.

Table 15.1 Achievement of specific milestones in automotive hydrogen fuel cell technologies in the United States and Europe, in 2009 and 2014

Technology	U.S. DOE's milestones 2015	European "Snapshot 2020 (2030)"	Situation in 2009	Current status 2014
Hydrogen cost at the station—untaxed	\$2–3/kg	<2.5€/kg[a]	\$3–9/kg[b]	\$2.75–5.7/kg[c] (\$7.7–12.9/kg)
Onboard hydrogen storage	\$2/kWh	10 (5)€/kWh	\$15–18/kWh[d,e] >\$60/kWh[e]	\$15–19/kWh[d,e,f]
Fuel cell cost (PEM-FC system)	\$30/kW	<100 (<50)€/kW	\$60/kW[d] >\$500/kW	\$51/kW[d,f] \$230/kW[f]
FC energy efficiency	60%	–	53–58%	53[f]–59%[g]
Durability	5,000 h	5,000 h	2,000 h	2,521 h[g] 2,200–3,600 h[h]

[a]Between 2 and 5€/kg for hydrogen produced from sustainable sources, such as solar and wind (HFP, 2007).
[b]\$3/kg for onsite production from the reforming of natural gas; \$9/kg for large-scale production from renewable sources (transport costs included).
[c]Lower-bound refers to onsite natural gas reformation and upper-bound to onsite electrolysis. Early market in brackets, assuming 1,500 kg/day.
[d]For an annual production of 500,000 units.
[e]5 kg compressed hydrogen at 350–700 bar (35–70 MPa).
[f]National Research Council (2013, Appendix F: Vehicles).
[g]U.S. DOE (2013).
[h]Defined as the average projected time of the stack before losing 10% of its original voltage (Kurtz et al., 2013).
Source: NRC (2013, 2008), DOE (2009, 2013), Kurtz et al. (2013), HyWays (2008), Roads2HyCom (2009), and HFP (2007).

American and European goals, especially when hydrogen is produced from renewable energies. Fuel cells, on the other hand, have undeniably made progresses over the past five years in terms of performance, operating life of equipment, and costs. However, improvements are still needed for fuel cells to become competitive in transportation.

The most important news for the hydrogen transition is that the cost of fuel cells is still high but has significantly decreased in recent years. The cost of the Proton Exchange Membrane (PEM) fuel cell, the most promising for mobile applications, reduced from \$273/kW in 2002 to \$51/kW in 2013 (NRC, 2013; DOE, 2009), assuming large-scale production (500,000 units)—otherwise the cost would be \$230/kW. In addition, the actual durability of fuel cells doubled from 940 h to almost 2,000 h between 2006 and 2009 (DOE, 2009). Although in the following five years the durability increased less rapidly to 2,521 h (DOE, 2013), which is still half the minimum operating time required for commercialization (i.e., 5,000 h). More advances are therefore needed to solve the cost and durability issues before fuel cells will be ready for commercialization. Several lines of research have been followed such as minimizing or even eliminating catalysts platinum loading of fuel cells and reducing the cost of balance of plant (Sharaf and Orhan, 2014; DOE, 2013). Toyota recently announced that has cut by half the size of the fuel cell system, and significantly reduced its cost to $50,000 (Reuters, 2014; Fairley, 2012). The company began the commercialization of Mirai in Japan in December 2014, and plans to start retail sales in the US and in Europe (Germany and Denmark) in late 2015.[7]

Figure 15.2 compares the average cost reductions as technology manufacturers gain experience, i.e., learning rates, estimated for PEM fuel cells and reported for several energy technologies: onshore wind power in Denmark, 1981–2009; cars in the US (model T), 1907–1927; solar PV world average prices, 1975–2007; and CFLs world average prices, 1988–2006. Results shows that the targets and goals for fuel cell costs are feasible since they compare with the decrease in the cost of solar PV or cars in the past (learning rates of 20%).[8] However, two other historical examples are shown, i.e., wind power and CFLs, for which the speed of cost reduction was slower (learning

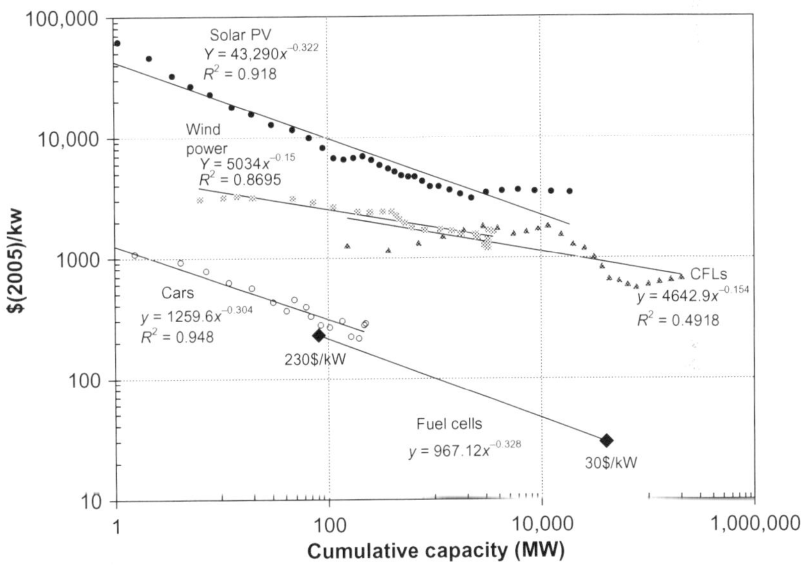

Figure 15.2 Learning curves for selected energy technologies and PEM fuel cells (projected). Source: NRC (2013) [PEM-FC], Grubler et al. (2012) [Wind], Weiss et al. (2010) [CFLs], and Nemet (2009) [Solar PV].

[7]http://www.toyota.com/mirai/fcv.html (last accessed 11.06.15).

[8]The learning rate (LR) is the indicator of cost reductions whenever production doubles. This measure can be computed from the exponential (β) of the trend line (potential regression) as follows: $LR = (1 - 2^{\beta}) \times 100$. In the case of fuel cells, β takes the value -0.328 (see Figure 15.2) and thus $LR = (1 - 2^{(-0.328)}) \times 100 = 20\%$: See more details and results for several energy technologies in McDonald and Schrattenholzer (2001).

rates of 10%). Therefore, fuel cells could become competitive through major cuts in production costs like in the case of the automobile, or R&D advances, which both cut costs and improve performance as in the case of solar PV.

Finally, the capacity to store enough hydrogen on the vehicle to drive 500 km without refueling is another key design issue because of the complexity and the bulkiness of the gas. Moreover, the method of hydrogen delivery depends on the technique used for storing it onboard, and must allow refueling in a few minutes with the same convenience as a gasoline car. Such hydrogen tanks already exist but are of a size, weight, and cost deemed excessive (NRC, 2013). In fact, the cost of the tanks is still an order of magnitude higher than the target for commercialization (up to $19/kWh vs. $2/kWh of the US DOE's objective for 2015), even when high-volume manufacturing is assumed. Major technological advances are occurring in several hydrogen storage technologies (e.g., compressed, liquid, solid materials) to resolve the challenge of onboard storage. Compression at very high pressure 700 bar (70 MPa) appears to be the preferred choice of the industry in the medium term in anticipation of the development of solid materials that would allow greater efficiency and storage capacity in the longer term (NRC, 2013; Bakker, 2010b).

In summary, further progress needs to be made in hydrogen production (especially from renewable sources), PEM fuel cell cost and durability, and hydrogen storage onboard. Two factors influence the ongoing dynamics. First, as is the case in any technological advances, the penetration of hydrogen cars on the market will accelerate technical improvements and cost reductions due to economies of scale and learning. Second, fuel cells face the competition of other alternative technologies (e.g., BEVs, plug-in hybrid vehicles, hybrid electric vehicles) to substitute the conventional internal combustion engine and petroleum fuels. Under these circumstances, the penetration of FCVs will also depend on the evolution of these technologies, and thus, a significant delay in commercial launch could undermine their market potential (Bento, 2010b). This raises additional uncertainties concerning the evolution and uptake of existing alternative technologies that affects the prospects of the investment in the infrastructure.

15.2.3 Importance of coordination among stakeholders

The previous sections discussed how commercialization of hydrogen-powered cars depends critically on the existence of an infrastructure, but investment in high-cost stations is risky as uncertainties persist about technology and demand behavior. In this context, demonstration projects have been organized to bridge the gap between the developments that FCVs have seen in the laboratory and the commercial launch expected to occur after 2015. These projects depend on a sufficient number of HRSs and cars. While the gains for companies with experience in commercializing FCVs and refueling equipment are likely to give them an advantage compared to their competitors (e.g., learning, cost reductions, reputation), these first-mover benefits must be weighed against the costs of such projects—involving early FCVs that still cost more than a $100,000 each and multimillion dollars HRS (Reuters, 2014). Hence, there is a high level of dependency between different actors' decisions, the main feature of this "chicken or egg" problem.

It is well-known in economics that whenever the transaction needs the establishment of a long-term, very specific investment, a number of moral hazard issues arise as the group of intervening firms is reduced and a party may be dependent upon its contractual counterparty (Laffont and Tirole, 1993). For instance, the decision to deploy hydrogen vehicles becomes subject to the existence of a local infrastructure, and the viability of a certain hydrogen station will always depend on the number of cars that carmakers make available to that location. This dependency amplifies the risk of opportunistic behavior, which increases the cost of transactions (Williamson, 1985). In this context, a cooperative engagement is often put in place to manage the moral hazard and uncertainty problems (Ménard, 2004). In the case of hydrogen, hybrid organizations started with some time-limited agreements, such as research collaborations between manufacturers and public laboratories, and gradually moved toward the formalization of partnerships as the technology approached the commercialization stage and agents' strategies relied more on each others' decisions.[9]

The transition toward a hydrogen economy in transports is therefore only possible if decisions are coordinated among actors in terms of RD&D, market entry, and infrastructure implementation (Konrad et al., 2012; Nygaard, 2008). This requires collaboration between the actors in activities that go beyond their core business to enable the formation of the first blocks of the future infrastructure. Several actors are currently involved in activities that aim to build infrastructure and bring hydrogen-powered cars into the market. Automakers' interest in hydrogen FCVs is explained by the raising regulatory pressures to reduce emissions and the fact that those vehicles do not oblige to a complete overhaul of the car (Bakker and Budde, 2012). Industrial gas suppliers and energy companies are also actively motivated to preserve their prominent role in the mobility system, in the case of oil companies, or to enter into new markets in areas related to their core business, in the case of industrial gas companies (Bento, 2010c). Fuel cell manufacturers are particularly motivated by the market potential of fuel cells in mobile applications. And national governments, who often facilitate the administrative procedures and provide incentives and subsidies for the construction of the first stations, seek to both reduce carbon emissions and petroleum use in transportation while at the same time stimulate the economy and jobs creation.

[9]In the United States, the US Department of Energy conducts R&D programs in collaboration with industry, universities, and national laboratories (e.g., National Renewable Energy Laboratory, or NREL) to overcome technical barriers to the commercialization of hydrogen FCVs. The Presidential Initiative for the Promotion of Hydrogen in Transport invested 1.5 billion dollars between 2004 and 2009. In 2009, the energy secretary at the time decided to cut funding for hydrogen FCV R&D under the argument that hydrogen needed four miracles to happen (in production, distribution, onboard storage, and to be used in more economical fuel cells) (Bullis, 2009). The budget for the fiscal year 2010 was finally restored by Congress, despite the initial proposition by the Obama Administration, but this situation created more uncertainties about the emergence of hydrogen in transport and reinforced the perception of its decline after the "hype" earlier in the decade.

The plans of carmakers have in particular evolved in recent years and are a good indicator of the activities and general expectations regarding the arrival of FCVs. Table 15.2 presents the number of FCVs involved in main demonstration projects as well as current plans for commercialization. Tens of hundreds of vehicles are on the road today with Daimler, Honda, and Toyota offering more than a hundred FCVs each. Hyundai is rapidly catching up with this group lately. The Korean company reaffirmed its intention to produce a thousand vehicles between 2013 and 2015.[10] The largest part of the vehicles is equipped with a fuel cell and uses gaseous hydrogen (compressed at 35 or 70 MPa). BMW demonstrated a small fleet based on its 7-Series line that carried cryogenic hydrogen to burn in a dual-fuel internal combustion engine, but this famous exception was not followed by others (Bakker et al., 2012). The commercialization of FCVs is announced for 2015 in Japan, United States, and Europe. Honda, Hyundai, and Toyota present the earliest plans for commercial launch. They also expect a growth in demand that make possible mass production of FCVs by the end of the decade. Toyota has released more details about its FCV that will be commercialized in 2015 in the US and Europe (sales started in Japan in December 2014), which cost should reduce from the current $100,000 (under mass production conditions) to be priced between $50,000 and $100,000 (Reuters, 2014).

These announcements may give the impression that fuel cell cars are "just around the corner," but the recent experience with this technology is full of advances and setbacks. In the early 2000s, when the "hype" around hydrogen was at its peak (FCT, 2013; Bakker and Budde, 2012; Romm, 2004), several auto companies were announcing the commercialization of thousands FCVs in a matter of years. In the late 1990s, Daimler was already seeing the commercialization of 40,000 units in 2004. In the following years other companies made similar statements about the introduction of hydrogen cars into the market by the same date (Bakker, 2010a).

More recently, in the highly competitive context of the auto industry, many auto companies started to form multiple partnerships to accelerate the development of fuel cell technology and reduce costs (Piper, 2014). In 2013, Honda and GM announced a partnership to collaborate on the next generation of fuel cell systems and hydrogen storage technologies within the 2020 time frame. Renault–Nissan has also signed an agreement with Daimler and Ford to join the Automotive Fuel Cell Cooperation (AFCC), which may result in the production of a new FCV by 2017. Toyota and BMW have entered into a strategic partnership to share a number of technologies in order to jointly develop a new FCV platform by 2020. In addition to the development of FCVs, Toyota has also become directly involved in the establishment of the early hydrogen network in California by investing $7.2 million in First Element Fuel,

[10]Hyundai, "Hyundai's ix35 Fuel Cell Receives Technology Award At Annual U.K. Fleet World Honours 2013," Hyundai Motor News, 22 May, 2013. Available at: http://www.hyundaiglobalnews.com/prCenter/news/newsView.do?dID=1757.

Table 15.2 **Carmakers' plans for demonstration and commercialization of hydrogen fuel cell vehicles.**

Companies	Current plans			First announcement	
	Demonstration	Launching	Mass production	Year	Commercialization
BMW (Germany)	2009 *100 7-Series ICE-LH_2*				
Daimler (Germany)	2013 *200 B-Class*	*2017*		1997	2004 *40,000 units*
Ford (United States)	2005-2010 *30 Focus*	*2020*		2002	2004 (launching) 2010 (comm.)
GM (United States)	Since 2007 *120 Chevy Equinox*	No plans announced	No plans announced	2001	2004
Honda (Japan)	2008-2012 *200 FCX Clarity*	2015 *5,000 units*	2018 *Target $65,000*	2001	2004
Hyundai (South Korea)	Since 2011 *48 ix35*[a] 2013 *Added 15 ix35*	2013-15 *1,000 units*	2015 *10,000 units* *Target $50,000*		
Nissan (Japan)		2015-16			
Toyota (Japan)	2013 *100 units*[a]	2015 *Current cost:* *$100,000* *Targets:* *Cost $50,000* *Price $50,000-$100,000*	2020 *X0,000 units*	2001	2004

[a]Demonstration vehicles on the road in the United States, cf. IPHE (2013).

Source: Carter and Wing (2014), Fuel Cell Today (2013, 2009), Automotive News (2014); Reuters (2014), official carmakers' websites.

a company created to open and operate the first hydrogen stations in the state, which also receives financial support from the California Energy Commission.[11]

Automakers have also started to collaborate with infrastructure providers and governments to prepare market introduction of hydrogen vehicles. In September 2009, a group of six automotive original equipment manufacturers (OEM)—Daimler, Ford, General Motors, Honda, Hyundai–Kia, and Toyota—signed a memorandum of understanding (MoU) addressed to oil and energy companies and governments. In this joint letter of intent the automakers restated their plans to commercialize FCVs by 2015 and called for the development of a network of HRS. Shortly after, a group of German industrial companies (Linde, Daimler, EnBW, OMV, Shell, Total, Vattenfall), Daimler, NGO (NOW GmbH National Organization Hydrogen and Fuel Cell Technology), and government representatives signed another MoU marking the beginning of the "H2 Mobility" initiative in Germany. This initiative aims to evaluate the setup of a hydrogen infrastructure in the country that eventually supports "the introduction of series produced hydrogen powered vehicles in Germany around 2015." In January 2011, the three largest Japanese carmakers—Honda, Nissan, and Toyota—and 10 Japanese oil and energy companies signed an agreement supported by the Japanese Ministry of Economy, Trade and Industry (METI) to cooperate with the introduction of FCVs into four major urban areas (Tokyo, Nagoya, Osaka, and Fukuoka), starting in 2015.[12] The carmakers agreed to reduce the costs of the future FCVs to increase sales in the second half of the decade, and the hydrogen fuel suppliers agreed to construct a network of a hundred HFS in Japan by 2015. A year and a half later, in October 2012, a MoU was signed between automakers (Toyota, Nissan, Honda, and Hyundai) and organizations of the Nordic countries to deploy FCVs and hydrogen refueling infrastructure there during the period of 2014–2017. The signature of joint letters of intent became more frequent with the approximation of the date expected for the start of FCVs commercialization, around 2015.[13,14]

Several countries are currently working with carmakers to establish the early hydrogen infrastructure to prepare the arrival of the first FCVs. Table 15.3 shows

[11]"Toyota Joins California Hydrogen Push in Station Funding," Bloomberg, May 2, 2014. Available at: http://www.bloomberg.com/news/print/2014-05-01/california-awards-46-6-millionfor-hydrogen-car-stations.html.

[12]Toyota Motor Corporation, "13 Japanese Companies Eye Smooth Domestic Launch of FCVs," 13th January 2011. Available at: http://www2.toyota.co.jp/en/news/11/01/0113.html.

[13]These agreements reveal a positive attitude toward the collaboration and expansion of the use of hydrogen FCVs. However, it is worthwhile to note that similar pledges were made in the past which were later postponed or even abandoned. Indeed, as early as September 2006, BMW, DaimlerChrysler, Ford, General Motors, MAN, Shell, Total, and Volkswagen issued a common position paper urging authorities to deploy the infrastructure in Europe "more quickly in order to allow the beginning of commercialization of hydrogen vehicles around 2015 (or earlier)." A couple of years later several of these automakers dropped their expectations and became considerably more skeptical about the role that hydrogen could have in the future.

[14]"Next Steps for the Development of a Hydrogen Infrastructure for Road Transport in Europe—A common position paper of BMW Group, DaimlerChrysler AG, Ford Motor Company, General Motors Europe AG, MAN Nutzfahrzeuge AG, Shell Hydrogen B.V., Total France, and Volkswagen AG," September 2006.

Table 15.3 **Number of fuel cell cars and hydrogen stations involved and projected in several demonstration projects around the world**

Project	Region	Period	Deployment plans	Target stations	Actors involved	Sources
California Fuel Cell Partnership (CaFCP)	California (US)	Current 2015 2023	200 FCVs Not specified Not specified	9 (+47 planned) 68 stations 100 stations	Air Liquide, Air Products, Ballard, Chrysler, Daimler, GM, Hyundai, Honda, Hydrogenics, Linde, Nissan, Toyota, VW, …	Office of Governor (2013) FCT (2013) CaFCP (2009)
Fuel Cells and Hydrogen Joint Undertaking (FCH JU 2)	EU	2008-2012 2013 2015 2020	>150 FCV, >45 buses 31 buses 5 K FCV, 500 buses 500 K FCV, 1 K buses	>20 stations Not specified <300 stations >2,000 stations	Air Liquide, Air Products, Ballard, BMW, Ceramic Fuel Cells, Daimler, Honda, Hyundai, Linde, Nissan, Opel, Shell, SOFCpower, Toyota, Vattenfall, Volkswagen, Volvo, …	FCH JU (2014) Carter and Wing (2014) FCH JU (2011)
HySUT	Japan	Current 2015	n.a. Commercialization	23 (+19 planned) 100 stations	(19 companies:) JX Nippon Oil & Energy Corp., Iwatani, Toyota, Nissan, Honda, Mitsubishi Kawasaki, …	H2stations.org http://hysut.or.jp/ Hara (2013) Carter and Wing (2014)
Scandinavia (MoU'12)	Norway Sweden Denmark Iceland	Current 2014-17	19 FCVs Deployment (500 FCVs[a])	9 (+2 planned) Not specified (30 + 15 stations[a])	Toyota, Nissan, Honda, Hyundai, infrastructure companies, HyNor, H2 Link, H2 Sweden, SHHP, Icelandic New Energy, …	News release from automakers [a]Vision of the SHHP by '15 (scandinavianhydrogen.org)

Continued

Table 15.3 **Continued**

Project	Region	Period	Deployment plans	Target stations	Actors involved	Sources
Korean program and roadmaps	Korea Republic	Current 2015 2030	100 FCVs (2013) 1,000 FCVs Not specified	10 stations 43 stations 500 stations	Local infrastructure companies, Hyundai, government	H2stations.org Kim (2013)
H_2 Mobility Initiative	Germany	Current 2015 2018 2023 2025	n.a. 5,000 FCVs Not specified Not specified Not specified	25 stations 50 stations 100 stations 400 stations 1,000 stations	Air Liquide, Daimler, Linde, OMV, Shell, Total, BMW, Honda, Hyundai, Intelligent Energy, Nissan, Toyota, VW	H2stations.org Press releases available at now-gmbh.de FCT (2013)
Clean Energy Partnership (CEP)	Berlin, Hamburg (Germany)	Current (2011–2016)	>100 FCVs, 30 buses Not specified	7 stations (+5 planned)	BMW, Daimler, Ford, GM/Opel, Honda, Hyundai, Toyota, VW, Air Liquide, Berlin BVG, EnBW, Hamburg Hochbahn, Linde, Shell, Siemens, SSB, Total, Vattenfall	H2stations.org cleanenergypartnership.de
UK H_2 Mobility	United Kingdom	Current 2015 2020 2025 2030	n.a. Not specified 10 K FCV sales/year Not specified 1.6 millions FCVs (300 K annual sales)	13 (+12 planned) 65 stations Not specified 330 stations 1150 stations	UK Government Departments, Greater London Assembly, FCH JU, Air Liquide, BOC, Daimler, Hyundai, Intelligent Energy, ITM Power, Johnson Matthey, Nissan, Toyota, …	H2stations.org ukh2mobility.co.uk Carter and Wing (2014)

n.a., not available.

key data from some of the national hydrogen programs concerning both light-duty vehicles and buses. While this information is not exhaustive, it does cover the most representative projects currently in progress around the world. The most active regions are still found in Japan, Europe, and United States (especially, California). Other countries have also started important programs in recent years, such as Korea.

The number of refueling stations is expected to grow everywhere until 2015 to provide a minimum infrastructure for the commercialization of FCVs, which would increase the number of vehicles on the road from hundreds to several thousands of units. There are already plans in Europe and Korea for a significant increase in the number of refueling stations subject to the start of mass-commercialization of these vehicles around 2020. In the European Union, for instance, the European Commission announced in January 2013 a package of measures to enable the build-up of alternative fuel infrastructure across Europe.[15] The proposal, which has been discussed at the European Parliament and the European Council, includes "binding targets on Member States for a minimum level of infrastructure." The Commission intends "to form a network with common standards ensuring the mobility of Hydrogen vehicles" among the 14 Member States that currently have refueling stations. For that, the EU allocated almost €3.5 million from the TEN-T infrastructure programme to examine the feasibility of an interconnected hydrogen network.[16] In addition, the Commission has recently renewed the main European RD&D program on fuel cells and hydrogen, the Fuel Cells and Hydrogen Joint Undertaking (FCH JU2), with a budget of €665 million—that must be complemented by an equivalent amount from industrial and research partners—from the Framework Program "Horizon 2020."[17]

All these initiatives contributed to the increase in the total number of hydrogen stations in service worldwide, which was 185 (not all with public access) in May 2014, according to H2stations.org developed by the LBST and TÜV SÜD. Most of the stations in service are located in Europe (72), North America (67), and Asia (44, mostly in Japan and Korea). A third of the HRS installed in Europe is situated in Germany with seven of them included in the demonstration project Clean Energy Partnership (CEP). This project tests hydrogen vehicles and filling station technologies in Hamburg, Berlin, and Dusseldorf (more details in Section 15.3.3). In September 2013, Air Liquide, Daimler, Linde, OMV, Shell, and Total agreed to expand the network of stations in Germany to 50 by 2015 in the framework of the "H2 Mobility Initiative".[18] The Initiative is funded jointly by the Germany's federal government

[15] "EU launches clean fuel strategy," European Commission, IP/13/40, 24/01/2013. Available at: http://europa.eu/rapid/press-release_IP-13-40_en.htm.

[16] "Hydrogen on TEN-T road network closer to reality thanks to EU grant," European Commission, Innovation and Networks Executive Agency, 16 January 2013. Available at: http://inea.ec.europa.eu/en/news__events/newsroom/hydrogen_on_ten-t_road_network_closer_to_reality_thanks_to_eu_grant.htm.

[17] "Green light for FCH 2 JU," Horizon 2020 Projects, 7 May 2014. Available at: http://horizon2020projects.com/industrial-leadership/green-light-for-fuel-cells-and-hydrogen-joint-undertaking/.

[18] "H2 Mobility initiative: Leading industrial companies agree on an action plan for the construction of a hydrogen refuelling network in Germany," Stuttgart, 30 September 2013. Available at: http://www.now-gmbh.de/en/presse-aktuelles/2013/h2-mobility-initiative.html.

and the industry, which together will invest 40 million EUR for the establishment of a nationwide infrastructure.[19] The current state of progress of the initiative is globally in line with the plans, but early ambitions of deploying hundreds of refueling stations to assist in the diffusion of one million FCVs by 2015 had to be revised (Stiller and Wurster, 2010). Similarly, the initial plans had to be reformulated in other areas, such as in California where more than 4,000 FCVs were previously expected on the roads between 2012 and 2014 (CaFCP, 2009) and in Europe where the partnership between the industry and the European commission lowered its initial previsions of almost two million cars in 2020 (HFP, 2007) to a more realistic 500,000 (FCH JU, 2014). These revisions were the result of the delays taken in the progress of the FCV as well as the over-optimistic assumptions concerning the development of the technology in late 2000s.

Hence, the first commercial fuel cell vehicles are likely to be launched in regions where plans for the development of a network of HRS are the most advanced. These regions historically comprise California, Germany, and Japan. The next section provides an overview of the hydrogen programs in these areas, their current targets, and instruments deployed to promote the penetration of hydrogen in mobile applications.

15.3 Implementation of the early infrastructure: case studies

15.3.1 Californian regulatory initiatives to promote FCVs commercialization

California has a long tradition of promoting the commercialization of new zero-emission technology in transportation. In 1990, the state began setting more advanced vehicle emission standards, such as the Zero-Emission Vehicle (ZEV) mandate, to solve severe air quality problems (Collantes and Sperling, 2008). The ZEV program requires automakers to produce a certain percentage of zero-emission vehicles (ZEVs) for sale in California, such as hydrogen fuel cell and BEVs. The California Air Resources Board (CARB) is in charge of the ZEV program, and typically updates it every 3 years. However, this regulation has faced strong resistance from automakers, which over time has delayed and greatly restricted its implementation in practice.[20]

The early version of the ZEV regulation required 2% of vehicles for sale in California in 1998 and 10% of vehicles in 2003 to be ZEVs. The reaction from carmakers as well as concerns about the readiness of technology lead the CARB in 1996 to remove the intermediate 1998 mandate, but left the 10% ZEV requirement for 2003. At the same time, the CARB allowed credits for partial zero-emission vehicles (PZEVs) that were not 100% ZEVs, such as hybrid electric vehicles (HEVs). The

[19] "50 hydrogen filling stations for Germany," Press/News 20.06.2012. Available at: http://www.now-gmbh.de/en/presse-aktuelles/2012/50-hydrogen-filling-stations-for-germany.html.

[20] For a good overview of the origins of the ZEV mandate, see: Collantes and Sperling (2008). The CARB website also provides detailed information about the changes of the ZEV program over time. Available at: http://www.arb.ca.gov/msprog/zevprog/zevregs/zevregs.htm.

following years saw the introduction of BEVs, including the GM EV1 and Toyota EV RAV4, and the HEVs Honda Insight and Toyota Prius. In 2001, the CARB again changed the ZEV regulation to allow automakers to meet the 10% requirement through more environmentally efficient (conventional) gasoline cars (6%), and the remaining 4% distributed equally between pure ZEVs and advanced technology PZEVs. Even so, the automakers pressured the program to stop its application, and in 2003, the CARB was forced to transform it into a complex system that allows the banking of credits. It also created an Alternative Path requiring significantly fewer FCVs. In 2012 the CARB completely reformulated the ZEV program, which is now part of the Advanced Clean Cars standards regulation focusing on California's long-term global warming goals.[21] This new emission-control program also includes the control of smog, soot, and global warming gases. In the end, the ZEV mandate led to the growth of less radical innovations (e.g., HEV) but was unsuccessful at promoting the commercialization of BEVs and FCVs. Still, it stimulated the development of several FCV programs and the expansion of the number of hydrogen stations across California.

The "California Fuel Cell Partnership" (CaFCP) was formed in the context of the ZEV program. The partnership was established in January 1999 between two public agencies (California Air Resources Board and California Energy Commission) along with six private companies (Ballard, DaimlerChrysler, Ford, BP, Shell Hydrogen, and Chevron) to promote the commercialization of FCVs in transportation. Since then, this initiative has supported the installation of refueling stations and the demonstration of hundreds of hydrogen-powered cars.

In 2012, the California Governor Edmund G. Brown's signed an executive order[22] that urged state government to assist the commercialization of ZEVs (including FCVs and BEVs) in California. This order and the "2013 Zero Emission Vehicles (ZEV) Action Plan" contain much of the current hydrogen policies of the state. The action plan includes a roadmap for the goal of reaching 1.5 million ZEV on the Californian roads by 2050. It also mandates that by 2020 "the State's zero-emission vehicle infrastructure will be able to support up to one million vehicles," and incorporates the suggestion of the CaFCP's roadmap to build a network of 68 hydrogen stations to allow FCVs launch in 2015. In September 2013, Governor Brown received legislative authorization to spend $20 million a year for 10 years in the construction of 100 stations.

In May 2014, the California Energy Commission awarded $46.6 million for the installation of 28 new refueling stations to add to the 9 existing and the 17 currently planned.[23] If the construction of all the projects is confirmed, these 54 refueling

[21]In 2006, the California Global Warming Solutions Act (AB 32) established the goal of reducing greenhouse gases to 1990 levels by 2020. An executive order issued by Governor Schwarzenegger and reinforced by Governor Brown called for reducing GHGs a total of 80% by 2050. According to the CaFCP (2009), the CARB estimates that meeting the 2050 goal will require nearly 100% of passenger vehicles sold by 2040 to be ZEVs.

[22]Executive order B-16-2012, Governor Edmund G. Brown, State of California, 2012.

[23]"California Investing Nearly $50 Million in Hydrogen Refueling Stations," The California Energy Commission, 1/5/2014. Available at: http://www.energy.ca.gov/releases/2014_releases/2014-05-01_hydrogen_refueling_stations_funding_awards_nr.html.

stations would represent progress toward the goals of 68 refueling stations by 2015 and the 100-station network to support the commercialization of FCVs in California in the time frame of a decade. In particular, more than $27 million went to the startup FirstElement Fuel—the startup that is also backed with $7.2 million from Toyota—for the construction of 19 new refueling stations. To be eligible for this grant, a third of the hydrogen sold by the refueling stations must come from renewable energies. Funding is derived from taxes in vehicle and boat registrations, as well as smog check and license plate fees.

The history of the Californian initiatives to promote FCV is at the same time pioneer in the world and illustrative of the succession of ups and downs that hydrogen energy went through in the past two decades. In addition to the problems with the implementation of the ZEV program, the official expectations had to evolve to reflect the real progress of the technology. Over a decade ago, in April 2004, Governor Schwarzenegger announced his vision for the "California Hydrogen Highway Network" initiative, which comprised the installation of 50–100 refueling stations along California's major highways by 2010. These refueling stations would support "automakers [which] have indicated that "tens of thousands" of fuel cells vehicles will be commercially available, provided there is fueling infrastructure in place." Meanwhile, the CaFCP have released several surveys of automakers' estimates for FCV roll-out in California. In the 2009 study, carmakers estimated that more than 700 hydrogen FCVs would be on the road by 2011, increasing to over 4,000 by 2014 and reaching about 50,000 vehicles by 2017 (CaFCP, 2009). In the 2012 Action Plan, the projections were more cautious suggesting a much slower ramp up of sales with around 1,300 vehicles in 2014 and 5,000–10,000 in 2015 (CaFCP, 2012).[24] These more modest plans may be a sign of realism and of a more pragmatic approach. Indeed, California has recently joined the H2 USA project, a public–private partnership led by the US Department of Energy focused on studying the establishment of a nationwide infrastructure following the example of current projects in Germany and the UK.

[24]In the early 2000s, the perspectives of market development for hydrogen-powered cars in California were often much more ambitious, illustrating the general context of "hype" that hydrogen and fuel cells were experiencing. In an often-cited article entitled "An Integrated Hydrogen Vision for California, " Lipman et al (2004, p. 36) suggested a "A California Hydrogen Strategy" that anticipated the begin of roll-out "to be at least through 2008, with up to 1,000 hydrogen-powered vehicles in California and perhaps 50–60 refueling locations." The commercialization was expected to start around "2008 through at least 2011 with from 1,000 to 20,000 hydrogen powered vehicles in the state and 100 or more refueling stations of various sizes." Finally, the growth accelerated "Post-2012 with over 20,000 hydrogen-powered vehicles in California and hydrogen fuel becoming widely available." In Europe, there were echoes of these high expectations. For instance, the high-level group formed by the European Commission to study the development of the hydrogen economy published in 2003 (HLG, 2003) its vision about the penetration of FCVs in transport: 5% of new cars in 2020 (2% of the fleet), 25% by 2030 (15% of the fleet), and 35% by 2040 (32% of the fleet). All these overoptimistic projections added to the usual humor around hydrogen FCVs that it would always be "ten years away," as noted by Dan Carter in the "The last Analyst View from Fuel Cell Today" (January, 2014, http://www.fuelcelltoday).

15.3.2 Japanese association for infrastructure development

The coordination of the infrastructure deployment in Japan is ensured by an industry group, called HySUT,[25] the Research Association of Hydrogen Supply/Utilization Technology, comprising 19 private companies and organizations. The association started to operate in 2009 with 13 private companies (mostly oil and gas utilities, as well as industrial gas suppliers) to prepare for the construction of the hydrogen supply infrastructure for FCVs launch in 2015—the target year was set by the Fuel Cell Commercialization Conference of Japan (FCCJ, 2008). In January 2011 13 companies—including automakers and energy firms—signed a MoU confirming 2015 as the year of introduction of FCVs and agreeing on the installation of 100 HRSs across four major metropolitan areas prior to commercialization. This goal became a national target with the "Japan Revitalization Strategy" in June 2013, at the same time that Toyota and Honda reasserted their commitment to launch FCVs in 2015 (Hara, 2013).

Government agencies are collaborating with automakers and infrastructure companies in Japan to streamline regulations on refueling stations and FCVs, reduce vehicle costs, and create early demand for the cars. A $46 million government subsidy was made available in 2013 and $82.5 million was requested in 2014 to support the construction of HRSs in Japan (Hara, 2013).[26] The subsidy covers up to 50% of the station's capital costs with a maximum of $2.5 million per project. HySUT estimates the current costs per station around $5 million (Hara, 2013). This enabled the construction of 19 new refueling stations in 2013 that enlarged the existing network of 23 refueling stations in operation. If the budget requested in 2014 for HRS construction is confirmed, the subsidy could support 34 more refueling stations, bringing the number of refueling stations in Japan close to 76. The subsidy is provided for early dissemination of HRS and may be reduced over time in order to incentivize private companies to accelerate investments and put the 100 refueling stations goal within reach. Indeed, JX Nippon Oil & Energy Corp. and Iwatani announced plans to construct 60 of the 77 refueling stations needed to meet that target (Wing, 2013). Some of the new HRS may be built at existing gasoline stations, thus reducing capital needs.

In Japan there is a tradition of collaboration in the field of fuel cells—even if the companies then compete fiercely between each other to be on the edge of the technology. In fact, the developments mentioned above appear in the sequence of the "Japan Hydrogen and Fuel Cell Demonstration Project" (JHFC). This landmark project was launched in 2002 by the Ministry of Economy, Trade and Industry (METI) in partnership with major Japanese and foreign carmakers (Toyota, Honda, Nissan, Daimler, and GM), and energy companies (e.g., Tokyo Gas and Shell). The program supported the opening of 12 hydrogen stations, which assisted the demonstration of nearly 60 hydrogen cars until 2010 (Aki, 2009).

[25]Available at: http://hysut.or.jp.

[26]Considering an exchange rate: ¥100 = $1.00.

15.3.3 CEP and H2 mobility initiative in Germany

As the technology progresses into the deployment phase, there is a growing need for more formal and specific commitments. This point concerns, in particular, concealing different elements such as: demand growth and public incentives, number and location of refueling stations from fuel providers, number of vehicles from carmakers, and costs and performance from fuel cell manufacturers.

The German collaboration between public and private actors, called "H2 Mobility," seeks to promote the commercialization of FCVs in 2015 through the deployment of the hydrogen infrastructure in Germany. Established in September 2009 between the National Association for the Advancement of Hydrogen and Fuel Cells ("NOW") and seven private companies (Daimler, Linde, EnBW, OMV, Shell, Total, and Vattenfall), the "H2 Mobility" initiative aims to develop a nationwide network of refueling stations starting as early as 2011. This program comes as part of the agreement signed in 2009 between the car manufacturers Daimler, GM, Honda, Toyota, Ford, and Hyundai for the commercial launch of FCVs by 2015. The initial plans included the construction of hundreds of hydrogen stations in Germany before 2015 (Daimler, 2009), but that number was significantly reduced to 50, which was certainly influenced by a less favorable environment for FCVs. In June 2012, the federal government signed a joint letter of intent agreeing to support the expansion of the German's network of HRS from 15 to 50, in both metropolitan areas and major interconnections between these regions. Similar to the Japanese case, the government subsidizes half of the stations' capital cost—estimated at 40 million Euros—through the National Innovation Programme for Hydrogen and Fuel Cell Technology (NIP). The construction of the refueling stations is coordinated by the "NOW" program, and will serve the 5,000 FCVs expected to be on the German roads by 2015.[27] The industry partners of the "H2 Mobility" initiative recently called on the government to enlarge the network to 100 refueling stations between 2015 and 2017, and to 400 by 2023.[28] As of May 2014, there are 25 hydrogen stations in operation and more three dozens are planned in Germany what is in line with the objectives for 2015.[29,30] A large part of the refueling stations is operated in the context of the CEP.

[27]"50 hydrogen filling stations for Germany," Press/News 20.06.2012. Available at: http://www.now-gmbh.de/en/presse-aktuelles/2012/50-hydrogen-filling-stations-for-germany.html.

[28]"H2 Mobility initiative: Leading industrial companies agree on an action plan for the construction of a hydrogen refuelling network in Germany," Stuttgart, 30 September 2013. Available at: http://www.now-gmbh.de/en/presse-aktuelles/2013/h2-mobility-initiative.html.

[29]cf. H2stations.org (accessed in 26.05.14).

[30]Moreover, the industry initiatives, the National Organization Hydrogen Fuel Cell Technology (NOW), the Clean Energy Partnership (CEP), H2Mobility and Performing Energy (PE), reaffirmed their commitment to market hydrogen and FCVs in Germany by signing a declaration to invest €2 billion (US$2.7 billion) over the next 10 years. This commitment is subject to the continuation of the NIP funding of hydrogen and fuel cells R&D by the Federal government. See: "Industry initiatives reinforce market introduction of hydrogen and fuel cell technologies—locations for H2 refuelling stations are confirmed," NOW Press release, 08/04/2014. Available at: http://www.now-gmbh.de/en/presse-aktuelles/2014/visit-state-secretary-reiche.html (last accessed 24.05.14).

The CEP is one of the largest demonstration projects in the world and a lighthouse initiative of the German programme (NIP), under the coordination of "NOW."[31] The CEP was formed in 2002 as a joint political initiative lead by the German Ministry of Transport and industry. Partners are currently composed of technology, oil and energy companies, major carmakers, public transport systems, government agencies, and German states. All these actors work together to test the deployment of hydrogen as a fuel in everyday use, contributing to increase its social acceptance. This initiative comprises the use of hydrogen-powered vehicles as well as the operation of the hydrogen chain (production, storage, and distribution). The project runs 7 refueling stations in Berlin and Hamburg that serve a fleet of more than 100 cars and 30 buses. By testing and optimizing vehicles and infrastructure, the CEP intends to become both a predecessor and a facilitator of the national infrastructure for hydrogen. Similar lighthouse projects have been organized in other European regions, such as in London with the local Hydrogen Partnership, or in Scandinavian cities under the Scandinavian Hydrogen Highway Partnership.

15.3.4 Synthesis of the cases

The three cases analyzed above have several characteristics in common when it comes to promoting the use of hydrogen and fuel cells in transport. On the one hand private companies are increasingly working with public authorities in an effort to build the infrastructure and prepare the arrival of FCVs in the target date. On the other hand, the promotion of economic competitiveness and jobs creation are important drivers for public action in all three cases. Other factors more specific to each case are also important such as: the reduction of local pollution in California, the concerns about fuel import dependency in Japan, and the maintenance of a strong position in the auto industry in Germany. However, there are differences in the approaches followed by each region in the preparation of the hydrogen energy transition.

California has the longest experience with these technologies starting as far back as the 1990s. The state took a more radical legislative approach by setting command and control type of policies such as the ZEV mandate requiring auto manufacturers to sell a certain percentage of specific technologies (BEVs or FCVs). As a consequence, the state had to face a strong resistance by the carmakers in the courts, which blocked most applications of the law. Even though the legislation was unable to spur a massive adoption of ZEVs, it created a favorable environment for the penetration of more incremental innovations in the auto industry, such as hybrid vehicles and plug-in hybrid vehicles.

In Japan there is a strong commitment to develop FCVs and the hydrogen infrastructure at the national level. The interest in these technologies started early and was resilient against the waves of "hypes" and disappointments that surrounded hydrogen energy over the past decade. The government has played a central role in the R&D and demonstration stages, and its collaboration with the industry is still essential in this phase of infrastructure deployment.

[31] http://www.cleanenergypartnership.de.

In Germany the strategy is more concerned with not lagging too far behind the more advanced areas, such as California and Japan, than finding leadership in the field. This is illustrated by the fact that, for a long time, the main hydrogen project was the CEP, which is a more regional organization (around Berlin and Hamburg) and directed toward demonstration and testing of vehicles and infrastructure. The recent decision to scale up the partnership to a national level was made possible due to the support of the federal government and the mobilization of organizations already involved with the existing projects.

15.4 Future trends

The recent announcements concerning the development of networks of HRSs raise the question of the possible trends they may experience in the next few decades. It is reasonable to assume that future trends will be influenced, to a certain extent, by the trends of the past, especially in terms of the attention/visibility of hydrogen cars and the readiness of fuel cell technologies.

On the one hand, hydrogen and FCVs give increasing signs of returning to the energy agenda. Commercial launch is again expected to start within 1 or 2 years as it was a decade ago when expectations were at their highest. Thus, very soon it will be possible to know whether these are serious statements or just a new episode of "overexpectations." A clear hype-disappointment cycle such as described by the Gartner curve was noted a few years ago. Figure 15.3 shows the number of times the term "fuel cell vehicles" appeared in books between 1990 and 2008 (available in Google Books Ngram Viewer).[32] This is compared with the number of times the same phrase was searched for in Google during the past 10 years.[33] In both cases, the raw data is indexed to the maximum value in the sample (which is shown by the number 100) to ensure comparability of the results. The figure shows on the one hand a marked increase in the number of times the term "fuel cell vehicles" appeared in published books after 1999. The trend of the number of searches in Google reveals on the other hand multiple peaks in 2004 and 2008, followed by a major drop afterward. Leaving aside the discussion about the potential pitfalls of using search results and book appearances, the analysis seems to confirm the findings of previous studies that documented the hype over hydrogen in the last decade (Bakker and Budde, 2012; Konrad et al., 2012; Bakker, 2010a). The question now is to understand to what extent the technology is really improving and reaching the plateau of productivity or entering into a new phase of "hype" that will again turn into disappointment (or even abandon). To answer this question, we need to analyze with special attention the recent developments around fuel cell technology.

[32] https://books.google.com/ngrams (analysis performed in 5/6/2014 for the number of times the phrase "fuel cell vehicles" (case-insensitive) occurred in the English corpus of books).

[33] http://www.google.com/trends (analysis performed in 5/6/2014 for the number of times the phrase "fuel cell vehicles" was searched in Google in the past decade).

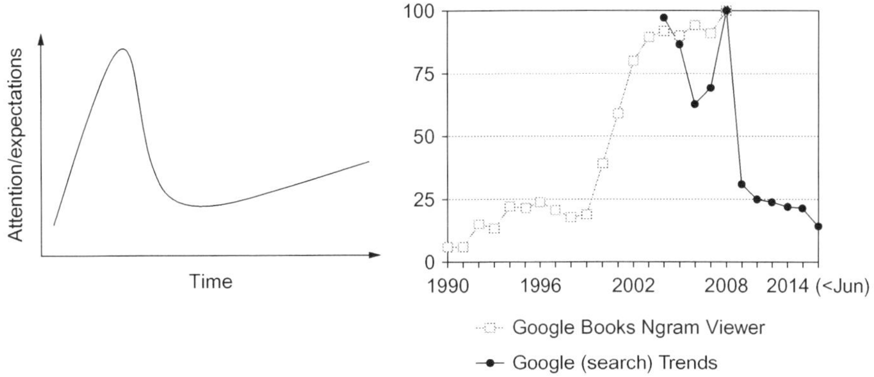

Figure 15.3 Gartner hype cycle (left-hand) and the number of times the term "fuel cell vehicles" appeared in books and was searched for in Google (right-hand), between 1990 and 2014.

On the other hand, fuel cell costs have declined significantly over the past decade, despite the increase in the cost of platinum. The US Department of Energy (DOE) recently upgraded its cost projections to high-volume manufacturing of 80-kW automotive PEM fuel cell system to take into account the new price of platinum (from $1100 to $1500 per troy ounce) and the new DOE requirements regarding heat management. This increased the projected costs from $47 to $55/kW in 2013, which assumed a production of 500,000 units per year (Spendelow and Marcinkoski, 2013). Figure 15.4 shows the decline in costs according to the old and the updated methodology. In the first years of commercialization, the number of cars produced should stay around a thousand units, which would raise the cost of fuel cells to $285/kW (Spendelow and Marcinkoski, 2013).

The reduction in fuel cell costs appears to be confirmed in the last reports from the industry. Toyota announced a significative reduction of the cost of its fuel cell propulsion system from $1 million a decade ago to $50,000 today—this translates into $500/kW or $625/kW depending on the system power considered, 100 kW or 80 kW, which was not specified by the company. Among other improvements, the new fuel cell would use around 30 g of platinum, down from the 100 g previously reported (Reuters, 2014).

Even though the ultimate targets for the cost of fuel cells seem within reach, the key challenge remains financing the first tens of thousands vehicles. The roll-out of these more expensive units is essential to begin the transition towards commercialization, but the total cost to buy down FCVs to competitive levels—through economies of scale and learning—could amount to tens of billions of dollars. Automakers should take a part in these learning costs, motivated by the possibility of increasing their share in the future market as did Toyota with the Prius in the past. Nevertheless, further support may be necessary, especially in case the most optimistic predictions of cost reductions are not realized. Therefore, the actual evolution of fuel cell costs will be primordial for the commercial success of FCVs, namely against other new alternative technologies.

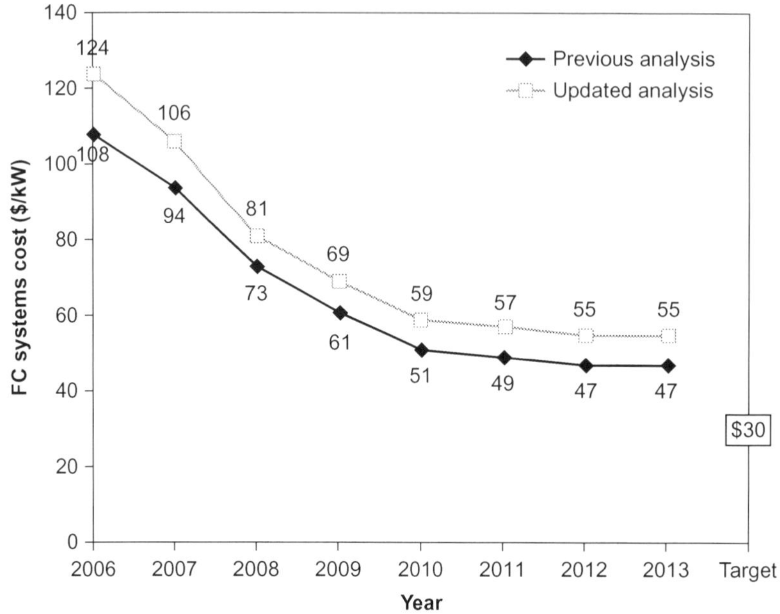

Figure 15.4 Cost projections from the DOE for an automotive 80-kW PEM fuel cell, assuming mass-production (500,000 units per year).
Source: Spendelow and Marcinkoski (2013).

Hydrogen and fuel cells are in competition with other alternative fuel vehicles for the car of the future. However, they can also benefit from the increasing electrification of vehicles to penetrate more rapidly in the market. For example, the new hydrogen-powered car announced by Toyota for 2015 uses spare parts of other gasoline-electric hybrids produced by the company (Reuters, 2014). In addition, some lessons can be drawn from the successful commercialization of BEVs to inform the strategies to boost the commercialization of FCVs. In Norway, for instance, tax exemptions in acquisition and several other factors, such as driving in bus lanes or free parking, have contributed to the rapid increase in the market for electric cars—which already accounted for 15% of new immatriculations in the first semester of 2014.[34]

15.5 Conclusions

The hydrogen economy in mobile applications will only become possible if there is a sufficient network of refueling stations to serve adopters of hydrogen-powered cars (which most likely will be equipped with fuel cells). Hydrogen and fuel cells are

[34] "Au pays du pétrole, la voiture électrique va bien," Le Monde, 17/07/2014.

resurging from a period of generalized disappointment after the hype in the early 2000s. Investment in the early infrastructure for such an emergent innovation is therefore risky and surrounded by multiple uncertainties. The choice of the infrastructure's configuration on the one hand raises important questions as the least expensive "pathways" for production and distribution of the fuel (e.g., onsite small methane reforming) to supply the (weak) initial demand are unlikely to produce the cheapest hydrogen. Furthermore, the choice of the infrastructure depends on the type of technique adopted to store hydrogen onboard, but this seems more of a consensus as automakers are increasingly adopting higher gas compression. The adoption of FCVs will on the other hand depend on the price of the car, which is still dominated by the fuel cell cost. However, this has significantly declined in the past few years and independent projections indicate that FCV costs can further reduce to competitive levels in the future, assuming high-volumes manufacturing. Yet the production of half a million units per year is unlikely to happen in the early years of commercialization, and it is still unclear how the additional costs can be shared among the stakeholders (e.g., automakers, users, governments, infrastructure providers, etc.). Additionally, the durability of fuel cells remains an issue that must be rapidly solved to enable the commercialization of FCVs.

Under these circumstances, the first networks of refueling stations are being built in several points of the world to prepare for the commercial launch of hydrogen-powered cars. These projects often involve the collaboration of fuel suppliers, fuel cell providers, local authorities and automakers, which in one case (California) financially helped the construction of the first refueling stations. Local efforts to build the infrastructure and commercialize FCVs were analyzed more in detail for three leading cases: United States (California), Japan, and Germany. Some lessons can be derived from the study of these cases. The Californian zero-emission vehicle mandate started in the 1990s. Even though it was never applied in its original form, the mandate put the hydrogen car on the agenda of the automobile industry. The stable support provided after the "hype" by the governments in Japan and Germany was essential to preserving the knowledge created during the previous phases and gradually move the technology forward. Nevertheless, the success of hydrogen and FCVs seems to depend mostly on the reduction of fuel cell costs and the readiness of the technology. This would allow the commercialization of the car on time, which could ameliorate the general credibility of the technology and consequently help in gradually establishing the infrastructure.

15.6 Sources of further information and advice

A nonexhaustive list of websites that provide updated and useful information in the field of hydrogen and fuel cells can be found here:

http://www.iphe.net—International Partnership for the Hydrogen Economy
http://iahe.org—International Association for Hydrogen Energy
http://www.ieahia.org—International Energy Agency (IEA) Hydrogen Implementing Agreement
http://www.fch-ju.eu—European "Fuel Cells and Hydrogen" Joint Undertaking
http://www.h2euro.org/—European Hydrogen Association (EHA)

http://www1.eere.energy.gov/hydrogenandfuelcells—Fuel Cell Technologies Office of the U.S. Department of Energy (DOE).
http://www.hydrogen.energy.gov—Hydrogen and Fuel Cells Program of the DOE
http://www.fchea.org—Fuel Cell and Hydrogen Energy Association (US)
http://www.cafcp.org—California Fuel Cell Partnership (CaFCP)
http://hysut.or.jp—Japanese Research Association of Hydrogen Supply/Utilization Technology (HySUT)
http://www.now-gmbh.de—German National Organization for Hydrogen and Fuel Cell Technology (NOW)
http://www.hfcletter.com—The Hydrogen & Fuel Cell Letter
http://www.fuelcells.org/—Fuel Cells 2000
http://www.fuelcelltoday.com—Fuel Cell Today
http://www.nrel.gov—U.S. National Renewable Energy Laboratory
http://steps.ucdavis.edu—Sustainable Transportation Energy Pathways research program from the University of California, Davis
http://www.h2mobility.org—LBST database on hydrogen vehicles and stations
http://www.cleanenergypartnership.de—Clean Energy Partnership (CEP)
http://www.scandinavianhydrogen.org—Scandinavian Hydrogen Highway Partnership

References

Adamson, K.-A., Crawley, G., 2006. Fuel cell today 2006 worldwide survey. Fuel Cell Today, January.
Aki, H., 2009. Country update—Japan. In: Presented at the 12th IPHE Steering Committee Meeting, December, Washington, DC.
Automotive News, 2014. Honda, Toyota plan to launch fuel cell vehicles in 2015. Retrieved at: http://www.autonews.com/apps/pbcs.dll/article?AID=/20140326/OEM04/140329915&gt (26.03.14.).
Bakker, S., 2010a. The car industry and the blow-out of the hydrogen hype. Energy Policy 38 (11), 6540–6544.
Bakker, S., 2010b. Hydrogen patent portfolios in the automotive industry—the search for promising storage methods. Energy Policy 35, 6784–6793.
Bakker, S., Budde, B., 2012. Technological hype and disappointment: lessons from the hydrogen and fuel cell case. Tech. Anal. Strat. Manag. 24 (6), 549–563.
Bakker, S., Van Lente, H., Meeus, M., 2011. Arenas of expectations for hydrogen technologies. Technol. Forecast. Soc. Chang. 78 (1), 152–162.
Bakker, S., Van Lente, H., Meeus, M.T., 2012. Dominance in the prototyping phase—the case of hydrogen passenger cars. Res. Policy 41 (5), 871–883.
Bento, N., 2008. Building and interconnecting hydrogen networks: insights from the electricity and gas experience in Europe. Energy Policy 36, 3009–3018.
Bento, N., 2010a. Investing in the hydrogen delivery infrastructure: methodology for a public policy. Energy Stud. Rev. 17 (2), Article 1.
Bento, N., 2010b. Is carbon lock-in blocking investments in the hydrogen economy? A survey of actors' strategies. Energy Policy 38, 7189–7199.
Bento, N., 2010c. Dynamic competition between plug-in hybrid and hydrogen fuel cell vehicles for personal transportation. Int. J. Hydrog. Energy 35, 11271–11283.
Bento, N., 2010d. La transition vers une économie de l'hydrogène: infrastructures et changement technique. Ph.D. thesis, Université de Grenoble, Grenoble, March, 410 p.

Borup, M., Brown, N., Konrad, K., Van Lente, H., 2006. The sociology of expectations in science and technology. Tech. Anal. Strat. Manag. 18 (3–4), 285–298.

Bullis, K., 2009. Q&A: Steven Chu. MIT Technology Review, May 14, 2009. Retrieved at: http://www.technologyreview.com/business/22651.

Bullis, K., 2013. Why Toyota and GM are pushing fuel-cell cars to market. MIT Technology Review, July 5. http://www.technologyreview.com/news/516711/why-toyota-and-gm-are-pushing-fuel-cell-cars-to-market/.

CaFCP, 2009. Hydrogen fuel cell vehicle and station deployment plan: a strategy for meeting the challenge ahead. California Fuel Cell Partnership, February. www.cafcp.org.

CaFCP, 2012. A California road map: bringing hydrogen fuel cell vehicles to the golden state. California Fuel Cell Partnership, July. www.cafcp.org.

Carter, D., Wing, J., 2014. The fuel cell industry review 2013. Fuel Cell Today. www.fuelcelltoday.com.

Collantes, G., Sperling, D., 2008. The origin of California's zero emission vehicle mandate. Transp. Res. A 42, 1302–1313.

Daimler, 2009. Initiative "H2 Mobility" - Major companies sign up to hydrogen infrastructure built-up plan in Germany. Berlin, September 10, 2009. http://media.daimler.com/dcmedia/0-921-656186-1-1236407-1-0-0-0-0-0-0-0-1-0-0-0-0-0.html (last accessed 11.06.15.).

Fairley, P., 2012. Hydrogen cars: a dream that won't die. MIT Technology Review, October 8. http://www.technologyreview.com/news/429495/hydrogen-cars-a-dream-that-wont-die/.

FCH JU, 2011. Multi-annual implementation plan 2008–2013. Adopted by the FCH JU Governing Board on 22nd November 2011.

FCH JU, 2014. Programme review 2013. Fuel Cells and Hydrogen Joint Undertaking.

Fenn, J., Time, M., 2007. Understanding Gartner's hype cycles. Gartner Research.

Fuel Cell Commercialization Conference of Japan (FCCJ), 2008. Commercialization of fuel cell vehicles and hydrogen stations to commence in 2015. Press Release, July 4.

Fuel Cell Today, 2013. Fuel cell electric vehicles: the road ahead. www.fuelcelltoday.com.

Grubler, A., Aguayo, F., Gallagher, K., Hekkert, M., Jiang, K., Mytelka, L., Neij, L., Nemet, G., Wilson, C., 2012. Policies for the Energy Technology Innovation System (ETIS). In: Global Energy Assessment—Toward a Sustainable Future. Cambridge University Press/The International Institute for Applied Systems Analysis, Cambridge, UK, New York, NY, USA/Laxenburg, Austria, pp. 1665–1744 (Chapter 24).

Hara, I., 2013. Current status of H_2 and fuel cell programs of Japan. In: 20th IPHE Steering Committee Meeting, 20 November, Fukuoka.

High Level for Hydrogen and Fuel Cells—HLG, 2003. Hydrogen energy and fuel cells: a vision for our future. Summary report, European Commission, Brussels, June.

Hoffmann, P., 2012. Tomorrow's Energy: Hydrogen, Fuel Cells, and the Prospects for a Cleaner Planet. revised and expanded edition, MIT Press, USA.

HyWays, 2008. The European hydrogen energy roadmap. Final report, February. http://www.hyways.de.

IPHE, 2013. 2012 hydrogen and fuel cell global commercialization & development update. International Partnership for Hydrogen and Fuel Cells in the Economy, May 2013. www.iphe.net.

Kim, J.W., 2013. Recent achievements in hydrogen and fuel cells in Korea. In: International Hydrogen Energy Development Forum 2013, January, Fukuoka, Japan.

Konrad, K., Markard, J., Ruef, A., Truffer, B., 2012. Strategic responses to fuel cell hype and disappointment. Technol. Forecast. Soc. Chang. 79 (6), 1084–1098.

KPMG, 2014. KPMG's global automotive executive survey 2014. Strategies for a Fast-Evolving Market. KPMG International, UK. http://kpmg.com/automot.

Kurtz, J., Dinh, H., Sprik, S., Saur, G., Ainscough, C., Peters, M., 2013. V.D.4 analysis of laboratory fuel cell technology status—voltage degradation. DOE Hydrogen and Fuel Cells Program, Excerpt from the 2013 annual progress report, December 2013. http://www.hydrogen.energy.gov/pdfs/progress13/vii_7_kurtz_2013.pdf.

Laffont, J.-J., Tirole, J., 1993. A Theory of Incentives in Procurement and Regulation. MIT Press, USA.

Lipman, T., Kammen, D., Ogden, J., Sperling, D., 2004. An integrated hydrogen vision for California. White Paper/Guidance Document, July.

McDonald, A., Schrattenholzer, L., 2001. Learning rates for energy technologies. Energy Policy 29 (4), 255–261.

Ménard, C., 2004. The economics of hybrid organizations. J. Inst. Theor. Econ. 160 (3), 345–376.

National Research Council (NRC), 2013. Transitions to Alternative Vehicles and Fuels. The National Academies Press, Washington, DC.

Nemet, G.F., 2009. Interim monitoring of cost dynamics for publicly supported energy technologies. Energy Policy 37 (3), 825–835.

NRC, 2008. Transitions to Alternative Transportation Technologies: A Focus on Hydrogen. Committee on Assessment of Resource Needs for Fuel Cell and Hydrogen Technologies, The National Academies Press, Washington, DC.

Nygaard, S., 2008. Co-evolution of Technology, Markets and Institutions: The Case of Fuel Cells and Hydrogen Technology in Europe. Lund University, Sweden.

Office of Governor Edmund G. Brown Jr., 2013. ZEV action plan—a roadmap toward 1.5 million zero-emission vehicles on California roadways by 2025. Governor's Interagency Working Group on Zero-emission Vehicles, February. www.opr.ca.gov.

Ogden, J., Yang, C., 2009. Build-up of a hydrogen infrastructure in the U.S. In: Ball, M., Wietschel, M. (Eds.), The Hydrogen Economy Opportunities and Challenges. Cambridge University Press, Cambridge (Chapter 15).

Ogden, J., Yang, C., Cunningham, J., Johnson, N., Li, X., Nicholas, M., Parker, N., Sun, Y., 2011. The hydrogen fuel pathway. In: Ogden, J., Anderson, L. (Eds.), Sustainable Transportation Energy Pathways—A Research Summary for Decision Makers. University of California, Davis, pp. 64–94 (Chapter 3).

Piper, J., 2014. Is 2014 the year of the fuel cell car? Scientific American, January 23, 2014. http://www.scientificamerican.com/article/is-2014-the-year-of-the-fuel-cell-car/.

Reuters, 2014. Insight: in green car race, Toyota adds muscle with fuel-cell launch, 17/4/2014. Retrieved at: http://www.reuters.com/assets/print?aid=USBREA3F1UN20140417&gt.

Roads2HyCom, 2009. Fuel cells and hydrogen in a sustainable energy economy: final report of the Roads2HyCom project, April, R2H8500PU.6.

Romm, J., 2004. The Hype about Hydrogen—Fact and Fiction in the Race to Save the Climate. Island Press, USA.

Ruef, A., Markard, J., 2010. What happens after a hype? How changing expectations affected innovation activities in the case of stationary fuel cells. Tech. Anal. Strat. Manag. 22 (3), 317–338.

Sharaf, O.Z., Orhan, M.F., 2014. An overview of fuel cell technology: fundamentals and applications. Renew. Sust. Energ. Rev. 32, 810–853.

Spendelow, J., Marcinkoski, J., 2013. Fuel cell system cost—2013. DOE Fuel Cell Technologies Office Record, October 6.

Stiller, C., Wurster, R., 2010. Build-up strategies for a hydrogen supply and refueling infrastructure including a comparative outlook on battery-electric vehicles and their infrastructure requirements. In: Proceedings of the Second International Conference on Sustainable Automotive Technologies. Springer, Berlin, Heidelberg.

The European Hydrogen and Fuel Cell Platform (HFP), 2007. Implementation plan—status 2006. http://www.hfpeurope.org.

U.S. Department of Energy (DOE), 2013. 2013-technology validation: summary of annual merit review of the technology validation program. 2013 merit review and peer evaluation report. Office of Energy Efficiency and Renewable Energy, Washington, DC, December, 2013. http://www.hydrogen.energy.gov/pdfs/review13/60260-08_tech_val.pdf.

U.S. Department of Energy—DOE, 2009. Multi-Year Research, Development and Demonstration Plan: Technical Plan—Technological Validation. Office of Energy Efficiency and Renewable Energy, Washington, DC. http://www1.eere.energy.gov/hydrogenandfuelcells/mypp.

Van Lente, H., 1993. Promising Technology. The Dynamics of Expectations in Technological Developments. Twente University, The Netherlands.

Van Lente, H., Rip, A., 1998. The rise of membrane technology from rhetorics to social reality. Soc. Stud. Sci. 28 (2), 221–254.

Weiss, M., Jungingere, M., Patel, M.K., Blok, K., 2010. A review of experience curve analyses for energy demand technologies. Technol. Forecast. Soc. Chang. 77, 422–428.

Williamson, O.E., 1985. The Economic Institutions of Capitalism. The Free Press, New York.

Wing, J., 2013. The propagation of hydrogen refuelling stations. Analyst View, April, Fuel Cell Today. www.fuelcelltoday.com.

Index

Note: Page numbers followed by *f* indicate figures and *t* indicate tables.

D

Q

R

S

T

Printed in the United States
By Bookmasters